Introduction to Strain-Based Structural Health Monitoring of Civil Structures

Introduction to Strain-Based Structural Health Monitoring of Civil Structures

Branko Glišić
Princeton University
Princeton
United States

Registered Offices
John Wiley & Sons, Inc., 111 River Street, Hoboken, NJ 07030, USA
John Wiley & Sons Ltd, The Atrium, Southern Gate, Chichester, West Sussex, PO19 8SQ, UK

For details of our global editorial offices, customer services, and more information about Wiley products visit us at www.wiley.com.

Wiley also publishes its books in a variety of electronic formats and by print-on-demand. Some content that appears in standard print versions of this book may not be available in other formats.

Library of Congress Cataloging-in-Publication Data is applied for

Hardback ISBN 9781118495353

Cover Design: Wiley
Cover Image: © Reproduced by permission of Branko Glišić

Set in 9.5/12.5pt STIXTwoText by Straive, Chennai, India
Printed and bound by CPI Group (UK) Ltd, Croydon, CR0 4YY

C9781118495353_030624

To my spouse Tanja and daughter Lana, to our extended family in Serbia, Croatia, North Macedonia, and Sweden, to our friends, and to all SHM enthusiasts.

Contents

Preface *xiii*
Foreword (Jerome P. Lynch) *xv*
Foreword (Daniele Zonta) *xvii*
Acknowledgments *xix*
About the Author *xxv*

1   **Introduction** *1*
1.1   Structural Health Monitoring – Basic Notions, Needs, Benefits, and Challenges *1*
1.2   SHM Process *5*
1.2.1   General Observations *5*
1.2.2   Definition and Four Dimensions of SHM *5*
1.2.2.1   Monitoring Parameters *6*
1.2.2.2   Timeframe and Pace *6*
1.2.2.3   Level of Sophistication *7*
1.2.2.4   Scale of Application *8*
1.2.3   SHM Core Activities *9*
1.2.4   Parties Involved in SHM *12*
1.3   Gota Bridge – Example of an SHM Project *13*

2   **SHM System Components and Properties** *19*
2.1   SHM System *19*
2.2   Principles of Functioning of an SHM System *21*
2.3   Properties of SHM System *22*
2.3.1   General Observations *22*
2.3.2   Basic Specifications of Measuring Subsystem *22*
2.3.3   Reading Type and Frequency *25*
2.3.4   Data Analysis and Management *29*
2.3.5   Interference with Structure and Environment, Operational Conditions, and Maintenance *32*
2.3.6   Reliability of SHM System *33*
2.3.7   Summary *34*
2.4   Example of a Heterogeneous SHM System: Tacony–Palmyra Bridge *34*

3   **Strain Sensors**   *39*

3.1   Importance of Strain Monitoring and Strain-Based SHM   *39*

3.2   Strain – Definition and Use of Term   *41*

3.3   Historic Summary   *42*

3.4   Desirable Properties of Strain Sensors   *46*

3.5   Examples of the First-Generation Strain Sensors   *48*

3.5.1   General Advantages and Challenges of the First-Generation Strain Sensors   *48*

3.5.2   Vibrating Wire (VW) Strain Sensor   *49*

3.5.3   Resistive Strain Gauge   *52*

3.5.4   Best Performances of the First-Generation Strain Sensors   *55*

3.6   Examples of the Second-Generation Strain Sensors   *56*

3.6.1   General Advantages and Challenges of the Second-Generation Strain Sensors   *56*

3.6.2   Interferometric or SOFO Sensors – Strain Sensors Based on Michelson and Mach–Zehnder Interferometry   *59*

3.6.3   Fiber Bragg-Grating (FBG) Strain Sensors   *61*

3.6.4   Distributed Sensors – Strain Sensors Based on Brillouin and Rayleigh Scattering   *65*

3.6.5   Best Performances of the Second-Generation Strain Sensors   *69*

3.7   Example of Research on the Third-Generation Strain Sensor   *71*

3.7.1   Direct and Indirect Detection of Unusual Structural Behaviors, General Advantages and Challenges   *71*

3.7.2   Example of a Contact-Based Third-Generation Strain Sensor: Sensing Sheet   *73*

3.7.3   Example of a Non-contact Based Third-Generation Strain Sensor: Camera Vision and Convolutional Neural Networks   *74*

3.8   Closing Remarks   *75*

4   **Sources of Error, Uncertainty, and Unreliability in Strain Measurements, and Fundamentals of Their Analysis**   *79*

4.1   General Observations   *79*

4.2   Errors and Uncertainties Related to SHM System   *80*

4.2.1   Measurement Error: Relative Error, Absolute Error, and Limits of Error   *80*

4.2.2   Random (Accidental) Error and Systematic Error (Bias)   *80*

4.2.3   Limits of Error, Standard Uncertainty, and Standard Error of Measurements   *82*

4.3   Fundamentals of Error and Uncertainty Analysis   *83*

4.3.1   Error and Uncertainty Propagation   *83*

4.3.2   Rounding and Significant Figures   *86*

4.4   Thermal Compensation of Sensor as a Source of Error or Uncertainty   *88*

4.5   Gauge Length and Spatial Resolution of Sensors as a Source of Error   *92*

4.5.1   Gauge Length of Discrete Sensors as an Inherent Source of Error   *92*

4.5.2   Contribution to the Error Inherent to Gauge Length Due to Homogeneous Part of Material   *94*

4.5.3   Contribution to the Error Inherent in Gauge Length Due to Inclusions in Material   *96*

4.5.4   Contribution to the Error Inherent in Gauge Length Due to Discontinuities in Material   *98*

4.5.5   Example of Influence of Gauge Length to Strain Measurement   *99*

4.5.6   Recommendation for Choice of the Type of Gauge Length Depending on Scale of Monitoring   *102*

4.5.7   Recommendation for Choice of Gauge Length for Monitoring of Beam-Like Structures at Local or Global Structural Scale   *104*

4.5.8   Spatial Resolution of Distributed Sensors as an Inherent Source of Error   *104*

4.6   Errors in Strain Measurement Inherent to Installation of Sensor and Sensor Packaging   *109*

4.6.1   Strain Transfer from Structure to Sensing Element of the Sensor   *109*

4.6.2   Error Due to Installation of Sensor   *111*

4.6.3   Error Due to Geometrical and Mechanical Properties of Sensor Packaging   *115*

4.6.4   Error Due to Compromise Between Survivability of Sensor and Quality of Strain Transfer   *117*

4.7   Fundamental Assessment of Reliability of Strain Measurement   *120*

4.7.1   Cleansing of Data   *120*

4.7.2   Examples of Identification and Correction of Outliers   *121*

4.7.3   Examples of Handling Missing Data   *123*

4.7.4   Examples of Evaluations of Reliability of Measurement Values   *124*

5   **SHM at Local Scale: Interpretation of Strain Measurements and Identification of Unusual Structural Behaviors at Sensor Level**   *131*

5.1   Introduction to SHM at Local Scale   *131*

5.2   Strain Tensor Transformations   *132*

5.3   Principal Strain Components   *134*

5.4   Reference Time and Reference (Zero) Measurement   *137*

5.5   Strain Constituents and Total Strain   *138*

5.6   Mechanical Strain and Stress   *140*

5.6.1   Scope   *140*

5.6.2   Instantaneous Uniaxial Stress–Strain Relationship and Comparison with Ultimate Values   *140*

5.6.3   Instantaneous Multidirectional Stress–Strain Relationship and Comparison with Ultimate Values   *145*

5.6.4   Analytical Expressions for Stress and Strain in Beams   *152*

5.7   Thermal Strain and Thermal Gradients   *156*

5.7.1   Thermal Strain at a Point   *156*

5.7.2   Cross-Sectional Thermal Gradients   *158*

5.7.3   Longitudinal Thermal Gradients   *166*

5.8   Rheologic Strain – Creep and Shrinkage   *169*

5.8.1   Creep   *169*

5.8.2   Shrinkage   *174*

5.9   Physics-Based Interpretation of Strain Measurements and Identification of Unusual Structural Behaviors in Steel and Concrete Structures   *176*

5.9.1   Example of Physics-Based Modeling of Total Strain and Stress   *176*

5.9.2   Comparison of Strain Measurements with Corresponding Model Values   *179*

5.9.3   Comparison of Stress, Derived from Strain, with Corresponding Model Values   *184*

5.9.4   Case Study on Analysis of Strain in Steel Structure   *187*

5.9.5   Case Study on Analysis of Strain in Concrete Structure   *203*

6   **SHM at Global Scale: Interpretation and Analysis of Strain Measurements from Multiple Sensors and Identification of Unusual Structural Behaviors at Structural Level**   *213*

6.1   Introduction to SHM at Global Scale   *213*

6.2   Global-Scale SHM Approach   *213*

6.2.1   Implementation Guidelines   *213*

6.2.2   Application Example of Global-Scale SHM on a Pile Subjected to Axial Loading   *215*

6.2.3   Application Example of Global-Scale SHM on Streicker Bridge   *216*

6.3   Physics-Based Interpretation of Parameters Derived from Sensor Topologies and Identification of Unusual Structural Behaviors   *218*

6.4   Defining Topology of Sensors for Instrumentation of Observed Cell   *219*

6.5   Simple Topology and Evaluation of Corresponding Cell Parameters   *221*

6.5.1   Implementation of Simple Topology   *221*

6.5.2   Change in Length of the Cell, $\Delta L_{cell}(t)$ *222*

6.5.3   Axial Stiffness of the Cell, $E_{cell}A_{cell}(t)$ *223*

6.5.4   Average Normal Force in the Cell, $N_{z,cell}(t)$ *225*

6.5.5   Examples of Analysis of Simple Topology Sensor Measurements and Their Derivatives   *225*

6.6   Parallel Topology and Evaluation of Corresponding Cell Parameters   *231*

6.6.1   Implementation of Parallel Topology   *231*

6.6.2   Average Curvature $\kappa_{cell,x,sensor}$ and Change in Relative Rotation $\Delta\varphi_{cell,x,sensor}$ *231*

6.6.3   Average Locations of Neutral Axis $d_{NA,cell,btm}(t)$ and $d_{NA,cell,top}(t)$ and Centroid of Stiffness $d_{cell,btm}(t)$ and $d_{cell,top}(t)$ *233*

6.6.4   Average Strain at the Centroid $\varepsilon_{cell,ctr,sensor}(t)$ and the Change in Length of the Cell $\Delta L_{cell}(t)$ *235*

6.6.5   Average Axial and Bending Stiffness of a Cell, $E_{cell}A_{cell}(t)$ and $E_{cell}I_{cell}(t)$ *236*

6.6.6   Average Normal Force $N_{z,cell}(t)$ and Bending Moment $M_{x,cell}(t)$ in the Cell   *238*

6.6.7   Examples of Analysis of Parallel Topology Sensor Measurements and Their Derivatives   *239*

6.7   Crossed Topology and Evaluation of Corresponding Cell Parameters   *251*

6.7.1   Implementation of Crossed Topology   *251*

6.7.2   Average Shear Strain $\gamma_{cell,zy,sensor}(t)$ and Change in Average Shear Stress $\Delta\tau_{cell,zy,sensor}(t)$ *252*

6.7.3   Average Shear Force $S_{y,cell}(t)$ and Shear Stiffness $G_{cell}A_{cell}(t)$ *254*

6.7.4   Average Strain $\varepsilon_{C_s,z,tot,sensor}(t)$ and Average Normal Force $N_{z,cell}(t)$ *256*

6.7.5   Example of Analysis of Crossed Topology Sensor Measurements and Their Derivatives   *256*

6.8   Examples of Global SHM Analysis   *261*

6.8.1   General Observations   *261*

6.8.2   Evaluation of Change in Average Frictional Stress Distribution $\Delta\tau_{sfr,cell,i/i+1}$ between Pile and Soil   *262*

6.8.3   Evaluation of Deformed Shape and Deflection of a Prismatic Beam Subjected to Bending   *264*

7   **SHM at Integrity Scale: Interpretation and Analysis of Sensor Measurements at Location of Damage**   *275*

7.1   Introduction to SHM at Integrity Scale   *275*

7.2   Crack Identification: Detection, Localization, and Quantification at Discrete-Sensor Level *276*
7.2.1   Crack Detection and Localization Using Soft-Packaged Short-Gauge Strain Sensors Installed Via Adhesive *276*
7.2.2   Crack Detection and Localization Using Short- and Long-Gauge Strain Sensors Installed Via Clamping or Embedding *281*
7.2.3   Crack Quantification Using Short- and Long-Gauge Strain Sensors Installed Via Clamping or Embedding *284*
7.2.4   Crack Diagnosis and Prognosis Using Short- and Long-Gauge Strain Sensors Installed Via Clamping or Embedding *286*
7.3   Distributed Sensing for Integrity Monitoring *291*
7.3.1   Brief Overview on 1D Distributed Sensing *291*
7.3.2   Sensor Packaging, Installation, Survivability of Sensor, and Thermal Compensation *291*
7.3.3   Sampling Interval and Coordinates Mapping *294*
7.3.4   Spatial Resolution and Crack Identification *296*
7.3.5   Implementation Examples *298*

8   **Closing Remarks and Future Perspectives** *301*

          References *303*
          Appendix A: Structural Health Monitoring Glossary *317*
          Index *333*

Preface

Importance of safe and reliable civil structures and infrastructure, or simply "structures," has been well ascertained in every society throughout human history, which is confirmed by many still-standing and still-in-use historic and old structures. Experiences teach us that many centuries-old structures survived to present day thanks to, of course, the quality of their design and construction, but also proportionately, if not more critically, thanks to preservation activities performed throughout their lifetime, including inspections, maintenance, repairing, and retrofitting or repurposing.

Over the centuries, in-person inspections have been, and still are, an invaluable source of information on structural performance and condition and have been often used as the trigger for informed preservation activities. Modern Structural Health Monitoring (SHM) was born about a century ago with the idea to complement in-person inspections and address their limitations using advanced technologies. And while progress in applications was slow during the twentieth century, revolutions in informatics, communication, and computing technologies enabled research, development, and more widespread application of SHM in the last 30 years and made SHM in general mature, viable, and, with some limitations, mostly related to cost-benefit evaluation policies, accepted among practitioners.

Strain-based SHM has been at the forefront of SHM applications on real structures due to importance, quality, and versatility of information provided by this type of monitoring. Yet, at the time of the inception of this book, both researchers and practitioners whose work involves strain-based SHM could not find fundamentals or comprehensive overview of the technique in a single reference; rather, they would have to search the information in multiple sources – books, manuals, guidelines, scientific articles, etc. – that are often presented at higher level and written with focused technical language, which can often be non-intuitive and difficult to understand for wider populace of readers, especially for beginners but frequently for advanced readers too.

Hence, the aim of this book is to address this issue, i.e., to provide a single reference book with the fundamentals of strain-based SHM that can be used by both beginners and advanced users interested in strain-based SHM. That is the reason why the book builds its contents from very broad and introductory topics toward very specific and practical subjects.

Chapters 1 and 2 of the book provide a general introduction to SHM, SHM systems and subsystems, and their specifications; while the readers experienced in SHM can skip these two chapters, they can still find a good overview of logistics of implementation of SHM that they might find useful. Chapter 3 is dedicated to strain sensors; an overview of the most frequently used strain sensors is provided, along with their advantages, shortcomings, and summaries of best performances. Hopefully, this chapter helps the reader identify technologies suitable for their applications.

Chapter 4 deals with the errors and uncertainties in strain measurements and provides basic methods for their assessment and evaluation. Errors and uncertainties due to properties of measuring subsystem are presented first, followed by presentation of errors and uncertainties introduced by specific sensor features such as thermal compensation, gauge length, spatial resolution, sensor packaging, and the manner of installation of sensor. Depending on the case, these errors and uncertainties are analyzed either quantitatively or qualitatively.

Chapters 5, 6, and 7 provide practical approaches for model-based analysis of monitored strain at local, global, and integrity scales, respectively. Chapter 5 gives first an overview of strain, sources of strain such as loading, temperature, and rheological effects, constitutive equations for typical construction materials, and analytical models for strain distributions in beams under the assumptions of linear theory. Then, the same chapter provides analytical expressions and criteria for detection of unusual structural behaviors, such as damage or deterioration, and applies them at the local scale. Chapter 6 expands the expressions and criteria to groups of sensors and global scale, and Chapter 7 to integrity scale while mostly focusing on crack detection and characterization as the most fundamental type of damage. The book closes with Chapter 8 that provides a summary and future perspectives.

Chapters 4, 5, 6, and 7 are extensively illustrated with examples taken from real-world applications. Particularly in Chapters 4, 5, and 6, there are numerous tables with data taken from real projects that are given alongside the equations and algorithms for data analysis, so the readers can apply the equations and algorithms, perform the analysis on their own, and compare it with the results presented in the book. In my opinion, the possibility for readers to practice data analysis using real-world examples makes this book especially useful and probably unique on the market. However, as a disclaimer, while all problems presented in the book were solved and the solution verified at least once, there may still be some errors present; in that case, please contact me so I can make corrections and make available corrected solutions to the readers.

For readers interested in more advanced topics of strain-based SHM, I strongly recommend the book I co-authored with one of my mentors, my former employer, and dear friend, Dr. Daniele Inaudi, *Fibre-Optic Methods for Structural Health Monitoring*. While that book has some common points with this one, it contains principles for creation of monitoring strategies for numerous types of structures and extremely rich set of real-world applications.

My involvement in strain-based SHM started at the Swiss Federal Institute of Technology in Lausanne (EPFL) when I built my first batch of fiber-optic sensors and embedded them in hybrid, steel-concrete specimen; since then, my passion for strain-based SHM has only grown, and my awe for its performance has been fulfilling my professional life. I hope that this book, which encompasses experience from more than a quarter of a century of my work in strain-based SHM, will transmit my admiration for the technique and serve not only educational and professional purposes but also inspire future research and applications.

Princeton (New Jersey, USA),
Valjevo (Serbia), and Rijeka (Croatia)
July 2023

Sincerely,

Branko Glišić
bglisic@princeton.edu

Foreword

I am honored to have been invited to write the Foreword to *Introduction to Strain-Based Structural Health Monitoring* by Dr. Branko Glišić of Princeton University. It is not every day you get asked to write a foreword by a luminary like Branko, so of course, I said yes! Not only is this foreword an opportunity to introduce an outstanding book on strain and its role in assessing structural health, but it is also a way for me to pay Branko back for everything I have learned from him over the years.

I was first introduced to the magical world of structural health monitoring when I was a graduate student at Stanford University in the late 1990s. At the time, the nascent field and the small research community forming around it were just beginning to define what we today would define as "structural health monitoring." A quarter of a century later, I am amazed by the dramatic progress this field has made with structural health monitoring, now a well-established, interdisciplinary field devoted to developing sensing and decision-support systems that can be used to detect deterioration in structures to ensure their safety. For me, the beauty of structural health monitoring lies in how it can be generalized and applied to so many types of "structure": some obvious like airplanes and bridges, and some not so obvious like metallic implants in human bone. Today, the field is vibrant and alive with innovation. Essential to building momentum in structural health monitoring is the diverse group of young students, researchers, and practitioners across the globe entering the field to contribute their talents to ensure safe and efficient structures for societal use. This future generation will be well served by Branko's *Introduction to Strain-Based Structural Health Monitoring* which provides a comprehensive and elegantly designed overview of strain sensors and their use to assess structural health and performance.

Those readers who enjoy music will appreciate the term "oldies but goodies" which refers to legendary hits that remain very much relevant and popular to millions. The topic of this book is strain, which is the most basic form of structural response measurable – for certain, an "oldy but goody" measurement. While the piezoresistivity of conductive materials has been known since the 1850s, the era of strain measurements really launched in 1938 with the invention of the strain gauge by Simmons and Ruge. Strain is a fundamental structural response to its environment and an essential modeling parameter in the field of engineering mechanics. Ironically, structures are designed based on concepts of stress, yet stress is a conceptual abstraction that is immeasurable. In contrast, strain can be measured with stress inferred from strain using constitutive models. Hence, strain has a special place in the field of structural health monitoring given its relationship to estimating the stress used to determine structural performance relative to engineering limit states. Strain has been a challenging measurement to reliably collect in structures over decades of service, especially when using low-cost metal foil gages that can only measure strain at single points where damage may or may not be detectable. This has encouraged the field to explore other sensing modalities for structural health monitoring. More recently, exciting innovation in strain sensing has once again renewed dramatic interest in strain-based structural health monitoring. Innovations like distributed strain sensing using fiber optics and smart appliques that can map strain over large surfaces are driving that interest.

Branko's *Introduction to Strain-Based Structural Health Monitoring* provides readers with a complete overview of strain and its potential to empower practical structural health monitoring solutions for real-world structures. The book's greatest feature is how it offers readers a complete structural health monitoring framework in which strain sensors can be applied. His review of the available strain-sensing technologies that have evolved over three generations of innovation offers readers a complete understanding of the strengths and weaknesses of these sensors when applied to real-world systems. Equally novel is the articulation of how to incorporate strain measurements into damage detection methods at various length scales ranging from the local to the global and integrity scales.

Without question, Branko is offering readers one of the most comprehensive books I have ever read on strain measurements and their use in diagnosing structural health. He impressively provides a delicate balance between theory and practical application, a skill he uniquely possesses due to his leading innovation as both an academic and practitioner. His years serving as lead engineer at SMARTEC SA, which at the time was one of the global leaders in long-gauge and distributed fiber-optic sensors, provide Branko with a unique understanding of how structural health monitoring works in complex operational structures. For those readers who are fans of *Fibre Optic Methods for Structural Health Monitoring* co-authored by Branko with another field luminary, Dr. Daniele Inaudi, you are sure to not be disappointed with *Introduction to Strain-Based Structural Health Monitoring*. I am sure you will enjoy the read and learn as much as I did from this titan of the structural health monitoring field!

Durham, North Carolina
July 2023

Jerome P. Lynch, Ph.D., F.EMI
Vinik Dean of Engineering
Fitzpatrick Family University
Distinguished Professor of Engineering
Duke University

Foreword

Retrospectively, I must recognize that a proper textbook on strain health monitoring of civil structures was long due, particularly one addressing the matter with method and clarity. I have been offering a SHM course at the University of Trento since 2013. While preparing the teaching material for the first year, I remember struggling to develop a consistent syllabus suitable for civil engineers: at that time, SHM, and particularly civil SHM, was still a discipline composed of a vast, heterogeneous, sparse material with no all-encompassing reference textbook that students could use as a guide.

Of all the available bibliography, recommended readings to the students were, for instance, Ewins' "Modal Testing" for vibrational methods; Wenzel's "Health Monitoring of Bridges" for case studies; Melchers' "'Structural Reliability" for evaluating structural safety based on monitoring information. Then there was Glišić and Inaudi's "Fibre Optic Methods," much more than a book on fiber-optic sensors, as the title would suggest. What made Prof. Glišić's book different was its rigorous quantitative approach, based on continuous mechanics, to data interpretation and monitoring design – a germinal version of what You will find, fully developed and matured, in the present book.

So, 10 years on, has the civil SHM panorama changed? In terms of number of textbooks: a lot. In terms of method: not that much, to be honest. Still today, the layman tends to see SHM as a magic tool that You install on a bridge or a building and wondrously tells You whether this bridge or building is safe thanks to unspecified technological wizardry.

Even practitioners and parts of the academic community do not always get the logic of monitoring right. In a 2014 paper, Prof. Glišić and I observed a civil engineer paradox: engineers use a rigorous quantitative approach when dimensioning structures but usually the rule of thumb when designing monitoring systems. It should come as no surprise that a civil engineer is (normally) very good at designing a bridge or a building. The objective of structural design is probably obvious to most: dimension structural members to ensure stability under design loads. Designing a structure, the good engineer sticks to a well-established process, often acknowledged in standards and codes: define the design loads; calculate the member's stress demand by structural analysis, using a structural model (e.g., a finite element model); dimension the members to the required demand; assess that the capacity is greater than the demand.

Let the same good civil engineer design a monitoring system, and most likely, their approach will be heuristic, based on common sense or experience rather than on quantitative analysis. It looks like the average engineer does not grasp the objective of monitoring and does not master its underlying logic.

Effectively, SHM is about acquiring data using sensors to understand the condition state of a structure. A monitoring is well designed when it allows inferring the state of the structure with few or no uncertainty. Uncertainty management is the keyword: reducing the uncertainty of structural state knowledge is the goal of SHM.

Monitoring design should follow a rigorous design process similar to structural design: define the target accuracy of the structural information meant to be learned through monitoring; calculate the required accuracy (demand) of instrumental data using uncertainty propagation analysis; choose the sensor technology to meet the required demand for accuracy; assess whether the expected uncertainty (capacity) is better than the target demand.

One of the breakthroughs of this book is to finally frame the problem of monitoring performance under the right light. Whereas the structural design objective is to achieve stability with an appropriate level of safety, the object of monitoring is to learn the state of the structure with an appropriate level of confidence. Uncertainty management and structural analysis play a crucial role in this process.

Notably, Chapter 4 is entirely dedicated to uncertainty analysis and error propagation. In logical terms, structural health monitoring (SHM) is formally identical to the metrology problem of indirect measurement, where the measurand is indirectly estimated based on observation of other physical quantities linked to the measurand. Similarly, Chapter 5 approaches the problem of monitoring data interpretation from a rigorous mechanics-based standpoint.

Another side of this book that deserves praise is its practical cut. For the very few who don't know him, Prof. Glišić is much more than the clichéd Princetonian academic. He features a unique industrial expertise in SHM, having matured initially as an R&D manager at Smartec SA between 2000 and 2008 and then as one of the most dynamic applied scientists in our community. In his successful career, he accomplished hundreds of SHM projects worldwide, addressing bridges, buildings, historic construction, dams, and lifeline facilities. His unparalleled first-hand experience is evident in this book, where every theoretical step is supported and made clear with plenty of case studies and real-life examples.

During my past visits to Princeton University, I have had the privilege to personally appreciate Prof. Glišić's rare teaching capabilities. His Structural Health Monitoring course, offered since 2009, is possibly the first regular graduate course on SHM for civil engineering launched in the US. By reading this book, practitioners, academics, and students alike will all acknowledge Prof. Glišić's extraordinary communication skills: He is a truly inspiring educator, capable of conveying his genuine enthusiasm for SHM with rigor and ease.

Even if there is still a long way to go before consolidating a proper common syllabus in Civil Structural Health Monitoring, this book, and the message it carries, is a fundamental milestone toward this goal. In short, I very much welcome a textbook on civil SHM that finally gets it right – Good job, Branko!

University of Trento
Italy

Daniele Zonta, Ph.D.
Professor, Department of Civil,
Environmental, and
Mechanical Engineering

Acknowledgments

I would like to acknowledge and deeply thank numerous selfless individuals – students, colleagues, friends, and family members – as well as various institutions – professional associations, agencies, schools, and companies – whose availability, support, collaboration, professionalism, kindness, and patience, in one way or another, influenced the creation of this book.

First and foremost, I would like to thank my spouse, Tanja, and daughter, Lana, for their endless love, encouragement, understanding, and patience, without which I could never complete this book. I would also like to acknowledge my extended family in Serbia, Croatia, North Macedonia, and Sweden and all my friends for their continuous support, comprehension, kindness, and laughter, which made right ambience for this book to happen.

My career in Structural Health Monitoring (SHM) would probably not happen, and thus, probably, neither this book, without generous support, great mentorship, and dear friendship offered by Prof. Jean Claude Badoux, former President of the Swiss Federal Institute of Technology in Lausanne (EPFL); Prof. Leopold Pflug, my PhD adviser in the Laboratory of Stress Analysis (IMAC) at the EPFL; Nicoleta Casanova and Dr. Daniele Inaudi, the founders of SMARTEC, Switzerland, my former employer; and Princeton University, my current employer. The extraordinary intellectual vigor, exceptional collegiality, and highly professional yet relaxed settings at IMAC–EPFL, SMARTEC, and Princeton University, and in particular, at Princeton's Department of Civil and Environmental Engineering, resulted in a productive ambit for this book to germinate and grow.

I would like to make a very special thanks to the former and current members of my research group, SHM*lab* at Princeton University, i.e., undergraduate, graduate, and visiting students, as well as visiting scholars, whose work has been related to SHM and who have been essential in performing the research and realizing several applications shown in this book:

Undergraduate students: George Lederman, Kenneth Liew, Jeremy Chen, Patrick Park, Tiffany Hwang, Julie Ditchfield, Prof. Anjali Mehrotra, Ellen Tung, Prof. Katherine Flanigan, Jose Alvrez, Matt Gerber, Jett Stearns, Rachel Marek, Corrie Kavanaugh, Anna Blyth, Mitchel Hallee, Camille Heubner, Elizabeth Keim, Yolanda Jin, Tessa Flanagan, Gabbie Acot, Jessica Chen, Maximilian Garlock, Melanie McCloy, Michaela Hennebury, Bryan Boyd, Anne Grinder, Tiffany Agyarko, and Daniel Trujillo.

Graduate students: Dr. David Hubbel, Prof. Dorotea Sigurdardottir, Dr. Yao Yao, Dr. Hiba Abdel-Jaber, Michael Roussel, Xi Li, Dr. Kaitlyn Kliewer, Dr. Jack Reilly, Prof. Rebecca Napolitano, Prof. Isabel Morris, Zeyu Xiong, Vanessa Notario, Dr. Vivek Kumar, Prof. Shengze Wang, Dr. Antti Valkonen, Mauricio Pereira, Moriah Hughes, Yitian Liang, and Kent Eng.

Visiting students: Kai Oberste-Ufer, Pedro Afonso Souza, Dr. Denise Bolognani, Dr. Carlo Cappello, and Dr. Daniel Tonelli.

Visiting scholars: Prof. Pedro Calderon, Polytechnic University of Valencia, Spain; Prof. Marco Domaneschi, Polytechnic University of Turin, Italy; Prof. Byung Kwan Oh, Yonsei University, Korea; Prof. Antonio Maria D'Altri, University of Bologna, Italy; Prof. Daniele Zonta, University of Trento, Italy, and Prof. Hui Li, Harbin Institute of Technology, China.

Additional special thanks go to Prof. Daniele Zonta, University of Trento, Italy, and Prof. Jerome Lynch, Duke University, USA, for their enormous influence on my academic career and thinking and for being such great role models and dear friends.

Concepts and ideas developed in this book could not be generated without interactions, discussions, and friendship with many dear colleagues whose research, presentations, talks, workshops, and wisdom helped shaping ideas and advancing the field of SHM. Thus, in addition to all the abovementioned individuals, I would like to deeply thank (in no particular order) the following researchers:

Prof. Billie Spencer Jr., University of Illinois, Urbana Champaign, USA
Prof. Fu-Kuo Chang, Stanford University
Prof. Kenichi Soga, University of California, Berkeley, USA
Prof. Hoon Sohn, KAIST, Korea
Prof. Satish Nagarajaiah, Rice University, USA
Prof. Raimondo Betti, Columbia University, USA
Prof. Farhad Ansari, University of Illinois, Chicago, USA
Prof. James Brownjohn, Exeter University, UK
Prof. Neil Hoult, Queens University, Canada
Prof. Wieslaw Ostachowicz, Polish Academy of Sciences, Poland
Prof. Susan Taylor, Queen's University, Belfast, UK
Prof. Kenneth Loh, University of California, San Diego, USA
Prof. Yang Wang, Georgia Tech, USA
Prof. Ming Wang, Northeastern University, USA
Prof. Dionisio Bernal, Northeastern University, USA
Prof. Genda Chen, University of Missouri S&T, Rolla, USA
Prof. Nicos Makris, Southern Methodist University, USA
Prof. Yi-Qing Ni, The Hong Kong Polytechnic University, USA
Prof. Kara Peters, North Carolina State University, USA
Prof. Maria Feng, Columbia University, USA
Prof. Aftab A. Mufti, University of Manitoba, USA
Prof. Necati Catbas, University of Central Florida, USA
Prof. Filippo Ubertini, Perugia, Italy
Prof. Campbell Middleton, Cambridge University, UK
Prof. Emin Aktan, Drexel University, USA
Prof. Franklin Moon, Drexel University and Rutgers University, USA
Prof. Matthew Yarnold, Auburn University, USA
Prof. Ali Maher, Rutgers University (CAIT), USA
Prof. Nenad Gucunski, Rutgers University (CAIT), USA
Prof. Hani Nassif, Rutgers University, USA
Prof. Thomas Schumacher, Portland State University, USA
Prof. Simon Laflamme, Iowa State University, USA
Prof. Mohammed Pour-Ghaz, North Carolina State University, USA
Prof. Rolands Kromanis, University of Twente, Netherlands

Prof. Ian Smith, Swiss Federal Institute of Technology in Lausanne (EPFL), Switzerland
Prof. Eugen Bruhwiler, Swiss Federal Institute of Technology in Lausanne (EPFL), Switzerland
Prof. Eleni Chatzi, Swiss Federal Institute of Technology in Zurich (ETHZ), Switzerland
Prof. Didem Ozevin, University of Illinois, Chicago, USA
Prof. Maria Giuseppina Limongelli, Polytechnic University of Milan, Italy
Prof. Bozidar Stojadinovic, Swiss Federal Institute of Technology in Zurich (ETHZ), Switzerland
Prof. Austin Downey, University of South Carolina, USA
Prof. Sigurd Wagner, Princeton University, USA
Prof. James Sturm, Princeton University, USA
Prof. Naveen Verma, Princeton University, USA
Prof. Amir Gandomi, University of Technology Sydney, Australia
Prof. Dryver Huston, University of Vermont, USA
Prof. Ignacio Paya-Zaforteza, Polytechnic University of Valencia, Spain
Prof. Haeyoung Noh, Stanford University, USA
Prof. Anne Kiremidjian, Stanford University, USA
Prof. Armen Der Kiureghian, University of California, Berkeley, USA
Dr. Alexis Mendez, MCH Engineering LLC, Alameda, CA, USA
Prof. Tulio Nogueira Bittencourt, University of Sao Paolo, Brazil
Prof. Werner Lienhart, Graz University of Technology, Austria
Prof. Dan Frangopol, Lehigh University, USA
Prof. Matteo Pozzi, Carnegie Mellon University, USA
Prof. Piervincenzo Rizzo, University of Pittsburgh, USA
Dr. Wolfgang Habel, BAM, Germany
Prof. Zhishen Wu, Southeast University, China, and Ibaraki University, Japan
Prof. Andrea Del Grosso, University of Genoa, Italy
Prof. Tribikram Kundu, Arizona State University, USA
Prof. Oral Buyukozturk, Massachusetts Institute of Technology, USA
Prof. Admir Masic, Massachusetts Institute of Technology, USA
Prof. Sebastian Thons, Lund University, Sweden
Prof. Xin Feng, Dalian University of Technology, China
Prof. Xuefeng Zhao, Dalian University of Technology, China
Prof. Jeffrey Weidner, University of Texas, El Paso, USA

… and to anyone else who I may have omitted unintentionally.

Also, I would like to acknowledge professional associations that enabled networking and vibrant exchange of knowledge and experiences:

- ISHMII – International Society for Structural Health Monitoring of Intelligent Infrastructure
- IWSHM/EWSHM/APWSHM – International/European/Asia-Pacific Workshops on Structural Health Monitoring
- SPIE – International Society for Optics and Photonics
- EMI – Engineering Mechanics Institute of American Society of Civil Engineers (ASCE)
- ACI – American Concrete Institute
- WCSCM – World Conference on Structural Control and Monitoring

… and, again, any other association that I may have omitted unintentionally.

I would like to greatly thank entire teams of SMARTEC, Switzerland, and Roctest Ltd., Canada, for their continuous and unselfish support, help, and friendship, which enabled many applications presented in this book. In addition, I would like to thank other SHM companies who helped with material in this book: RST Instruments, Canada; Telemac, France; Advantech Engineering Consortium, Taiwan; Newsteo, France; Micron Optics, USA; Fiber Sensing, Portugal; Omnisens, Switzerland; fibrisTerre, Germany; and Marmota Engineering, Switzerland.

Part of the material presented in this book is based on work supported by several funding agencies:

- National Science Foundation (NSF), USA
- United States Department of Transportation (USDOT) Office of the Assistant Secretary for Research and Technology, USA

Any opinions, findings, conclusions, or recommendations expressed in this book are those of the author and do not necessarily reflect the views of any of the above agencies.

The following list is an acknowledgment to the agencies, companies, institutions, and individuals who have contributed to the application examples presented in this book:

Streicker Bridge Project:

- NSF Grants No. CMMI-1362723, CMMI-1434455
- USDOT Grants No. DTRT12-G-UTC16/4650, DTRT13-G-UTC28/5237, CAIT-UTC-REG13/0615
- NSF Graduate Research Fellowships, USA
- IBM Fellowship, USA
- Leifur Eiriksson Fellowship, USA
- Turner Construction Company, Somerset, NJ, USA
- HNTB Corporation, New York, NY, USA
- A.G. Construction Corporation, Lincoln Park, NJ, USA
- Vollers Excavating & Construction, Inc., North Branch, NJ, USA
- SMARTEC SA, Switzerland
- Micron Optics, Inc., Atlanta, GA
- Princeton University, Princeton, NJ: Department of Civil and Environmental Engineering, Department of Electrical and Computer Engineering, Department of Physics, Facilities, Office of Design and Construction, Office of Sustainability

Tacony-Palmyra Bridge Project:

- NSF Grant No. EEC-0855023
- Drexel University, Philadelphia, PA, USA
- Process Automation Corporation (PAC), NJ, USA
- The Burlington County Bridge Commission, Burlington, NJ, USA
- Intelligent Infrastructure Systems, Philadelphia, PA, USA
- Pennoni Associates, Philadelphia, PA, USA
- National Instruments, Austin, TX, USA

US202/NJ23 Highway Overpass (Wayne Bridge) Project:

- USDOT Grants No. DTRT12-G-UTC16/4650
- Drexel University, Philadelphia, PA, USA
- New Jersey Department of Transportation (NJDOT), Lawrenceville, NJ, USA
- USDOT: Long-Term Bridge Performance (LTBP) Program
- PB Americas, Inc., Lawrenceville, NJ, USA
- Center for Advanced Infrastructure and Transportation (CAIT), Rutgers University, Piscataway, NJ, USA
- SMARTEC SA, Switzerland

Sensing Sheet Project:

– USDOT Grants No. DTRT12-G-UTC16/4650, DTRT13-G-UTC28/5237

Historic Staircase:

– The Museum of the City of New York, New York, NY, USA
– Guy Nordenson and Associates, New York, NY, USA

Gota Bridge Project:

– Traffic Authority of Gothenburg (Trafikkontoret), Sweden
– Norwegian Geotechnical Institute (NGI), Oslo, Norway
– Omnisens, SA, Morge, Switzerland
– Royal Institute of Technology (KTH), Stockholm, Sweden
– SMARTEC SA, Manno, Switzerland

Punggol Building Project:

– Housing Development Board (HDB), Singapore
– Sofotec, Singapore
– SMARTEC SA, Switzerland

Pile foundations:

– Route Aero, Taipei, Taiwan
– Fu Tsu Construction Co., Taipei, Taiwan
– Bovis Lend Lease, Taipei, Taiwan
– SMARTEC SA, Switzerland

About the Author

Prof. Branko Glišić received his degrees in civil engineering and theoretical mathematics from the University of Belgrade, Serbia, and PhD from the EPFL, Switzerland. After eight years of experience at SMARTEC, Switzerland, where he was involved in numerous Structural Health Monitoring (SHM) projects, he has been employed as a faculty member at the Department of Civil and Environmental Engineering of Princeton University, where he is currently serving as the Chair of the Department. His research is in the areas of SHM, smart structures, heritage structures, and engineering and the arts.

Prof. Glišić's research includes advanced sensing techniques: long-gauge and distributed fiber-optic sensors, 2D sensors based on large area electronics, and 3D sensors based on radio-frequency devices and ground penetrating radar; advanced data analysis for diagnostics, prognostics, and decision-making based on structural analysis, machine learning, and hybrid physics-informed machine learning; documentation, integration, and visualization using virtual tours, information modeling, and augmented reality; smart, kinetic, deployable, and adaptable structures; holistic analysis of heritage structures; and engineering and the arts in general. His application domains include concrete, steel, and masonry structures: bridges, buildings, pipelines; smart structures for coastal protection; and historical buildings, monuments, a university course and sites.

Prof. Glišić is the author and coauthor of more than 100 published papers, short courses on SHM, and the book *Fibre Optic Methods for Structural Health Monitoring*. He is a Council Member and Fellow of ISHMII, voting member of ACI Committee 444, and member of several other professional associations and journal editorial boards. Prof. Glišić is recipient of several awards, including the prestigious SHM Person of the Year Award, the ASCE Moisseiff Award, and Excellence in Teaching by E-Council of Princeton's School of Engineering and Applied Science.

1

Introduction

1.1 Structural Health Monitoring – Basic Notions, Needs, Benefits, and Challenges

Civil structures and infrastructure (simply referred to as "structures" in the further text) form our built environment and affect human, social, ecological, economical, cultural, and aesthetic aspects of societies. They are essential for the well-being and security of the people; vitality of the economy; and prosperity, sustainability, and resilience of society. This is especially emphasized in the twenty-first century, as for the first time in history, more than half of the world's population lives in urban areas, while climate change threatens with more frequent and more devastating hazardous events. Consequently, not only resilient design and quality construction are required from engineers but also sustainable management and durable and safe exploitation of structures.

In the course of their lives, structures are subjected to adverse changes in their structural health conditions and performances due to potential damage or deterioration induced by environmental degradation, wear, fatigue, errors in design and construction, and episodic events such as earthquakes, floods, strong winds, or impacts. Damage and deterioration can lead to malfunction and in extreme cases failure of the structures, which in turn can have significant adverse consequences in terms of life losses and injuries, worsening of general public well-being and security, and material losses for individuals, society, and economy. For instance, the collapse of the I35W Minneapolis Bridge is a sad reminder of the catastrophic consequences of structural failure: the loss of 13 lives while 145 people were injured; the unavailability of the river crossing generated economic losses of US\$ 400,000 per day for road users. In addition, losses for the Minnesota economy were estimated at US\$ 17 million in 2007 and at US\$ 43 million in 2008 (DEED 2009). The cost of rebuilding the bridge was approximately US\$ 234 million (MnDOT 2009).

The aging of infrastructure is, in general, a major concern for societies. For example, the American Society of Civil Engineers (ASCE) estimated that if the deterioration trends related to surface transportation infrastructure continue, annual costs imposed on the US economy will increase by 351%, i.e., to \$520 billion by 2040, and it will cost the national economy more than 400,000 jobs (ASCE 2011).

Structures are subjected to adverse changes, and their integrity and performance may be compromised over the time for some of (but not limited to) the following reasons (Glisic 2009):

- There is no ideal construction material: initial defects always exist, and they represent potential initiation points for damage and deterioration.
- Damage and deterioration induced by wear and environmental degradation, such as

- o Excessive external static or dynamic loadings (e.g., the material reaches critical levels of stress or strain or critical levels of buckling stability).
- o Repetitive or cyclic dynamic loading (e.g., the material is exposed to fatigue).
- o Stress concentrations (e.g., at the location of abrupt cross-section changes, dents, grooves, inclusions, welds, forging flaws, material porosities, and voids).
- o Excessive or repetitive thermally generated loads (i.e., stresses due to temperature variations and gradients).
- o In concrete: Damage due to early age deformation, drying shrinkage, freeze–thaw cycles, sulfate attack, alkali–aggregate reaction, chloride penetration and corrosion in reinforcement bars (rebars), etc.
- o In steel: Local buckling (bowing), cracking due to stress corrosion or fatigue, loss of material (and capacity) due to corrosion, etc.
- o Changes in the condition of foundations due to scour, erosion, liquefaction, differential settlements, etc.
- o Etc.
- – Natural or human-induced episodic events such as
  - o Earthquakes, storm surges, and tsunamis.
  - o Strong winds (e.g., hurricanes, tornados, and typhoons).
  - o Accidents involving fire, impact, or explosion.
  - o Etc.
- – Changing operational and environmental conditions:
  - o Old structures were not designed for modern load demands, and the latter can result in damage and deterioration of the former.
  - o Changes in the environment may impose new loads that did not exist at the time of the design and construction of the structures (e.g., strong wind); these loads can result in damage and deterioration of the structures.
- – Unintentional design, construction, and maintenance imperfections or errors.

Thus, it is desirable to assess the health condition and performance of structures in order to mitigate risks, prevent disasters, and plan maintenance activities in an optimized manner. For this purpose, ideally, a modern structure should be able to "generate" and "communicate" information concerning the changes in its health condition and performance to responsible operators and decision makers, in-time, automatically or on-demand, and reliably. To achieve this, a modern structure should be equipped with a "nervous system," a "brain" and "voice," i.e., it has to be subjected to structural health monitoring (SHM), which is continuously in operation and able to sense structural conditions.

The concept of SHM can be understood from comparison with the nervous system of the human body. An unhealthy condition of the body or exhaustion of performance is detected by the nervous system in the form of pain or tiredness. Nerves in the involved areas are activated, and the information is transmitted to the brain, which analyzes the data. A person realizes that they are ill or exhausted and addresses a doctor in order to prevent further degradation of health or performance. The doctor undertakes detailed examinations, establishes a diagnosis, and proposes a cure.

The SHM, similar to the nervous system of the human body, should be able to automatically detect unusual structural behaviors (e.g., damage, deterioration, and lack of performance), characterize them (ascertain the times of occurrence, localize them and quantify them, or rate them), and report them, providing an important and actionable information for engineers and managers who are responsible for the monitored structure. The similarity of SHM and the nervous system of humans is schematically presented in Figure 1.1.

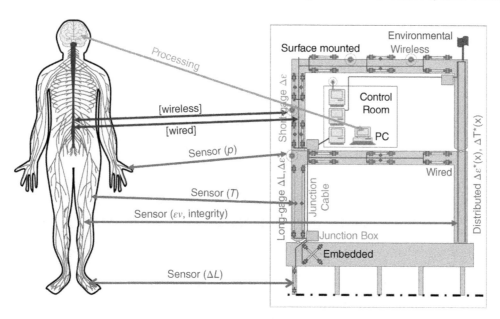

Figure 1.1 SHM as a nervous system of structures (Glisic 2009).

The potential benefits of SHM could be manifold:

- Improved safety: Unusual structural behaviors can be detected at an early stage; therefore, preservation and maintenance actions can be undertaken on time, the risk of collapse can be minimized, and human lives, nature, and goods can be preserved. In addition, the monitored structure is perceived as safe, which improves the general public's feeling of safety and security.
- Improved management: Objective and reliable information can allow for planning and reducing life-cycle operation costs and decrease of economic losses for the owner of the structure (e.g., due to untimely, premature or tardive, preservation actions, maintenance, repair, retrofit, and reconstruction), but also for users (e.g., longer and more expensive travel and delivery times due to unavailability of deficient bridge); early detection of structural malfunctioning allows for an in-time intervention that involves limited maintenance costs (Radojicic et al. 1999); well-preserved and maintained structures are more durable and resilient, and increase of durability and resilience decreases the direct and indirect economic losses (Frangopol et al. 1998).
- Improved knowledge: As new materials, new construction technologies, and new structural systems are used more and more in practice with no extended experience, it is necessary to increase the knowledge of their long-term structural behavior on-site in order to verify the design, confirm structural safety, and calibrate and improve numerical models (e.g., Bernard 2000).
- Improved use of existing structures: Unexpected structural reserves discovered in functionally obsolete structures lead to better exploitation of traditional materials and better use of existing structures; if the structure can accept a higher load, increased performance is obtained without additional (e.g., replacement or retrofit) costs (Glisic and Inaudi 2007).
- Improved resilience of society: At the local geographical level, SHM can help evaluate the pre-event state of the structure and indicate actions that, in turn, can help prevent adverse social, economic, ecological, and aesthetic impacts that may occur in the case of structural deficiency caused by a natural or human-induced disaster (event); at the wider geographical

level, SHM can help with post-event assessment of structural health and performance and inform restoration planning of manifold structures.

– Improved sustainability: All the above benefits lead to less use of construction materials and a reduction of unnecessary detours for users, which in turn eventually leads to less adverse interference with the environment, less gas emissions, and less public and private spending.

Thus, SHM can be defined (based on experience, e.g., Glisic et al. 2010) as a process aimed at providing accurate, reliable, and actionable information concerning structural health condition and performance in quasi-real-time (Glisic and Inaudi 2007).

The information obtained from monitoring is used to yield one or a combination of the above benefits. In a broader sense, SHM helps prevent the adverse social, economic, ecological, and aesthetic impacts that may occur in the case of structural deficiency and is critical to the emergence of sustainable and resilient civil engineering.

Structures have been inspected (and maintained and repaired) on a periodic basis since ancient times. The first qualitative (quasi-continuous) monitoring of various machines was performed in the late nineteenth century. Sensor-based SHM of civil structures started in the 1920s (Glisic 2022), with applications on tunnels and dams under construction. Due to their importance and devastating consequences of failure, dams were among the first structures that were continuously monitored in the long term, while more detailed and systematic inspection of bridges started in the 1960s and 1970s. Until the early 1990s, SHM involving sensors was relatively sporadically performed on bridges and buildings, mainly on those with identified deficiencies, in short to medium terms. The main challenges during these times were technological limitations related to the long-term reliability of SHM systems operating in real-life settings, remote access to SHM systems (readings were often performed manually), and the management and analysis of big SHM data. Technological developments during and after the 1990s in the domains of new materials, metrology, informatics, and telecommunications made possible the large-scale application of SHM for structures. Since then, the number of structures equipped with SHM is rapidly growing, as the research, development, and applications push the boundaries of this new multidisciplinary branch of engineering. Some of the critical challenges being addressed by research are listed below:

– Selection of appropriate damage-sensitive features: a damage-sensitive feature is a quantifiable property or a data pattern sensitive to unusual structural behavior (e.g., damage, deterioration, or lack of performance), which can be either directly monitored or determined by analyzing monitoring data in order to detect (ascertain the existence) unusual structural behaviors; typical damage-sensitive features are strain, stiffness, natural frequency, damping, position of center of stiffness, deformed shape, etc.; challenges related to selection of appropriate damage-sensitive features are in its potentially small sensitivity to unusual structural behaviors (e.g., cracking at locations distant from the strain sensor might not be detected by that sensor) and in variabilities of usual structural behaviors that, in some cases, may be indistinguishable from the influence of unusual structural behaviors (e.g., usual variability of strain due to temperature might be wrongly interpreted as unusual or might "mask" the unusual structural behavior, which would then remain undetected).

– Reliable detection and characterization of unusual structural behaviors in real-life settings (on-site conditions): this challenge results from a various forms of manifestation of damage (e.g., cracking , corrosion, etc.), large volume of the structure (i.e., numerous points at which the damage can occur), uncertainties related to loads and construction materials, temperature and humidity variations, lack of data on the undamaged state, uncertainties in setting the thresholds for damage-sensitive features, etc.

- Management of big SHM data (e.g., processed and unprocessed measurements, static and dynamic measurements, camera images, etc.) and metadata (e.g., technical drawings of structure, properties of construction materials, specifications of SHM systems, etc.), including data storage, documentation, accessibility, and visualization; the challenges originate from SHM data and metadata heterogeneity and from the need to provide meaningful access to them by diverse groups of users (e.g., managers, engineers, and inspectors) in a simple and intuitive manner.
- SHM data analysis, including the transformation of data into a meaningful and actionable information on current structural health condition and performance and establishing the prognosis of future structural behaviors; challenges result from the need to create data analysis algorithms that are robust in real-life settings and numerical models that accurately reflect the behavior of the structure.
- Evaluation of SHM benefits: the challenge is that SHM involves immediate measurable initial costs (e.g., cost of hardware and installation on structure), but promises benefits only in long term, and the monetary value of benefits is difficult to evaluate in advance due to a lack of experience and evaluation models; challenge in evaluation of SHM benefits may lead to reluctance to implement SHM in long term, despite the benefits that it enables.
- Education gap between researchers and practitioners makes the latter unaware of the potential of SHM techniques; publishing successful SHM applications and introducing SHM in university curricula may help bridge the gap and educate a new generation of engineers.

In spite of its importance, the SHM is not yet widely applied, mostly due to the challenges presented above. However, a growing body of research outcomes combined with an increased number of successful applications in the first two decades of the twenty-first century are visibly accelerating the maturation and general adoption of SHM by practitioners.

1.2 SHM Process

1.2.1 General Observations

The SHM is a complex process, full of delicate phases, and only proper and detailed planning of each of its steps can lead to its successful and maximal performance (Glisic and Inaudi 2007). There are several approaches in the literature to defining and describing the SHM process, depending on the level of sophistication and applications (e.g., Glisic and Inaudi 2003a, Ansari 2005, Wenzel and Pichler 2005, Andersen and Fustinoni 2006, Balageas et al. 2006, Farrar and Worden 2007, Glisic and Inaudi 2007, Karbhari and Ansari 2009, Wenzel 2009, etc.). In this book, a combination of various aspects of the abovementioned approaches is compiled and modified to adapt to specific requirements related to civil structures and infrastructure.

1.2.2 Definition and Four Dimensions of SHM

Two definitions of SHM found in literature that are equally concise and accurate are given below.

- SHM is a process aimed at providing actionable and reliable quasi-real-time information concerning structural health conditions and performance (adapted from Glisic and Inaudi 2007).
- The process of implementing an identification strategy for unusual structural behaviors (e.g., damage) is referred to as SHM (adapted from Farrar and Worden 2007).

The main difference between the two above definitions is that the first is targeting structural health condition and performance as the primary aim, with assumption that the unusual structural behaviors would then be identified as the alteration of health condition and performance, while the second is targeting unusual structural behaviors as the primary aim, with the assumption that this will lead to ascertaining the health condition and performance. In both cases, the SHM process consists of permanent, continuous, periodic, or periodically continuous recording of parameters that, in the best manner, reflect the health condition and performance of the structure (Glisic and Inaudi 2003a).

Thus, it involves repetitive quantification (measurement or rating) of monitored parameters over a certain period (or periods) of time and transformation of the collected data into an actionable information on structural health condition and performance. Here, "actionable" means that the provided information can be used by the manager of the monitoring structure to perform an informed action toward improvement of structural health condition and performance.

The following four "dimensions" of the SHM process can be identified:

- Monitored parameters
- Timeframe and pace
- Level of sophistication
- Scale of application

1.2.2.1 Monitoring Parameters
The selection of parameters to be monitored, that are representative of structural health and performance, depends on the aims of SHM and several structure-specific factors, such as its type and the purpose, construction materials, expected loads, environmental conditions, and expected damage and degradation phenomena. The monitored parameters could be either damage-sensitive features themselves (e.g., strain and deflection) or they are used to extract damage-sensitive features (e.g., modal characteristics determined from vibration monitoring or position of centroid of stiffness of a cross section determined from strain monitoring). In general, the monitored parameters can be mechanical, geometrical, physical, chemical, or environmental, and they can be monitored directly or indirectly. Commonly monitored parameters are given in Table 1.1 (the list is not exhaustive).

1.2.2.2 Timeframe and Pace
Depending on the aims of a specific SHM project, the type of monitored structure, its exposure to natural and human-induced hazards, and its perceived condition, the SHM can be performed in the short term (typically from few hours to several days), medium term (few weeks to few months), long

Table 1.1 The most frequently monitored parameters.

Type	Parameter
Mechanical	Strain, acceleration, deformation, displacement, crack opening, wave propagation, force,[a] stress,[a] natural frequency,[a] mode shape,[a] damping,[a] centroid of stiffness,[a] etc.
Geometrical	Survey, camera image, laser scanning, photogrammetry, etc.
Other physical	Temperature, humidity, pore pressure, electrical conductivity, etc.
Chemical	Chloride penetration, sulfate penetration, pH value, carbonatation penetration, oxidation of rebars, oxidation of steel, decaying of timber, etc.
Environmental	Ambient temperature, ambient humidity, wind speed, loading

a) These parameters are determined indirectly from strain, acceleration, etc.
Source: Adapted from Glisic (2009).

term (typically several months to several years), or during the whole lifespan of the structure. When deciding on implementation of SHM, it is important to recognize that SHM should not be applied only to structures with recognized deficiencies for the following reasons (Glisic and Inaudi 2007):

– Structure with recognized structural deficiency performs with limited capacity; thus, the economic losses are already generated, and the owner and the users are already adversely affected.
– The history of processes and events that lead to structural deficiency is not registered, which makes it difficult to identify them, understand their impact on structure, and establish a diagnosis.
– The information concerning the healthy structural behaviors is very important as a reference (baseline), notably for structures whose structural behavior is complex and unusual structural behaviors might not be always ascertained by direct comparison with design and numerical models due to their inherent large uncertainties (e.g., long-term monitoring of concrete structures, where rheological effects make an accurate modeling very challenging).

Consequently, whole lifespan monitoring, which includes all the important phases in the structure's life, i.e., construction, testing (if any), service, refurbishment (if any), and dismantling (if any), is highly recommended (Glisic et al. 2002a).

Recording monitoring parameters can be static (e.g., strain and temperature), dynamic (e.g., strain and acceleration), tomographic (e.g., wave propagation and digital image correlation), or combined. Different pace of recording parameters results in different data analysis and management requirements, as dynamic monitoring can result in data sets that are several orders of magnitude larger than those of static monitoring, and tomographic monitoring can result in data sets that are several orders of magnitude larger than those of dynamic monitoring.

1.2.2.3   Level of Sophistication

Depending on the level of sophistication, the SHM may be able to perform the following actions:

– Detect unusual structural behaviors in the structure; in literature, the capability of SHM to detect unusual behaviors (e.g., damage, deterioration, or lack of performance) is often referred to as "Level 1 monitoring" (Rytter 1993).
– Identify the time or timeframe of occurrence of unusual structural behavior; the former is for quasi-instantaneous occurrence (e.g., cracking), while the latter is for slowly developing behaviors (e.g., settlement of foundations and corrosion); to keep consistency with literature, the capability of SHM to identify the time or timeframe of occurrence of unusual structural behaviors could be referred to as "level 1.5 monitoring."
– Indicate the physical position (geographical coordinates) of the detected unusual behavior on the structure; in the literature, the capability of SHM to localize unusual behaviors is often referred to as "level 2 monitoring" (Rytter 1993).
– Quantify or rate the detected unusual structural behavior (e.g., size of crack and depth of chlorine penetration); in the literature, the capability of SHM to quantify or rate the severity of the unusual structural behavior is often referred to as "level 3 monitoring" (Rytter 1993).
– Evaluate or predict structural safety and performance and provide actionable information; in the literature, the capability of SHM to estimate residual capacity and lifespan of the structure (prognosis based on which the actions can be executed) is often referred to as "level 4 monitoring" (Rytter 1993).

Note that, in general, SHM is not supposed to make a diagnosis (ascertain the origin or cause of the unusual structural behavior) and propose remedy actions against the detected unusual structural behavior; to make a diagnosis and propose the remedy, it might be necessary to perform additional site inspections and perform further analyses.

In this book, the combination consisting of ascertaining the timeframe of occurrence (Level 1.5 SHM), the localization (Level 2 SHM), and the quantification of unusual structural behavior (Level 3 SHM) is referred to as characterization of unusual structural behavior.

While Level 1 SHM could be achieved in real applications, other levels are more challenging, and difficulty rises with the targeted level of sophistication. Levels 1.5, 2, and 3 SHM (i.e., characterization of unusual structural behavior) can be achieved with an appropriate strain-based approach (limited to those unusual structural behaviors that are perceivable by strain-related parameters). Level 4 SHM is "grand challenge" of SHM as it requires not only the achievement of Level 3 SHM but also numerical modeling of structure and damage and degradation phenomena, which all involve important measurable and non-measurable (epistemic) uncertainties. Nevertheless, strain-based SHM approaches often successfully reach Level 4 SHM.

1.2.2.4 Scale of Application

Strain-based SHM can be carried out at the local scale, the global scale, or the integrity scale. Monitoring at the local scale can target local materials or local structural properties. As the damage frequently has local characteristics, it helps detect the damage in the monitored area of the structure. However, local-scale monitoring gives limited information concerning the behavior of the structure as a whole, and its damage detection capability is challenged if damage occurs in a non-monitored area. Monitoring at the global level involves large parts of structures or the entire structure and provides better information related to the global behavior of structures as a whole, and indirectly, through the changes in structural behavior, it may provide some information related to material performance. Integrity monitoring encompasses certain aspects of both local and global monitoring and combines damage detection capability with an assessment of global structural behavior. The local, global, and integrity monitoring are presented in more detail in Chapters 5–7, respectively. The four dimensions of SHM are shown in Figure 1.2.

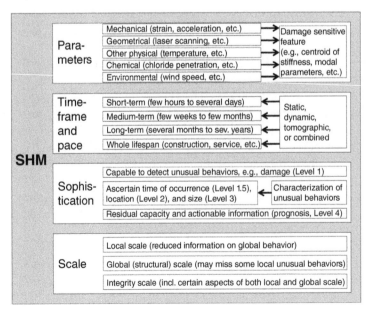

Figure 1.2 Four dimensions of SHM.

As a conclusion of this subsection, it is important to highlight that for each dimension of SHM, there are challenges that are related to technical limitations and the cost. For example, digital image correlation using cameras can be challenging in real-life settings due to technical limitations (e.g., unfavorable lighting conditions and obstructed line of sight), while integrity monitoring can be challenging due to elevated costs (e.g., to instrument a significant volume of structure with sensors). Achievement of long-term Level IV SHM can be challenging due to both technical limitations (e.g., to achieve prognostication) and elevated costs (e.g., potential need for a large number of sensors).

1.2.3 SHM Core Activities

There are five core activities of the SHM process (Glisic and Inaudi 2003a, 2007):

(i) Selection of the monitoring strategy.
(ii) Installation of the monitoring system.
(iii) Maintenance of the monitoring system.
(iv) Data analysis and management.
(v) Closing activities (in the case of interruption of monitoring).

Each of these activities can be further split into sub-activities, as presented in Figure 1.3.

Each of the core activities is very important, but the most important is to create an appropriate monitoring strategy. The monitoring strategy is influenced by each of the other core activities and sub-activities and consists of

– Establishing the aims of SHM.
– Identifying and selecting representative parameters to be monitored.
– Selecting appropriate monitoring systems.

Monitoring strategy	Installation of SHM system	Maintenance of SHM system	Data analysis & management	Closing activities
Establishing aims of SHM	Installation of sensors	Collecting data (reading of sensors)	Data processing (cleansing, etc.)	Interruption of monitoring
Selection of monitored parameters	Installation of accessories (boxes, cables...)	Providing for electrical supply	Data interpretation	Dismantling of monitoring system
Selection of monitoring systems	Installation of reading units and firmware	Providing for communication lines (wired or wireless)	Data analysis	Storage of monitoring components
Design of sensor network	Installation of data analysis software	Implementation of maintenance plans for hardware	Storage of all SHM data and metadata	
Schedule of monitoring	Interface with users (software)	Firmware and software upgrades	Integration of data and metadata	
Data exploitation plan		Repairs and replacements	Visualization of data and metadata	
Costs			Accessibility to data and metadata	
			Export of data and metadata	
			The use of data	

Figure 1.3 SHM core activities and sub-activities. Source: Adapted from Glisic and Inaudi (2007).

- Designing the sensor network.
- Establishing the pace of monitoring (monitoring schedule).
- Planning data analysis, management, and exploitation.
- Costing the SHM.

To start an SHM project, it is important to define its aims and to identify SHM dimensions related to the aims, i.e., to select parameters and damage-sensitive features to be monitored, timeframe and pace (long term, medium term, short term, or whole lifespan; static, dynamic, tomographic, or combined), level of sophistication (1–4), and scale of SHM (local, global, or integrity). Any combination of SHM dimensions is characterized by advantages and challenges, and which one (or ones) will be used depends mainly on the expected structural behavior and the aims of monitoring.

The parameters to be monitored must be properly selected in a way that reflects the structural behavior. Each structure has its own particularities and, consequently, its own set of indicative parameters to be monitored. In some cases, the monitoring parameters are "obvious" (e.g., acceleration or dynamic strain monitoring for evaluation of behavior during an earthquake). However, identification (detection and characterization) of unusual behaviors may require the selection of a damage-sensitive feature that is not directly measurable but requires extraction from the data (e.g., determination of centroid of stiffness based on strain measurements).

SHM involves measuring the magnitude of the monitored parameters and recording the time and value of the measurement. In order to perform a measurement and register it, one can use different types of apparatus. The set of all the means destined to carry out a measurement and register it is called an SHM system (Glisic and Inaudi 2007). Nowadays, there are many types of SHM systems, based on different functioning principles. In general, however, they all have similar components: sensors, carriers of information, reading units, interfaces, and data analysis and management subsystems (e.g., software), see Chapter 2.

The selection of an SHM system mostly depends on the monitoring aims, dimensions of SHM related to the monitoring aims (selected parameters, timeframe, level of sophistication, and scale of SHM), required specifications of the SHM system such as accuracy measures (resolution, precision, and limits of error), frequency of reading, compatibility with the environment (sensitivity to electromagnetic interference, temperature variations, humidity, etc.), installation procedures for different components of the monitoring system, possibility of automatic functioning, remote connectivity, data analysis and management, and the available funds.

For example, long-term strain monitoring of a new concrete structure subjected to dynamic loads at the global (structural) scale should be performed using sensors that are not influenced by local material defects or discontinuities (such as cracks and inclusions). Since short-gauge sensors are subject to local influences, a good choice is to use a monitoring system based on long-gauge or distributed sensors. In addition, the sensors should be embeddable in the concrete, insensitive to environmental influences (for long-term reliability), and the reading unit must be able to perform both static and dynamic measurements with a predefined frequency and accuracy.

Several parameters are often required to be monitored, such as average strains and curvatures in beams, slabs, and shells; average shear strain; deformed shape and displacement; acceleration; and crack occurrence and quantification. That is why the ease of integration of various SHM technologies should also be considered when making a selection.

In order to extract valuable information from the system, it is necessary to place the sensors in representative positions on the structure. The sensor network to be used for monitoring mostly depends on the monitoring aims, geometry, the type of structure to be monitored, selected parameters, the scale and level of sophistication of the SHM, and the costs of hardware and installation.

The schedule and pace of monitoring depend on the timeframe of monitoring and on the expected rate of change of the monitored parameters. The schedule can be modified and adapted to various phases of the structure's life (construction, testing, service, repair and refurbishment, and dismantling). Also, the schedule can be modified upon detection of an unusual structural behavior (e.g., structure that may collapse shortly after unusual behavior is detected should be monitored continuously, with maximum frequency of measurement). For some applications, periodic monitoring gives satisfactory results, but information that is not registered between two measurement sessions is lost forever. Only quasi-continuous monitoring during the whole lifespan of the structure can completely register its history, help understand its real behavior, and fully exploit the monitoring benefits.

The installation of the SHM systems is a particularly delicate phase (Glisic and Inaudi 2007). Therefore, it must be planned in detail, seriously considering on-site conditions and notably the structural component assembly activities, sequences, and schedules. The components of the monitoring system can be embedded (e.g., into the fresh concrete or between the composite laminates) or installed on the structure's surface using fastenings, clamps, or adhesives. The installation may be time-consuming, and it may delay construction work if it is to be performed during construction of the structure. For example, components of a monitoring system that are to be installed by embedding in fresh concrete can only be safely installed during a short period between rebar completion and pouring of concrete. Hence, the installation schedule of the monitoring system has to be carefully planned to take into account the schedule of construction works and the time necessary for the system installation. At the same time, one has to be flexible in order to adapt to work schedule changes, which are frequent on building sites.

When installed, the monitoring system has to be protected, notably if monitoring is performed during the construction of the structure. The protection has to prevent accidental damage during the construction and ensure the longevity of the system. Thus, all external influences, periodic or permanent, have to be taken into account when designing protection for the monitoring system.

Data analysis and management is a particularly important SHM core activity. It is not rare in practice to find SHM only partially used or even completely abandoned due to improper analysis or management of large amounts of data (Glisic and Inaudi 2007). That is why data analysis and management should be addressed properly when establishing a monitoring strategy. Data analysis and management practically starts with data collection, although the data collection technically falls into maintenance activity of SHM process. It can be performed manually, semi-automatically, or automatically, on site or remotely, periodically or quasi-continuously, statically or dynamically. These options can be combined in different ways depending on SHM aims and schedule; for example, during testing of a bridge, it is necessary to perform measurements semi-automatically (i.e., the measurements are triggered by the operator once the loads are properly placed on the structure), on site, and periodically (after each load step). For long-term in-use monitoring, the maximal performance is automatic, remote (from the office), and quasi-continuous collection of data, without human intervention. Besides the SHM data, every project also has SHM metadata – dimensions of structure, locations of sensors (i.e., their geographical coordinates), specifications of SHM system components (e.g., resolution and manufacturer information), etc. Data management should enable integration, correlation, and easy visualization of SHM data and meta-data.

Data are usually automatically processed (cleansed from outliers, filtered for specific purpose, compensated for environmental influences, and normalized for specific parameters) and stored, for example, in the form of numbers, arrays of numbers, strings (words), images, or video streams on different types of data supports, such as electronic files (on hard disk, CD, etc.) or hard versions (printed on paper). The manner of storage of data has to ensure that data will not be lost (data stored in a "central library" with backups) and that prompt access to any selected data is possible (e.g., one can be interested to accessing only data from one group of sensors during a selected period of time). Modern SHM systems generate large amounts of data, and in order to easily and promptly access the data, the latter should be stored in a single database. For the storage of large amounts of data, there are several options in the market, and they have different properties and payment models. In order to contain the costs, data may also be compressed and stored on local hard disk. The software that manages the collection and storage of data is to be a part of the monitoring system (see Chapter 2). Otherwise, data management can be difficult, demanding, and expensive.

Collected data frequently consist of a huge amount of numbers (e.g., time series containing dates and magnitudes of monitoring parameters) and have to be transformed into actionable information concerning the structural behavior, enabled by data analysis. This transformation depends on the established SHM strategy and algorithms that are used to interpret and analyze the data. This can be performed manually, semi-automatically, or automatically. Automatic data analysis is the most convenient approach. Finally, based on the information obtained from data analysis, planned actions can be undertaken (e.g., warnings can be generated and the use of the structure restricted or stopped in order to guarantee safety). The data exploitation has to be planned along with the selection of the monitoring strategy, and appropriate algorithms and tools compatible with the established SHM system have to be selected.

The monitoring strategy is often limited by the budget available. From a monitoring performance point of view, the best is to use versatile monitoring systems, dense sensor networks (many sensors installed in each part of the structure), and software allowing remote and automatic management and analysis. On the other hand, the cost of such an SHM can be very high and unaffordable. That is why it is important to develop an optimal monitoring strategy, providing appropriate overall performance of the SHM (four dimensions of SHM) and which is also affordable in terms of costs. For example, based on experience, an estimated budget for SHM of a new structure typically ranges between 0.5% and 1.5% of the total cost of the structure (Glisic and Inaudi 2007).

1.2.4 Parties Involved in SHM

The main parties (entities) involved in SHM are the monitoring authority, the consultant, the monitoring companies, and the contractors (Glisic and Inaudi 2007). These parties should collaborate very closely with each other, so an efficient and performing SHM can be created and implemented.

The monitoring authority is the entity that is interested in and decides to implement SHM. It is typically the owner of the structure or the entity that is responsible for management, safety, or security of the structure (e.g., public agency). The monitoring authority finances the monitoring and benefits from it. It is responsible for establishing the monitoring aims and for approving the proposed monitoring strategy. The same authority is later responsible for maintenance of SHM and data management (directly or by subcontracting to the monitoring company or contractor).

The consultant's role is to propose a monitoring strategy to the monitoring authority based on analysis of the structural system, estimated loads, numerical modeling, risk evaluation, etc. The consultant may propose several SHM options and create an alternative monitoring strategy if the initial one is rejected by the monitoring authority. The consultant may perform supervision of the production, delivery, installation, and commissioning of the monitoring system.

The monitoring company is usually a manufacturer of monitoring systems or integrator of monitoring solutions. It is essentially responsible for delivery of the monitoring system, and it guarantees the SHM system performance.

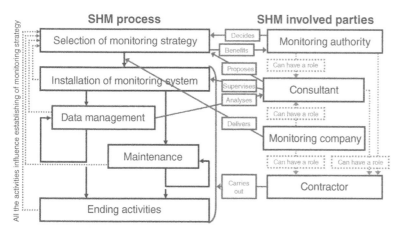

Figure 1.4 The interaction between the SHM involved parties and SHM core activities. Source: Adapted from Glisic et al. (2010).

The contractor is responsible for installation of the monitoring system. In order to perform timely and quality installation, the support of the monitoring company and the monitoring authority are often necessary.

The SHM entities need not necessarily be mutually exclusive (different); for example, if a monitoring authority has engineering and contractor sections (e.g., traffic authorities or departments of transportation), then it can also have a role of a consultant or contractor. Similarly, if the monitoring company is a system integrator, then it is likely that it will also have an engineering section and the capability to deploy the SHM system (i.e., it can perform the roles of consultant and contractor). The interaction between the parties involved in SHM and SHM core activities is given in Figure 1.4 (modified from Glisic et al. 2010).

To illustrate the SHM process on a real application, an example of properly managed SHM project is briefly presented in the next section.

1.3 Gota Bridge – Example of an SHM Project

The Gota Bridge (Götaälvbron) in Gothenburg, Sweden, built in 1936–1939, consisted of a concrete deck slab supported by nine continuous steel girders (Enckell et al. 2011). The bridge was approximately 950 m long, and the girders with deck were supported on columns over more than 50 spans of variable length (see Figure 1.5). The bridge had a 20-m-long bascule that opened on demand to allow passage for tall ships.

The Gota Bridge represented the main communication line across the Göta River that connected Hisingen and Gothenburg City. It carried both the road and the light-rail traffic and had pedestrian and bicycle lanes as well. The traffic volume was expected to increase from 26,000 vehicles per day in 2008 to 40,000 per day in 2020 (Myrvoll et al. 2009).

Trafikkontoret, the Gothenburg City Traffic Authority, recognized that the bridge's service life was close to its theoretical design life of 80 years, and some deterioration or damage was expected to occur. A bridge inspection carried out in 1999 revealed cracks in the flanges of steel girders above the support columns (see Figure 1.6), that were caused by the combined effect of fatigue over 60 years of service and low quality of steel (Myrvoll et al. 2009). The bridge was repaired (cracks were closed), and additional structural improvements were made with the aim of keeping the bridge

Figure 1.5 View of Gota Bridge (Glisic and Inaudi 2007). Source: Courtesy of SMARTEC.

Figure 1.6 Example of the damaged flange of the Gota Bridge. Source: Reproduced by permission of Norwegian Geotechnical Institute (NGI).

in service until 2020, the time estimated as necessary to plan a new crossing (bridge or tunnel) and replace the existing bridge. Finally, a new bridge was opened in 2021 (Hisingen Bridge), and Gota Bridge was decommissioned in the same year. Its dismantling was completed in 2022.

Restriction of acceptable axle load of vehicles was imposed, and comprehensive investigation and analysis was carried out by the Traffic Authority. The steel quality was found to be variable along the length of the bridge and more brittle than expected (Thomas steel). Low and variable steel quality as well as unfavorable column–girder connections resulted in risk of further cracking of the bridge girders, due to cyclic traffic and temperature loads, and corrosion (Myrvoll et al. 2009, Enckell et al. 2011). In addition, bascule opening caused unsymmetrical static loads during ship transits, especially in rush hours. The risk of further cracking was particularly emphasized during cold temperatures in the winter. In order to mitigate the risk of further cracking, the Trafikkontoret (monitoring authority) decided to equip the bridge with SHM.

The overall objective of the SHM, set by the Trafikkontoret, was to automatically detect, localize, and report excessive elastic strains and occurrence of cracks in the steel girders.

Trafikkontoret, as the monitoring authority in the project, hired Norwegian Geotechnical Institute, Norway (NGI) as a consultant who proposed monitoring of the bridge at integrity scale, using
distributed fiber optic sensors (see more on distributed fiber optic sensors in Chapter 3 and on
integrity monitoring in Chapter 7). The monitored parameters, selected by NGI, were (i) strain,
(ii) cracking, indirectly through strain changes, and (iii) temperature, for thermal compensation
purposes, as well as to discriminate thermal strain from elastic strain. Thus, the damage-sensitive
feature in this project was mechanical (elastic) strain. The timeframe of SHM was set to long
term, i.e., from 2007 to at least 2020 (de facto remained until 2021), and the level of sophistication
included detection and characterization of high strain values by SHM system (Level 3), detection
and localization of cracks by SHM system (Level 2), and evaluation of capacity of the bridge under
high strain or cracking, or upon any other event detected by SHM (Level 4).

The SHM system that could fulfill these requirements was identified to be a distributed fiber
optic sensing system based on stimulated Brillouin scattering (Glisic et al. 2007). The monitoring
company selected for the project was SMARTEC SA, Switzerland, with its supplier Omnisens SA,
Switzerland.

The specifications of the system were developed by NGI (consultant) in collaboration with
SMARTEC (monitoring company) and approved by Trafikkontoret (monitoring authority). These
specifications are given in Table 1.2.

Besides requirements presented in the table, the SHM system had to have self-monitoring capabilities, i.e., the capacity to report any malfunctioning or failure of the SHM system itself.

The sensor network was determined by NGI in collaboration with Trafikkontoret and SMARTEC;
it was decided to install distributed strain sensors (see Section 3.6.3) along the upper flange of the
five main girders (see Figure 1.7). This solution was judged as optimal, taking into account the aims
of the SHM and the available budget.

The lengths of individual sensors were approximately 90 m. Two reading units were needed for
the project, one on the north side and one on the south side. Each reading unit had a 32-channel
switch, and the sensors were read sequentially (one after another) in a quasi-static mode (the
duration of each measurement was slightly below 10 minutes). Based on the number of individual

Table 1.2 Specifications of the SHM system as applied to Gota Bridge (Glisic et al. 2007, Myrvoll et al.
2009).

Property	Values
Strain resolution	$\pm 3\,\mu\varepsilon$
Strain limits of error	$\pm 21\,\mu\varepsilon$
Strain range	-5000 to $+10\,000\,\mu\varepsilon$
Crack detection and localization	0.5–5 mm (over 0.1 m) with 99% probability
Temperature limits of error	$\pm 1°C\ (\pm 1.8°F)$
Temperature range	-30 to $+85°C\ (-22$ to $+185°F)$
Spatial resolution	1 m (3′–3″)
Sampling interval	0.1 m (4″)
SHM coverage	5 girders $\times \sim 1000$ m $= \sim 5000$ m ($\sim 16,405'$)
Measurement time per sensor	<10 min
Measurement time for whole system	<2 h

Figure 1.7 Left: View of girders equipped with sensors; right: positions of strain and temperature sensors on a girder and deck respectively. Source: Courtesy of SMARTEC.

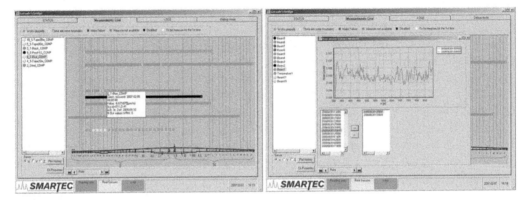

Figure 1.8 Examples of data visualization on a central server dedicated to SHM. Source: Courtesy of SMARTEC.

sensors and the range of measurements, the schedule (pace) of measurements was set to two hours (one reading over all the sensors takes a little less than two hours).

The installation was performed in several phases by Bemek, Sweden; Royal Technical University (KTH), Stockholm, Sweden; NGI; and SMARTEC (all having the role of contractors). The strain sensors were bonded to the upper flange of the girders, and the adhesive used was selected based on a series of tests.

The conformity of the SHM system to specifications was tested in a series of factory acceptance tests (FAT) before delivery and with site acceptance tests (SAT) once the system was installed onto the bridge. All FAT procedures were designed by SMARTEC in collaboration with NGI and were performed under NGI supervision.

The collected data were transmitted and stored on a dedicated server. Data visualization and analysis software were installed on the server, accessible via the internet by authorized personnel. Data visualization was simple and intuitive, as shown in Figure 1.8. In the event of a high strain or crack occurrence, an "alarm" message would be generated, while in the case of the SHM system malfunctioning, an "alert" message would be generated. The messages were sent in the form of e-mails and mobile phone messages (for redundancy purposes) to responsible personnel in NGI, Trafikkontoret, and SMARTEC, who could then react upon the data exploitation plan developed by Trafikkontoret. Alarm messages included the type of event (high strain or crack), location on the bridge, and for high strains the magnitude and whether the threshold was exceeded due to a

Figure 1.9 Left: Four specimens equipped with sensors and connected to the SHM of Gota Bridge; right: detail of the gap. Source: Courtesy of SMARTEC.

short-term event (e.g., overloading and plastic deformation) or long-term event (e.g., slow settlement of foundations and rheological redistribution of the loads). The alert message would point to the issue on the SHM system (e.g., unintentional restart of the system, interruption of power, and malfunction of the communication line).

Upon the installation of the system, several SAT were performed before official commissioning. SAT were designed by NGI in collaboration with SMARTEC and performed under the supervision of NGI, Trafikkontoret, and an independent reviewer hired by Trafikkontoret.

For SAT purposes, four steel specimens were manufactured, painted with the same paint as the bridge, and equipped with distributed strain sensors in the same manner as on the bridge, in order to simulate real conditions (see Figure 1.9). Each specimen consisted of two aligned 1.25-m-long steel beams with a U-shaped cross section. Specimens were equipped with screws on the wooden support, so the distance between the two parts of each specimen could be controlled. A gap opening,

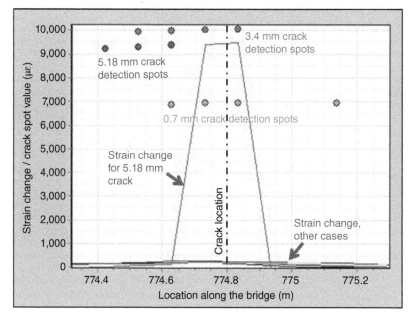

Figure 1.10 Example of successful crack detection and localization on specimens tested during the Gota Bridge SAT (Enckell et al. 2011). Source: Adapted from Enckell et.al., 2011.

simulating a crack, was created by a screw. The gap opening was measured using a mechanical caliper attached to metallic bolts welded on both sides of the gap. The view of the closed gap is shown in Figure 1.9.

The four specimens were connected to the bridge SHM system. Comprehensive crack opening scenarios were simulated in order to evaluate the performance of the monitoring system (Enckell et al. 2011). The capacity of the system to detect the cracks and send the correct messages was then tested. In addition, malfunctioning of the system was also tested by artificially introducing problems (e.g., disconnection of a sensor and interruption of power), and the capability of the system to report these problems by sending correct messages was verified. An example of the results is presented in Figure 1.10.

Upon successful SAT, the SHM was commissioned and put into service. After a one-year testing period, dedicated to debugging and proofing, the system was fully handed over to Trafikkontoret. Under the contract, NGI, as the consultant, supervised the function of SHM system, analyzed the data, and provided periodical reports to Trafikkontoret. The SHM system was continuously in use until the end of service life of the bridge, i.e., over a 14-year period (2007–2021).

2

SHM System Components and Properties

2.1 SHM System

In order to implement structural health monitoring (SHM), it is necessary to instrument the structure with an SHM system. There are several definitions of SHM systems found in literature (e.g., Andersen and Fustinoni 2006, Glisic and Inaudi 2007, Wenzel 2009), and they are all similar and valid. For the purposes of this book, we will simply define an SHM system as a *totality of means used to achieve SHM*.

While there is a large diversity of SHM systems available on the market, they all have similar subsystems and components, as shown in Figure 2.1.

It is common in real projects to combine several SHM systems in order to achieve the aims of monitoring. The combination of the SHM systems is referred to as heterogeneous SHM system. While in some cases it is possible to interface different SHM systems, this is commonly not done, and they are kept independent from each other. However, the data they collect is typically integrated, analyzed, and managed holistically.

The two main subsystems of an SHM system (and consequently any heterogeneous SHM system) are measuring subsystem and data analysis and management subsystem. The measuring subsystem aims at acquiring quantitative or qualitative data related to the observed parameters (often called measurands) of the structures. The data analysis and management subsystem is responsible for all activities related to data and metadata, i.e., data processing, analysis, interpretation, storage, integration, visualization, accessibility, export, and the use of the data. Depending on how the SHM system is conceptualized, the processing of the data, i.e., transformation of the encoding parameter into the measurand (see the next section for clarification), can be either part of measuring subsystem or the data management subsystem.

The typical components of the measuring subsystem are sensors, carriers of information (wired or wireless), accessories (including, but not limited to cable extensions, connection or junction boxes, and protective ducts and containers), reading units (including channel switches and/or wireless nodes if any), firmware, user interface, and self-monitoring software (with or without an associated computer or server). If the infrastructure for power supply and communication is not available on-site, then it should be provided and considered as a set of components of SHM system. The components of SHM system are schematically illustrated in Figure 2.2.

The aim of the sensor is to measure the magnitude of monitored parameter (measurand) and transform it into a signal readable by operator or reading unit. The carriers of information transfer the signal from the reading unit to the sensor and vice-versa. Reading unit decodes the signal,

Introduction to Strain-Based Structural Health Monitoring of Civil Structures, First Edition. Branko Glišić.
© 2024 John Wiley & Sons Ltd. Published 2024 by John Wiley & Sons Ltd.

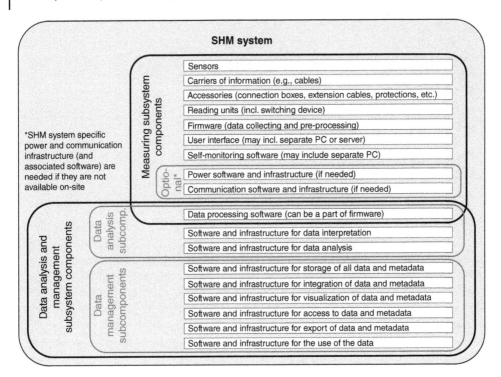

Figure 2.1 Typical subsystems and components of an SHM system.

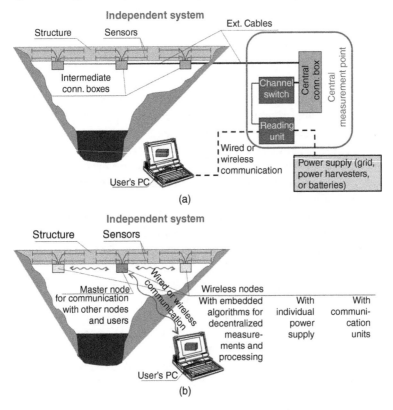

Figure 2.2 Schematic examples of typical components of SHM measuring subsystem with (a) wired and (b) wireless data transfer. Source: Modified from Glisic (2000, 2009).

retrieves the magnitude of the monitored parameter, and transforms it into presentable information. It is frequently equipped with a channel switch for reading multiple sensors. Firmware (internal reading-unit software) enables programming of the internal electronics, reading, and processing of the signals. Wired SHM systems usually have one or a few reading units, while wireless SHM systems can have large number of reading units, i.e., each wireless node can have a reading unit capable of data processing and partial data management. The aim of user interface software is to make interaction between the system and the user possible. It can be integrated into the reading unit (wired or wireless) or installed on a separate server or computer. Finally, the self-monitoring capability (i.e., the capacity to report any malfunctioning or failure of the SHM system itself) can be either embedded in the firmware or require separate external software installed on a dedicated server or personal computer.

The data analysis and management subsystem consists of hardware infrastructure (servers, computers, and global and local communication lines, e.g., internet and local area networks) and various software components (for processing, interpretation, and analysis, i.e., for transformation into actionable information), storage (database), integration, visualization, accessibility, export, and the use of data (sending messages, executing actions, etc.). Data analysis and management subsystem can be fully centralized or partially distributed. The former is frequently a property of wired SHM systems where all the data is managed from one center (central measurement point), while the latter is a feature of wireless systems where each individual wireless node can serve as a local data analysis and management subsystem.

The schematic presentations of SHM systems and subsystems shown in Figures 2.1 and 2.2 are typical but not exhaustive. For example, hardware and software components of an SHM system can be completely separate but can also often be combined in the same physical device or software (e.g., sensor and connection cable are frequently one physical device, user interface and self-monitoring software are frequently combined and installed on the same server or computer). Thus, a good understanding of a specific SHM system and its components is necessary before its implementation onto the structure.

2.2 Principles of Functioning of an SHM System

The principle of functioning of an SHM system is determined by the physical principle that enables the measurement to be taken. In general, a measurement can be taken directly, by comparison with a reference standard, or indirectly, by first converting a measured property (measurand) into an encoding parameter and then decoding that encoding parameter to get the value of the measurand. An example of the former is measurement of length using meter-tape (unknown length is directly compared with the reference standard, calibrated meter-tape), and an example of later is a classical mercury-thermometer where the temperature is encoded by the height of the mercury in a calibrated cylinder, see Figure 2.3.

Modern SHM systems are mostly based on indirect measurements. The nature of encoding parameter determines the functional principle of the SHM system, and in general, we can classify SHM systems as electrical, optical, acoustic (electro-mechanical), etc. Electrical systems can further be classified as resistance-based, impendence-based, inductance-based, etc. Optical systems can be fiber-optic-based, optical-image-based, laser-scanning-based, etc. Acoustic systems can be based on acoustic emission, wave propagation, ultrasonic scanning, etc.

The same measurand can be monitored using SHM systems with different principles of functioning; for example, the strain can be monitored using classical electrical resistive strain gauges or

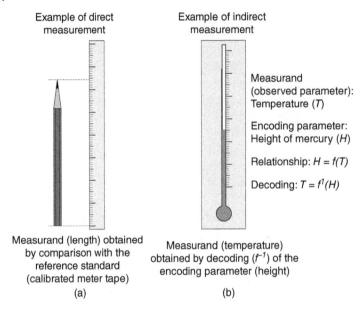

Example of direct
measurement

Example of indirect
measurement

Measurand
(observed parameter):
Temperature (*T*)

Encoding parameter:
Height of mercury (*H*)

Relationship: *H = f(T)*

Decoding: *T = f¹(H)*

Measurand (length) obtained
by comparison with the
reference standard
(calibrated meter tape)

Measurand (temperature)
obtained by decoding (*f⁻¹*) of the
encoding parameter (height)

(a)

(b)

Figure 2.3 Schematic representation of: (a) Direct and (b) indirect measuring system.

recently matured fiber optic strain sensors (see Chapter 3). The suitability of a principle of functioning of SHM system for application in a specific SHM project depends on the project requirements and the properties of the SHM system.

2.3 Properties of SHM System

2.3.1 General Observations

The main properties of an SHM system are measuring specifications, reading type and frequency, data analysis and management properties, interference with environment and maintenance, and reliability. These properties determine the SHM system's applicability to a specific project, and they are presented in more detail in the next subsections.

2.3.2 Basic Specifications of Measuring Subsystem

For the purposes of this book, the terminology used in this section, which has its source in metrology, is slightly modified and adapted to practical civil engineering vocabulary. More accurate vocabulary is given in ISO/IEC (2007).

A measurable quantity, physical quantity, observed parameter, or simply measurand is a property of phenomena, bodies, or substances that can be defined qualitatively and expressed quantitatively (Rabinovich 2005). Measurement is the process of determining the true value of measurand using measurement instruments (Rabinovich 2005), i.e., the SHM measuring subsystem. The basic axiom of metrology states that measurand has a true value; however, due to imperfections of measurement instruments, we cannot determine the exact true value of measurand (Kirkup and Frenkel 2006), but only its approximate value (as the result of measurements).

Accuracy of measurement is the closeness of the agreement between the result of a measurement and the true value of the measurand (Kirkup and Frenkel 2006). As per definition, it is not a property of measuring subsystem but a property of individual measurement. It is important to highlight that the accuracy cannot be quantified, as it is impossible to know whether the value of measurement is close to its true value or not. However, it is possible to evaluate whether one measurement system is likely to have better accuracy than the other. For example, the measurement of length in Figure 2.3a is likely to be more accurate if divisions on the meter-tape are given in millimeters than if they are given in centimeters.

Precision describes the closeness of the agreement between replicated measurements of the constant (unchanged) measurand under specified conditions (e.g., repeatability or reproducibility conditions).

Repeatability conditions of measurement include the replication of measurements on the same or similar objects by the same operators, under the same operating conditions (e.g., constant temperature and humidity), at the same location, using the same measuring subsystem and the same measurement procedures over a short period of time. Measurement repeatability is precision under repeatability conditions (ISO/IEC 2007).

Reproducibility condition of measurement includes the replication of measurements on the same or similar objects by different operators under different operating conditions (e.g., variable temperature and humidity), at different locations, using different measuring subsystems, and using different measurement procedures. Measurement reproducibility is precision under the reproducibility conditions (ISO/IEC 2007).

As per definition, precision (and consequently repeatability and reproducibility) is not a property of measuring subsystem but a property of measurement (do not confuse with instrumental measurement uncertainty, see below). In the past, precision used to be considered as qualitative term (Kirkup and Frenkel 2006), which actually gives a sense of how much the values of the repeated measurements are scattered. More precise measurements are less scattered and vice-versa. Recently, precision was defined as quantitative term "usually expressed numerically by measures of imprecision, such as standard deviation, variance, or coefficient of variation under the specified conditions of measurement" (ISO/IEC 2007). Thus, repeatability and reproducibility are quantitative terms expressed in the same way as precision. Note that the repeatability conditions provide the most favorable circumstances to carry out measurements, while the reproducibility conditions provide the most unfavorable circumstances to carry out measurements, and thus, the repeatability precision is expected to be better than reproducibility precision.

Precision and accuracy are frequently confused in SHM practice, and the following example is given to better understand the distinction between them. Let two teams of surveyors, *Team A* and *Team B*, measure the coordinates of the geometric center (centroid) of a field shown in Figure 2.4. Each team repeats the measurements four times and plots the results onto the plan of the field as shown in the figure. For reference, the true centroid is also shown in the figure.

Based on results presented in Figure 2.4, one can conclude that *Team A* performed more accurate measurements than *Team B*, while *Team B* performed more precise measurements than *Team A*. It is important to highlight that while the measurements of *Team B* were precise, they were not accurate! If *Team A* could improve the accuracy of measurement, it would also improve its precision, but *Team B* could not improve the accuracy by improving the precision. Hence, high accuracy implies high precision, but the opposite is not true. The difference between the mean (average) of repeated measurements and the true value is called bias (i.e., an estimate of the systematic error, see Chapter 4).

Stability is qualitative property of measuring subsystem, which indicates that the metrological properties of the measuring subsystem remain constant in time (ISO/IEC 2007), i.e., no change of encoding parameter would happen without a change of measurand. The stability can be quantified

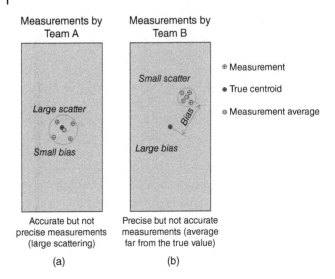

Figure 2.4 Accuracy vs. precision – Schematic comparison.

in several manners, the most characteristic being: (i) in terms of the duration of a time interval over which a specification (e.g., precision or resolution) changes by a stated amount and (ii) in terms of the change of a specification over a stated time interval (ISO/IEC 2007).

Drift is related to a loss of stability of measuring subsystem and represents change in measurement (encoding parameter) due to changes in metrological properties (specifications) over time, i.e., it is not consequence of the change of the measurand or any other recognized influence (e.g., change in temperature, and humidity). This is illustrated in the following example. Let teams *A* and *B* measure the position of the centroid of the field based on single measurements, and let them repeat the measurement once a year over five years. The hypothetical results are shown in Figure 2.5.

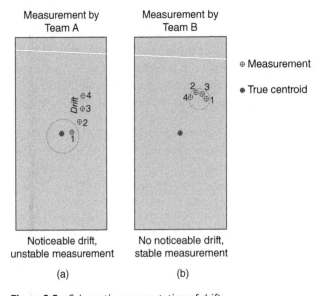

Figure 2.5 Schematic representation of drift.

The measurement of *Team A* shown in Figure 2.5a drifts over time more than the measurement of Team B (which appears to be within the range of precision). Thus, *Team B* has a measuring subsystem that is more stable than the measuring subsystem of *Team A*.

Figures 2.4 and 2.5 show that it is very easy to confuse the terms "accuracy," "precision," and "stability," and that is why it is important to understand them properly. Actually, these terms are frequently confused in SHM practice with another term: "limits of (permissible) error" and "uncertainty" (see the further text).

The resolution of measuring subsystem is the smallest change in a measurand Δx that causes a perceptible change in the encoding parameter Δy (ISO/IEC 2007). It can depend on noise (internal, instrumental, external, or environmental), magnitude of measurand, etc. Resolution of a displaying device can be different from the resolution of a measuring subsystem, and it represents the smallest difference between displayed encoding parameters that can be meaningfully distinguished. The resolution of display device cannot improve the resolution of the measuring subsystem, but it can worsen it. For example, the resolution of displaying device in Figure 2.3b corresponds to smallest division in the scale, while the resolution of the measurement is smaller – we can perceive the change in height of mercury column, which is equal to half or the third of the division.

Sensitivity of a measuring system is the ratio dy/dx between the change of the encoding parameter y and the measurand x (ISO/IEC 2007); it is quantifiable property that shows how much the encoding parameter varies as the measurand varies, and higher variation in encoding parameter indicates higher sensitivity. In the example shown in Figure 2.3b, the sensitivity is represented by the ratio between the variations in the mercury column height over the variation of temperature; for the same temperature variation, the instrument with higher sensitivity will have higher change in height of mercury column.

Limit of error or maximum permissible measurement error is extreme value of measurement error with respect to a known reference quantity value, permitted by specifications for a given measuring subsystem. Limit of error provided in measuring subsystem specifications by the manufacturer could be arguable if there is a lack of information on bias and known reference quantity. Limit of error is defined when two extreme values can be identified, and it is frequently (but not always) expressed as a positive value with the meaning of a plus–minus interval (e.g., "limit of error is 5 µε" actually means "±5 µε").

Instrumental measurement uncertainty is component of measurement uncertainty (see Chapter 4) created by a measuring subsystem (ISO/IEC 2007). It is determined through calibration and is usually included in measuring subsystem specifications in form of standard deviation or multiples of standard deviation.

Measuring interval, or working interval (but also measuring range, or measurement range, or dynamic range), is the set of measurand values that can be measured by the measuring subsystem with specified instrumental uncertainty, under defined conditions (ISO/IEC 2007). For example, the interval for the instrument in Figure 2.3b is defined by the smallest and largest values on the scale.

2.3.3 Reading Type and Frequency

To complete a measurement on one sensor, the measuring subsystem requires a certain amount of time, informally called sensor measuring time or sensor reading time. This time may include several actions included in sensor reading process, which depend on the principle of measuring subsystem and its hardware and firmware performances. The sensor measuring time is defined as the minimum time that can possibly elapse between two successive measurements, and it depends

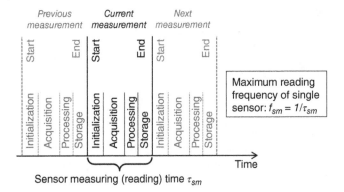

Figure 2.6 Schematic example of actions included in sensor reading process, sensor measuring (reading) time, and a sequence of successive readings with maximum reading frequency.

on the duration of the actions needed to complete the measurement. Reciprocal value of sensor measurement time is informally called maximum reading frequency of measuring subsystem over one sensor. An illustration of the actions included in the sensor reading process, sensor measuring time, and sequence of sensor readings performed with the maximum reading frequency are schematically shown in Figure 2.6.

Civil structures are frequently equipped with a large number of sensors (sensor networks), and all these sensors can be read simultaneously (sensors read all at once), sequentially (sensors read one after another), or combined (sequence of simultaneous measurements). Set of measurements performed over all the sensors in the network within the shortest possible time interval is called measuring or reading session. The session measuring (reading) time is defined as the minimum time that elapses between two successive measuring sessions, as shown in Figure 2.7. In the case of simultaneous measuring session, the measuring time is typically equal to sensor measuring time (see Figure 2.7).

For sequential measurement, the session measuring time is equal to sensor measuring time increased by switching time (if any) and then multiplied by the number of sensors (see Figure 2.8).

Finally, the session measuring time for combined session is equal to the sensor measuring time increased by switching time (if any) and then multiplied by number of switching actions (see Figure 2.9).

Regardless of the type of session reading (simultaneous, sequential, or combined), the reciprocal value of session measuring time is informally called maximum reading frequency of measuring subsystem over the sensor network.

Typical first natural frequencies in civil structures range between 0.1 and 10 Hz, and this range is used to approximately delimit static and dynamic measurement capability of an SHM system. To avoid aliasing, measuring subsystem has to be faster than Nyquist frequency, which is theoretically twice the value of the measurand's frequency. Measuring subsystems requiring more than one-twentieth of a second to complete a measurement session can be considered as static, while those requiring less than one twentieth of a second can be considered as dynamic. However, this delimitation should not be considered as strict, as the applicability of static or dynamic measuring subsystems largely depends on the specific project requirements; for example, if higher modes of vibration have to be captured by the measurement, then monitoring systems with an appropriate, much higher frequency of reading should be used.

Simultaneously performed session contains measurements that are fully synchronized since the time stamp of the measurements from all the sensors is identical. Sequentially performed sessions

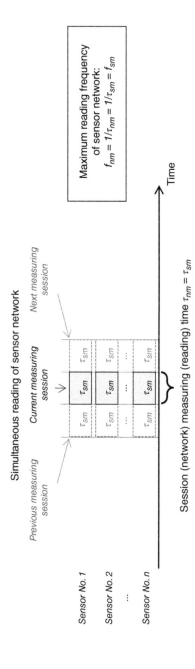

Figure 2.7 Schematic representations of session measuring time for simultaneous readings of sensors.

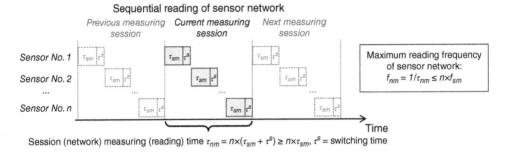

Figure 2.8 Schematic representations of session measuring time for sequential readings of sensors.

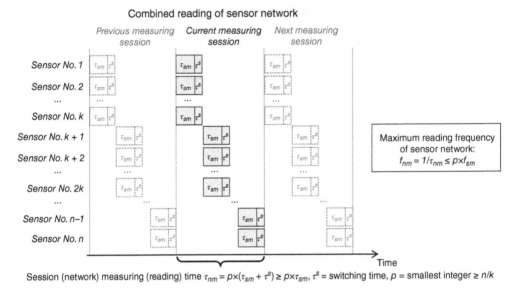

Figure 2.9 Schematic representations of session measuring time for combined readings of sensors.

as well as sessions performed with combined (simultaneous and sequential) reading contain measurements that are not fully synchronized due to switching time.

As the measurand can change during the sensor and/or session measuring times, the values of some measurements, while commonly assigned to the session time stamp, may actually not correspond to that specific time. For example, the values of measurands may be averaged over sensor/session measuring time, and in case of sequential and combined sessions, different sensors may have different time stamps, and thus measurements performed within the same session may be related to different events on the structure. Thus, technically, the measurements are either quasi-static or quasi-dynamic (depending on their maximum frequency). However, if the measurand does not change significantly during the measurement session, then we can consider measurements as "truly" static or "truly" dynamic.

In the case of wireless sensing network, additional potential issue regarding the timestamp of measurements is synchronization of wireless nodes, i.e., ensuring that the internal clocks in all wireless nodes are synchronized.

2.3.4 Data Analysis and Management

In this book, the term "SHM data" refers to the unprocessed data collected by the SHM system and all its derivatives, such as processed and analyzed data. Term "SHM metadata" refers to any data that describes the SHM system (e.g., positions of sensors, plan of sensor network, type and brand of reading unit, and various specifications) and the monitored structure (e.g., geographical location, dimensions, materials, and numerical models). To ease narration, in the further text, the term "data" will be used instead of "SHM data" or "SHM data and metadata" when appropriate, without introducing ambiguity.

Data analysis and management represent an important component of the SHM process. Inability to analyze and/or manage big SHM data properly may represent a barrier to its implementation and fruitful use. Partial or full abandonment of an already implemented SHM due to improper analysis or management of data is not unusual (Glisic and Inaudi 2007). Data analysis and management have two subcomponents: the data analysis subcomponent, which includes data processing, interpretation, and analysis; and the data management subcomponent, which includes storage, integration, visualization, accessibility, export, and the use of data. Thus, SHM data analysis and management include processing, interpretation, analysis, storage, integration, visualization, access, export, and the use of data, as shown in Figure 1.3 (Glisic and Inaudi 2007). Data analysis and management practically follow the collection of data; although the latter, being performed by hardware of the SHM system, is actually considered a part of maintenance of SHM system (see Figure 1.3). Both the measuring subsystem and the data management subsystem can carry out a part of the data management, as shown in Figure 2.1. In general, the data analysis and management can be basic or advanced, and they can be performed manually, semi-automatically, or automatically, as shown schematically in Figure 2.10. Basic data analysis and management are frequently automated and provided with measuring subsystems (except, maybe, data interpretation), while advanced data analysis and management may be project specific and may require additional hardware and software development to achieve full automation. However, both data analysis and the use of the data require human intervention; the former requires at least

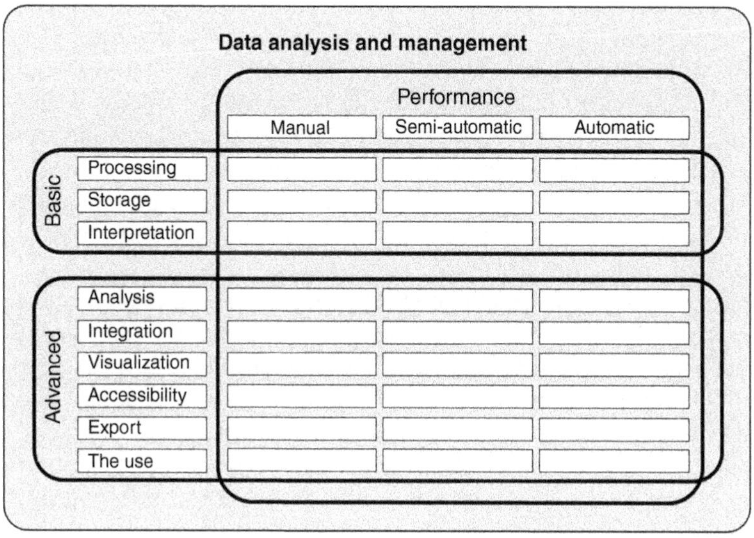

Figure 2.10 Basic and advanced SHM data analysis and management.

professional oversight (e.g., by consultant), while the latter requires the final decision on how to use the data, which is to be made by a decision maker (e.g., consultant or owner).

Data is usually collected in the form of time series of an encoding parameter. This raw (unprocessed) data is very useful for measurement validation purposes, and it should be stored. If the validity of processed data is questioned in the future, access to raw data and its analysis can help validate it.

The raw data should be converted into processed data, which is then used in the data interpretation and analysis. The processing commonly consists of decoding (transformation of encoding parameter into measurand, e.g., by calibration equation), elimination of outliers (measurements with excessive error due to malfunctioning of the system, undesired interference with environment, etc.), compensation (e.g., for temperature, and known drift), resolving the issues related to missing data (e.g., due to malfunction of the system or due to removal of outliers), and filtering (e.g., removal of noise or extracting only the data of specific interest).

Interpretation of data includes understanding the meaning of the data (e.g., correlation of unusual changes in data to real events, for example, correlation of unusually high strain with temperature variations or post-tensioning of the structure, or correlation of slow changes in long-term with rheologic effects). While the two processes above, processing and storage, are frequently made automatically without human intervention, interpretation frequently requires the participation of an engineer (consultant) to establish interpretation procedures for a specific project. Once the procedures are established, the process can be automated.

The critical parts of data management are data analysis and the use of the data. The former represents the transformation of the data into actionable information for the end user, and the latter consists of pre-set or ad-hoc actions to be taken upon reception of the information. Typical data analysis approaches are model-based, data-driven (model-free), or combined. They will be discussed in more detail in Chapters 5–7.

All data, including raw, processed, and analyzed data, as well as metadata, should be stored in a manner that ideally allows easy access, retrieval, export, and creation of reports.

Data management is particularly challenging when heterogeneous data are collected. Various sensors based on different technologies can measure many parameters related to the structure in many formats, such as strain, tilt, and weather; various tomographic responses (e.g., changes in electrical resistivity, wave propagation patterns, laser scanning outcomes, and results of digital imaging correlation); live cameras can be used to both visualize traffic action and structural response; and more recently, advancements in augmented reality were made in order to collect both data and metadata (Moreu et al. 2017). Moreover, depending on the type of sensors and specific project requirements, the data can be registered both statically and dynamically.

Data management may become even more complex if multiple users have to consult data and especially if they have diverse backgrounds and interests (e.g., the owner/manager of the structure, consultants, technology providers, contractors, inspectors, and researchers).

Integration, visualization, accessibility, and export of data are very important as they allow the users to rapidly assess the status of the structure and monitoring system; hence, the components of SHM that address data integration, visualization, accessibility, and export should be interactive, automatic, and both intuitively simple to navigate and comprehensively informative for all user profiles. They might be delivered as parts of SHM system or specifically designed by a consultant, depending on the specifications of the SHM project. Modern data management components for integration and visualization of data are based on 2D approaches (see Glisic et al. 2014), while 3D approaches such as those based on BIM (building information modeling), virtual tours, augmented reality, and digital twins are emerging (e.g., Moreu et al. 2017, Napolitano et al. 2017, 2018). The general principles for data visualization and accessibility are presented in Glisic et al. (2014).

The data analysis and management subsystem consists of infrastructure and software. The infrastructure mostly (but not exhaustively) consists of dedicated spaces (e.g., rooms), computers, servers, mobile devices, screens, communication network, and, still in the experimental phase, virtual and augmented reality sets. It can have components installed on-site, "in-house" (in the user premises), and/or in third party premises, with a known location (e.g., monitoring company, and consultant) or a virtual location (in the informatics "cloud"). The software is in general installed on computers, servers, and mobile devices, while access to it is available on-site, in-house, or remotely via mobile network or internet, automatically or on-demand. The software may include the components delivered as a part of SHM system, components developed ad-hoc, for the specific SHM project, and third party components that in one way or another complement the data management (e.g., BIM, and FEM – Finite Element Modeling). The components of data management subsystem are schematically presented in Figure 2.11.

It is desirable for the data analysis and management subsystem to be as automated and standardized as possible. Typically, manufacturers of measuring subsystems also provide standardized flexible software packages for basic data analysis and management. Also, they may offer general software packages for data analysis and/or services for advanced data analysis and management via remote internet access (in the cloud). However, taking into account project requirements and specifics, some parts of software (especially regarding the data analysis) often have to be developed ad-hoc and/or interfaced with other software (e.g., with software provided by manufacturers of SHM systems, and with software for structural analysis).

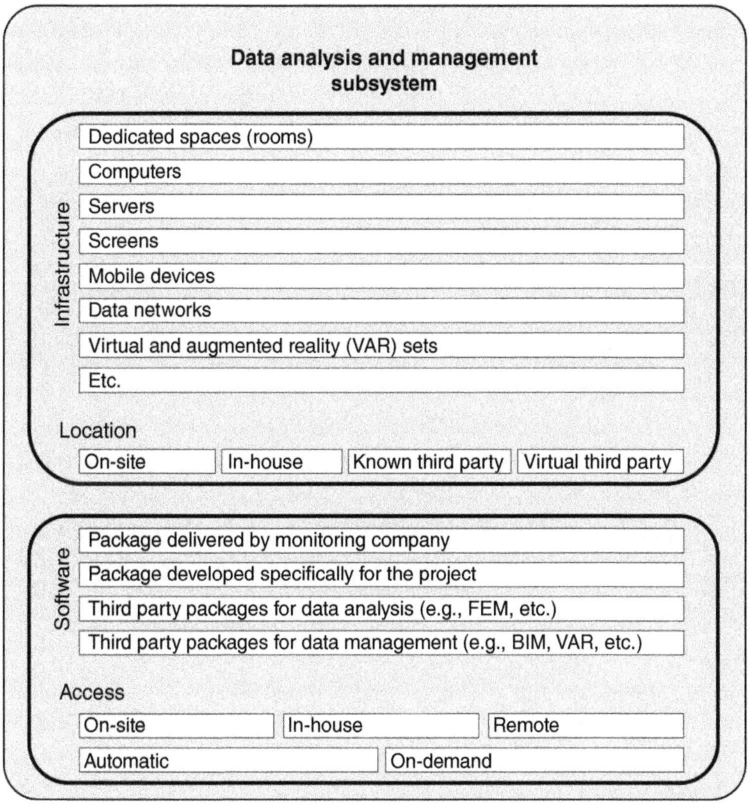

Figure 2.11 The components of data analysis and management subsystem.

2.3.5 Interference with Structure and Environment, Operational Conditions, and Maintenance

SHM components (e.g., sensors and accessories) are frequently in physical contact with the structure, and they mutually interfere with each other, chemically and mechanically. In order to avoid adverse chemical interference and preserve the specified performance of both the SHM system and the structure, it is important for materials used in SHM components to be compatible with material of host structure, i.e., the SHM components and the structure must be mutually non-intrusive. For example, aluminum parts of SHM components should not be embedded in concrete or in direct contact with concrete surfaces that may be wet for extended period; this might create chemical reactions that could damage the concrete and/or aluminum component. Another example: ordinary steel should not be in contact with stainless steel – it is likely that the former will corrode the latter. To avoid adverse mechanical interference, SHM system components, and in particular sensors, should not affect the structure nor the measurand. For example, if a strain sensor is significantly stiffer than the structural material to which it is attached, it may perturb the strain field of the structure at the location of measurement and thus lead to misinterpretation of SHM data. Another example: drilling the holes or welding with the purpose of installing the anchors for sensors may reduce the cross-section of the structure or make it prone to fatigue damage.

The environment whether natural or human-made, also interferes with SHM components. Thus, the SHM components and their installation method must be designed to operate and withstand extreme site conditions while keeping their specified performance. Typical natural environmental influences are temperature and humidity variations, exposure to elements (dust, rain, lightning, snow, freezing, and wind), animals (mostly rodents), and in some areas, exposure to earthquakes and hurricanes. Typical human-made influences are exposure to vibrations (e.g., due to traffic), electromagnetic fields (e.g., due to closeness to power lines), workforce (e.g., during the construction and maintenance of the structure), and vandalism (all exposed visible components of SHM system are potential "victims" of theft or vandalism, but also data management infrastructure may undergo cyber-attacks).

Modern sensors and accessories commonly have excellent flexibility regarding operating conditions (for sensors and accessories exposure to elements is common; desirable and mostly achievable operating ranges are for temperature −40 to +80°C and for relative humidity 0–100% with condensation allowed in windy and dusty environments). However, the reading units and data management hardware usually have very restricted operating conditions in terms of temperature (typically 10–40°C) and relative humidity (no condensation allowed), while exposure to elements and excessive vibrations is mostly prohibited.

The components of SHM system have various electrical, mechanical, or optical components, and despite all precautions and proper design, they could be damaged or deteriorated due to environmental interference or due to natural aging of components. Typically, unshielded electrical sensors can be damaged by lightening; unprotected connection cables might be accidentally "cut" by rodents or intentionally removed by thieves; junction boxes might be filled with water; and computing performance will slow down after extended period of use. As a consequence of damage or deterioration, the SHM system may malfunction or function with limited performance, or completely cease functioning.

Hence, maintenance (including repairs, updates, and upgrades) of the SHM system is practically inevitable. The maintenance activities can be scheduled (e.g., cleaning of connectors every 5 years, and changing of computers every 10 years) or unscheduled, based on the need (mostly repairs). To preserve uninterrupted full performance of the SHM system, it is necessary to design it so that the maintenance is minimized by choosing robust and durable components.

Finally, most SHM systems need electrical power and communication lines. Electrical power can be supplied from the grid (if available), energy harvesters (e.g., solar panels, small wind turbines, or vibration-based harvesters), or by batteries (the least convenient power source, as it requires frequent replacements). In all cases, low-consumption systems are preferred, and back-up energy devices (e.g., uninterruptible power supply devices – UPS) are highly recommended. The communication lines are usually provided by a wired or mobile network, but in remote areas, the communication lines might need to be specifically built for the SHM system, or data must be physically transferred (e.g., on an external hard disk or USB memory stick, by contractor who travels on-site).

2.3.6 Reliability of SHM System

The reliability of SHM system can be defined as the ability of system to perform its required functions under stated conditions for a specified period of time. The required function of SHM system is in accordance with the properties and specifications presented in this section. However, there are two aspects that should be considered:

(1) Reliability of SHM system components, including both infrastructure (hardware) and software
(2) Reliability of data analysis algorithms, in particular identification of unusual structural behaviors (e.g., damage detection and characterization).

In general, the reliability of an SHM subsystem component can be defined as its mean time to failure. For example, for sensors embedded in concrete, it is frequently determined by the manufacturer and presented in form of "survival rate" (e.g., survival rate of long-gauge fiber optic strain sensors upon embedding in concrete is typically better than 95%). In addition, the loss rate over time can also be specified (e.g., 1% of sensors per year can experience malfunction, or loss of performance, or cease functioning).

Reliability of data analysis algorithms can be presented by expected numbers of false positive detections (SHM system indicates detection of an unusual structural behavior while actually there is no one on the structure) and false negative detections (SHM system does not detect an actual unusual structural behavior). While false positive detections are more tolerable than false negative ones, they are undesirable too, as they may uselessly trigger a series of actions. Reliability of data analysis algorithms in detection and characterization of unusual behaviors can be challenged by many factors, such as influence of environmental "noise," type of damage-sensitive feature, sensor network (number, specifications, and distribution of sensors), and type of algorithms used, just to name a few. Examples of environmental noise are temperature and humidity variations that can either make unusual behavior undetectable or make usual behavior appear unusual and trigger false alarm (e.g., temperature variations can change the deformed shape of a structure, which can be falsely interpreted as an unusual behavior; Koo et al. 2013). Damage-sensitive features are not universal: SHM systems designed for monitoring certain parameters are not able to detect unusual behaviors that do not manifest through changes in these parameters (e.g., vibrational methods are unlikely to detect prestress losses in non-cracked beams; see De Roeck 2003). Sensor networks, i.e., proximity of sensors to damage and their sensitivity, may have an important role in detection and characterization of unusual behavior (e.g., discrete strain sensor that is located only a few centimeters far from crack might not be able to detect it; see Glisic et al. 2016). The ability of SHM system to detect and characterize unusual structural behavior also depends on the type of data analysis algorithms that are deployed, as some algorithms are more effective and some less effective for some specific types of damage or deterioration (e.g., see Posenato et al. 2010).

2.3.7 Summary

In summary, an SHM system has two subsystems (measuring and data management), and each subsystem has several components, which in turn have specific properties regarding performance, interaction with the environment, and reliability. All these components and their properties must be properly analyzed when planning implementation of SHM.

2.4 Example of a Heterogeneous SHM System: Tacony–Palmyra Bridge

This section briefly illustrates the components and complexity of a heterogeneous SHM system deployed on a real bridge. The Tacony–Palmyra Bridge over the Delaware River (in the USA, between New Jersey and Pennsylvania) was constructed in 1929 by the renowned engineer RalphModjeski (see Ralph Modjeski et al. 1931). It is owned and operated by the Burlington County Bridge Commission (BCBC). The bridge consists of a 168 m (550 ft) steel tied arch span, a 79 m (260 ft) Scherzer, two-leaf rolling-lift-system bascule, and two 224 m (735 ft) half through truss three-span continuous units, totaling approximately 1115 m (3,659 ft). A photograph of the bridge is shown in Figure 2.12.

The deployment of the SHM system was planned in several stages, and the one including application on the bascule span is presented. The bascule span is instrumented due to the pressing need to assess the behavior of this complex part of the bridge. The owner considers the bridge irreplaceable, and thus, the bridge must be efficiently managed to ensure adequate performance into the distant future. As a consequence, the SHM system is deployed with the objective of preserving the bridge indefinitely. The SHM system was designed so that it addresses several needs:

(i) Operational (safe opening and closing of the bascule).
(ii) Structural (safe levels of strain, deformation, and forces).
(iii) Traffic safety (assessing weather conditions, icing of the road, and congestion).

To address the above needs, the measuring subsystem consists of a combination of various technologies, including dynamic strain sensors (resistive strain gauges), static strain sensors (vibrating wires – VW), tilt-meters (inclinometers), temperature sensors, weather station, and video surveillance. More detail about the SHM systems and the structure itself can be found in Modjeski et al. (1931), Yarnold et al. (2012, 2015), Yarnold (2013), Glisic et al. (2014), and Yarnold and Weidner (2016).

The measurements were collected and recorded with the use of several technology-specific reading units and transferred to a Central Server via a local area network (LAN). Dynamic strain sensors,

Figure 2.12 Photograph of Tacony–Palmyra Bridge.

temperature sensors, and tiltmeters were connected to and read by two reading units (i.e., two client microcomputers running a real-time version of Windows XP) installed at Pennsylvania and New Jersey sides of the bascule. Static strain sensors are connected to a separate reading unit (VW Spectrum Analyzer) through a channel multiplexer. The VW Spectrum Analyzer reads the sensors and sends the data to the client microcomputer at Pennsylvania side of the bascule through the serial port. Video surveillance combines four axis video cameras installed at various locations on the bridge with one 180° panoramic video camera installed at high point next to the bridge, all connected to the Central Server via LAN. A weather station is installed in the proximity of the bridge and collects data related to temperature, humidity, pressure, dew point, and heat index, as well as various rain and wind data. It is provided with its own internet interface, which is used to connect to Central Server via LAN. The SHM measuring subsystem components, their connections, and data flow on the bridge are schematically shown in Figure 2.13 (Glisic et al. 2014). A Detailed description of all the instruments deployed at Tacony–Palmyra Bridge is given in the literature (Yarnold 2013).

The data management infrastructure consists of a Central Server, installed in the Burlington County Bridge Commission premises, and LAN that connects the measuring subsystem with the server. The data management was enabled by creating a two-component software package. Due to heterogeneous nature of the measuring subsystem, it was necessary to create System Control Component (SCC) of the software that enables the configuration and operation of various components (sensors, video camera images, weather station) from a unique center with a unique user interface. The SCC also provides all types of data (metadata, configuration data, and SHM data) and performs data analysis. The Data Management Component (DMC) of the software provides live and historic SHM data accessibility and visualization. The two-component software package was split into three parts:

(1) Main System Software
(2) Live Portal Software
(3) Playback Software

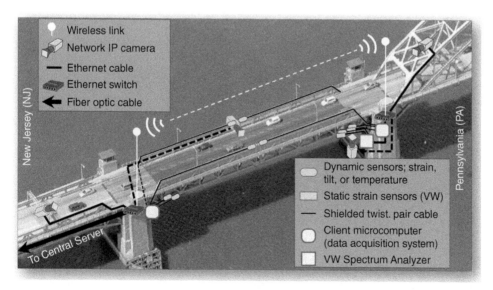

Figure 2.13 Schematic representation of measuring subsystem components, their connections and data flow on the bridge (Glisic et al., 2014 / JOHN WILEY & SONS, INC.).

The Main System Software represents the SCC, while the two other software parts represent the DMC. The Main System Software and the Live Portal Software are installed on the Central Server, while the Playback Software can be downloaded and installed on the personal computers of interested users (e.g., engineers who want to perform more detailed data analysis, and researchers).

The software package was developed by professional informatics company and uses a LabVIEW based Supervisory Control and Data Acquisition (SCADA) platform developed by the same company (PAC 2011). It contains (i) a real time database for acquisition and storage of instantaneous values of SHM data and (ii) a data historian that processes, compresses, stores, retrieves, and displays the current and past states of the stored data. The SHM system installed on the bridge generates large amounts of data in very different formats: text files with scalar time series, text files with matrix time series, images (.jpg format), and video streams (.mpg format). To properly store and allow for easy and prompt access to data, it was necessary to develop data storage strategies and decide on the type of database to be used. In order to contain costs, data storage in this project was made possible by data compression and local hard disk storage. Basic threshold data analysis was put in place (see Chapter 5), and it was judged sufficient at the current stage of the project (enabling control of proper opening and closing of the bascule). Data analysis software for detailed health analysis was planned to be developed at a later stage (no major structural issues were expected at the time the bascule was instrumented since the bridge was undergoing rehabilitation), and important work in creating data analysis algorithms was well advanced (see Yarnold 2013).

Depending on their level of clearance, a licensed user can connect to the Central Server and interact with the Main System Software to change configuration of the system, access live data (through the Live Portal Software) or download the data of interest to their PC and visualize it using the Playback Software. Figure 2.14 shows the interactions between the data flow, the software components, and the user.

To enable comprehensive yet easy and intuitive visualization and access to data, each software part consisted of several tabs that could be selected by users. Examples of screenshots are given in Figures 2.15–2.17 for illustrative purposes only, and more details about their functionality are found in the literature (Glisic et al. 2014).

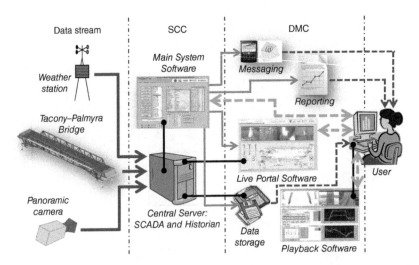

Figure 2.14 Schematic representation of interactions between the data flow, the software package components, and the user (Glisic et al. 2014 / JOHN WILEY & SONS, INC.).

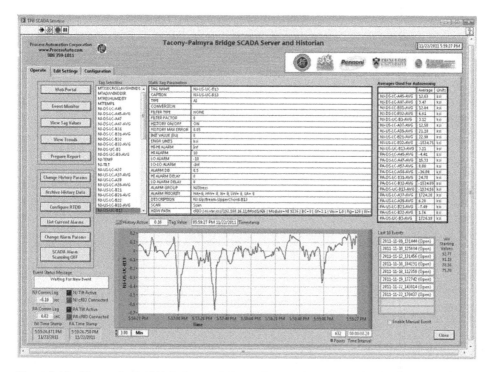

Figure 2.15 "Operate" tab of Main System Software for metadata and configuration parameters accessibility and visualization (Glisic et al. 2014 / JOHN WILEY & SONS, INC.).

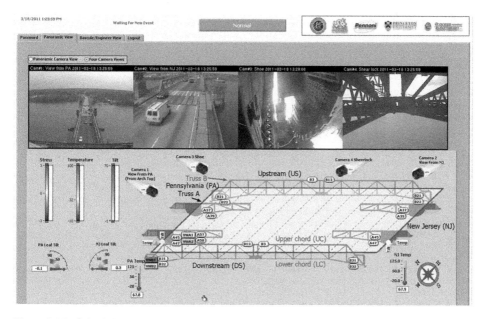

Figure 2.16 "Live Bridge View" tab of Live Portal Software (Glisic et al. 2014 / JOHN WILEY & SONS, INC.).

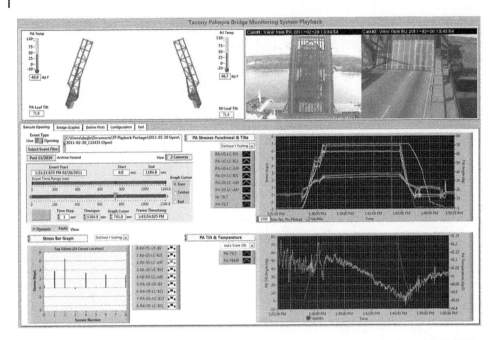

Figure 2.17 "Dynamic View" of Playback Software (Glisic et al. 2014 / JOHN WILEY & SONS, INC.).

The example briefly presented in this section demonstrates the complexity of implementation of a heterogeneous SHM system in a real structure and illustrates some of challenges that must be addressed when selecting and designing appropriate SHM system components.

3

Strain Sensors

3.1 Importance of Strain Monitoring and Strain-Based SHM

Strain-based structural health monitoring (SHM), in this book, is referred to as SHM that uses strain as a primary directly monitored parameter or strain "derivatives" (i.e., parameters derived from strain such as stress, curvature, deformed shape, and prestress force) as primary indirectly monitored parameters, combined with some secondary monitored parameters (e.g., temperature, and tilt) and associated data analysis algorithms, to achieve the objectives of SHM.

Structures are commonly designed based on structural analysis and codes that include a series of procedures and criteria for structural safety and serviceability. These procedures and criteria deal with the stress state in the structure to ensure safety, and deflections (deformations) to ensure serviceability (e.g., AISC 2005, ACI 2008). Hence, assessment of these two parameters represents important objectives of SHM, as they enable identification (i.e., detection and characterization) of unusual structural behaviors. Assessments of stresses and deflections, as well as identification of unusual structural behaviors, are not simple tasks; however, strain monitoring can successfully address all of them.

Characterization of unusual structural behaviors is challenging in real-life settings due to environmental noise and strongly depends on the selection of damage-sensitive features and corresponding data analysis algorithms; strain and strain derivatives have been proven to be robust and reliable damage-sensitive features. In addition, there are no effective means to directly monitor stress in on-site conditions, except in some very specific applications (e.g., monitoring tension in cables using an elastomagnetic sensor; see Jarosevic et al. 1996, Wang et al. 2005, Cappello et al. 2018), while direct monitoring of deflections in real-life settings is also a challenge (see below). However, stresses and deflections are frequently monitored indirectly, i.e., through strain monitoring, as briefly described below (for details, see Chapters 5 and 6).

The first signs of unusual behaviors (e.g., damage, deterioration, or loss of performance) are frequently reflected through strain field anomalies. Cracking, bowing, excessive rheological strain, differential settlement of foundations, and reduction (loss) of cross-sections due to corrosion, spalling, or alkali–silica reactions, are just some examples of damage or deterioration that manifest as an alteration of the strain field (i.e., in the form of a strain field anomaly). Thus, strain can be and was successfully used as a damage-sensitive feature for the characterization of unusual structural behaviors (e.g., see Glisic et al. 2002b, 2013, Enckell et al. 2011, Glisic and Yao 2012, Yao and Glisic 2015a). In addition, many other strain derivatives can be used as a damage-sensitive feature for characterization of unusual structural behaviors and evaluation of the structural health condition and performance, such as (but not limited to): the position of the neutral axis (e.g., Sigurdardottir and Glisic 2013), curvature (e.g., Kliewer and Glisic 2017), prestressing force (in

Introduction to Strain-Based Structural Health Monitoring of Civil Structures, First Edition. Branko Glišić.
© 2024 John Wiley & Sons Ltd. Published 2024 by John Wiley & Sons Ltd.

concrete structures, e.g., Abdel-Jaber and Glisic 2019), cross-sectional stiffness (e.g., Sigurdardottir and Glisic 2015), cross-sectional integrity (e.g., Abdel-Jaber and Glisic 2015), and thermal "signatures" (e.g., Yarnold et al. 2015). The fact that strain-based SHM can provide damage detection and characterization represents an important reason for monitoring strain.

It is commonly assumed in engineering practice that a construction material fails at a point when some stress parameter at that point reaches a certain threshold value. That threshold value is usually related to the uniaxial strength or yielding stress of the material. For example, a threshold value can be given in terms of the maximum allowable normal component of stress (i.e., limit uniaxial stress or strength, which is based on a single component of the stress tensor) or the yield criterion (for example Von Mises or Tresca yield criterion), which is based on a combination of components of the stress tensor (e.g., see Malvern 1969).

Mechanical strain is a parameter that is directly correlated to the stress, and any change in the stress field causes a change in the mechanical strain field. Since the stress–strain relationship can be used to infer the stress from mechanical strain, and given that there are no effective means to monitor stress on-site, the strain has emerged and became established as an important parameter to monitor for evaluation of stress (e.g., Davidenkoff 1934, Galambos and Armstrong 1969, Bridges et al. 1976, Altounyan 1981, Murnen and Laubenthal 1985, Widow 1992, Inaudi et al. 1999a, Omenzetter and Brownjohn 2006, Glisic and Inaudi 2007). The structures built of the most frequently used structural materials – steel, concrete, wood, and fiber-reinforced thermoplastic or thermoset composite materials – are usually designed to be in a linear regime, in which case the stress–strain relationship reduces to the evaluation of Young (elastic) and shear moduli; this evaluation can be made directly, from for-the-purpose performed laboratory tests, or indirectly, from some other tests performed in the laboratory or on-site. As an example of direct evaluation, a concrete specimen can be subjected to axial force while simultaneously the strain or relative displacement between the extremities of the specimen are measured, which enables the evaluation of the Young modulus. As an example of indirect evaluation of the stress–strain relationship, Young modulus of concrete can be estimated using the results of the compressive strength tests (e.g., determined from standard cylinder tests) combined with empirical formulae provided in codes and manuals (e.g., ACI318-08). Another approach to evaluating stress–strain relationships is structural identification, i.e., by applying controlled load tests to the monitored structure and performing data analysis of the test results that yields a Young modulus that fits analytical or numerical (structural) models (e.g., see Li and Wu 2005, Catbas 2013, Yarnold et al. 2015). Thus, strain-based SHM has the potential to provide an evaluation of stresses and an assessment of structural safety, and this is another important reason for monitoring strain.

Contrary to stresses, deflections can be directly monitored on real structures using linear variable differential transformers – LVDTs, hydrostatic systems (e.g., Burdet 1993, Li et al. 2012), cameras (e.g., Jauregui et al. 2003, Tian et al. 2013), global positioning systems – GPS (e.g., Roberts et al. 2004, Figurski et al. 2007), laser-based sensors (e.g., Psimoulis and Stiros 2007, Fang and Yunfei 2012), and radar-based sensors (e.g., Gentile and Bernardini 2010). Their main advantages are direct deflection measurement (error or uncertainty in measurements depends only on SHM system specifications), universal application to a variety of types of structures, and the fact that knowledge of boundary conditions is not required. Nevertheless, they have important limitations – they frequently depend on unobstructed lines of sight, stable reference points, and need clear targets (e.g., Choi et al. 2013, Guan et al. 2014) – which makes them difficult and challenging to implement and use in medium- and long-term SHM applications. Deformed shape, as an important component of deflection, can be successfully monitored indirectly by double-integration of curvature, the latter being derived from strain measurements (e.g., Vurpillot et al. 1998, Glisic et al. 2002b, Sigurdardottir et al. 2017). Hence, strain-based SHM can provide an evaluation of deformed shape and an assessment of structural serviceability, and this is the third important reason for monitoring strain.

3.2 Strain – Definition and Use of Term

The reader of this book is assumed to have fundamental knowledge of the strain in structures. Nevertheless, to ease reading, important details about the definition of strain and strain sources and components are given in Chapter 5. In addition, to ease understanding of strain sensor measurement, some important strain properties are highlighted below (e.g., see Malvern 1969):

1. Strain state (or simply strain) is described by a tensor (and not a scalar!), which contains two types of components: normal strain component (or simply normal strain), commonly denoted with ε, and shear strain component (or simply shear strain), commonly denoted with γ, see Figure 3.1 and Equation 3.1 (an alternate notation of strain components is also given in Equation (3.1); note that $\frac{1}{2}\gamma$ is tensor component, not γ, e.g., see Malvern 1969); shear strain is conjugated by definition, i.e., $\gamma_{xy} = \gamma_{yx}$, $\gamma_{yz} = \gamma_{zy}$, and $\gamma_{zx} = \gamma_{xz}$.
2. Strain state and strain components are defined at a mathematical (dimensionless) point; the infinitesimal "squares" (dimensions dx and dy) shown in Figure 3.1 should be perceived as "boundaries" of a point.
3. While the strain state at a point is independent from the selected coordinate system, the strain tensor components depend on the selection of the coordinate system, i.e., on the directions (angles) of its axes (in this book, mostly Cartesian coordinate system is used); for any state of the strain (any strain tensor), there is a coordinate system with directions of axes such that shear strain vanishes and only normal strain components exist (strain tensor matrix becomes diagonal); these axes are called principal axes or principal directions, and the corresponding normal strain components are called principal strains (see more details in Chapter 5).
4. In the most general, three-dimensional (3D) case, strain is a tensor that has nine components, and it is frequently represented by a 3×3 matrix denoted with **E**, see Equation 3.1; in the 3D case, only six components out of nine are independent as the shear strain components are conjugate by definition (see Point 1 above); in the two-dimensional (2D) case, the strain tensor reduces to four components (e.g., by removing one row and conjugate column from the matrix shown in Equation 3.1), three of which are independent; and in the uniaxial (one-dimensional – 1D) case, the strain tensor reduces to only one component (e.g., by removing two rows and two conjugate columns from the matrix shown in Equation 3.1).

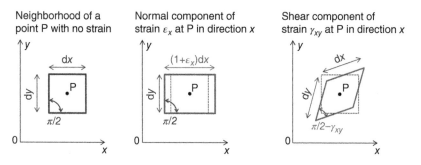

Figure 3.1 Simplified elementary definitions of strain components.

Simplified elementary definitions of strain components ε_x and γ_{xy}, in the plane $0xy$ defined with axes x and y, are illustrated in Figure 3.1 (for details, see Malvern 1969). These definitions are simply extended to define the strain components in planes $0yz$ and $0zx$ by appropriately substituting the axe labels in subscripts. Figure 3.1 shows that normal strain reflects the change in the "length" along one axis (direction) of an infinitesimal square ("boundary" of a point), while shear strain reflects the change in the angle between the two axes of the square. Thus, elementary definitions show that normal strain components, ε_x, ε_y, and ε_z, are related to directions of axes x, y, and z, of the coordinate system $0xyz$ (e.g., ε_x is normal strain in direction x), while shear strain components are related to planes $0xy$, $0yz$, and $0zx$ defined by the axes of coordinate system x and y, y and z, and z and x respectively (e.g., γ_{xy} is shear strain in plane $0xy$).

Two most common notations for the components of the strain tensor $\mathbf{E}$ are given in Equation 3.1.

$$\mathbf{E} = \begin{bmatrix} \varepsilon_x & \frac{1}{2}\gamma_{xy} & \frac{1}{2}\gamma_{xz} \\ \frac{1}{2}\gamma_{yx} & \varepsilon_y & \frac{1}{2}\gamma_{yz} \\ \frac{1}{2}\gamma_{zx} & \frac{1}{2}\gamma_{zy} & \varepsilon_z \end{bmatrix} = \begin{bmatrix} \varepsilon_{xx} & \varepsilon_{xy} & \varepsilon_{xz} \\ \varepsilon_{yx} & \varepsilon_{yy} & \varepsilon_{yz} \\ \varepsilon_{zx} & \varepsilon_{zy} & \varepsilon_{zz} \end{bmatrix} \tag{3.1}$$

In this book, the first notation, i.e., the matrix with "epsilons" (ε's) and "gammas" (γ's), will be used to simplify presentation and emphasize the difference between normal strain components (ε's) and shear strain components (γ's).

In engineering practice, term "strain" is frequently simplified and used to describe only the normal strain component at an observed point in a given direction (and not the strain tensor), which sometimes can lead to confusion. Nevertheless, unless specified differently, this book will follow the use of term "strain" as presented above, i.e., to describe normal strain component at observed point in a given direction. The reason for adopting this simplified notation is to improve readability of the book. Commercial strain sensors used in SHM actually measure only one normal strain component, and thus the term "strain measurement," which aligns with this simplified notation, can be used instead of correct term "measurement of normal strain component," which is, however, too long. To avoid confusion with terms "strain state" or "shear strain component," full term "normal strain component" will also be used when needed, depending on the context.

3.3 Historic Summary

Importance of measuring the strain in structures was identified far back in the nineteenth century, and it is not surprising that strain sensors were among the first to be invented. So far, strain sensing technologies can be classified into three generations. The first-generation includes electrical sensors (discrete, short-gauge), the second includes fiber-optic sensors (discrete, short- and long-gauge, as well as 1D distributed), and the third-generation includes a variety of technologies and materials for 2D strain sensing (quasi-distributed or truly distributed) or strain-based detection of unusual structural behaviors.

The development of first-generation of strain sensors practically started in 1856, when Lord Kelvin, in his Bakerian Lecture "On the Electro-dynamic Qualities of Metals," demonstrated to the Royal Society of London that metallic conductors exposed to mechanical strain undergo a change in their electrical resistance (Perry and Lissner 1955). However, this physical principle could not be applied in practice at that time due to the difficulty of bonding the conductor to the structure. Resistive strain sensors (commonly known as "strain gauges") became reality in 1938 when Edward

E. Simmons and Arthur C. Ruge, simultaneously and independently from each other, invented the strain gauge in 1938 (Perry and Lissner 1955). They represent an important step forward in strain-based SHM, as they enabled high-frequency dynamic strain measurements and, thus, were used in numerous field projects.

However, while the sensing principle of a resistive strain gauge was the first to be identified, it was not the first to be transferred into a real strain sensor. To the best of the author's knowledge, some 19 years before Simmons and Ruge, i.e., in 1919, German researcher Otto Schaefer discovered the vibrating wire (VW) sensing principle and translated it into VW strain sensor (Schaefer 1919). The work of Schaefer was known to Soviet professor N. Davidenkoff, who created his own VW sensor ("teletensometer") in 1926, made it embeddable in concrete, and applied it in Tunnel of Zoragetstroi in 1931 (Davidenkoff 1934). Furthermore, in 1931, in France, André Coyne patented his VW strain sensor ("témoin sonore"), applied it in dam on the river La Bromme (Coyne 1938), and funded company Télémac (Télémesures Acoustiques) in 1947 that manufactured and applied a wide range of sensors based on VW, including strain sensors (Rosin-Corre et al. 2011). Potocki reinvented embeddable VW sensor (Potocki 1958), and the number of real-life applications of medium- and long-term strain monitoring was growing through second half of the twentieth century. Due to their excellent performance, the VW sensors became an established tool for long-term strain monitoring.

While VW sensors and their applications were being developed in Europe and the Soviet Union, the pioneering of strain sensors also happened in the United States. Burton McCullom and O.S. Peters from the former U.S. Bureau of Standards (present-time the National Institute of Standards and Technology – NIST) created, in 1924, a resistive strain sensor called "electrical telemeter," see Figure 3.2 (McCollum and Peters 1924). It was made of a stack of carbon discs, which would change their electrical resistance if exposed to a change in compression.

These sensors were applied to Stevenson Creek Experimental Dam, near Fresno, California, in order to measure strain in concrete during the test performed in 1925–1927 (Rogers 2010). Stevenson Creek Experimental Dam was one of the first structures instrumented with sensors and monitored remotely. Figure 3.3 shows a schematic of the remote monitoring of the dam. To the

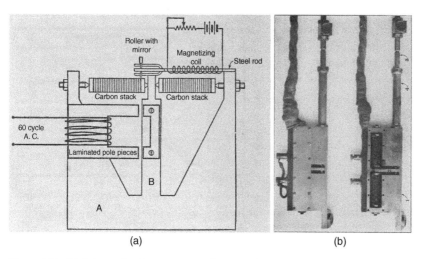

Figure 3.2 (a) Schematic of components of an electrical telemeter and (b) photograph of closed and open packaging of electrical telemeter (McCullom and O.S. Peters., 1924 / National Bureau of Standards).

Figure 3.3 Schematic of the implementation of remote monitoring in Stevenson Creek Experimental Dam. Source: The Stevenson Creek test dam (1925). Southern California Edison Collection of Photographs (photCL SCE), The Huntington Library, https://go.exlibris.link/461Zn0Q0, last accessed on February 28, 2022.

best of the author's knowledge, it was the first known schematic used to demonstrate remote monitoring of civil structures.

Electrical telemeters were not stable in long-term and were adversely affected by changes in humidity. To address these issues, Roy Carlson developed during the 1930s a new embeddable resistive strain sensor called "strain meter." The sensors consisted of two coils of unbonded carbon steel wire mounted on steel rods connected to anchors at each extremity of the sensor. The change in the distance between the anchors would increase the length of one coil and decrease the length of the other coil. The ratio of resistance between the two coils was used to measure the strain, while their total resistance was used to measure temperature. To protect the coils from humidity, the sensor packaging was filled with oil. This design of sensors enabled excellent long-term performance of sensors, and Carlson created a company for their commercialization. These sensors have been successfully applied in Hoover Dam (Rogers 2010) and in numerous other civil and geotechnical structures.

The second-generation strain sensors are based on fiber-optic technologies. Various sensing principles were explored in 1970s (e.g., see Hill 1974, Cielo 1979, Culshaw et al. 1981) and the first application of discrete fiber optic sensors (FOS) for strain measurement appeared in 1980s (e.g., see Asawa et al. 1982, Meltz et al. 1987, Hogg et al. 1989, Udd et al. 1991). Commercial discrete FOS for strain monitoring in civil structures became available on market in 1990s by various existing companies such as FiberMetrics Corporation (Mason et al. 1992) and especially by newly funded companies such as Blue Road Research, FISO, and SMARTEC. Commercial availability, in turn, enabled the first large-scale applications for SHM (e.g., see Inaudi et al. 1999a, Choquet et al. 2000, Udd and Kunzler 2003). The first important paradigm change brought about by FOS into strain measurement was availability of long-gauge sensors, which opened new paths for strain-based SHM (e.g., see Inaudi et al. 1999a, Glisic and Inaudi 2003b, Xu et al. 2003). The second important paradigm change was availability of distributed (or more precisely truly distributed) FOS. Exploration of distributed FOS started in 1980s, first for temperature sensing (e.g., see Hartog 1983, Kingsley 1984, Kikuchi et al. 1988), followed by strain sensing (e.g., see Dunphy et al. 1986), but true breakthrough was made with developments of sensing techniques based on Brillouin and Rayleigh scattering. Brillouin effect was researched mostly in 1990s, again first for temperature monitoring (e.g., Culverhouse et al. 1989, Bao et al. 1993), followed by strain monitoring (e.g., Horiguchi et al. 1993, Nikles et al. 1994). Development in strain sensing systems based on Rayleigh scattering mostly happened in parallel with development of Brillouin-based monitoring, with temperature sensing first (e.g., Rogers and Handerek 1992, Wait and Newson 1996), followed by strain sensing (e.g., Liu and Yang 1998, Posey et al. 2000). Commercialization of distributed FOS for strain monitoring started at the end of the twentieth and beginning of the twenty-first century with companies providing reading units, such as Ando, Omnisens, Luna Innovations, Marmota and Oz Optics, and the first commercially available distributed FOS for strain monitoring appeared on the market in the first decade of twenty-first century (e.g., Inaudi and Glisic 2005, 2006). The first large-scale applications of distributed strain monitoring using FOS were also realized in the first decade of twenty-first century (e.g., see Glisic et al. 2007, Mohamad et al. 2007, Hoult et al. 2009, Inaudi and Glisic 2010). While FOS market for strain monitoring is well established, the research is still ongoing in order to push the boundaries in various aspects of performance, e.g., Rinaudo et al. (2016) or Bao et al. (2017) to achieve high-temperature sensors; Maraval et al. (2017) or Zhou et al. (2018) to achieve dynamic Brillouin-based sensing system, Xu et al. (2014) to improve spatial resolution of Brillion-based sensing systems; Feng et al. (2017), Bado et al. (2018), or Fan et al. (2018) to explore new applications for distributed FOS.

The development of third-generation strain sensors started approximately with the end of the twentieth and beginning of the twenty-first century, and it has been targeting mostly (but not exclusively) 2D distributed and quasi-distributed sensors, whose spatial coverage is large, and whose purpose is mostly (but not exclusively) characterization of unusual structural behaviors rather than accurate strain monitoring, i.e., to perform SHM at integrity scale rather than global structural scale. Examples of ongoing research are various forms of quasi-distributed sensors such as expandable sensor networks (e.g., see Lanzara et al. 2008, Salowitz et al. 2014) and sensing sheets (e.g., see Glisic and Verma 2011, Glisic et al. 2016), and truly distributed sensors such as photonic crystals (e.g., see Zonta et al. 2009, Zur et al. 2017), sensing skins (e.g., see Loh et al. 2009, Pyo et al. 2011, Laflamme et al. 2013, Hallaji and Pour-Ghaz 2014, Schumacher and Thostenson 2014), and nano-paints and nano-based materials and adhesives (e.g., see Lim et al. 2011, Withey et al. 2012, Ladani et al. 2017).

2D sensing techniques described above require contact between sensors and the structure. Alternative techniques that emerged in the second decade of the twenty-first century are based on non-contact camera imaging approaches. In these approaches, digital images of structural members are taken over time and compared to reference photographs or to each other using various algorithms. Modern, powerful computers combined with various machine learning techniques

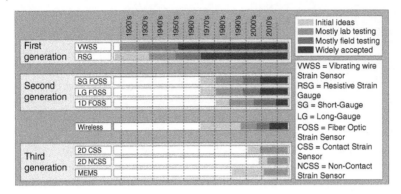

Figure 3.4 Timeline for the development of strain sensors and strain-based sensing techniques for civil structures and infrastructure (Glisic 2022/ MDPI / CC BY 4.0.).

enabled digital image processing at a very fine (pixel) scale, and extraction of strain fields from photographed structural members (e.g., see Nonis et al. 2013) or strain-based damage-sensitive features (e.g., see Yang et al. 2018).

Besides the above presented 2D sensors, development of discrete sensors based on Micro-Electro-Mechanical Systems (MEMS) has also been ongoing since 1990s (e.g., see Jacobsen et al. 1991). MEMS-based strain sensors have found applications in bio-medical industry, mechanical engineering, or as a sensing element in other types of sensors (e.g., force sensors), but they are not yet mature for applications in civil structures, and the research is still ongoing (e.g., see Pozzi et al. 2011, Saboonchi and Ozevin 2015).

Finally, in between the developments of the second and the third-generation of strain sensors, fast progress in wireless technologies brought a paradigm change on how the SHM data can be collected, processed, and communicated. While early experimentation on wireless strain sensing dates back to 1970s (e.g., Adler 1971), booming in development of wireless capabilities for strain sensing actually started in 1990s (e.g., see Varadan et al. 1997, Straser et al. 1998, Lynch et al. 2002, Kurata et al. 2003). Wireless nodes enable reading and processing data from strain sensors and their transmission to storage media or the users (e.g., see Lynch et al. 2003). The technology matured rapidly with large-scale applications starting in first decade of twenty-first century (e.g., see Hou and Lynch 2006, Jang et al. 2010, Torfs et al. 2013).

To conclude this historical review of strain sensors and strain-based sensing techniques, an illustrative timeline is given in Figure 3.4 (Glisic 2022).

3.4 Desirable Properties of Strain Sensors

In general, the main parts of strain sensors are the sensing element, packaging, and connecting cables. Sensing element is the part that converts measurand (strain) into encoding parameter (e.g., electrical, or optical signal). The purpose of packaging (or housing) is to protect sensing element and to enable transfer of measurand (strain) from the structure to sensing element through physical contact with the structure (frequently realized by embedding, gluing, clamping, bolting, welding, etc.). Finally, cables enable transfer of information from sensing element to reading unit or wireless node. The sensor parts are presented in more detail in Sections 3.5 and 3.6.

Regardless of type, physical principle, or generation, the following properties are in general desirable for strain sensors used in SHM

1. The sensors should be sensitive to the dedicated measurand (i.e., strain sensor should be sensitive to strain); this is the ideal that is usually well achieved in practice.

   Important: In the case of strain sensors, while sensitivity mostly depends on the properties of sensing element, the final sensitivity also depends on the type of packaging and the quality of strain transfer from the monitored material to the sensor, i.e., on the manner in which the sensor is attached to the structure (e.g., embedded, glued, clamped, bolted, or welded).

2. To the extent possible, the sensor should be insensitive to any other parameter (other than strain), property, or influence; this ideal is not easy to achieve as sensing elements themselves, besides the strain, may be, in one way or another, unintentionally (but inevitably) sensitive to variations in temperature, humidity, or pressure, and electrical sensors may be also affected by electro-magnetic interferences.

   Important: In the case of strain sensors, the undesired sensitivity to many other parameters (other than strain) can be minimized by using appropriate packaging, protection, and shielding; however, in the most of the cases, the sensitivity to temperature cannot be eliminated given that temperature variations influence not only the dimensional changes of sensor and reading unit components, but also modify physical properties related to encoding parameter (e.g., electrical resistivity in strain gauges, and refraction index in optical fibers); hence, in the most of the cases, a temperature sensors must be installed in parallel with strain sensors, and thermal compensation of each strain sensor must be performed by using temperature measurements (e.g., see Section 4.3).

3. The sensors that are in contact with the monitored structure should not derange neither measurand nor structure, i.e., they should neither influence or alter the measurand (strain) at and around location of sensors, nor adversely affect the monitored structure in terms of performance, safety, and durability; to achieve this ideal, it is important to properly select the geometrical, mechanical, and chemical properties of the sensor packaging (e.g., dimensions, stiffness, thermal expansion coefficient, and chemical composition), and the manner of sensor installation (embedding, gluing, clamping, bolting, welding, etc.); for example, strain sensor encapsulated in stiff steel packaging and embedded in fresh concrete will alter strain field in concrete at very early age as the latter is much softer than the former (e.g., see Glisic and Simon 2000, Calderon and Glisic 2012); another example: attaching very stiff sensor to (tinny) structural member with cross-section comparable to that of the sensor, or attaching it to the material which stiffness is significantly lower than stiffness of the sensor (e.g., attaching sensor in stiff steel packaging to wooden members), would result in redistribution of stress and strain fields in the monitored structural member, which in turn will provide results that do not reflect intended strain measurement.

   Important: Methods of installation such as gluing, clamping, bolting, and welding involve minor modifications to monitored structures, such as removal of protective paint, grinding of surface, drilling the holes, heating, and deformation, just to name a few; these modifications, while usually small by extent and harmless to structure, may be, in some cases, excessively intrusive and may damage monitored structure with adverse consequences to both the performance of sensor measurement and the structure itself (e.g., as mentioned in Chapter 2, drilling holes in very thin webs, embedding the sensors in very shallow concrete cross-section); similar would happen in the case of chemical incompatibility between sensor packaging and structure (e.g., as mentioned in Chapter 2, embedding aluminum packaged sensor in concrete or installing aluminum brackets to potentially wet concrete will result in chemical reactions that may damage the structure or sensor or both; putting in contact stainless steel with ordinary steel might have similar outcome).

4. The preferred (but not necessary) relation between measurand and encoding parameter should be linear; linear relation simplifies process of calibration of sensors, as well as the process of installation in the cases where reference measurement should be set within certain limits (see Chapter 5); moreover, linear relation between measurand and encoding parameters enables faster data validation.

5. Sensors for SHM should be reliable in long-term; this means that sensors themselves, as well as the manner of installation of sensors, must be durable and exhibit a high level of resistance to aging, i.e., to any adverse mechanical or chemical process that may affect performance of sensors and lead to their malfunction or failure; typical adverse mechanical and chemical processes that may influence sensor parts or installation are fatigue, rheologic effects (e.g., creep, shrinkage, and relaxation), thermal instability, chemical reactivity (e.g., corrosion), electro-magnetic interference, etc. (e.g., see Habel et al. 1994).

   Important: To achieve long-term reliability of sensors, it is crucial to use adequate quality materials for sensor parts and controlled processes for their assemblage; also, installation procedures should be followed as strictly as possible; finally, if needed, sensors should be protected from any external influence that can affect them once installed (rain, snow, wind, vandalism, rodents, unintentional human-made damage, electro-magnetic interference, chemicals, etc.); the most frequent reasons for premature sensor malfunction or failure are installation errors and manufacturing errors (see Section 4.6).

6. Packaging (or housing) of sensors should allow or enable all the above points 1–5 and provide physical robustness for safe handling and easy installation of sensors.

   Important: Design of packaging is very important as the packaging ultimately influences all the properties described in the points 1–5 above; packaging should respond well to both the application requirement and site conditions (e.g., see Potocki 1958, Glisic et al. 1999, Inaudi et al. 1999b, Lin et al. 2005).

Due to large number of manufacturers of strain sensors, reading units, and wireless nodes, and large number of varieties in terms of strain sensor configurations, packaging, and reading schemes, there is even larger number of potential strain monitoring solutions that could be applicable to a specific SHM project. The selection of a suitable solution requires informed decision-making, i.e., good knowledge and understanding of the strain-sensing technologies. While presenting all existing technologies in detail largely exceeds the scope of this book, the next few sections present examples of technologies that are widely accepted in practice, i.e., the examples of strain sensors of the first- and second-generation. These examples illustrate and emphasize the need for complete understanding of strain sensor components and functional principles, as prerequisites to correctly selecting and deploying sensors, as well as to correctly process and interpret (and subsequently analyze) the data they provide. Hence, simply considering strain sensing technology as a "black box" that only provides time series of strain would be wrong in many aspects.

3.5 Examples of the First-Generation Strain Sensors

3.5.1 General Advantages and Challenges of the First-Generation Strain Sensors

The first-generation strain sensors extensively used in SHM of civil structures and infrastructure include VW strain sensors, resistive strain gauges, and Carlson strain meters. Besides them, strain sensors based on piezo-electric sensing elements have also been developed. However, they did not find extensive application in strain-based SHM of civil structures and infrastructure, mostly due to limitations resulting from transverse sensitivity and challenge in performing an accurate true static measurement. Hence, they are not presented in more detail. However, sensors with piezo-electric sensing elements have found extensive applications in acoustic and wave-propagation based SHM.

Being electrical sensors based on their sensing principle, the VW strain sensors, resistive strain gauges, and Carlson strain meters have some common attributes in terms of advantages and challenges.

General advantages of the first-generation strain sensors are listed below.

- Long tradition: There is extensive experience and sustained improvement in performances and applications of the first-generation sensors, which makes them easy to understand, use, and operate, and brings confidence when applying them.
- Affordable cost: Due to the simple functioning principle, inexpensive sensor parts, and widespread use, the cost of the first-generation sensors is lower than cost of the second-generation sensors; however, their cost is still strongly correlated with the quality, i.e., inexpensive sensors are likely to lack the quality in some way.
- High measurement resolution and repeatability precision: The first-generation strain sensors have a resolution typically in the range of $1\,\mu\varepsilon$ ($1\,\mu\varepsilon = 1$ microstrain $= 1\,\mu m/m = 1\cdot10^{-6}\,m/m$) and precision typically between 0.5% and 2% of sensors dynamic range (or full scale – F.S.), which is suitable for SHM of civil structures and infrastructure.
- Easy repair of cables in case of damage: Being simple electrical cables, the carriers of information of the first-generation sensors are easy to repair and replace in the case of damage.
- Simple interface to wireless nodes: Being encoded by an electrical signal, the first-generation sensors are simple to interface, read, and analyze by wireless nodes.

General challenges of the first-generation strain sensors are listed below.

- Electromagnetic interference (EMI): Various sources of electromagnetic field may be found nearby monitoring structures (e.g., lightning and thunders, radio waves, and proximity of electrical power conductors), which can interfere with sensors, affect the measurement, and in some cases even result in sensor malfunction or damage; depending on the strength of EMI, in some cases mitigation can be performed by shielding, which, however, incurs additional material and installation cost, and may even result in an additional weight to the structure.
- Temperature compensation: Sensing elements of the first-generation sensors are sensitive to temperature, and hence the measurements must be thermally compensated, which affects the accuracy of the measurement and incurs additional material and installation costs.

The advantages and challenges related to VW strain sensors and resistive strain sensors (strain gauges) are presented in the corresponding subsections below. Given that Carlson strain meter does not have such a broad applicability as the two other groups of sensors, its detailed presentation is omitted in this book, but for the completeness in presenting the first-generation sensors, its appearance is shown in Figure 3.5, and its performances are summarized in Table 3.1 (Section 3.5.4). More details on the Carlson strain meter can be found in RST Instruments (2022) and Glisic (2022).

3.5.2 Vibrating Wire (VW) Strain Sensor

Functioning principle of a VW strain sensor is based on the known relation between the first fundamental frequency of a tensioned wire and the mechanical strain generated in the wire by the

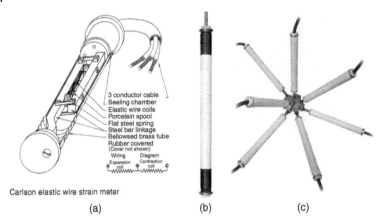

Carlson elastic wire strain meter

(a)   (b)   (c)

Figure 3.5 (a) View of the components of the Carlson strain meter. Source: Reprinted from Carlson (1939), with permission of Massachusetts Institute of Technology – MIT. (b) Modern packaging of the sensor and (c) so-called "spider configuration" with multiple sensors enabling assessment of a 3D strain tensor. Source: Courtesy of RST Instruments Ltd.

force that tensions the wire. The relation between fundamental frequency and force in the wire is given in Equation 3.2.

$$f_w = \frac{1}{2L_w}\sqrt{\frac{F_w}{m_w}} = \frac{1}{2L_w}\sqrt{\frac{F_w}{\rho_w A_w}} \tag{3.2}$$

where f_w is fundamental (resonant) frequency of the wire, F_w is tensioning force, L_w is the length of the wire, and m_w is linear density of the wire (the mass per unit length), i.e., $m_w = \rho_w A_w$, where ρ_w is volumetric density of the wire and A_w is the area of wire's cross-section.

If the area of cross-section is constant along the wire, the mechanical strain in the wire is uniform in the direction of the wire and can be derived from Equation 3.2 as follows:

$$\varepsilon_w = \frac{F_w}{E_w A_w} = 4\frac{L_w^2 m_w}{E_w A_w}f_w^2 = 4\frac{L_w^2 \rho_w}{E_w}f_w^2 \tag{3.3}$$

where E_w is Young modulus of wire.

WV sensing element includes metallic wire, anchors, an excitation coil, and a pick-up coil, as shown in Figure 3.6.

The purpose of anchors is to ensure tension in the wire (wire is pretensioned between anchor points, so that the sensor can measure both elongation and contraction), provide mechanical contact with monitored structure, and enable strain transfer from the structure to the wire. Two electrical coils, the excitation coil and the pick-up coil, are placed close to the midpoint of the wire. The excitation coil creates an electromagnetic field that forces wire into vibration, while the

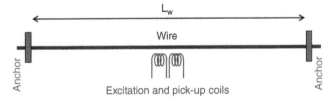

Figure 3.6 Schematic representation of a WV sensing element (Glisic 2009).

pick-up coil, inversely, converts the frequency of VW into an electrical signal. Frequency of the VW is decoded from an electrical signal, either in the reading unit or in the wireless node, depending on the architecture of the SHM system. Then the strain in the wire is determined using Equation 3.3.

Note that the strain in the wire depends on relative displacement of anchors, which in turn depends on the deformation of monitored structure. While the strain in the wire is uniform, and can be described with a single number, this might not be the case with the strain in the monitored structure, which may vary between the anchors (e.g., depending on loading). Hence, the strain in wire, i.e., the strain measured by VW sensing element, is equal to an average value of strain in the monitored structure, where averaging is performed between anchors and includes the normal component of strain in the direction of wire. The distance between anchors, i.e., the length over which the strain is averaged and measurements are taken, is called the gauge length of the sensor. More detail about gauge lengths for sensors is given in Section 4.4.

It is important to note that for long-term reliable functioning of VW strain sensors, the strain change in the wire should only be introduced by deformation of monitored structure. Given that the wire used in VW element is made of metal so that it can be operated by electrical energy, it has negligible rheological strain. Tension in wire is kept within adequate limits, so that fatigue in wire should not occur. Packaging is usually designed to prevent any chemical degradation of wire. However, one influence on wire that cannot be prevented, but can be mitigated, is temperature. Due to thermal expansion of wire, temperature changes will result in dimensional changes in wire, which in turn will change tension in wire and affect the measurement, and thus the VW strain sensors must be thermally compensated. This is usually performed by integrating an electrical temperature sensor into the packaging of the VW sensor and automatically applying compensation formulae. Given that the majority of adverse influences on sensor reliability could be prevented or mitigated, VWs have very good long-term performance (e.g., see Rosin-Corre et al. 2011). Practically, the only adverse influence that represents a challenge to VW sensors are electromagnetic interferences that may occur due to thunder, proximity of power lines, or similar.

Packaging of VW strain sensor has evolved over several decades, and there are various designs that target diverse applications, e.g., embedding in concrete, and surface mounting. This makes VW strain sensors versatile. In all cases, packaging provides excellent protection for the sensing element and guarantees good strain transfer from the monitored structure to sensing element. Figure 3.7 presents several examples of VW strain sensors that are commercially available.

VW sensors are accessed by a reading unit whose purpose is to send the signal to the excitation coil, receive the encoded signal from the pick-up coil, and decode the received signal into the strain. Reading unit can be single-channel or multi-channel, portable or rack-mounted, wired or wireless. Examples of commercially available reading units are shown in Figure 3.8.

Figure 3.7 Examples of commercially available VW strain sensors; left: embeddable in concrete; middle: two types of surface mountable sensors; and right: surface mountable (up) and embeddable (down) sensors. Source: Courtesy of Roctest / Telemac.

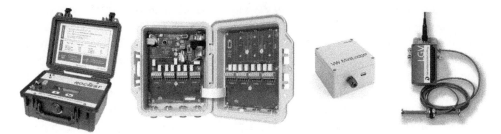

Figure 3.8 Examples of commercially available VW reading units. Source: Courtesy of Roctest / Advantech Engineering Consortium, and Newsteo.

VW strain sensors are either surface-mounted on the structure or embedded in concrete. The surface mounting is usually performed by using special brackets provided by the monitoring company (manufacturer of the sensors). The brackets must guarantee good strain transfer from the structure to the sensor. Thus, they are typically installed on the structural member by gluing, bolting, fastening, or welding; depending on the dimensions and material properties of structural member and the allowed procedure (installation of sensor should not adversely affect the safety of the structural member or the strain field in it). Many types of surface-mounted VW strain sensors and their accessories might not need additional protection from the elements (wind, rain, snow), rodents, and mechanical damage; some of them might steel need it, depending on manufacturing specifications. Also, additional shielding against EMI might be needed. Any additional protection (mechanical or EMI) should not affect the strain field in the structure or strain transfer from the structure to the sensor. Embedding in concrete is performed by simply attaching the sensor to rebars or by placing it in-between the rebars using plastic ties or ordinary wire.

Besides the general advantages of the first-generation sensors presented in previous subsection, an important specific advantage of VW strain sensors is the use of frequency of VW to encode the strain. This makes them insensitive to signal degradation that can occur due to long cable runs, potential changes in the electrical properties of the cables (e.g., due to damage, and environmental influences), and results in an excellent long-term stability and high precision of measurements. The main challenges specific to VW technology are the relatively limited number of sensors that can be attached to the reading unit (max. 32), short-gauge length, and the dimensions and stiffness that might be limiting for certain applications. Summary of best performances of VW strain sensors is given in Table 3.1 (Section 3.5.4).

3.5.3 Resistive Strain Gauge

Fundamental physical law behind a resistive strain sensor (in practice often referred to as strain gauge or strain gage or foil strain gauge or foil strain gage) relates the electrical resistance of a linear electric resistor with constant cross-section to its dimensions, as shown in Equation 3.4.

$$R_r = \rho_r \frac{L_r}{A_r} \tag{3.4}$$

where R_r is electrical resistance in resistor, L_r is the length of resistor, A_r is the area of cross-section of the resistor, and ρ_r is specific electrical resistance of resistor.

Functioning principle of a resistive strain sensor is based on the relation between the change in strain of a linear electrical resistor and change in its electrical resistance, derived from Equation 3.4

and presented in Equation 3.5.

$$GF = \frac{\Delta R_r / R_r}{\Delta L_r / L_r} = \frac{\Delta R_r / R_r}{\Delta \varepsilon_r} \Rightarrow \Delta \varepsilon_r = \frac{\Delta R_r / R_r}{GF} \tag{3.5}$$

where ΔR_r is change in electrical resistance of resistor, ΔL_r is change in length of resistor, $\Delta \varepsilon_r$ is change in strain in resistor, and GF is a constant called gauge factor.

Sensing element of strain gauge consists of serpentine-shaped resistor, which is patterned over a substrate (i.e., the packaging of the sensor), see Figure 3.9. Due to resistor shape, the gauge length of sensor L_s is significantly different from total length of resistor L_r. The resistor is serpentine-shaped to increase the length of the resistor, which in turn increases sensitivity of sensing element and improves measurement accuracy by averaging measurement over the gauge length.

Depending on application, resistors can be made of metals such as Constantan (Cu—Ni alloy), Nichrome V (Ni—Cr alloy), Platinum alloys (usually tungsten), Isoelastic (Ni—Fe alloy), Karma-type alloys (Ni—Cr alloy), etc., or semiconductors for an extended dynamic range of sensors. Substrate is typically polyimide (Kapton), but other materials are used too depending on application (e.g., it can be metallic for installation by welding).

The gauge factor GF depends on the resistor material and geometrical properties. It can have different values for sensing elements with different materials and geometries, but the typical value is approximately equal to two (i.e., GF $\approx$ 2).

Due to their small size, simple packaging, and variety of materials that can be used in their manufacturing, the strain gauges are available in numerous forms and configurations: with a single resistor for uniaxial strain measurement, with three resistors combined in "rosette" for monitoring several components of strain tensor, with four resistors combined in Whitestone Bridge for uniaxial strain measurement self-compensated to temperature, with or without lead wires (carriers of information), etc. Some examples are shown in Figure 3.10.

Reading units for strain gauges can be based on different principles, and which one will be used depends on application. The Chevron Bridge enables multiple channel arrangement and is frequently applied in SHM of rotating machines (Omega 1998). Four-Wire Ohm Circuit provides automatic voltage offset compensation, and the lead wires do not affect measurement; however, it cannot be used for high-frequency dynamic measurements (Omega 1998). Contrary, Constant Current Circuit can be used for high-frequency dynamic measurements and features low drift due to temperature (Omega 1998). The Wheatstone Bridge enables various configurations of up to four sensing elements or sensors to be accessed and read from a single channel. Examples of commercially available reading units are given in Figure 3.11.

Strain gauges are usually surface-mounted onto a structural member by gluing. The adhesive selected for gluing must be compatible with the material of the monitored structural member

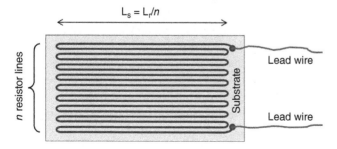

Figure 3.9 Schematic representation of sensing element and packaging of strain gauge (Glisic 2009).

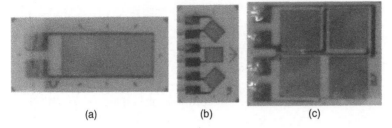

Figure 3.10 Examples of strain gauges with different configurations: (a) Strain gauge with single resistor; (b) "rosette," a system of three strain gauges for inferring a 2D strain tensor; and (c) full-bridge strain gauge, a system consisting of four resistors that specifically use differential measurement to perform thermal self-compensation. Source: Reproduced from Glisic., 2009 / PRINCETON UNIVERSITY.

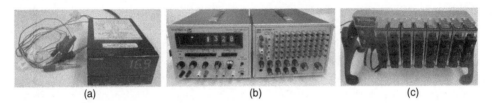

Figure 3.11 Examples of commercially available reading units for strain gauges: (a) For reading a single sensor; (b) older type of reading unit (left side) with a switch for reading multiple sensors (right side, switch for 25 sensors); and (c) computer operated reading unit, for reading multiple sensors (modern type, switch for 96 sensors).

(e.g., concrete, steel, timber, or composite material), and it is commonly suggested and provided by the monitoring company (i.e., sensor manufacturer). To achieve durability of installation and adequate strain transfer from the monitored structure to the sensor, a care must be taken in preparation of the surface of the structural member to which the sensor is to be glued, i.e., the surface must be properly smoothened and well cleaned before the gluing. In addition, the layer of the glue should be as evenly spread over the surface as possible and kept as thin as possible. Other means of surface mounting include welding (e.g., see HBM 2018) or magnetic attachment to steel structural elements (e.g., see TML 2017). Once installed on the surface of structural member, the strain gauges should be protected from the accidental damage, vandalism, rodents and other animals, elements (sun, snow, rain, and wind), and EMI by using various types of enclosures and shielding, which, in turn, should not affect neither the sensor measurement nor the strain field or integrity of the monitored structural member.

In general, strain gauges are not designed to be embedded in concrete; however, they can be installed on the rebars before pouring concrete by following the same procedures as for surface mounting. In this case, protection is needed to prevent damage from actions resulting from pouring and vibrating the concrete.

Advantages specific to strain gauges are: small size, suitable for monitoring strain concentrations; light weight, low stiffness, and varieties in substrate material, all resulting in applicability to virtually any structural member regardless of its dimensions or material; diversity of possible configurations that makes them flexible and tunable to specific applications; and high precision and resolution. Moreover, reading unit set-up is relatively simple, and practically any device that measures voltage with high resolution and long-term stability can have a role of reading unit.

Despite their long tradition and extensive use, stain gauges still face important challenges when applied for SHM purposes. Besides general challenges of the first-generation sensors presented in the introductory part of Section 3.5, the main challenges specific to strain gauges are: transverse sensitivity (strain component in the direction perpendicular to orientation of the strain gauge results in change in length of curved endings of "serpentines" and thus affects the measurement); tedious installation (quality installation requires surface smoothening and cleaning, correct application of adhesive, and mounting of suitable protection, which is all time-consuming); temperature compensation (change in temperature of monitored structure, but also current of air around unprotected sensor will both affect the measurement); sensitivity of lead wires (change in strain and/or temperature of lead wires of ordinary strain gauge may affect accuracy of measurement); potentially long lengths of cable extensions in wired SHM applications (resulting in high cumulative resistance and sensitivity similar to that of lead wires); and long-term instability (drift frequently occurs in long-term applications).

Hence, applying strain gauges for long-term static monitoring is challenging and difficult to realize in the field, on real structures. Nevertheless, in the applications where the comparison of strain measurements with an absolute reference is not necessary, such as periodic dynamic measurements repeated over long-term (e.g., to determine amplitude of strain due to live load or to monitor natural frequency derived from strain measurements), strain gauges can still be used with a satisfactory degree of precision.

Given that strain gauges can be provided with numerous types of substrates, resistors, and resistor configurations (e.g., see Figure 3.7), and can be read using several internal readout schemes, a good knowledge and understanding of technology is crucial for selecting the appropriate solution for a specific application. Summary of best performances of resistive strain gauges is given in Table 3.1 (Section 3.5.4).

3.5.4 Best Performances of the First-Generation Strain Sensors

Table 3.1 presents the best performances of the first-generation strain sensors that were commercially available at the time when this section was written (Glisic 2022). Thus, they are subject to future changes. The information provided in the table was taken, as is, from datasheets issued by various sensor manufacturers. Each field in the table contains the best achievable performance regardless of the potential trade-offs with performances presented in the other fields in the table (e.g., resolution and frequency of measurements are for some technologies correlated; improving one would diminish the other). Also, the best performances presented in the table are achieved in some idealized conditions (e.g., in the lab), while the actual performances in real-life settings may be altered depending on the application (e.g., quality of installation, and environmental conditions).

Table 3.1 is intended to help in the process of selecting a strain-based SHM system. However, it is somehow incomplete and contains some inconsistencies that result from information found in datasheets. The most important shortcoming is a lack of some important information regarding measures of accuracy, and lack of clarity in terminology in general. For example, majority of strain gauge datasheets do not contain any direct, practical information regarding repeatability, precision, or frequency of measurements – the user is supposed to calculate this information based on the chosen configuration of sensors and specifications of reading unit. In the case of VW strain sensors, the repeatability precision is given in datasheets through term "linearity," i.e., in relative, rather than absolute value. Inconsistency in the table is visible in dynamic range of sensors: while for resistive strain gauges this value is given as plus/minus value, for VW sensors this is not the case, which leaves doubts on true limits of the dynamic range. Finally, important shortcoming of the first-generation sensor datasheets reflected in Table 3.1 is the lack of any information regarding reproducibility precision.

Table 3.1 Best performances of the commercially available first-generation strain sensors (Glisic 2022).

Property	Vibrating wire strain sensor	Resistive strain gauge	Carlson strain meter
Gauge length	Typical: 50–150 mm Special (long): 300 mm	Typical: 0.3–20 mm Special (long): 150 mm	Typical: 203–508 mm Special (short): 102 mm
Multiplexing (the way multiple sensors are read)	Sequential (one by one)	Sequential (one by one)	Sequential (one by one)
Maximum number of sensors per reading unit (with a channel switch)	32[a]	1200[a]	24[a]
Stability	Long-term stable	Long-term drift (typical)	Long-term stable
Resolution	Typical: 1 με Best: 0.35 με	Typical: 1 με Best: 0.5 με	Typical: 2–3 με Best: 1.5 με
Linearity (repeatability precision)	Typical: ±0.5% Full Scale Best: ±0.1% Full Scale	0.2–2%	N/A
Sensor range	Typical: 3000 με Extended: 10,000 με	Typical: ±10,000 με Extended: ±100,000 με	Typical: 2000 με Extended: 3900 με
Temperature sensitivity	Compensated with an integrated temp. sensor	Compensation needed	Self-compensated (by measurement principle)
Measurement frequency	Typical: 0.25–1 Hz Dynamic: 20–333 Hz Max.: 1 kHz[a]	Typical: 100–200 Hz Max. 100 kHz	N/A

a) Unconfirmed.

3.6 Examples of the Second-Generation Strain Sensors

3.6.1 General Advantages and Challenges of the Second-Generation Strain Sensors

Second-generation strain sensors based on fiber-optics (Fiber-Optic Strain Sensors – FOSS) became widely accepted among practitioners in the second decade of the twenty-first century. Besides the short-gauge sensors, with applications similar to those of the first-generation sensors, they changed the paradigm in strain-based SHM by offering two new types of sensors: long-gauge and truly distributed sensors. These two types of sensors, practically unique to commercially available fiber optic technologies, expanded the scale of SHM from local-material (enabled by short-gauge sensors) to local-structural, global, and integrity monitoring (enabled by long-gauge and distributed sensors respectively). Global and integrity monitoring, in turn, enabled affordable spatial coverage of structures with sensors (e.g., see Inaudi et al. 1999a, Glisic et al. 2007) and created conditions for reliable achievement of Level III and Level IV SHM (see Figure 1.2) in real-life settings (e.g., see Glisic et al. 2013, Abdel-Jaber and Glisic 2015).

Optical fibers were originally designed for telecommunications industry in order to reliably transmit information over long distances. Ordinary optical fibers are made of fused silica, and they consist of three main components: core, cladding, and coating, see Figure 3.12.

The core is made of silica, and its function is to transmit the light. Its diameter is 5–10 μm for single-mode fibers and around 50 μm for multimode fibers. Cladding surrounds the core and is also

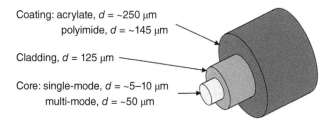

Coating: acrylate, d = ~250 µm
  polyimide, d = ~145 µm

Cladding, d = 125 µm

Core: single-mode, d = ~5–10 µm
  multi-mode, d = ~50 µm

Figure 3.12 Typical components and dimensions of an ordinary optical fiber (d denotes outer diameter). Source: Modified from Glisic (2009).

made of silica, but with refraction index different from that of the core. Its function is to physically support the core and to control the optical losses (i.e., to keep the light inside the core). External diameter of the cladding is 125 µm. Coating function is to provide robustness to fiber against physical actions (e.g., to enable safe handling and protect against humidity). Coating of ordinary optical fibers is made of acrylate, with external diameter of 250 µm. However, acrylate does not provide good strain transfer, and acrylate-coated fibers should not be used for strain sensors. Other typical coating material is polyimide, with an external diameter of 145 µm. Polyimide-coated fibers provide excellent strain transfer, but they are more expensive and more difficult to handle. Nevertheless, they are a much better solution for high-quality sensors. Optical fibers with less typical coatings such as aluminum, carbon and gold are also manufactured, and these coatings are used in special applications.

Light propagating through the optical fibers is in the infrared range, with wavelengths of approximately 1310–1590 nm for single-mode and 850–1300 nm for multi-mode fibers.

In addition to silica-based FOSS, polymer-based FOSS are also being developed, and they are targeting geotechnical applications where strain is large and exceeds the dynamic range of silica-based fibers (Liehr et al. 2008). However, their applicability in strain-sensing is still being researched, and therefore they are beyond the scope of this section.

As opposed to the first-generation sensors, where only two main physical principles were used to create strain-sensing elements, the second-generation sensors, FOSS, offer multiple physical principles that enable strain-sensing elements. In the case of discrete sensors, those are: (i) Extrinsic Fabry–Perot Interferometry (EFPI), (ii) Fiber Bragg-grating spectrometry (FBG), (iii) Michelson and Mach–Zehnder Interferometry (SOFO – French acronym for "surveillance d'ouvrages par fibres optiques" meaning "structural monitoring by optical fibers"), and (iv) intensity of light (micro-bending); in the case of distributed sensors, those are: (v) Brillouin scattering, and (vi) Rayleigh scattering. Commercially available SHM systems based on Brillouin scattering have several subdivisions: Brillouin Optical Time Domain Analysis (BOTDA), Brillouin Optical Frequency Domain Analysis (BOFDA), Brillouin Optical Time Domain Reflectometry (BOTDR), and Brillouin Optical Frequency Domain Reflectometry (BOFDR), where the first two are considered stimulated Brillouin scattering techniques, and the last two are considered as spontaneous Brillouin scattering techniques. Note that Raman scattering is also used in distributed sensing, but given that it can be used only for temperature monitoring, it is beyond the scope of this book. The first four physical principles are used in discrete sensing, where EFPI and FBG are used in short-gauge sensors, while FBG, SOFO, and Intensity-based sensors are used in long-gauge sensors (note that FBG can be used for both short- and long-gauge sensors). Stimulated Brillouin scattering (BOTDA and BOFDA), spontaneous Brillouin scattering (BOTDR and BOFDR), and Rayleigh scattering are used in 1D distributed strain sensing. Figure 3.13 presents the classification of FOSS based on their physical principles and gauge length.

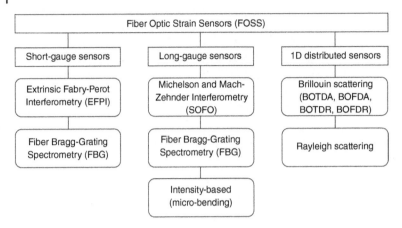

Figure 3.13 Classification of FOSS based on physical principle and gauge length (Adapted from Glisic 2009).

Having optical fiber sensing element, the FOSS have some common attributes in terms of advantages and challenges, regardless the physical principle behind their functioning.

General advantages of the second-generation strain sensors are listed below.

- Durability and long-term stability: FOSS are made of standard optical fibers used in the telecom industry; thus, they are designed to be chemically stable and inactive in the long-terms, which in turn provides FOSS with durability and long-term stability.
- Electrical passivity and reliability: Optical signal, which carries the encoding parameter, is insensitive to electro-magnetic interference; this property combined with chemical stability results in high long-term reliability of sensor measurements; in addition, the electrical passivity of light makes FOSS applicable in environments where the use of electrical sensors is restricted due to potential sparking (e.g., in the gas and oil industry).
- High measurement resolution and repeatability precision: The resolution and repeatability precision of FOSS varies depending on physical principle; in general, FOSS have resolution in the range of 0.2–5 με and repeatability precision typically ranges between 1 με (for discrete FOSS) and 20 με (distributed FOSS), which is suitable for SHM of civil structures and infrastructure.

General challenges of the second-generation strain sensors are listed below.

- Cost: FOSS are mature, yet their cost is still higher than the cost of the first-generation strain sensors; this is a consequence of expensive components and manufacturing; however, while FOSS cost did not decrease over the last decade, it did not rise either, and the discrepancy in cost between the second- and first-generation sensors, which was diminishing over time due to inflation, is expected to further diminish in the next decade.
- Tedious repair on-site: Optical fibers require special equipment in order to be re-connected (spliced) after the breakage; such equipment might not be easy to handle and operate in on-site conditions.
- Interfacing to wireless nodes: Being encoded by optical signal, the second-generation sensors cannot be directly interfaced, read, and analyzed by wireless nodes; wireless communication can be established with the reading unit, but sensors have to be connected to the reading unit via cable extensions.

The advantages and challenges related to the specific technology behind SOFO, FBG, BOTDA, and Rayleigh scattering are presented in the corresponding subsections below, as these technologies are frequently used in practice. More details related to other technologies can be found in Measures (2001) and Glisic and Inaudi (2007). Tables 3.2 and 3.3 (Section 3.6.5) summarize the best performances of discrete and distributed FOSS respectively (including technologies that are not presented in detail in the next subsection i.e., EPFI and intensity-based sensors).

3.6.2 Interferometric or SOFO Sensors – Strain Sensors Based on Michelson and Mach–Zehnder Interferometry

Sensors based on intrinsic interferometry are often called SOFO sensors, which originate from the French abbreviation "surveillance d'ouvrages par fibres optiques" meaning "structural monitoring by optical fibers" (Inaudi 1997). The sensing element of interferometric (SOFO) sensors consists of two optical fibers of different length. While Michelson interferometry is used for static measurement and Mach–Zehnder interferometry is used for dynamic measurement, both techniques are based on the analysis of the time delay between optical signals sent through the two optical fibers. The time delay is correlated to difference in length of the two optical fibers, which in turn is correlated to the strain in the monitored structure (see Figure 3.14).

In the case of Michelson interferometry, chemical mirrors are silvered at the ends of the two optical fibers; they reflect the encoded optical signal (light) back to a photo detector which is situated in the reading unit (see Figure 3.14). In the case of Mach–Zehnder interferometry, the fibers are looped back toward the reading unit, creating two direct outputs (instead of mirrors). Nevertheless, to improve the practicality of sensor implementation, a sensor based on Mach–Zehnder interferometry can be modified by adding mirrors, thus making its appearance the same as the sensor based on Michelson interferometry. Given that the sensor has the same appearance, the main difference between the two techniques is in the opto-electronics of the reading unit, which dictates the performance. A reading unit based on Michelson interferometry enables static measurements that are very stable in the long-term. Contrary, a reading unit based on Mach–Zehnder interferometry enables dynamic measurements that feature very high resolution but should be carried out in short-term, as the reference drifts over long-terms. The sensing element and functional principle of a standard SOFO sensor are shown in Figure 3.14.

Two distinct zones can be noticed in Figure 3.14, active and passive zone. The active zone, delimited with anchors, contains the actual sensing element, and its purpose is to interact with structure and measure the strain. The purpose of the passive zone is to bring an optical signal from the reading unit to the sensor (optical signal is denoted by "1") and send the encoded signal back to the reading unit for analysis (denoted by "5").

The active zone contains two optical fibers: the measurement fiber and the reference fiber; both fibers have mirrors at their ends. Part of the active zone that interacts with the structure is delimited by two anchors, whose purpose is to ensure mechanical contact with the monitored structure and strain transfer from structure to the sensor. Hence, the distance between the anchors defines

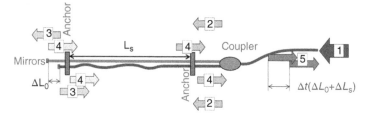

Figure 3.14 Sensing element and functioning principle of interferometric (SOFO) sensor (Glisic 2009).

the gauge length of the sensor. The passive zone consists of one single-mode fiber, connector, and coupler.

The measurement fiber is pre-tensioned between the anchors to allow measurement of both shortening and elongation. The reference fiber is purposely made loose (strain-free) so the strain from the structure cannot be transferred to it. The purpose of reference fiber is to perform thermal compensation: both the measurement and the reference fiber are close to each other and have the same spatial extent, thus they will have approximately the same temperature; given that the measurement is related to delay in time of flight in two fibers, the part of delay induced by temperature is canceled.

The measurement is performed as follows (see Figure 3.14): (i) light is sent from the reading unit into the sensor through the passive zone; (ii) coupler splits the light into two identical signals and inserts them in fibers of active zone; (iii) the signal from shorter reference fiber reflects back from the mirror, while it is still propagating along the longer measurement fiber; (iv) the light reflects back from longer measurement fiber, and due to difference in fiber lengths, a delay in time of flight between the two signals is created; (v) the two signals are combined in coupler, and the combined signal is brought to reading unit via passive zone; reading unit identifies the time delay and determines the strain in sensor. At rest (before the strain is transferred from the structure to sensor), the time delay between the two fibers in the sensor is determined using Equation 3.6 (Inaudi 1997).

$$\Delta t_r = \frac{1}{k_{cut}} \Delta L_0 \tag{3.6}$$

where Δt_r is delay in time of flight in sensor created by initial difference in length of two fibers ΔL_0 and $k_{cut} \approx 103\,\mu m/ps$ is a constant describing the optical properties of the fibers ($k_{cut} = c/(2n)$, where $c \approx 300\,\mu m/ps$ is speed of the light in a vacuum and $n \approx 1.46$ is the refractive index of the core of optical fibers).

Deformation of monitored structure will change the length of measurement fiber (i.e., the strain in measurement fiber), but not the length of reference fiber. This will result in a change in time delay, as described by Equation 3.7 (Inaudi 1997).

$$\Delta t_\varepsilon = \frac{1}{k_{cut}} \Delta L_0 + \frac{1}{k_{stress}} \Delta L_s = \frac{1}{k_{cut}} \Delta L_0 + \frac{1}{k_{stress}} \varepsilon L_s \tag{3.7}$$

where Δt_ε is delay in time of flight in sensor created after the average strain ε, generated in structure between sensor anchors and transferred from structure to sensor, caused a change in length of the sensing element of ΔL_s ($\Delta L_s = \varepsilon L_s$), and $k_{stress} \approx 130\,\mu m/ps$ is another constant describing optical properties of the fibers ($k_{stress} \approx k_{cut}/0.79$).

Finally, average strain in the structure (between anchors) is determined in the reading unit by the conversion given in Equation 3.8, which results from the subtraction and rearrangement of terms in Equations 3.6 and 3.7.

$$\varepsilon = k_{stress} \frac{\Delta t_\varepsilon - \Delta t_r}{L_s} \tag{3.8}$$

Any change in temperature will change the value of ΔL_0, which is canceled in Equation 3.8. The resolution of interferometric sensors, in terms of the difference in length between two fibers (measured in micrometers), depends on the reading unit but is independent on the difference in length of the two fibers; thus, the resolution in terms of strain is higher if the length of the two fibers between anchors (i.e., the gauge length of the sensor) is longer. Consequently, interferometric (SOFO) sensors are mostly conceived to have long-gauge-length (Inaudi 1997). These sensors are considered to be a "true" long-gauge sensors because they "integrate" the strain along the measurement fiber between the anchors (i.e., along the gauge length), and their measurement represents the average strain between anchors (see more details on long-gauge measurement in Section 4.4). Interferometric (SOFO) sensors were successfully implemented in numerous projects (e.g., Glisic et al. 2010), with various sensor packages (e.g., Glisic et al. 1999, Glisic and Inaudi 2003b). An example of a standard sensor with soft polyethylene packaging is shown in Figure 3.15a.

The static reading unit for SOFO sensors is based on a Michelson interferometer (see Figure 3.15b–d), while the dynamic reading unit is based on Mach–Zehnder interferometer (see Figure 3.15e). Static reading unit can be accompanied by a channel switch that can enable sequential reading of a practically unlimited number of sensors. Note that the reading units shown in Figure 3.15c,d enable the reading of both SOFO and FBG sensors (see the next subsection).

In addition to the general advantages of FOSS presented in the previous subsection, the other specific advantages of interferometric sensors are the following: insensitivity to temperature (they are self-compensated by reference fiber), compatibility with channel switches with practically any (unbounded) number of channels for static measurements, extremely high resolution (2 μm for static and 10 nm for dynamic measurements over any gauge length), possibility of very long-gauge length (up to 20 m, note that 20-m long sensor would have resolution of 0.1 microstrain (με) in static mode and 0.5 nanostrain (nε) in dynamic mode), and low stiffness which makes them both embeddable in concrete and surface mountable to wide range of materials.

In addition to the general challenges of FOSS, the major challenge specific to interferometric sensors is the current need for two separate reading units – Michelson-interferometer-based for static, and Mach–Zehnder-interferometer-based for dynamic readings. Other important challenge is limited number of channels that can be read by a dynamic reading unit (max. eight channels, as opposed to the practically unlimited number of channels in case of static reading unit).

The best performances achievable with SOFO sensors are given in Table 3.2 (Section 3.6.5). More information about technology can be found in Inaudi (1997).

3.6.3 Fiber Bragg-Grating (FBG) Strain Sensors

Sensing elements of a Fiber Bragg-grating (FBG) strain sensor consist of intentionally created periodical changes of the refractive index along the short length (typically a few to several millimeters) of the fiber core (e.g., by controlled exposure to ultraviolet light, Dadpay et al. 2008).

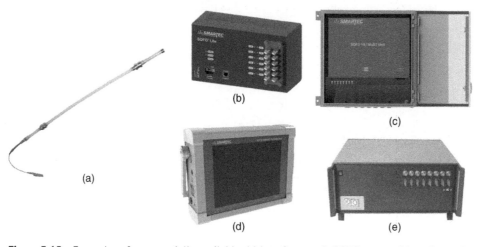

(b)

(c)

(a)

(d)

(e)

Figure 3.15 Examples of commercially available: (a) Interferometric SOFO sensor; (b) static reading unit; (c) rack-mounted static reading unit; (d) portable static reading unit; and (e) dynamic reading unit; reading units in figures (c) and (d) also enable reading of FBG sensors. Source: Courtesy of Roctest/SMARTEC.

An FBG acts as a partial mirror that reflects light with a specific wavelength; if no strain is present in the FBG, the reflected wavelength is called the reference wavelength. The reference wavelength is set during the manufacturing process by appropriately tuning the optical properties of the FBG. A light containing a range of wavelengths that travels along the optical fiber and arrives at an FBG will have the part of the light with the reference wavelength reflected back toward the source. The reminder of light containing all the other original wavelengths will propagate through the FBG and continue traveling along the fiber, as shown in Figure 3.16. Hence, if another FBG is inscribed on the fiber and its reference wavelength is different from the wavelength of the first FBG, then the part of the light that did not reflect off the first FBG will contain a wavelength that will reflect off the second FBG. This means that several FBGs with mutually different reference wavelengths can be inscribed on the optical fiber and read quasi-simultaneously with the same signal of light, i.e., FBG allows in-line multiplexing.

The refractive index and geometrical properties of FBG depend on strain and temperature; consequently, a change in strain or a change in temperature in FBG with respect to the reference state will result in a change in the wavelength of reflected light. The dependence of change is given in Equation 3.9.

$$\Delta\lambda_i(\varepsilon, \Delta T) = \lambda_i(\varepsilon, \Delta T) - \lambda_{i,r} = C_{\varepsilon,i}\varepsilon + C_{T,i}\Delta T \tag{3.9}$$

where $\Delta\lambda_i(\varepsilon, \Delta T)$ is change in wavelength of light reflected off FBG due to change in strain ε and change in temperature ΔT, $\lambda_i(\varepsilon, \Delta T)$ is wavelength of light reflected off FBG after ε and ΔT are applied, $\lambda_{i,r}$, is reference wavelength, "i" denotes the label of FBG, $C_{\varepsilon,i}$ is strain coefficient of FBG, and $C_{T,i}$ is temperature coeffect of FBG.

Strain and temperature coefficients are dependent on the reference wavelength and, thus, must be determined for each FBG individually through the calibration process. Their approximate values are $C_{\varepsilon,i} \approx 1.2\,\text{pm}/\mu\varepsilon$ and $C_{T,i} \approx 10.4\,\text{pm}/°\text{C}$, but these values should only be considered as estimates.

Given that an FBG itself cannot be temperature-free, FBG strain sensors must be compensated to temperature. Thermal compensation can be made by measuring the temperature change at the location of the FBG, either by using a strain-free FBG or by using any other temperature sensor. An FBG can be made strain-free (e.g., by leaving it loose within a packaging, similar to the reference fiber of the SOFO sensor); thus, if placed close to the FBG, which is coupled to the structure, the former can be used to compensate for the latter. This scheme is shown in Figure 3.17. Change in temperature measured by strain-free FBG is determined by setting strain equal to zero

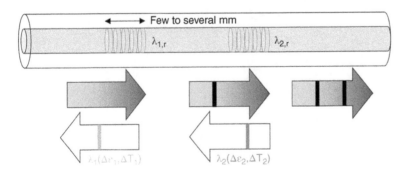

Figure 3.16 Schematic representation of two FBG sensing elements (reference wavelengths $\lambda_{1,r}$ and $\lambda_{2,r}$) inscribed along the same optical fiber (Glisic 2009).

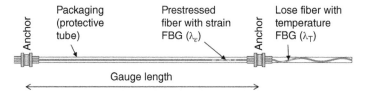

Figure 3.17 Schematic example of FBG strain sensor with thermal compensation (Glisic 2009).

in Equation 3.9 and rearranging it into Equation 3.10.

$$\Delta T = \frac{1}{C_{T,T}} \Delta \lambda_T \qquad (3.10)$$

where index "T" denotes FBG dedicated to thermal compensation ($i = T$ in Equation 3.9).

Strain is now determined by applying Equation 3.9 to FBG, which is coupled to monitored structure (the strain is transferred from structure to that FBG), Equation 3.10 to strain-free FBG, and combining the two equations into Equation 3.11.

$$\varepsilon = \frac{1}{C_{\varepsilon,\varepsilon}}(\Delta \lambda_\varepsilon - C_{T,\varepsilon} \Delta T) = \frac{1}{C_{\varepsilon,\varepsilon}} \left(\Delta \lambda_\varepsilon - \frac{C_{T,\varepsilon}}{C_{T,T}} \Delta \lambda_T \right) \qquad (3.11)$$

where index "ε" denotes FBG coupled to structure, i.e., the FBG dedicated to strain measurement ($i = \varepsilon$ in Equation 3.9).

Since several FBGs can be placed along the same optical fiber, strain and temperature FBG are frequently placed on the same fiber within the sensor. Figure 3.17 shows a schematic example of a strain sensor containing two FBGs, one coupled with structure, intended to measure strain (reference wavelength λ_ε), and the other strain-free, intended to measure temperature and be used for thermal compensation (reference wavelength λ_T).

Physical length of an FBG is typically a few to several millimeters. Hence, depending on packaging, FBG strain sensors can be manufactured and implemented as short-gauge (considered as fiber-optic equivalent of a VW-based strain sensor or resistive strain gauge) or long-gauge (e.g., by having the same packaging as a SOFO sensor where one FBG is placed on the part of the fiber that is pre-tensioned between the anchors). FBG strain sensors are very popular among practitioners, and there are several manufacturers worldwide offering a large variety of both short- and long-gauge sensors. Some examples are given in Figure 3.18.

The FBG reading unit sends through an optical fiber a light containing a spectrum of wavelengths. Light reflected back, toward the reading unit, is analyzed (e.g., using a tunable laser with a wavelength filter or spectrometer), and each reflected wavelength contained in the light is identified. As mentioned earlier in the text, several sensors can be connected in-line, one after another. Figure 3.19 emphasizes the advantage of in-line multiplexing (see the next subsection). If two sensors have wavelengths that are similar in value, then it might be difficult or even impossible to identify which wavelength comes from which sensor. To avoid confusion, it is necessary to define the reference wavelengths of all sensors properly so that the readings from different sensors can be clearly distinguished. Typically, the sensor wavelength should be set a part for at least 3–10 nm depending on the range of strain and temperature to be measured and any unintentional pre-tension or pre-compressing that the sensor can experience during installation. Total number of sensors that can be connected in-line depends on the dynamic range of the reading unit, which is typically 80–160 nm. Hence, typically, 5–10 compensated strain sensors (i.e., including

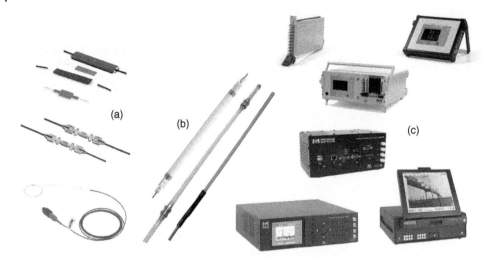

Figure 3.18 Examples of commercially available (a) short-gauge FBG strain sensors, (b) long-gauge FBG strain sensors, and (c) reading units. Source: Courtesy of Roctest / SMARTEC.

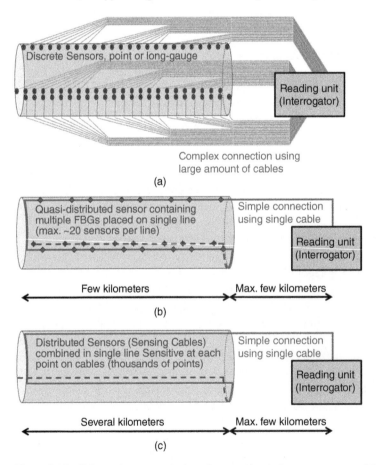

Figure 3.19 Schematic representation of connection between sensors and the reading unit for (a) typical discrete sensing (short-gauge sensors shown in figure with each sensor connected directly to reading unit), (b) FBG sensors interconnected over single line (long-gauge sensors shown in figure), and (c) distributed sensors. Source: Modified from Glisic (2009).

temperature FBG for each strain FBG) can be connected in-line to the reading unit with a range of 80 nm, and double this value to the reading unit with range of 160 nm. Several sensors can be connected in-line to a single channel of the reading unit, which in turn can have several (typically four) channels. Number of channels can be augmented by a channel switch (typically enables 16 channels). While sensors connected in-line to the same channel are read simultaneously, the lines of sensors (channels) are read one after another, i.e., sequentially. Examples of commercially available reading units are given in Figure 3.18, but also in Figure 3.15c,d, in which they are combined with reading units for SOFO sensors.

Besides the general advantages of FOSS, the principal additional advantages of FBG sensors are in-line multiplexing, the possibility to create both short-gauge and long-gauge sensors, and ability to monitor both static and dynamic strain using the same reading unit. All the above enables great flexibility in designing the strain network and may significantly lower the cost of complex SHM projects. The challenge specific to FBG strain sensors is need for thermal compensation, which requires an additional FBG per strain sensor, resulting in increased monitoring costs of and decreased accuracy of strain measurement.

3.6.4 Distributed Sensors – Strain Sensors Based on Brillouin and Rayleigh Scattering

Sensing element of a distributed sensor is the optical fiber itself, i.e., virtually each point along the length of the optical fiber. Packaging of a distributed sensor is practically an additional coating on optical fiber, which gives the distributed sensor a form of cable. Hence, a distributed sensor (frequently called a sensing cable) can be depicted as a cable sensitive to strain at each point of its length (Glisic and Inaudi 2007). Given sensitivity at each point, one distributed sensor is equivalent to a very large number of discrete sensors (long- or short-gauge).

Distributed sensors have a connecting scheme that is significantly different from that used for discrete sensors: while most of discrete sensors require one cable extension for each sensor, with the exception of FBG sensors that can be connected in-line and require one cable extension for every 5–20 sensors, a distributed sensor, which covers the same amount of measuring points, would require a single cable extension to communicate the measurements from all points to the reading unit. This is schematically shown in Figure 3.19 (Glisic 2009). This simplified connection is particularly advantageous in the cases where very large structures are monitored, such as pipelines, long bridges, tunnels, and dams.

Distributed sensor is sensitive to strain at each point along its length, which means that it theoretically provides with infinity of measurement points. For practical reasons, the set of all measurement points is reduced to a finite number of equally spaced points. The distance between the measurement points is called the "sampling interval," and is set by the user depending on project requirements. Similar to long-gauge sensors, at every measurement point, the distributed sensor provides an average strain measured over a predefined length of optical fiber, which is called "spatial resolution" (Lanticq et al. 2009). Spatial resolution is predefined by the user depending on project requirements. For example, if the length of the sensor is 1 km, the sampling interval is set to 10 cm, and the spatial resolution to 0.5 m, then the strain measurements will be taken at 10,000 points spaced every 10 cm along 1 km, and at each point, the value of the measurement will represent the strain averaged over 0.5 m. Hence, one measurement over a distributed sensor executes measurement over all measurements points, yielding a discretized 1D average strain distribution along the sensor.

It is important to highlight that for Brillouin-based SHM systems, sampling interval and spatial resolution of measurement are correlated with the precision and resolution of the strain measurement, and with the measurement time. For example, a short sampling interval and/or spatial resolution would require a longer time to execute measurements with high resolution and precision, but if the measurement time is to be kept short, then the measurements will be less resolved and less precise; fast and more precise measurements would require less measurement points (larger sampling interval) and averaging strain over larger lengths (larger spatial resolution). Consequently, trade-offs between the measurement parameters must be carefully decided for each application, depending on the project requirements.

The two main physical principles that enable distributed strain measurements in optical fibers are Brillouin scattering (e.g., Horiguchi et al. 1993) and Rayleigh scattering (e.g., Posey et al. 2000). In addition, sensors based on Raman scattering exist, but they are sensitive to temperature only (e.g., Kikuchi et al. 1988). Light with a single frequency is sent from the reading unit into the optical fiber. By propagating through the core of the optical fiber, the light encounters various imperfections, and a very small amount of light scatters around these imperfections in all directions. This is schematically shown in Figure 3.20 (modified from Glisic and Inaudi 2007). Figure 3.20 shows that Raman- and Brillouin-scattered light have different wavelengths (and consequently different frequencies) from the light originally inserted in optical fiber, which is not the case with Rayleigh-scattered light. Raman- and Brillouin-scattered lights are dependent on strain and temperature, and the latter two quantities modify the former. Part of the scattered light returns to the reading unit, where it is analyzed, and strain and temperature decoded. Both Rayleigh and Brillouin scattering are sensitive to temperature too, and thus, distributed sensors based on those physical principles need temperature compensation, which is frequently performed by a parallel lose optical fiber that is mechanically uncoupled from the monitored structure, i.e., strain-free.

Each physical principle defines corresponding measurement technique that decodes the relation between the strain (and temperature) and the change in properties of the scattered light, as illustrated in Figure 3.20. Intensity of Raman-scattered light is sensitive to temperature only; while this technique cannot be used to monitor strain, it can be used to monitor temperature and is finding applications where temperature is a critical parameter, such as fire detection in tunnels, leakage detection in dams, dykes, and pipelines.

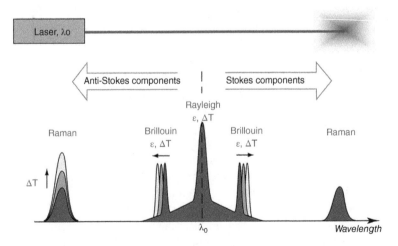

Figure 3.20 Scattered light as an encoding parameter for strain and temperature measurements in distributed fiber optic sensors. Source: Modified from Glisic and Inaudi (2007), courtesy of SMARTEC SA.

Rayleigh-scattering sensors are based on the dependence of the shifts in the local Rayleigh backscatter pattern on the change in strain (and temperature) of optical fiber. Rayleigh-scattering decoding technique enables very fast, dynamic measurement, high resolution of strain (and temperature), short sampling interval, and short spatial resolution. However, the length of sensor is relatively limited (maximum length of sensor is approximately 100 m). Hence, this system is best used for projects where monitoring localized strain changes over relatively short distances are of interest. Best performances of the Rayleigh-scattering technique are given in Table 3.3 (in the next subsection).

Brillouin-scattering sensors are based on the dependency of the change in frequency of Brillouin scattered light on change in strain (and temperature) in optical fiber, as shown in Equation 3.12 (note similarity with Equation 3.9 for FBG sensors).

$$\Delta f_i(\varepsilon, \Delta T) = f_i(\varepsilon, \Delta T) - f_{i,r} = K_{\varepsilon,i}\varepsilon + K_{T,i}\Delta T \tag{3.12}$$

where $\Delta f_i(\varepsilon, \Delta T)$ is change in frequency of Brillouin scattered light due to strain ε and change in temperature ΔT, $f_i(\varepsilon, \Delta T)$ is the frequency of Brillouin scattered light after ε and ΔT are applied, $f_{i,r}$, is reference frequency of Brillouin scattered light, i.e., the inherent frequency shift upon manufacturing of optical fiber, before ε and ΔT are applied (see Figure 3.17), "i" denotes the optical fiber of distributed sensor (strain or temperature), $K_{\varepsilon,i}$ is strain coefficient of optical fiber and $K_{T,i}$ is temperature coeffect of optical fiber.

The approximate values of strain and temperature coefficients in Equation 3.12 are $K_{\varepsilon,i} \approx 0.05\,\text{MHz/}\mu\varepsilon$ and $K_{T,i} \approx 1\,\text{MHz/°C}$. Exact values of coefficients should be determined by calibration.

As mentioned above, thermal compensation of distributed sensors must be performed using strain-free optical fiber (distributed temperature sensor) installed in parallel and very close to the strain sensor. For practical purposes, some manufacturers package both strain and temperature sensing fibers in the same packaging, so that only one physical sensing cable (containing both optical fibers) has to be installed on the structure. Equation 3.13 presents the determination of temperature change using a distributed temperature sensor, and Equation 3.14 presents determination of strain after thermal compensation (note similarity to Equations 3.10 and 3.11).

$$\Delta T = \frac{1}{K_{T,T}\Delta T}\Delta f_T \tag{3.13}$$

where index "T" denotes optical fiber dedicated to thermal compensation ($i = T$ in Equation 3.12).

$$\varepsilon = \frac{1}{K_{\varepsilon,\varepsilon}}(\Delta f_\varepsilon - K_{T,\varepsilon}\Delta T) = \frac{1}{K_{\varepsilon,\varepsilon}}\left(\Delta f_\varepsilon - \frac{K_{T,\varepsilon}}{K_{T,T}}\Delta f_T\right) \tag{3.14}$$

where index "ε" denotes optical fiber coupled to structure, i.e., dedicated to strain measurement ($i = \varepsilon$ in Equation 3.12).

As mentioned in Section 3.6.1, the two most established groups of sensing techniques based on Brillouin scattering are spontaneous Brillouin sensing (based on BOTDR or BOFDR, e.g., see Wait and Hartog 2001 or Garus et al. 1997, and Ghafoori-Shiraz and Okoshi 1986, respectively), and stimulated Brillouin sensing (based on BOTDA or BOFDA, e.g., see Nikles et al. 1994, 1997 or Garus et al. 1997, respectively). Each group of techniques has advantages and limitations.

In the case of spontaneous sensing, the sensor is accessed only from one side, scattering occurs as the light propagates through the fiber, and reading unit picks up scattered light as it arrives from various locations of the fiber. This makes measurement of sensor independent of its far ending, and thus, if the damage occurs in the fiber, the sensor will still be operational from starting point to the point of damage; however, given that reading unit picks up spontaneous, weak, scattered light, the total measurable length of sensor (spatial range of sensor) can be limited and strongly depends

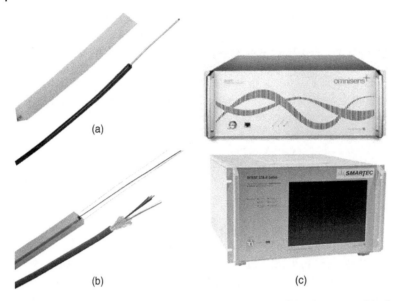

(a)

(b)   (c)

Figure 3.21 Examples of commercially available (a) distributed sensors with single fiber (needs separate temperature sensor for thermal compensation), (b) distributed sensors with multiple fibers for both strain and temperature monitoring, and (c) reading units. Source: Courtesy of Roctest / SMARTEC.

on cumulative optical losses that occur in the optical fiber due to manufacturing, installation, connections, and splicing.

In the case of stimulated sensing, the sensor is accessed from both sides (frequently, a temperature sensor is used to connect far end of strain sensor with the reading unit). The reason for this is to insert two signals in fiber – continuous, with constant frequency on one side and probe (pulsing), with variable frequency on the other side. The aim of pulse is to match the Brillouin frequency and amplify the signal. In addition, optical fiber is mechanically excited to further stimulate Brillouin scattering. This allows for exceptionally large spatial range of sensor (Thevenaz et al. 1999). However, given that the access from both sides of sensor is needed to perform measurement, damage to sensing optical fiber will result in malfunction of entire sensor. Examples of commercially available distributed sensors and reading units are given in Figure 3.21.

While Rayleigh-based systems have better strain resolution and shorter measurement time than Brillouin-based systems, the latter can cover significantly longer lengths (greater spatial range). Thus, the Brillouin-based systems are the most optimally used when deployed for monitoring global strain changes over large areas of structures. The best performances of commercially available Brillouin-based systems are presented in Table 3.3 (in the next subsection).

General advantages of distributed FOSS are their large spatial coverage, which enables instrumentation of large areas of structure and results in monitoring of 1D strain fields, which in turn enables direct damage detection and integrity monitoring (see Chapter 7). In the case of large structures, besides a significantly simpler connection scheme, distributed sensors are more economical to deploy and operate; however, this might not be valid in the case of moderately large and small structures.

The major challenge of distributed FOSS is a reduction of optical losses generated by sensor packaging and installation. Cumulative optical losses can significantly decrease the spatial range of the sensor and worsen the quality of the optical signal, which in turn will yield less precise measurements. The less difficult challenge is the thermal compensation of sensor, which requires accurate matching of coordinates of strain and temperature sensors (Equation 3.14 has to be applied to every point of strain and corresponding point of temperature sensor). Finally, while embedding in concrete or soil is relatively simple and easy, surface mounting of distributed sensors might be tiresome depending on site conditions and geometry of the monitored structure (i.e., surface on which the sensor is to be installed).

Specific advantages of Rayleigh-based technique are short measurement time (dynamic measurements are possible), high resolution of strain (0.1 με), and short sampling interval (under mm), and spatial resolution (under 10 mm). The main challenge is relatively limited spatial range of sensor (~70–100 m).

The main specific advantage of Brillouin-based techniques is large spatial range of the sensors (up to several kilometers for stimulated sensing), which results in a large monitoring coverage of the structure. The challenges are relatively long spatial resolution (>0.5 m), relatively low precision (~20 με), and slow measurement time (several minutes). Ongoing research is aimed at reducing the effects of these challenges, and it is very likely that new, more performant Brillouin-based techniques will be commercially available soon.

3.6.5 Best Performances of the Second-Generation Strain Sensors

Tables 3.2 and 3.3 present the best performances of the second-generation strain sensors that were commercially available at the time of writing this section. The information provided in tables

Table 3.2 Best performances of the discrete second-generation strain sensors used in SHM (Glisic 2022).

	Extrinsic interferometry (EFPI)	Intrinsic interferometry (SOFO)	Fiber Bragg Gratings (FBG)	Intensity-based (micro-bending)
Gauge length	51–70 mm	250 mm to 20 m	10 mm to 2 m	100 mm to 10 m
Multiplexing	Sequential (one by one)	Sequential (one by one)	Simultaneous and sequential	Sequential (one by one)
Max. no. of sensors per reading unit (with switch)	32	Static: unlimited Dynamic: 8	80–300 (depending on reading unit)	80
Stability	Long-term stable	Static: long-term stable Dynamic: short-term only	Long-term stable	Long-term drift (typical 2%)
Resolution[a]	0.01% full scale	Static: 2 μm/gauge length Dynamic: 10 nm/gauge length	0.2 με	1 μm/gauge length
Repeatability ("precision")	N/A[a]	Static: 0.2% full scale Dynamic: N/A[a]	1 με	1%
Sensor range[b]	±3000 με	Static: −5000 to +10,000 με Dynamic: ±5 mm/gauge length	−5000 to +7500 με	±5000 με
Temperature sensitivity	Insensitive (by measurement principle, but might be packaging dependent)	Self-compensated (by measurement principle)	Compensation needed	0.6 μm/°C
Measurement frequency	20 Hz	Static: 0.1 Hz Dynamic: 10 kHz	0.5 MHz	100 Hz

a) For SOFO sensors and intensity-based sensors, the strain resolution is determined by dividing the number provided in the table with gauge-length of the sensor.
b) For SOFO sensors measured with dynamic reading unit, the strain range is determined by dividing the number provided in the table with gauge-length of the sensor.

Table 3.3 Best performances of the distributed second-generation strain sensors used in SHM (Glisic 2022).

	Stimulated Brillouin	Spontaneous Brillouin	Rayleigh
Spatial resolution[a]	0.2–5 m	1 m	10 mm
Sampling interval[a]	100 mm	50 mm	0.65 mm
Max. no. of sensors per reading unit (with switch)	16	N/A	8
Stability	Long-term stable	Long-term stable	Long-term stable
Resolution[a]	2 µε	2 µε	0.1 µε
Repeatability	±2 to ±50 µε	±20 µε	±30 µε
Sensor range	±10,000 µε	±10,000 µε	±15,000 µε
Spatial range[b]	50 km	25 km	100 m
Temperature sensitivity	Compensation needed	Compensation needed	Compensation needed
Measurement speed[a]	10 s to 15 min	4–25 min	250 Hz

a) Parameters are mutually correlated, improving one requires diminishing the others.
b) Spatial range assumes minimal losses in optical fibers; these lengths are severely shortened if cumulative losses are present in sensor due to manufacturing, installation, connections, etc.

are taken, as is, from datasheets issued by various sensor manufacturers. Each field in the tables contain the best achievable performance, regardless of the potential trade-offs with performances presented in the other fields in the same table. For example, if channel switch is used to read multiple FBG sensors, then the maximal frequency of reading is diminished (see Table 3.2). Another example is correlation between measurement time, strain resolution, spatial resolution, and sampling interval in the case of distributed BOTDA sensors – improving one parameter requires diminishing the others (see Table 3.3). It is important to highlight that the presented best performances are achieved in idealized conditions, i.e., in the lab environment, while the actual performances in real-life settings may be altered depending on the application (e.g., quality of installation and environmental conditions).

Tables 3.2 and 3.3 are intended to help in the process of selection of strain-based SHM system for specific project. However, they are to certain extent incomplete and contain some inconsistencies that result from information found in datasheet. Similar to the first-generation sensors, the second-generation sensor datasheets lack some important information and lack clarity in terminology in general. Typically, there is confusion between resolution and repeatability precision, while reproducibility precision is not mentioned.

Table 3.2 presents the best performances of discrete FOS. While EFPI-based and intensity-based sensors were not presented in detail in this section, their performances are, however, presented in Table 3.2 for the purposes of completeness and comparison. Table 3.3 presents the best performances of distributed FOS.

3.7 Example of Research on the Third-Generation Strain Sensor

3.7.1 Direct and Indirect Detection of Unusual Structural Behaviors, General Advantages and Challenges

The first- and second-generation sensors provided tools to greatly address the two out of the three needs presented in Section 3.1, i.e., to assess safety and performance of structures through stress and deflection evaluations and beyond (see Chapters 5 and 6). In many cases, the third need, i.e., identification of unusual structural behaviors, can be successfully addressed using these sensors; however, identification of unusual behaviors remains very challenging in general, and this motivates research and development of the third-generation strain sensors.

Detection of unusual structural behaviors (e.g., damage, deterioration, or lack of performance) based on strain sensing can be performed based on two approaches: direct and indirect approaches (Yao and Glisic 2012, Glisic 2022).

The main assumption in indirect approach is that sensors are not necessarily placed at locations of occurrence of unusual structural behaviors, i.e., they are distant from these locations, and thus, they are not in direct contact with unusual structural behaviors when the latter occur. In that case, unusual structural behaviors may or may not affect the strain field at locations of sensors. As an illustration, Figure 3.22 shows an example of relationship between the magnitude of strain change and distance from the tip of the crack (see more details in Yao and Glisic 2015a, Glisic 2022). Note that strain change in the figure decreases rapidly with the distance from the damage, and sensors installed around 50 mm from damage are not likely to capture changes in the strain field.

If the strain field at locations of sensors is not affected, then it is practically impossible to detect existence of unusual structural behavior directly by strain measurement. One example is given in Figure 3.22 at a distance of around 50 mm. Another example would be localized damage in statically determined beam structures, which does not result in any perturbation of strain field at locations that are typically half to full cross-sectional depths distant from the damage. If the strain field at locations of sensors is, however, affected by occurrence of unusual structural behavior (e.g., due

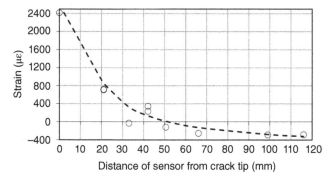

Figure 3.22 Experimental results showing a change in the strain at the location of the sensor due to cracking in metallic plate as a function of the distance between the sensor and the crack tip; markers represent measurements and dashed curve represents trendline (Glisic 2022).

to redistribution of stresses in indeterminate structures), then unusual structural behavior may be detected. Yet, this is also challenging and might be unreliable due to numerous factors. The strain field at locations at even modest distances from the origin of unusual structural behavior is typically only slightly affected, and the magnitude of resulting strain field anomaly is often difficult to discriminate from usual strain variations due to changes in loads and environmental conditions. This is even more challenging if the unusual structural behavior is in its early stages and its magnitude is small. Hence, in all those cases, the detection of unusual structural behaviors cannot be reliably performed by the simple use of thresholds (see Chapter 5), and consequently, advanced data analysis algorithms have to be implemented for this purpose. However, the implementation of these algorithms can be complex and expensive, yet there is no guarantee of their full reliability when applied to real structures. Thus, the indirect approach for detection of unusual structural behaviors, while successful in many applications, has important limitations that cannot be generally addressed.

The main assumption in the direct sensing approach is that the sensors are placed at the locations of potential occurrences of unusual structural behaviors. In that case, an unusual structural behavior will create a strain field anomaly that is detected by the sensors as an unusually high change in the measured strain, i.e., by simple thresholding. For example, for a strain sensor with a gauge length of 1 m a crack that opens for 0.05 mm would generate a strain change of $0.05/1000 = 0.00005$ m/m $= 50$ µε, which is easily detectable with sensors of the first- and the second-generation. This approach is illustrated in Figure 3.23 (Glisic 2022), which shows detection of early age cracks that occurred at locations of sensors and consequently resulted in unusual strain change measured by sensors (see more details in Hubbell and Glisic 2013, see also Section 4.4 and Figure 4.11).

Given high sensitivity to unusual structural behaviors of sensors that are in direct contact with them and simple thresholding algorithm for their detection, the direct detection approach features very high reliability of detection. However, the main challenge of the direct detection approach is

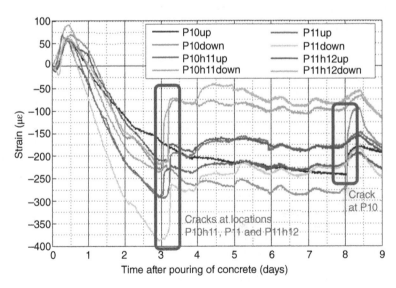

Figure 3.23 Example of direct detection of early age cracking. Source: Modified Glisic (2009, 2022).

that the locations of occurrence of unusual structural behaviors often cannot be exactly predicted. This challenge can be addressed by installing a large number of densely placed discrete sensors, but this would incur very high costs for the equipment and the installation. Implementation of DFOSS certainly addresses these concerns to a large extent, as it can continuously cover larger areas of structure. Nevertheless, they face limitations in the case of very complex geometries of structures that require dense 2D coverage of large areas and monitoring in multiple directions. This motivates development of quasi-distributed and truly distributed 2D sensors that use strain field anomalies to detect unusual structural behaviors. These sensors can be contact or non-contact, and they represent a vast majority of the third generation of strain sensors (see Section 3.3). While they are at different stages of research and prototyping, they are not yet fully commercially available.

The most important advantage of the 2D sensors is their ability to detect and characterize unusual structural behaviors over large structural surfaces by direct detection approach, which significantly improves the reliability of identification of unusual structural behaviors on real-life structures, i.e., under variable environmental conditions. Such an achievement can enable 2D integrity monitoring of large structures.

The main challenge for 2D strain sensing techniques is scaling to the size of the real structures in terms of deployment over large areas of structures, long-term reliability, and data management, but some of them (if not all of them) will certainly achieve maturity and find their way to real-life applications.

3.7.2 Example of a Contact-Based Third-Generation Strain Sensor: Sensing Sheet

Sensing Sheet is an example of a contact-based third-generation strain sensor. It is conceived to cover large areas of structures with an estimated cost of approximately $400/m^2 (Glisic et al. 2016). Sensing sheet consists of: (i) resistive strain sensors densely patterned over flexible organic substrate; (ii) integrated circuits that serve as intelligence units for smart sensing; and (iii) protective coating that can also serve for power harvesting, see Figure 3.24 (Glisic 2022).

Unit sensors (resistive strain gauges) are densely patterned over the large substrate, to increase probability of sensors being in contact with damage (see Yao and Glisic 2015b). Such an arrangement results in quasi-distributed 2D sensor. One integrated circuit is assigned to each dedicated group of sensors, and it controls measurements, performs data processing, optimizes the use of power, and enables wireless communication. Protective layer ensures longevity of installation and it can consist of flexible photovoltaic for power harvesting. Given that the sensor is flexible, it can be installed via gluing onto irregular surfaces as it will conform to their shape and texture. Thanks to affordable cost, it can be tailored to fit the shape of the monitored area by simply cutting it (see Figure 3.24).

If an unusual structural behavior occurs (e.g., crack, see Figure 3.24a), each sensor in direct contact with it will detect it (highlighted in Figure 3.24a). Involved integrated circuits (highlighted in Figure 3.24a) will identify geographical locations of these sensors and, through mutual communication, infer the location of the unusual structural behavior and its extent.

Sensing sheet was applied to Streicker Bridge on Princeton campus (Figure 3.24a,b) and results of crack monitoring are presented in Figure 3.24c. The readings were performed using external, non-integrated circuitry. The results confirmed that the sensor is able to detect, localize, and quantify the crack opening due to daily temperature variation. Research on Sensing Sheet continues in

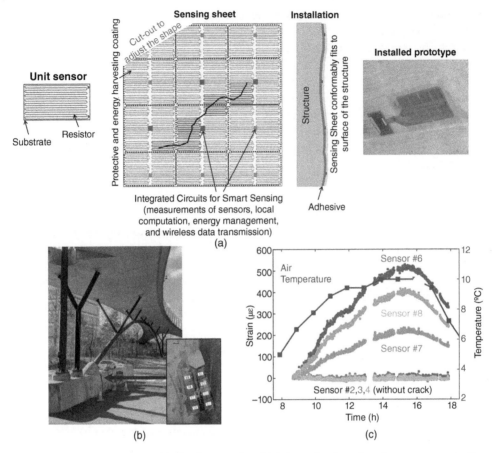

Figure 3.24 Sensing sheet and its implementation: (a) Schematic of sensing sheet components. Source: Modified from Glisic (2009). (b) Sensing sheet prototype installed over a shrinkage crack on the Streicker Bridge foundation. Source: Modified from Aygun et al. (2020). (c) Results of measurements showing the crack opening over time. Source: Reproduced from Aygun et al., 2020 / MDPI /CC BY 3.0.

order to build an integrated prototype and evaluate its long-term performance in real-life settings. More information on Sensing Sheet can be found in the references cited in this subsection.

3.7.3 Example of a Non-contact Based Third-Generation Strain Sensor: Camera Vision and Convolutional Neural Networks

Resolution of digital cameras has been rapidly improving over the last decade, and at present day even ordinary smartphone cameras can be used to acquire high resolution images. They can be used to record images of the structures, and these images can contain potential visible damage. Hence, these images can be considered as 2D non-contact "scans" of strain field anomalies. In order to automatically recognize the damage, various machine learning (ML) methods can be used. However, ML has to be trained for each specific type of damage. To illustrate this non-contact approach, one such method that targets cracks in masonry structures is briefly presented here. More detail on this method can be found in Hallee et al. (2021).

To create a training dataset, small-scale experimental walls were constructed in laboratory using miniature bricks with dimensions of $3.4 \times 1.7 \times 1.7$ cm. This made possible repeated cracking and

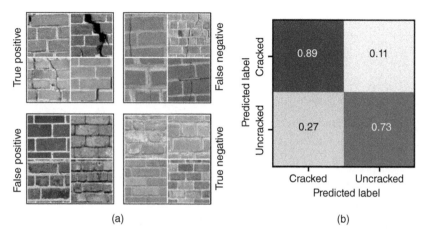

Figure 3.25 (a) Examples of randomly selected photographs and (b) confusion matrix. Source: Reproduced from Hallee et al., 2021 / MDPI / CC-BY 3.0.

rebuilding of the walls after the photographs of cracked walls were recorded. Walls typically contained 30–50 bricks, and total of 53 walls were built and cracked to create a training dataset. In total, 1068 usable images were generated, 642 of which were used to train ML model, 213 to tune hyperparameters, and 213 to test the results (i.e., to validate the model).

Nikon D90 digital camera (mirrorless) was installed on a tripod, and 4608×3456-pixel photographs were recorded with a two-second shutter delay. This was important to guarantee pixel-level sharpness of photographs, as the cracks were identified and labeled manually by zooming into sections of the photographs. The distance of the camera from the walls as well as the lightning conditions was varied in the tests in order to minimize the potential bias introduced by these two parameters. Besides the photographs taken during the tests, validation was also performed on real-world images collected from internet. ML method used was based on Convolutional Neural Networks (CNNs). It was implemented with the Keras deep learning library using the TensorFlow backend. Examples of images used in crack detection based on CNNs are shown in Figure 3.25a. The confusion matrix, given in Figure 3.25b, demonstrates the performance of the method.

The example presented here is just one of many that are currently being developed. Methods based on camera vision are rapidly maturing and will soon be available on market.

3.8 Closing Remarks

Strain sensors were used in civil SHM applications for more than 100 years. During this period, three generations of sensors were researched, developed, and implemented, and chronology of their evolution is shown in Figure 3.4. Each generation of sensors pushed boundaries of strain-based SHM and expanded its capabilities in terms of spatial coverage and identification of unusual structural behaviors, as shown in Figures 3.26 and 3.27, respectively (Glisic 2022).

The first-generation strain sensors are discrete (point), short-gauge, and based on electrical principles. They enabled fundamental evaluation of strain at a local material level and practically created strain-based SHM.

The second-generation sensors are discrete short-gauge, discrete long-gauge, or 1D distributed sensors, and they are based on optical fiber technologies. While short-gauge sensors brought

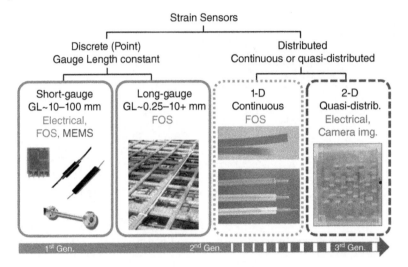

Figure 3.26 Progress in spatial coverage (gauge length) of strain sensors; outline and arrow coding: continuous outline and arrow = mature; short-dashed outline and arrow = mature in part, but research is still needed; long-dashed outline and arrow = under research and development. Source: Courtesy of SMARTEC.

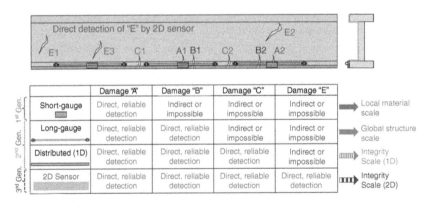

Figure 3.27 Progress in damage detection capabilities of strain sensors and their transformative impact on the scale of applicability of SHM; bracket and arrow coding: continuous bracket and arrow = mature; short-dashed bracket and arrow = mature in part, but research is still needed; long-dashed bracket and arrow = under research and development. Source: Modified from Glisic (2009, 2022).

about performance that was complementary to the first-generation sensors, long-gauge and 1D distributed sensors changed the paradigm of strain-based SHM by enabling monitoring of materials at a macro-scale, larger spatial coverage of structures, and improved capabilities of detection of unusual structural behaviors; this raised the scale of SHM to local and global structural, and integrity monitoring.

The third-generation of strain sensors addresses the need for improved reliability in the detection of unusual structural behaviors by enabling a direct detection approach. The third-generation of strain sensors is based on a variety of technologies, but two main subsets are contact and non-contact sensors. These sensors further change the paradigm of strain-based SHM, as they depart from the ideal of accurate strain measurement and put unusual structural behaviors in the focus, i.e., the strain becomes an indicator of unusual structural behavior rather than the principal observed parameter of interest.

"*In conclusion, one century after the development of the first strain sensors and almost one century after their first real-life applications, strain sensors based on several technologies are, at present day, commercially mature and are widely and successfully applied in many real-life SHM projects. They have enabled better understanding, optimized maintenance, and improved safety of civil structures worldwide. They gave birth to many companies and an entire industry sector. Regardless of their long tradition, strain sensors and strain-based SHM still represent vivid areas of research, development, and innovation.*" (Glisic 2022).

4

Sources of Error, Uncertainty, and Unreliability in Strain Measurements, and Fundamentals of Their Analysis

4.1 General Observations

Taking measurements represents important part of SHM, yet this process is unavoidably subjected to errors and uncertainties. The sources of errors and uncertainties are, in general, twofold: first are related to the measures of accuracy of SHM system itself (e.g., resolution, bias, and precision, see Chapter 2), and second are related to sensor settings (e.g., need for thermal compensation, gauge length, and installation procedure). Having knowledge of and, to the extent possible, accounting for all types of errors or uncertainties in strain measurement analysis is crucial for estimating error or uncertainty of strain measurements and their derivatives (i.e., other parameters determined from strain measurements, such as curvature, position of neutral axis, stress, deformed shape, and curvature). If the error or uncertainty in measurement of observed parameter (strain or its derivative) can be determined, then the thresholds could be meaningfully set for that specific parameter. Depending on the project requirements, estimating error or uncertainty of measurements can be performed using elementary approaches, but more often, estimating error or uncertainty can be very challenging and requires the use of advanced methods (e.g., Monte Carlo simulation).

In this chapter, first the errors and uncertainties related to SHM system are presented, followed by fundamentals of error and uncertainty analysis. Then, some frequent sources of errors related to sensor settings are presented, such as thermal compensation, gauge length and spatial resolution, sensor packaging and manner of installation. The chapter finishes with considerations related to evaluation of measurement reliability.

To ease the presentation in the further text, we refer to factors that influence the accuracy of monitoring (SHM) system, such as resolution, repeatability precision, reproducibility precision, bias, uncertainty, and limits of error, simply as the "measures of accuracy." The reason for this simplification is the fact that some monitoring systems under different conditions can have different accuracy of measurement.

Overall, the aim of this chapter is to introduce elementary approaches to estimating error or uncertainty in measurements and the propagation of error or uncertainty in evaluation of parameters that are derived from the measurements, so that the complexity of this process can be understood. However, this chapter should not be considered exhaustive. For example, there might be some other project-specific sources of errors or uncertainties related to sensor settings that are not presented in this chapter (e.g., variations in humidity, and aging of sensor components). In addition, elementary approaches presented here are applicable only under certain conditions and restrictions (as specified throughout this section), and in many real-life cases, more advanced techniques

Introduction to Strain-Based Structural Health Monitoring of Civil Structures, First Edition. Branko Glišić.
© 2024 John Wiley & Sons Ltd. Published 2024 by John Wiley & Sons Ltd.

should be used. More details about error and uncertainty analysis can be found in literature (e.g., Rabinovich 2005, Kirkup and Frenkel 2006, Ghanem et al. 2017).

4.2 Errors and Uncertainties Related to SHM System

4.2.1 Measurement Error: Relative Error, Absolute Error, and Limits of Error

Outcome of a measurement is an estimated value of measured parameter (measurand). Difference between the measurement outcome and the true value of measurand is called measurement error, see Equation 4.1.

$$\delta A = A_m - A_{true} \tag{4.1}$$

where A denotes measured parameter (measurand), δA is measurement error, A_m is measured value (outcome of measurement) of parameter A, and A_{true} is true value of parameter A.

Absolute value of measurement error is called absolute error, see Equation 4.2.

$$\delta A_{abs} = |\delta A| = |A_m - A_{true}| \tag{4.2}$$

where δA_{abs} denotes absolute error of measurement of parameter A.

Ratio between measurement error and the true value of measurand is called relative error, see Equation 4.3.

$$\delta^* A = \frac{A_m - A_{true}}{A_{true}} = \frac{\delta A}{A_{true}} \tag{4.3}$$

where $\delta^* A$ denotes relative error of measurement of parameter A.

Absolute value of relative error is called absolute relative error, see Equation 4.4.

$$\delta^* A_{abs} = \left| \frac{A_m - A_{true}}{A_{true}} \right| = \left| \frac{\delta A}{A_{true}} \right| \tag{4.4}$$

It is important to note that, in principle, we never know the true value of measurand; to the best of our abilities, we can only know approximate value of measurand with some (better or worse) degree of accuracy.

Limits of error represent interval bounded with minimal and maximal error values. This means that the outcome of individual measurements is not expected to be outside the bounds defined by limits of error. In other words, the error of an individual measurement is not expected to be greater than positive limit of error or smaller than negative limit of error. In the majority of available SHM literature and in the most of available datasheets with specifications of sensors used in SHM, the limits of error are wrongly called "accuracy." As per Chapter 2, accuracy is qualitative term, and limits of error are just one measure of accuracy.

4.2.2 Random (Accidental) Error and Systematic Error (Bias)

The experience has shown that repeated measurements of the same quantity under unchanging conditions will result in slightly different measurement realizations (note that described conditions correspond to repeatability conditions defined in Chapter 2). These differences in measurements can be attributed to influence of uncontrollable small changes in environment that affect measurement instruments, measurands, or both. That is the reason why the resulting measurement error is called a random or accidental error.

Random errors are usually determined through repeatability and/or reproducibility tests as differences between the values of measurements (measurement realizations) and mean value of all valid measurements performed during the test. As an example, let us assume that four strain measurements performed under unchanging (repeatability) conditions or under changing (reproducibility) conditions (measurand kept constant) yielded four different values, as shown in the middle column of Table 4.1. Statistical parameters, i.e., mean of the measurement values, as well as minimal error, maximal error, and standard deviation of the errors, are shown in the last row of the column. Random error for each measurement is then calculated as difference between the value of measurement and mean (average value) of all measurements, as shown in the right column of Table 4.1.

Error that remains constant when measurements are repeated under unchanged conditions is called systematic error or bias. It can be additive (adds a constant to measurement error) or multiplicative (adds a constant to relative error). An offset in instrument (typically results in additive systematic error), wrong calibration constants (typically results in multiplicative systematic error), or person executing measurement (operator) are the most frequent sources of systematic error. While random error can be inferred by repeating the measurements, this is not possible for systematic error. Practically, we know nothing about systematic error of measurements shown in Table 4.1. Given that statistical analysis of repeated measurements cannot help in identifying systematic error, an upgrade of experimental set-up might be required to detect it (e.g., exchanging reading units, having different operators of measurement, comparing with measurements from independent sensors with different physical principles). Evaluation of the systematic error can be performed by comparison with measurements obtained with a measurement system with certified traceability of calibration to primary standard (etalon).

Note that in the most of cases, the value of systematic error is not provided in datasheets with sensor specifications, which results in unknown systematic error in sensor applications and practically unknown limits of error. Given that the strain is measured with respect to some reference measurement, i.e., it is measured as a change with respect to reference value, additive systematic error is minimized by subtracting reference value from current value of measurement. Multiplicative systematic error can be minimized by confirmation or correction of calibration constants of measurement systems through comparison with an independent measurement system with different physical principles for which we have some knowledge regarding its multiplicative systematic error.

Table 4.1 Realization of fours strain measurements and random errors of measurements.

Measurement number	Value of measurement ($\mu\varepsilon$)	Random error ($\mu\varepsilon$)
1	152	2
2	139	−11
3	163	13
4	146	−4
Statistics	Mean = 150	Min. = −11; Max. = 13; St. Dev. = 10

4.2.3 Limits of Error, Standard Uncertainty, and Standard Error of Measurements

Table 4.1 presents four measurements performed under conditions of repeatability or reproducibility. The table can give us information about the accuracy of measurement system used in the experiment. This information can be inferred using deterministic or probabilistic approach.

In deterministic approach, we can simply find the minimum and maximum error and take these values as limits of (permissible) random error of our monitoring system (see Section 2.3.2). For example, based on Table 4.1, the limits of random error for monitoring system used in the experiment are −11 µε and +13 µε. In this case, we have to make sure that in our tests we performed enough measurements so that we were able to have these values as realizations of our measurements. For example, in Table 4.1, if we performed only the first two measurements, we would wrongly conclude that limits of error are −11 µε and +2 µε, which would not be satisfactory; however, performing only four measurements might not be satisfactory either, as this small number of measurements might miss some other larger errors.

In probabilistic approach, we are looking for a confidence probability that the random error does not exceed a specific value. This specific value is called uncertainty of measurement, and it is denoted with u(A). For example, in Table 4.1, the standard deviation of measurements is 10 µε. If we assume Gaussian distribution of measurements around the mean and adopt the standard deviation as measure of uncertainty, then our confidence probability that the measurement value will be within ±10 µε is 68.27%. This means practically that approximately one out of three measurements will be expected to be out of this bound (which indeed happens in Table 4.1). If we want to be more conservative, then we can use two (confidence probability of 95.45%) or three standard deviations (confidence probability of 99.73%) for uncertainty bounds. If only one standard deviation is used as measure of uncertainty, then this uncertainty is referred to as standard uncertainty.

In the two previously presented approaches, we assume that in real-life settings, we will perform exactly one measurement at each moment in time (time stamp) when we measure. However, in some cases, it is possible to perform multiple measurements for a single time stamp. This enables us to further reduce uncertainty in measurement by using standard error of the mean (standard error) as the measure of accuracy. Standard error is equal to standard uncertainty divided by root square of sample size. For example, standard error of measurements presented in Table 4.1 is $10/\sqrt{4} = 5$ µε. Interpretation of standard error is similar to interpretation of standard uncertainty, but the main difference is that the former deals with the mean of repeated measurements, while the latter deals with individual measurements.

Two important facts from the considerations above have to be highlighted. First, for all of the above approaches, an increase in number of repeated measurements improves the accuracy in estimation of limits of random error or of uncertainty of random error of the monitoring system. Second, tests presented above refer only to random error (and not systematic error) and only to repeatability or reproducibility precision conditions.

Estimation of measurement error in field conditions is very important in SHM. On the one hand, it enables setting of thresholds that guarantee safety, and on the other hand, it provides measures of accuracy for evaluation of structural health and performance. As an example, let us assume that yielding strain in structural steel is 2000 µε (typically, yielding is not allowed in structural members), and we have two candidates for monitoring strain, SHM system "A," with a limit of errors (both systematic and random errors included) of ±10 µε and SHM system "B," with a standard uncertainty (both systematic and random errors included) of ±10 µε. In addition, it is possible to perform only one measurement at each time stamp. If we want SHM system to alert us when strain in structural members is reaching yield value, then the threshold should not be set to 2000 µε, but rather to value that is at least safely lower than that. In the case of SHM system "A," the threshold should be set to max. 1990 µε, and in the case of SHM system "B," to max. 1970 µε (three standard deviations were used to cover more than 99% probability of not exceeding of 2000 µε). If the SHM system "B" is able to perform more than one measurement at each time stamp, then standard error can be used instead of standard uncertainty. Similar applies to SHM system "A" (standard error can be used instead of limits of error) if its random error follows Gaussian distribution (which is implied in system "B," as we use standard uncertainty for that system).

However, for most commercially available monitoring systems, limit of errors and/or uncertainty are provided based only on repeatability precision tests and, rarely, on both repeatability and reproducibility precision tests. Before applying the SHM in real-life settings, it is important to understand and evaluate all factors that can influence accuracy of strain measurements as well as of strain measurement derivatives: resolution, limit of errors/uncertainty under reproducibility precision conditions, systematic error, drift, temperature influence, location of sensors, etc. (see Section 2.3.2). To do that, it is necessary to apply error propagation formulae.

4.3 Fundamentals of Error and Uncertainty Analysis

4.3.1 Error and Uncertainty Propagation

Let y be a parameter of interest that cannot be directly measured by sensor (e.g., curvature in beam element) and $x_1, x_2, \ldots x_n$ input quantities that can be measured (e.g., strain and temperature in cross-section), so that $y = f(x_1, x_2, \ldots x_n)$, where f is differentiable function that defines relationship between input quantities and parameter of interest. Let $\pm \delta x_i$, $i = 1, 2, \ldots n$ be known estimations of limits of error in measurements x_i. Then, the limit of error δy of parameter y is determined using the following error propagation formula (Rabinovich 2005):

$$\delta y = \frac{\partial f}{\partial x_1} \delta x_1 + \frac{\partial f}{\partial x_2} \delta x_2 + \ldots + \frac{\partial f}{\partial x_n} \delta x_n \tag{4.5}$$

Few typical examples of error propagation formulae resulting from Equation 4.5 are given in Table 4.2.

Note that values of δx_i in Equation 4.5 (and consequently Table 4.2) can have both positive and negative values; thus, both values have to be tested in equations and the worst-case scenario (the largest absolute values) chosen for δy, using trial and error method. Similar applies if the limits of errors are not the same by absolute value for positive and negative limits of error (e.g., in Table 4.1, limits of error are different by absolute value: -11 µε and 13 µε).

Table 4.2 Examples of formulae for estimation of limits of error propagation.

Relationship	Estimated limits of error	Estimated limits of relative error
$y = a_1 x_1 + a_2 x_2$	$\delta y = a_1 \delta x_1 + a_2 \delta x_2$	$\delta^* y = \dfrac{a_1 \delta x_1 + a_2 \delta x_2}{a_1 x_1 + a_2 x_2}$
$y = x_1 x_2$	$\delta y = x_2 \delta x_1 + x_1 \delta x_2$	$\delta^* y = \delta^* x_1 + \delta^* x_2$
$y = x_1^n$	$\delta y = n x_1^{n-1} \delta x_1$	$\delta^* y = n \delta^* x_1$
$y = x_1 / x_2$	$\delta y = \dfrac{1}{x_2} \delta x_1 - \dfrac{x_1}{x_2^2} \delta x_2$	$\delta^* y = \delta^* x_1 - \delta^* x_2$

For example, let limits of errors of measurands x_1 and x_2 be $\pm\delta x_1 = \pm 3$ µε and $\pm\delta x_2 = \pm 2$ µε, respectively, and two derivatives of these measurands, y_1 and y_2 defined as $y_1 = x_1 + x_2$ and $y_2 = x_1 - x_2$. Then the limits of error in y_1 and y_2 are calculated as follows:

$$+\delta y_1 = 3 + 2 = +5 \,\text{µε}$$

$$-\delta y_1 = -3 - 2 = -5 \,\text{µε}$$

$$+\delta y_2 = 3 - (-2) = +5 \,\text{µε}$$

$$-\delta y_2 = -3 - (+2) = -5 \,\text{µε}$$

Thus, both y_1 and y_2 have the same limits of errors $\pm\delta y_1 = \pm\delta y_2 = \pm 5$ µε. Note that, in the above equations, signs "+" and "−" in front of symbol δ are symbolic, and denote positive and negative limits of error, respectively.

If the error in observed quantity $y = f(x_1, x_2, \ldots x_n)$ is analyzed in terms of standard uncertainty $u(y)$, then a different propagation formula should be used. Let us assume that $x_1, x_2, \ldots x_n$ are input quantities that can be measured and that are mutually uncorrelated; let $u(x_i)$, $i = 1, 2, \ldots n$ be known standard uncertainties in measurements x_i. Then, the uncertainty $u(y)$ of parameter y is determined using the following uncertainty propagation formula (Rabinovich 2005):

$$u(y)^2 = \left(\frac{\partial f}{\partial x_1} u(x_1) \right)^2 + \left(\frac{\partial f}{\partial x_2} u(x_2) \right)^2 + \ldots + \left(\frac{\partial f}{\partial x_n} u(x_n) \right)^2 \tag{4.6}$$

It is important to note that Expression 4.6 applies only if standard uncertainty (standard deviation) is an appropriate probabilistic measure of error for all input quantities x_i. Also, Expression 4.6 applies only if the quantities x_i are mutually non-correlated. If any of the above conditions are not fulfilled, then Expression 4.6 should not be applied.

Few typical examples of error propagation formulae resulting from Equation 4.6 are given in Table 4.3.

For example, let $u(x_1) = 1.5$ µε, $u(x_2) = 1$ µε, and $y_1 = x_1 + x_2$. Note that u is standard uncertainty, i.e., it represents standard deviation of measurements obtained under repeatability conditions. We can assume that plus/minus two standard deviations correspond to deterministic limits of error (this is reasonable assumption as there is more than 95% probability that error will not exceed that limit; in a more conservative assumption, three standard deviations can be used to cover more than 99% probability). Limit of errors calculated in this way is the same as in previous example that illustrated error propagation (± 5 µε), so we can compare errors of y obtained in both ways – using

Table 4.3 Examples of formulae for estimation of uncertainty propagation.

Relationship	Estimated uncertainty	Estimated relative uncertainty
$y = a_1 x_1 + a_2 x_2$	$u(y)^2 = a_1^2 u(x_1)^2 + a_2^2 u(x_2)^2$	$u^*(y)^2 = \dfrac{a_1^2 u(x_1)^2 + a_2^2 u(x_2)^2}{(a_1 x_1 + a_2 x_2)^2}$
$y = x_1 x_2$	$u(y)^2 = x_2^2 u(x_1)^2 + x_1^2 u(x_2)^2$	$u^*(y)^2 = u^*(x_1)^2 + u^*(x_2)^2$
$y = x_1^n$	$u(y) = n x_1^{n-1} u(x_1)$	$u^*(y) = n u^*(x_1)$
$y = x_1/x_2$	$u(y)^2 = \left(\dfrac{1}{x_2} u(x_1) \right)^2 + \left(-\dfrac{x_1}{x_2^2} u(x_2) \right)^2$	$u^*(y)^2 = u^*(x_1)^2 + u^*(x_2)^2$

error propagation formula and uncertainty propagation formula. In the latter case, the result is the following:

$$u(x_1)^2 = 1.5^2 + 1^2 = 3.25 \, \mu\varepsilon \Rightarrow u(x_1) = 1.8 \, \mu\varepsilon$$

Using the same logic of two standard deviations, this value corresponds to limits of error of $\pm 3.6 \, \mu\varepsilon$, which is smaller than $5 \, \mu\varepsilon$ determined using Expression 4.5. The reason for this difference lies in the fact that Expression 4.5 looks for the worst-case scenarios, while Expression 4.6 considers that the extreme differences (errors) in input quantities x_i are not likely to occur simultaneously, so some of them will cancel each other. Expressions for propagation of standard error $u_{SE}(y)$ are the same as for standard uncertainty, and thus, they are obtained from Expression 4.6 and Table 4.3 by substituting symbol "u" with symbol "u_{SE}."

Selection of formula for the use in a specific project depends on project requirements and available measures of accuracy of SHM system. Expression 4.5 is more conservative, but Expression 4.6 depends on type of distribution of random error of SHM system (if it is not Gaussian, then it makes no sense to use standard uncertainty or standard error) and mutual correlation of input parameters (if they are correlated than the Expression 4.6 cannot be used). If elementary propagation formulae presented here are not appropriate for specific project, then other, more elaborate methods should be used (e.g., see Ghanem et al. 2017).

Example 4.1 *Examples of Error and Uncertainty Analysis*

To illustrate the process of error and uncertainty analysis, illustrative examples are given in this section. Let assume that after a series of tests, the following properties of the monitoring system were found:

Resolution $= 1 \, \mu\varepsilon$
Repeatability (standard uncertainty) $= 2 \, \mu\varepsilon$
Reproducibility (standard uncertainty) $= 5 \, \mu\varepsilon$
Absolute value of drift $< \pm 1 \, \mu\varepsilon/\text{year}$
Systematic error can be neglected.

If the maximal frequency of measurements is 250 Hz, let's estimate limits of error of the monitoring system in the following situations:

Scenario 1: Static measurements in long-term (e.g., five years), in field conditions; at each time stamp, the measurement is performed 100 times, and average value is used as final value of measurement.

Scenario 2: Dynamic measurements in very short-term (e.g., one minute), in field conditions.

Scenario 3: Measurements in short-term (e.g., two hours), taken under slowly changing loads in controlled lab conditions. Load change will result in change of strain smaller than $1 \, \mu\varepsilon$ per minute.

The limits of error are deterministic values, so we can use Equation 4.5 to calculate them; however, repeatability and reproducibility are given in probabilistic terms, and thus, they should be converted to deterministic values.

For scenario 1, the following measures of accuracy influence the limits of errors: resolution ($\pm 1 \, \mu\varepsilon$), reproducibility precision ($5 \, \mu\varepsilon$), as field conditions imply uncontrolled environment, and drift ($\pm 1 \, \mu\varepsilon/\text{year}$). Given that each measurement is the average of 100 samples, it takes $1/2.5 = 0.4$ seconds to execute it. We can assume that during that very short time, the strain on the structure does not change (e.g., due to load, temperature, or rheological effects). Hence, we can use standard error $u_{SE} = 5/\sqrt{100} = 0.5 \, \mu\varepsilon$ to describe uncertainty of measurement and $2u_{SE} = 1 \, \mu\varepsilon$

to define limits of random error (we can use $3u_{SE}$ if for some reasons more conservative approach is required by the project). Finally, the limits of error can be calculated using Equation 4.5:

$$\delta\varepsilon_{max/min} = \pm(1\ \mu\varepsilon + 1\ \mu\varepsilon + 1\ \mu\varepsilon/\text{year} \cdot 5\ \text{years}) = \pm 7\ \mu\varepsilon$$

Note that in the middle part of equation above, the first term corresponds to resolution, the second to random error, and the third to drift; in some situations, resolution can be taken as contributing half-value to the error, as change from minimal to maximal value involves two resolution values, i.e., it might be detectable. However, this consideration should be evaluated on a case-to-case basis.

For scenario 2, the following measures of accuracy influence the limits of errors: resolution ($\pm 1\ \mu\varepsilon$) and repeatability precision ($2\ \mu\varepsilon$). Although the field conditions imply uncontrolled environment, the conditions around measurement in real-life settings (e.g., temperature and humidity) will not change in very short time (one minute), and that is the reason why repeatability precision can be used instead of reproducibility precision. In addition, drift can be neglected due to this very short-term. However, given that measurements are taken dynamically, only one measurement can be taken at each time stamp, and thus the limits of random error are calculated as two times (or three times for more conservative approach) standard uncertainty. Thus, based on Expression 4.5, the limits of error for this scenario can be calculated as follows:

$$\delta\varepsilon_{max/min} = \pm(1\ \mu\varepsilon + 2 \cdot 2\ \mu\varepsilon) = \pm 5\ \mu\varepsilon$$

In the middle part of the equation above, the first term corresponds to resolution and the second to random error (two times standard uncertainty in repeatability conditions).

For scenario 3, the following measures of accuracy influence the limits of errors: resolution ($\pm 1\ \mu\varepsilon$) and repeatability precision ($2\ \mu\varepsilon$). Since the conditions in lab are controlled, repeatability, and precision are used even though measurements are taken over two hours. However, given that strain is slowly changing, the measurement of strain at each time stamp can be taken as an average of 100 measurements, similar to scenario 1. This will take 0.4 seconds per measurement, and during that time the measured strain is quasi-constant, with an expected change that is far below the value of resolution of the system ($1\ \mu\varepsilon/60\ \text{seconds} \cdot 0.4\ \text{seconds} = 0.007\ \mu\varepsilon$), thus it can be neglected. Consequently, the limits of error can be calculated using standard repeatability error $u_{SE} = 2/\sqrt{100} = 0.2\ \mu\varepsilon$ (instead of standard uncertainty), i.e., as follows:

$$\delta\varepsilon_{max/min} = \pm(1\ \mu\varepsilon + 0.2\ \mu\varepsilon) = \pm 1.2\ \mu\varepsilon$$

The above example shows that the same monitoring system deployed in three different scenarios can have different limits of error, depending on surrounding conditions and the type of loading. This emphasizes the importance of a good understanding of the meanings of measures of accuracy of monitoring system and conditions under which the measurements are taken, so that they can be properly considered and implemented to evaluate limits of error or uncertainties in measurements or in their derivatives.

4.3.2 Rounding and Significant Figures

This section serves as a reminder of the elementary rules of rounding and consideration of significant figures related to measurement. When a strain measurement is performed, its outcome is presented by the monitoring system in the form of numeral with several figures (or digits). However, depending on measures of accuracy (e.g., resolution, precision, and limits of errors), some of these digits might be irrelevant and are results of spurious decoding of the encoding parameter.

Example 4.2 *Rounding and significant figures in strain measurement resulting from the properties of SHM system*

Let us assume that FBG-based strain sensing system with resolution of 1 $\mu\varepsilon$ provides values of encoding parameter, back-reflected wavelength of light expressed in nanometers (nm), in five-figure format. Let us assume that for the purposes of this example that outcome of measurement is 0.1235 nm. Let us assume that the sensor calibration constant is 834 $\mu\varepsilon$/nm. Then, the outcome of measurement is the following numeral: $834 \cdot 0.1235 = 102.9990$ $\mu\varepsilon$. However, the resolution of the system is 1 $\mu\varepsilon$, and thus most of the figures to the right of decimal point are spurious results of measurement and thus meaningless. To avoid unnecessary trailing of the spurious figures, the measurement can be rounded to 103 $\mu\varepsilon$.

The above example shows two important features of measurement. The first is that some figures are insignificant (meaningless) for evaluation of the measured quantity, and the second is that we can round the measurement so that, as a result of rounding, all figures are significant. Thus, we can define significant figures of a measurement as those that carry meaningful contribution to evaluation of measured quantity, and rounding is an action of reducing the number of significant figures.

If an outcome of measurement is given without any information regarding measures of accuracy, then the significant figures can be determined as follows:

- All non-zero values are significant; in the above example, measurement is 102.9990 $\mu\varepsilon$, and assuming that we do not know the resolution of the measuring system, significant figures would be "1," "2," "9," "9," and "9."
- In addition, all zero values in between significant figures are significant; in the above example, "0" between "1" and "2" is significant.
- All zeros in front of the most-left significant figure (leading zeros) are insignificant; in the above example, there are no zeros to the left of "1"; if the measurement was expressed in original unit – meter per meter – then the measurement value would be expressed as 0.0001029990, and in this number the four zeros to the left of "1" would be all insignificant.
- The zeros to the right of the most-right significant figure are ambiguous, and more information is to be known in order to determine their significance; in the above example, the zero after the third "9" is ambiguous.
- Spurious figures, if known, are insignificant and can be eliminated after rounding.

Rounding to significant figure is process of eliminating insignificant figures. What figures are significant is determined by the measures of accuracy of the monitoring system (resolution, precision, etc.). In the above example, the number 102.9990 (seven significant figures) is rounded to 103 (three significant figures) because the resolution of the monitoring system determined that the most-right significant figure is found in the place of units. Rules of rounding are the following (Rabinovich 2005):

- General rule: Measurement is rounded to closest number with required significant figures; for example, if the numbers 1.3498 and 1.3501 are to be rounded to two significant figures, then 1.3498 is rounded to 1.3 and 1.3501 is rounded to 1.4.
- Rounding "5" rule in science and engineering: If the value to be rounded has only one right-side insignificant figure and that figure is "5," then round the most-right significant figure to even value; for example, if the numbers 1.35 and 1.45 are to be rounded to two significant figures, then both are rounded to 1.4.

The fourth bullet point in the above rules for determination of significant figures mentions that zeros following the most-right non-zero significant figure are ambiguous. This is illustrated by the following example: in the number 300, "3" is significant figure, but the two following zeros are ambiguous: they might be placeholders without significance (i.e., to show the entire number, in which case it might be more appropriate to write $3 \cdot 10^2$ instead of 300) but they also may have significance and show the number rounded with respect to the rules above (e.g., measurement of 299.9 µε is rounded to 300 µε if the known resolution of the monitoring system is 1 µε). This illustrative example shows that determining the significance of zeros that follow the most-right non-zero significant value requires additional information on how the measurement is obtained (e.g., placeholder or rounding).

In science and engineering, the absence of explicit measures of accuracy implies that the error limits can be considered as plus/minus half range of values with an extra decimal place. For example, if no information is provided regarding measures of accuracy of a monitoring system, a measured value of 3.1 µε is considered as 3.1 µε ± 0.05 µε. However, this approach also has to be considered carefully: while rounding in science and engineering is performed in order to reflect measures of accuracy of the measurement, in everyday practice it can also be performed in a very approximate manner, to simplify presentation (for example statement that equatorial radius of earth is 6400 km without specifying significant figures would imply that its value ranges between 6399.5 and 6401.5 km (6400 ± 0.5 km), which is inaccurate; the statement that the number is rounded to two significant figures would be accurate as it would place the radius between 6350 and 6450 (6400 ± 50 km), while more accurate value is 6378 km (NASA 2020), which, in turn, should be interpreted as 6378 ± 0.5 km.

4.4 Thermal Compensation of Sensor as a Source of Error or Uncertainty

Changes in temperature in the environment surrounding the monitored structure influence sensor measurement in two ways: first, it generates strain changes in the structure (thermal strain and thermally generated mechanical strain, see Chapter 5); second, it generates dimensional changes of sensor parts and affects the physical properties of sensing elements that are related to the encoding parameter of the sensor (e.g., resistivity of electrical sensors, and refraction index in fiber-optic sensors). This second influence has adverse consequences for sensor measurement and results in errors or uncertainties. That is the reason why, this influence has to be accurately evaluated (to the extent possible) and removed from sensor measurement. This evaluation and removal of influence of change in temperature on sensor measurement is called thermal compensation of sensor. It is important to highlight that thermal compensation does not involve evaluation of thermally generated strain changes in the structure (see Chapter 5), but only the effects of the temperature change on the sensor measurement itself.

Example 4.3 *Thermal compensation vs. thermal strain*
To emphasize differences between the first and the second influence, let us observe a table made of steel that is exactly 1-m long at 0°C and has a thermal expansion coefficient of 10 µε/°C; the length of the table is measured using the rulers, the first is made of the same steel, the second is made of wood, with thermal expansion coefficient of 3 µε/°C, and the third of plastic, with thermal expansion coefficient of 100 µε/°C. To simplify the discussion, let us assume that all errors can be neglected, and that rulers can measure with resolution of 1 µε (which is acceptable for the illustrative academic purposes of this example).

If the temperature changes from 0°C to 40°C, and we assume that there are no constraints applied to table (i.e., it is free to expand), the strain in table changes for $40 \cdot 10 = 400$ $\mu\varepsilon$, change in length of the table is $400 \cdot 10^{-6} \cdot 1 = 0.000400$ m (0.4 mm), and the length of the table becomes 1.000400 m. This is an example of thermal strain, i.e., the strain generated in the structure due to thermal change. Now, let us try to measure this new length of the table using our three rulers.

Note that all three rulers will also undergo thermal change themselves. The distance between markings on rulers indicating zero and one meter will change as follows: wooden rulers will become 1.000030-m long (i.e., shorter than the table), steel rulers will become 1.000400-m long (i.e., the same length as the table), and plastic rulers will become 1.001000-m long (i.e., longer than the table). These changes in the lengths of rulers illustrate the influence of thermal change on the sensors (rulers) themselves. If we now attempt to measure the length of the table using these three rulers, without evaluation and removal (compensation) of this thermal influence, we will obtain three different values: wooden ruler will indicate that the table became longer for approximately 0.370 mm, a steel ruler will indicate no change in length of the table, and a plastic ruler will indicate shortening of the table for approximately 0.600 mm. Note that approximations are used in the above calculations to simplify the presentation; thus, the numbers obtained are not fully accurate, which does not affect the general conclusions from the discussion. We can see that the influence of temperature on the sensor can produce not only inaccuracy in measurement but also can obstruct understanding of observed phenomenon – steel ruler will indicate that there is no change in dimension of table under temperature change, and plastic ruler will indicate that the steel table contracts under positive temperature change; both indications are contradictory to proven physical phenomenon that increase in temperature increases the dimensions of steel objects.

In order to minimize the influence of temperature on the sensor, i.e., to perform thermal compensation on the sensor, it is necessary to establish a relationship between temperature and the encoding parameter in the sensor through an experimental set-up. Different types of sensors may require different setups.

As an example of a thermal compensation test, the sensor that has to be thermally compensated can be installed onto a fully unrestrained material (free to expand and contract under temperature change, e.g., a beam supported by two rollers) with a known thermal expansion coefficient $\alpha_{T,material}$, and subjected to temperature changes, ideally in the range of those expected in field. Temperature change should be applied in a controllable way, e.g., by placing the testing setup in a thermal chamber, so uniformity of temperature across material and sensor can be guaranteed. At least two temperature sensors should be used to measure temperature: one installed on the material, in proximity of the sensor ($T_{material}$), and one placed in the air (T_{air}). Uncompensated (as is) strain reading $\varepsilon_{uncompensated}$ should be taken after every increment i of temperature is applied ($i = 0, \ldots n$), but only when the measurements of the two temperature sensors are stabilized. Note that uncompensated strain reading does not have immediate physical meaning as it represents combination of strain in structure and thermal effects in sensor. Set of measurements ($T_{material}(i), \varepsilon_{uncompensated}(i)$) will represent the relationship between temperature and uncompensated strain readings. Note that the first reading ($T_{material}(0), \varepsilon_{uncompensated}(0)$) represents reference reading, and while $T_{material}(0)$ represents temperature in the material and sensors, $\varepsilon_{uncompensated}(0)$ has no physical meaning – it just shows a value reflecting an offset in sensor due to relationship between strain and encoding parameter and deformation of sensor due to installation. That is the reason why, that for $i = 0$, both uncompensated and compensated sensor readings would have the same reading, i.e., $\varepsilon_{uncompensated}(0) = \varepsilon_{compensated}(0)$. The above relationship between uncompensated strain reading and temperature can be approximated by some function f(T), typically polynomial

$$\varepsilon_{uncompensated}(i) \approx f(T_{matrial}(i)) \tag{4.7}$$

As an example, the approximation given in Expression 4.7 can be made by using least-squares fitting, and the goodness of approximation (uncertainty) evaluated using standard deviation.

Thermal strain change in material at temperature increment i reflects the expected strain measurement, as it would be performed by compensated strain sensor:

$$\varepsilon_{compensated}(i) - \varepsilon_{compensated}(0) = \alpha_{T,material}(T_{matrial}(i) - T_{matrial}(0)) \tag{4.8}$$

By combining Equations 4.7 and 4.8 and taking into account that $\varepsilon_{uncompensated}(0) = \varepsilon_{compensated}(0)$, we obtain thermal compensation function $c(T)$, as shown in Equation 4.9.

$$
\begin{aligned}
c(T(i)) &= \varepsilon_{uncompensated}(i) - \varepsilon_{compensated}(i) \\
&= f(T(i)) - (\alpha_{T,material}(T_{matrial}(i) - T_{matrial}(0)) + \varepsilon_{uncompensated}(0)) \\
&= (f(T(i)) - f(T(0))) - \alpha_{T,material}(T_{matrial}(i) - T_{matrial}(0))
\end{aligned}
\tag{4.9}
$$

Note that $c(T(0)) = 0$ as $\varepsilon_{uncompensated}(0) = \varepsilon_{compensated}(0)$.

It is important to notice that measures of accuracy of the function $c(T)$, such as limits of error or uncertainty, depend on measures of accuracy of each term found in the right-hand side of Equation 4.9. Hence, to determine the limits of error or uncertainty of function $c(T)$, it is necessary to determine these measures of accuracy for the function $f(T)$, (i.e., by using standard deviation of its fit), thermal expansion coefficient $\alpha_{T,material}$ (e.g., in a separate experiment or by using certified specifications provided by manufacturer of material), and temperature measurements (e.g., by using certified specifications provided by manufacturer of temperature sensors). Additional potential sources of inaccuracy in determination of function $c(T)$ are inherent to sensor and its installation, as presented in Sections 4.5 and 4.6. Once all measures of accuracy are found for each term, the measures of accuracy (limits of error or uncertainty) of function $c(T)$ can be estimated by applying a suitable error or uncertainty propagation formula (Equation 4.5 or 4.6).

Once the compensation function $c(T)$ is determined, along with its measures of accuracy, we can derive an expression for compensated strain measurement. Let us assume that in real-life settings time t_0 is considered as reference time (e.g., time of installation of strain sensor), then change in strain at some time $t > t_0$ can be found by using Equation 4.10, which is obtained by including reference time in Equation 4.9.

$$
\begin{aligned}
\Delta\varepsilon_{compensated}(t) &= \varepsilon_{compensated}(t) - \varepsilon_{compensated}(t_0) \\
&= \varepsilon_{uncompensated}(t) - c(T(t)) - (\varepsilon_{uncompensated}(t_0) - c(T(t_0))) \\
&= \Delta\varepsilon_{uncompensated}(t) - \Delta c(T(t))
\end{aligned}
\tag{4.10}
$$

where symbol Δ indicates the change with respect to reference time.

Expression 4.10 shows that measures of accuracy of compensated strain measurement depend on the measures of accuracy of uncompensated strain measurement, which are typically the ones provided by the strain sensor manufacturer from various lab tests, and the measures of accuracy of function $c(T)$. Thus, the accuracy of compensated strain measurement (which is the one we are interested in) is always worse than the accuracy of uncompensated strain measurement. Note that uncompensated strain measurement, in general, does not have physical meaning (as it represents a combination of strain in structure and thermal effects in sensor) except in cases where the temperature remains constant between reference and observed time (e.g., during repeatability precision tests in the lab).

In most of the cases function $f(T)$ is linear for strain sensors and can be represented in form of $f(T) = m_{comp}T + b_{comp}$, where constants m_{comp} and b_{comp} are determined using linear regression (see

Equation 4.7). In that case the function $c(T)$ is also linear, as per Equation 4.11.

$$c(T(i)) = \varepsilon_{uncompensated}(i) - \varepsilon_{compensated}(i)$$
$$= m_{comp}T(i) + b_{comp} - (\alpha_{T,matrial}(T_{matrial}(i) - T_{matrial}(0)) + m_{comp}T(0) + b_{comp})$$
$$= (m_{comp}T(i) - m_{comp}T(0)) - \alpha_{T,material}(T_{matrial}(i) - T_{matrial}(0)) \qquad (4.11)$$

Finally, Equation 4.11 transforms into Equation 4.12.

$$\Delta\varepsilon_{compensated}(t) = \Delta\varepsilon_{uncompensated}(t) - \Delta c(T(t)) = \Delta\varepsilon_{uncompensated}(t) - k_{comp}\Delta T(t) \qquad (4.12)$$

where $k_{comp} = m_{comp} - \alpha_{T,material}$ and $\Delta T(t) = T(t) - T(t_0)$. Recall that $\alpha_{T,material}$ is thermal expansion coefficient of material used in lab tests to determine the thermal compensation function $c(T)$.

Sensors with a linear thermal compensation function $c(T)$ are preferred in practice, as the outcome of compensation does not depend on initial temperature $T(t_0)$, registered at reference time t_0, but only on the change in temperature from that time and single compensation constant k_{comp}. An example of the influence of thermal compensation on the accuracy of measurement is given below.

Example 4.4 *Influence of thermal compensation on accuracy of strain measurement*
Let us observe FBG strain sensor presented in Section 3.6.3 and shown in Figure 3.17. The sensor contains two sensing elements, one pre-tensioned between the anchors, coupled with structure, and sensitive to both strain and temperature, and the other kept strain-free, sensitive to temperature only. Expression 3.11 (see Section 3.6.3) represents the thermal compensation formula equivalent to Expression 4.12, with corresponding terms given in table below.

Expression 4.12	Expression 3.11
$\Delta\varepsilon_{compensated}(t)$	ε
$\Delta\varepsilon_{uncompensated}(t)$	$\dfrac{1}{C_{\varepsilon,\varepsilon}}\Delta\lambda_{\varepsilon}$
$-k_{comp}\Delta T(t)$	$-\dfrac{C_{T,\varepsilon}}{C_{\varepsilon,\varepsilon}}\Delta T = -\dfrac{C_{T,\varepsilon}}{C_{\varepsilon,\varepsilon}C_{T,T}}\Delta\lambda_T$

Calibration constants $C_{\varepsilon,\varepsilon}$, $C_{\varepsilon,T}$, and $C_{T,T}$ depend on reference values $\lambda_{\varepsilon,r}$ and $\lambda_{T,r}$ of FBG sensing elements (see Section 3.6.3); however, to simplify the presentation, the following approximate values of calibration constants are used in this example: $C_{\varepsilon,\varepsilon} \approx 0.0012$ nm/$\mu\varepsilon$ and $C_{\varepsilon,T} \approx C_{T,T} \approx 0.010$ nm/°C. Based on specifications provided by manufacturer, the resolution for reading the wavelength of FBGs is 0.2 pm = 0.0002 nm, and repeatability precision is 1 pm = 0.001 nm. Given that resolution is significantly smaller than repeatability precision, the influence of the former is neglected, and only the latter is used as the measure of accuracy of the monitoring system, i.e., $u(\lambda_{\varepsilon}) = u(\lambda_T) = u(\lambda) \approx 1$ pm = 0.001 nm. Expression 3.11 contains calibration constants $C_{\varepsilon,\varepsilon}$, $C_{\varepsilon,T}$, and $C_{T,T}$, and changes in wavelengths of strain and temperature FBGs, $\Delta\lambda_{\varepsilon} = \lambda_{\varepsilon} - \lambda_{\varepsilon,r}$ and $\Delta\lambda_T = \lambda_T - \lambda_{T,r}$. Considering that measurements taken at reference time (subscript r) and observed time t (no specific subscript) are, for each FBG, mutually uncorrelated, the standard uncertainty in change in wavelength can be evaluated by using the propagation formula given in Expression 4.6, as follows:

$$u(\Delta\lambda_{\varepsilon}) = u(\Delta\lambda_T) = u(\Delta\lambda) = (1^2 + 1^2)^{0.5} = 1.41 \text{ pm} = 0.00141 \text{ nm}.$$

Then, the standard uncertainty in determination of $\Delta\varepsilon_{uncompensated}(t)$, $\Delta T(t)$, and $\Delta\varepsilon_{compensated}(t)$ is evaluated as follows:

$$u(\Delta\varepsilon_{uncompensated}(t)) = \frac{1}{C_{\varepsilon,\varepsilon}}u(\Delta\lambda_\varepsilon) = \frac{1}{0.0012}0.00141 = 1.2\,\mu\varepsilon$$

$$u(\Delta T(t)) = \frac{1}{C_{T,T}}u(\Delta\lambda_T) = \frac{1}{0.010}0.00141 = 0.14°C$$

$$u(\Delta\varepsilon_{compensated}(t)) = \sqrt{\left(\frac{1}{C_{\varepsilon,\varepsilon}}u(\Delta\lambda_\varepsilon)\right)^2 + \left(\frac{C_{T,\varepsilon}}{C_{\varepsilon,\varepsilon}C_{T,T}}u(\Delta\lambda_T)\right)^2}$$

$$= u(\Delta\lambda_T)\frac{1}{C_\varepsilon}\sqrt{1+\left(\frac{C_{T,\varepsilon}}{C_T}\right)^2} = 0.00141\frac{1}{0.0012}\sqrt{1+\left(\frac{0.010}{0.010}\right)^2} = 1.7\,\mu\varepsilon.$$

Thus, the example above illustrates the influence of thermal compensation on the precision of strain measurement and shows that thermal compensation of FBG strain sensors introduces an increase in standard uncertainty of more than 40% (i.e., from 1.2 to 1.7 μɛ). Similar approach can be used for evaluation of errors or uncertainties introduced by thermal compensation of other types of sensors.

4.5 Gauge Length and Spatial Resolution of Sensors as a Source of Error

4.5.1 Gauge Length of Discrete Sensors as an Inherent Source of Error

Chapter 3 presents various types of discrete and distributed sensors that are all designed to monitor strain (i.e., normal strain component of strain tensor) in a specific direction (in the direction of the sensor). The sensing element (e.g., electrical resistor, vibrating wire, and optical fiber) is a part of the sensor which is coupled with the monitored structure, to which the structure's strain is transferred, and in which the strain is converted into encoding parameter (e.g., voltage, and optical signal). For discrete sensors, the length of sensing element over which the discrete sensor is coupled to structure is called gauge length of sensor. In practice, discrete sensors with gauge length shorter or equal to 100 mm (~4 in.) are called short-gauge sensors, while those with the gauge length longer or equal to 250 mm (~10 in.) are called long-gauge sensors. However, depending on material properties (notably inhomogeneity), in some cases, sensors shorter than 250 mm can be considered as long-gauge and sensors longer than 100 mm as short-gauge.

Let us observe, as an example, an inhomogeneous material, such as concrete, consisting of a continuous matrix (e.g., cement paste), inclusions (e.g., aggregate or air pockets), and discontinuities (e.g., cracks), as shown in Figure 4.1 (Glisic 2011). Such material is inhomogeneous at meso-scale but in engineering practice, for the purposes of design and performance evaluation, the behavior of such material is frequently observed as homogeneous at macro-scale (e.g., Neville 1975). To illustrate this statement, let us assume that beam shown in Figure 4.1 has cross-sectional dimensions of 1 × 1 m and is subjected to uniaxial compressive force of 1 kN applied at the left and right ends of the beam. If we observe an arbitrary cross-section in the middle zone of the beam, based on engineering practice, we will consider material as homogenous at macro-scale and conclude that stress at any point of this cross-section is equal to 1 kN/m^2; however, in reality, the material is inhomogeneous, and stress at different points of the cross-section would be different depending on material properties at these points. For example, let us assume that one observed point happens to be a

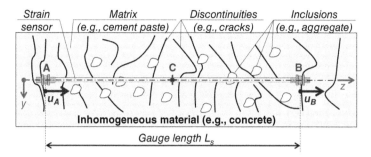

Figure 4.1 Schematic representation of a discrete (long-gauge) strain sensor coupled with an inhomogeneous material. Source: Modified from Glisic and Inaudi (2007), courtesy of SMARTEC SA.

part of matrix (e.g., cement paste), the other a part of inclusion (e.g., aggregate or air pocket), and another a part of discontinuity (e.g., crack or pore), where inclusion or discontinuity could be filled with air or with water. Then, the stress at each of these points will depend on mechanical properties of corresponding material, and these stresses will be, in general, different from each other. Thus, the value of $1\,kN/m^2$ is not exact; it is rather an average value, which is nevertheless, a good approximate for the purposes of design and performance evaluation. The reasoning used above for the analysis of stress can also be applied for analysis of strain. Thus, in the case of materials, such as concrete, that are inhomogeneous at material level (or meso-level) but for engineering purposes considered as homogeneous at structural level (or macro level), it would be reasonable to use sensors that directly provide the averaged value of strain, which reflects behavior at the structural level.

Strain is parameter defined at a mathematical, dimensionless point, while sensor gauge length has dimension. Thus, a discrete strain sensor does not measure strain at a point; it actually integrates strain along the gauge length and provides an average-integral value of strain (in further text, simply referred to as "average strain"). This value is usually attributed to point at the middle of the gauge length (see Figure 4.1). Thus, on one hand, regardless of their gauge length, discrete strain sensors always measure an average value of strain around their midpoint; on the other hand, a proper choice of gauge length can ensure that material will be monitored at desired scale: at material scale (meso-scale) or structural scale (macro-scale). The former is performed using short-gauge sensors (to enable monitoring at local, material scale), while the latter is performed using long-gauge sensors, and enables monitoring at both local and global structural scales and, to some extent, at the integrity (damage detection) scale (see Chapters 5–7). This is further explained through an analysis of strain measurement. Equation 4.13 describes measurement of the discrete sensor shown in Figure 4.1.

$$\varepsilon_{z,C,sensor} = \frac{\Delta L_s}{L_s} = \frac{u_B - u_A}{z_B - z_A} = \frac{1}{L_s}\int_{z_A}^{z_B} \varepsilon_{z,matrix}(z)dz + \frac{1}{L_s}\sum_i \Delta w_{d,i} \qquad (4.13)$$

where subscript z denotes direction of strain, A and B are start and end point of gauge length of the sensor, which coordinates are z_A and z_B; C is midpoint of the sensor to which the measurement is attributed; $z_C = (z_A + z_B)/2$ is coordinate of point C; $L_s = z_B - z_A$ is gauge length of sensors at reference time, i.e., before the change in strain occurred; u_A and u_B are components of displacements of points A and B in direction of z-axis (after change in strain occurred); $\varepsilon_{z,C,sensor}$ is average strain measured by sensor (attributed to point C); $\varepsilon_{z,matrix}(z)$ is strain distribution in continuous parts of matrix along the gauge length of sensor (i.e., in direction of z-axis); $\Delta w_{d,i}$ is dimensional change component of i-th ($z_i \in [z_A, z_B]$) discontinuity or inclusion (crack opening, inclusion dimensional change, etc.) in direction of sensor gauge length (z-axis) after the strain change occurred. To simplify presentation, coordinate y is omitted in Equation 4.13, as it is constant along the gauge length of the sensor.

Equation 4.13 shows that discrete strain sensor always measures an average strain value around point C, which is, in general case, different from true strain value at that point $\varepsilon_{z,C} = \varepsilon_z(z_C)$. Thus, the measurement of strain sensor contains an error inherent to gauge length of the sensor. This error can be represented by Equations 4.14.

$$\delta\varepsilon_{z,C,sensor}^{GL} = \varepsilon_{z,C,sensor} - \varepsilon_{z,C} \tag{4.14a}$$

$$\delta^*\varepsilon_{z,C,sensor}^{GL} = \frac{\delta\varepsilon_{z,C,sensor}^{GL}}{\varepsilon_{z,C}} = \frac{\varepsilon_{z,C,sensor} - \varepsilon_{z,C}}{\varepsilon_{z,C}} \tag{4.14b}$$

where superscript GL denotes the error inherent in gauge length.

Equation 4.13 indicates that the value of strain measurement $\varepsilon_{z,C,sensor}$ depends on the strain distribution $\varepsilon_{z,matrix}(z)$ between points A and B, the sum of dimensional changes of discontinuities (crack openings or inclusions) $\Delta w_{d,i}$ between points A and B, and the gauge length of sensor L_s. The former can be considered as contribution to strain measurement from homogeneous part of material (matrix, e.g., cement paste) and the latter as contribution from an inhomogeneous part of material (inclusions and discontinuities, e.g., aggregate, air pockets, and cracks). Both parts contribute to gauge-length induced inherent error in Equations 4.14, as per Equations 4.15.

$$\delta\varepsilon_{z,C,sensor}^{GL} = \delta\varepsilon_{z,C,sensor}^{GL,hom} + \delta\varepsilon_{z,C,sensor}^{GL,inhom} \tag{4.15a}$$

$$\delta^*\varepsilon_{z,C,sensor}^{GL} = \frac{\delta\varepsilon_{z,C,sensor}^{GL,hom} + \delta\varepsilon_{z,C,sensor}^{GL,inhom}}{\varepsilon_{z,C}} = \frac{\delta\varepsilon_{z,C,sensor}^{GL,hom}}{\varepsilon_{z,C}} + \frac{\delta\varepsilon_{z,C,sensor}^{GL,inhom}}{\varepsilon_{z,C}} \tag{4.15b}$$

where superscripts *hom* and *inhom* denote contributions to the error from homogenous and inhomogeneous parts of material.

4.5.2 Contribution to the Error Inherent to Gauge Length Due to Homogeneous Part of Material

Contribution to the error due to homogeneous part of material is discussed in this subsection. Let us assume that monitored structure is made of homogeneous material and that the strain distribution between endpoints of strain sensor can be considered a continuous function, as shown in Figure 4.2.

Let assume that Figure 4.2a shows the true strain distribution $\varepsilon_z(z)$, and the values of strain at points A, C, and B are $\varepsilon_{z,A} = \varepsilon_z(z_A)$, $\varepsilon_{z,C} = \varepsilon_z(z_C)$, and $\varepsilon_{z,B} = \varepsilon_z(z_B)$, respectively. Note that $\varepsilon_z(z)$ is equivalent to $\varepsilon_{z,matrix}(z)$ in Figure 4.1, but subscript "matrix" is omitted as the material is assumed to be homogeneous. Given that discrete strain sensor measures average strain, the value of strain measurement, $\varepsilon_{z,C,sensor}$, is equivalent to an imaginary constant strain distribution between points A and B, such that the shaded surfaces in Figures 4.2a and 4.2b have the same areas (by definition of average value of integral). Figure 4.2b shows that in general case, the values of $\varepsilon_{z,C,sensor}$ and $\varepsilon_{z,C}$ are different, and their difference is equal to inherent error of measurement $\delta\varepsilon_{z,C,sensor}^{GL,hom}$ (also shown in Figure 4.2b). This is expressed by Equation 4.16. Note that this error is null if the strain distribution along the gauge length of sensor is linear. It can also be null if strain distribution and gauge length are such that $\varepsilon_{z,C,sensor} = \varepsilon_{z,C}$ simply by accident.

$$\delta\varepsilon_{z,C,sensor}^{GL,hom} = \varepsilon_{z,C,sensor}^{GL,hom} - \varepsilon_{z,C}^{GL,hom} = \frac{1}{L_s}\int_{z_A}^{z_B} \varepsilon_z(z)dz - \varepsilon_z(z_C) \tag{4.16}$$

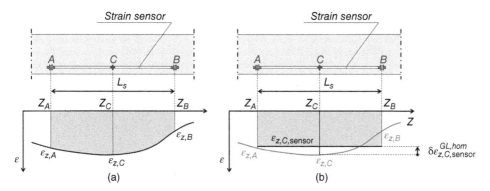

Figure 4.2 Schematic representation of (a) strain distribution along the sensor's gauge length and (b) average strain measurement at point C; the areas of shadowed surfaces in figures (a) and (b) are equal. Source: Modified from Glisic and Inaudi (2007).

Equation 4.16 shows that the error inherent to gauge length can be evaluated in homogeneous materials if strain distribution $\varepsilon_z(z)$ in the structure can be estimated, e.g., based on structural analysis. It also shows that with a decrease in the gauge length around point C (i.e., with a decrease in the distance between points A and B in Figure 4.2), the error decreases too, as mathematically, the limit of average integral of strain converges to the strain value at point C. Thus, for more accurate measurements, it is desirable to have shorter gauge lengths of sensors.

Analysis of typical strain distributions in beam-like structures along the gauge length of sensor parallel to centroid line shows that the error induced by gauge length can be estimated (see Glisic and Inaudi 2007, Glisic 2011 for more detail). Three typical examples for beams with invariable (constant) cross-section are given below:

- If the strain distribution between endpoints of sensor is linear, then the error is null.
- If the strain distribution between endpoints of sensor is parabolic (e.g., due to uniformly distributed load), then the relative error is proportional to the square of ratio of gauge length to the length of the beam; typically, sensors whose gauge length is six times shorter than the length of the beam will have a relative error in measurement smaller than 1%.
- If the strain distribution between endpoints of sensor is continuous and bilinear (i.e., shape of "broken line," e.g., due to concentrated force), then the relative error is proportional to the simple ratio of gauge length to the length of the beam; typically, sensors whose gauge length is 10 times shorter than the length of the beam will have a relative error in measurement smaller than 5%.

In the case of homogeneous materials, such as steel, the error induced by gauge length is only due to non-linear strain distribution, and thus, even sensors shorter than 250 mm can be considered as long-gauge sensors. Shorter sensors can be used for more accurate measurements, and this would be perfectly fine for monitoring at both the local material and structural scales, as well as at global scale. However, short sensors provide small spatial coverage over the structure, which may be less convenient for monitoring at integrity scale, i.e., for damage detection purposes. Longer sensors offer larger spatial coverage of structures and consequently, they are more likely to detect damage. Hence, if the monitoring is performed at integrity scale, longer sensors should be used, and the trade-off between accuracy in strain measurement (requiring shorter sensors) and sensitivity to damage (requiring longer sensors) should be made. This is discussed in more detail in Section 4.5.3 and Chapters 5–7, where monitoring at local, global, and integrity scales is presented (see also Glisic 2011).

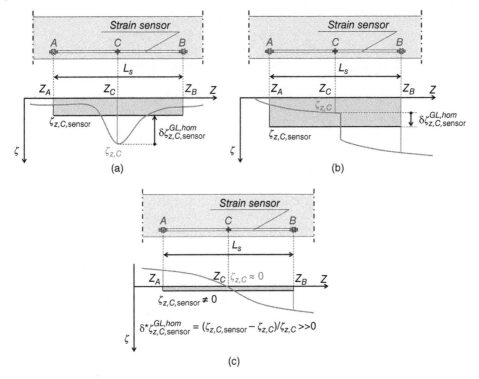

Figure 4.3 Three strain distributions along gauge length of sensor that may lead to potentially large errors: (a) Large strain gradient, (b) abrupt (discontinuous) strain change, and (c) change in sign of strain.

Before moving to the analysis of strain measurement in inhomogeneous materials, three specific strain distribution cases with a potential risk of large errors are presented in Figure 4.3 (compare to Figure 4.2). The first case (Figure 4.3a) is an abrupt strain field perturbation, e.g., due to proximity of sensor to the application point of concentrated force. In this case, strain gradients might be very high locally, and strain measurement may be prone to large absolute error ($\left| \delta \varepsilon_{z,C,sensor}^{GL,hom} \right| \gg 0$), depending on the gauge length of the sensor and its position. The second case (Figure 4.3b) is somewhat similar to the first ($\left| \delta \varepsilon_{z,C,sensor}^{GL,hom} \right| \gg 0$), except that the strain field is discontinuous, e.g., due to change in geometrical properties of structure or application of an external moment. The third case (Figure 4.3c) involves a sensor installed in proximity to the point where strain changes the sign, and thus strain measurement is close to zero, resulting in potentially large relative errors ($\left| \delta^{*} \varepsilon_{z,C,sensor}^{GL,hom} \right| \gg 0$). These cases must be carefully considered when deciding the gauge length and position of sensors, depending on the scale of monitoring (local, global, or integrity scale) and specific project objectives (see Chapters 5–7).

4.5.3 Contribution to the Error Inherent in Gauge Length Due to Inclusions in Material

Inherent error due to gauge length is larger for inhomogeneous materials than for homogeneous materials, as both components shown in Equations 4.15 are present. In addition, depending on the composition of the material, a component $\delta \varepsilon_{z,C,sensor}^{GL,inhom}$ can have several subcomponents. The case of concrete is presented in this section, while a similar approach can be taken for other materials.

Observed at meso-scale, concrete can be considered as to consist of a homogeneous matrix (cement paste), inclusions (aggregate, air pockets), and discontinuities (cracks). Component of error due to strain distribution is determined in the same way as for homogeneous material (Equation 4.16). Component of error due to inhomogeneity can be further subdivided into two components, see Equation 4.17

$$\delta\varepsilon_{z,C,sensor}^{GL,inhom} = \delta\varepsilon_{z,C,sensor}^{GL,inclus} + \delta\varepsilon_{z,C,sensor}^{GL,discon} \tag{4.17}$$

where $\delta\varepsilon_{z,C,sensor}^{GL,inclus}$ represents component of error due to inclusions and $\delta\varepsilon_{z,C,sensor}^{GL,discon}$ component of error due to discontinuities.

Evaluation of $\delta\varepsilon_{z,C,sensor}^{GL,inclus}$ is challenging as it depends on granulometric composition of the concrete and mechanical properties (notably elastic modulus and Poisson's ratio) of all concrete constituents (cement paste, aggregate). Tests performed in laboratory (Widow 1992) show that for uniaxially loaded concrete, relative error $\delta^*\varepsilon_{z,C,sensor}^{GL,inclus}$ depends on ratio between gauge length and diameter of greatest aggregate and approximately follows negative power rule, as per Equation 4.18. Example of test results is given in Figure 4.4.

$$\delta^*\varepsilon_{z,C,sensor}^{GL,inclus} = a\left(\frac{L_s}{D_{max}}\right)^{-b} \pm \delta_{fit}^* \tag{4.18}$$

where a and b are coefficients of trendline fit, L_s is gauge length of the sensor, D_{max} is maximal nominal diameter of aggregate, and $\pm\delta_{fit}^*$ is the limit of error of trendline fit; in the case of tests shown in Figure 4.4, $a = 32$, $b = 1.2$, and max. $\pm\delta_{fit}^* = \pm6\%$ (two relative standard uncertainties).

Tests shown in Figure 4.4 were performed using a vibrating wire strain sensor. They show that as the ratio between gauge length and maximal aggregate diameter increases, the error rapidly decreases and asymptotically approaches zero. It also suggests that the scattering of measurements around trendline is smaller for larger ratios (i.e., δ_{fit}^* decreases). While graph and trendline shown in Figure 4.4 are specific to tested concrete, they can be used as guidelines for determination of error component $\delta^*\varepsilon_{z,C,sensor}^{GL,inclus}$. For example, it is necessary to have ratio $L_s/D_{max} \geq 5$ to reduce the relative error to approximately 5% and $L_s/D_{max} \geq 8$ to reduce the relative error to approximately 3%. Using very short sensors, with $L_s/D_{max} \leq 2$, would result in relative error of 20% and higher.

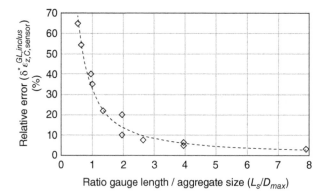

Figure 4.4 Example of dependence of relative error in strain measurement in concrete on ratio between sensor gauge length and maximal nominal diameter of aggregate. Source: Reconstructed from Widow (1992).

4.5.4 Contribution to the Error Inherent in Gauge Length Due to Discontinuities in Material

As mentioned earlier, concrete structures that are inhomogeneous at meso-level are designed and analyzed as homogeneous at macro-level. The average value of strain measurement provided by long-gauge sensors is in accordance with that principle as it reflects the concrete behavior as it were homogeneous and thus enables monitoring of concrete structures at local and global structural scales.

Reinforcing bars (rebars) provide reinforced concrete with tensional strength, but they also enable another component of inhomogeneity in form of tensional cracks. These cracks are small (typical opening is less than 0.1 mm, but larger openings can also be tolerated, depending on type of structure and conditions surrounding it), and they are not considered as damage, they are considered an integral part of material. The distance between these cracks can be variable, depending on the type of structure, and it is, typically ranged between 10 and 30 cm (4 and 12 in). Due to discontinuities (cracks), the sensors with short-gauge length cannot be used for monitoring of concrete structures at the structural scale (e.g., Glisic et al. 2002b, Chung et al. 2008). To illustrate this statement and evaluate errors in strain measurement due to discontinuity $\delta\varepsilon_{z,C,sensor}^{GL,discon}$, let us assume that three sensors with identical gauge lengths are embedded in a reinforced concrete beam with constant cross-section, subjected to uniaxial tensional stress, as shown in Figure 4.5. Let us assume that the position of the sensors is slightly different: Sensor 1 and Sensor 2 are placed exactly next to each other, i.e., their endpoints (anchors) belong to the same cross-sections, while Sensor 3 is shifted along the axis of the beam (e.g., due to imperfect embedding of sensor), see Figure 4.5.

Since the beam is subjected to tension, let us assume it cracked, and for the purposes of academic analysis, let us assume that all cracks opened for the same value and beam "split" into $n+2$ blocks of the same size (numbered from 0 to $n+1$ in Figure 4.5). Let us assume that one crack happens exactly in the cross-section with left endpoints of Sensors 1 and 2, but these endpoints became attached to different blocks, i.e., Block 1 and Block 0 respectively. The left endpoint of Sensor 3 is shifted axially but still belongs to Block 1, and the right endpoints of all three sensors belong to the same Block $n+1$.

If the reinforced concrete beam is observed as homogeneous at structural scale, the average strain field in the beam should be considered constant (i.e., equal average stress $\sigma_{z,average}$ divided by the modulus of elasticity E), and thus, all three of the sensors are expected to yield the same strain measurement, equal to theoretical average strain value $\varepsilon_{z,theoretical} = \sigma_{z,average}/E$. Yet, presence of cracks combined with imperfections in the positions of sensors' endpoints will result in measurements that are mutually different, and also different from the theoretical average strain $\varepsilon_{z,theoretical}$. Glisic and Inaudi (2007) analyzed in detail the differences in measurements and estimated limits of error inherent to gauge length, $\delta\varepsilon_{z,C,sensor}^{GL,discon}$, as given in Equation 4.19.

$$\delta\varepsilon_{z,C,sensor}^{GL,discon} \approx \varepsilon_{z,C,sensor} - \varepsilon_{z,C,theoretical} \approx \pm\frac{w_c}{L_s} \leq \pm\varepsilon_{z,C,sensor}\frac{L_c}{L_s} \tag{4.19}$$

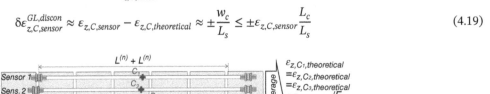

Figure 4.5 Idealized schematic of the influence of the discontinuities on a long-gauge sensor measurement. Source: Modified from Glisic and Inaudi (2007) and Glisic (2011).

where w_c is (average) crack opening, L_c is (average) length of cracked block (distance between the cracks), and L_s is gauge length of Sensors 1, 2, and 3.

Equation 4.19 shows that the measurement accuracy increases with the increase in gauge length. For example, to reduce relative error under ±10%, the gauge length of sensor should be 10 times longer than the distance between the cracks ($L_s \geq 10\ L_c$). To improve the durability of concrete structures, modern design favors a larger number of cracks with smaller crack openings, as opposed to fewer cracks with larger openings. A typical spacing between the cracks ranges between 100 and 300 mm (e.g., Piyasena 2002) and thus, to keep the relative error under ±10%, typical minimal gauge length of the sensor should range between one and three meters.

Based on Figure 4.5 and Equation 4.19, one can conclude that a measurement made by sensor with gauge length shorter than distance between the discontinuities (cracks) will be highly inaccurate. For example, let us assume a sensor with a gauge length of 100 mm, which is shorter than distance between the two cracks, is embedded in uniaxially loaded beam shown in Figure 4.5. If both endpoints of the sensor are within a single block of concrete (i.e., between two cracks), the measurement will be approximately equal to the local tensional strain in concrete, which is typically ranged between 50 and 120 µε. If the endpoints are slightly shifted, and are in two different adjacent blocks, the measurement will be approximately equal to ratio between the crack opening and the gauge length; for example, if crack opening is 0.1 mm, then the sensor of 100 mm will measure 1000 µε. Both of these measurements might provide some useful information at local material scale (e.g., ultimate strain in concrete or crack opening value), but they are both significantly different from the average strain in the beam observed at structural scale, as the former would be significantly lower, and latter significantly higher than the average. In addition, analysis of measurements from the sensor could be confusing because slight shift in the location of sensor along the axis of the beam may result in significantly different outcome of measurement (i.e., different in order of magnitude, as shown in the example above). Based on example above, for inhomogeneous materials, depending on geometrical distribution of discontinuities in specific case, even sensors with gauge length longer than 250 mm can be considered as short-gauge sensors (e.g., if the spacing between the discontinuities is larger than 250 mm).

4.5.5 Example of Influence of Gauge Length to Strain Measurement

This subsection uses real-life examples to illustrate how different gauge lengths of sensors influence the strain measurement. The structure used for this study is the Streicker Bridge, located at Princeton University campus (e.g., see Figure 3.24b and Chapter 6). The bridge consists of a main span and four curved continuous girders, called "legs." The main span is a deck-stiffened arch, where deck is made of post-tensioned concrete and arch and columns are weathering steel. Similarly, decks of all four legs are made of post-tensioned concrete and supporting columns of weathering steel. For the purposes of research and education, the bridge has been equipped with SHM system, consisting of long-gauge FBG sensors, Brillouin-based distributed fiber optic strain sensors, FBG displacement sensors, and Sensing Sheet (see Section 3.72). The overall aims of the project and details regarding the bridge and the SHM system are given in Chapter 6 and only a part relevant to influence of gauge length on measurement is presented here. Other results from the project are shown in Figure 3.23, Section 4.5.8, and Chapters 5–7 (see also Sigurdardottir and Glisic 2015, Reilly 2019).

Figure 4.6 (Glisic 2011) shows position of three sensors with different gauge lengths embedded in south-east "leg" of the bridge, close to abutment, where cross-section of the bridge changes the geometry. All three sensors are parallel to the centroid line of the beam and placed at the same distance from the centroid in the zone with a constant cross-section. They have the same

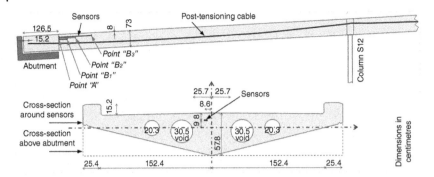

Figure 4.6 Position of the sensors in the structure, elevation view from the north side. Source: Upgraded from Glisic (2011).

starting point A, but different end points B_1, B_2, and B_3, and different mid-points C_1, C_2, and C_3 (see Figure 4.6). Respective gauge lengths of sensors AB_1, AB_2, and AB_3 are $L_{s,AB_1} = 0.3\,\mathrm{m}$, $L_{s,AB_2} = 0.6\,\mathrm{m}$, and $L_{s,AB_3} = 1.2\,\mathrm{m}$. Limits of error of sensors were estimated to be $\pm 4\,\mu\varepsilon$.

Note that point A is located in the cross-section close to abutment, where the geometry of the cross-section changes from quasi-rectangular to quasi-triangular (see Figure 4.6); consequently, based on Saint-Venant's principle, point A is situated in the zone where strain field is perturbed by the change in cross-sectional shape, the proximity of the support reaction from the abutment (concentrated force perpendicular to the deck of the bridge), and proximity of the zone where post-tensioning force was applied (concentrated axial force). However, the influence of the latter can be neglected, based on the theory of elasticity (Timoshenko and Goodier 1970), as the sensors are situated very close to the centroid of the cross-section. The influences of the abrupt change in cross-section and the transverse (shear) force could not be neglected. The length of the perturbed zone due to these influences was estimated to be not longer than approximately half of the depth of the cross-section, i.e., 0.289 m (Glisic 2011), and thus, end points "B" of all three sensors were assumed to be out of the perturbed zone.

The post-tensioning force was applied to the south-east leg in three steps. The first step introduced approximately 25% of the full post-tensioning force, the second step introduced 55%, and the third step introduced 100% of the full force. Strain measurements were repeated three times for the first two steps and four times for the third step. These measurements are presented in Figure 4.7 with discrete markers. Theoretical strain distributions along the gauge length of sensors are determined for each step based on simplified elastostatics theory (Timoshenko and Goodier 1970) and shown in Figure 4.7 as continuous or dashed lines.

Direct determination of error in strain measurement inherent to sensors' gauge lengths was not possible as there were no short-gauge sensors installed at the center points C of long-gauge sensors for comparison purposes. In addition, accuracy in evaluating the error depends on the accuracy of theoretically calculated strain distribution based on linear elastostatics theory, which is not known. Furthermore, accuracy in evaluating the error also depends on the accuracy with which the post-tensioning device introduced post-tensioning force (e.g., small force fluctuations were present at each step of post-tensioning), the friction and sliding of the concrete deck within the formwork, as well as tendons within the post-tensioning ducts, and the creep in concrete and relaxation in tendons. Nevertheless, while a direct and accurate evaluation of the error inherent to gauge length was not possible, the analysis presented in following text, based on consistency between theoretically predicted strain distribution and the measurements, serves the purpose of illustrating the influence of gauge length on strain measurement.

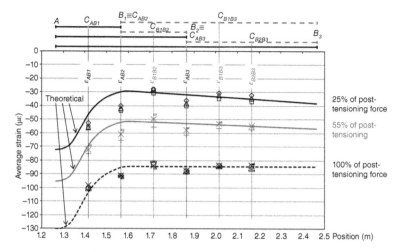

Figure 4.7 Results of measurements performed by sensors with different gauge length and theoretical strain distributions along their gauge lengths. Source: Modified from Glisic (2011).

Figures 4.6 and 4.7 show locations of sensors, their same starting point A, and their different ending points B and center points C. Besides the measurement of the three sensors, three additional strain values could be determined indirectly between the pairs of points B_1 and B_2, B_2 and B_3, and B_1 and B_3, using the following equations (see Figure 4.7):

$$\varepsilon_{C_{12}, \text{“sensor”}} = \varepsilon_{B_1 B_2} = \frac{\Delta L_{B_1 B_2}}{L_{B_1 B_2}} = \frac{\varepsilon_{AB_2} L_{AB_2} - \varepsilon_{AB_1} L_{AB_1}}{L_{AB_2} - L_{AB_1}}$$

$$\varepsilon_{C_{23}, \text{“sensor”}} = \varepsilon_{B_2 B_3} = \frac{\Delta L_{B_2 B_3}}{L_{B_2 B_3}} = \frac{\varepsilon_{AB_3} L_{AB_3} - \varepsilon_{AB_2} L_{AB_2}}{L_{AB_3} - L_{AB_2}}$$

$$\varepsilon_{C_{13}, \text{“sensor”}} = \varepsilon_{B_1 B_3} = \frac{\Delta L_{B_1 B_3}}{L_{B_1 B_3}} = \frac{\varepsilon_{AB_3} L_{AB_3} - \varepsilon_{AB_1} L_{AB_1}}{L_{AB_3} - L_{AB_1}} \qquad (4.20)$$

where quotes symbols around the word "sensor" indicate that those are not real sensors but rather "virtual" sensors, whose "measurements" are derived from real sensors.

Note that the endpoints of the three virtual sensors $B_1 B_2$, $B_1 B_3$, and $B_2 B_3$ are out of strain perturbation zone, and their "measurements" are not independent of each other: two measurements are independent, but the third can be expressed as a linear combination of the others.

Prior to post-tensioning, the concrete deck was supported by formwork, and strain in the direction of centroid line due to dead load was approximately null. After the first step of the post-tensioning, the bridge span lifted up, activating the dead load and generating reactions in the abutment and the columns. Based on Saint-Venant's principle and linear theory (see also Chapter 5), the combination of normal force and bending moments due to dead load and prestressing resulted in a parabolic strain distribution outside the perturbation zone. However, at the locations of sensors, this distribution could be approximated with a line, as the deviation from the parabola is smaller than limits of error of the monitoring system ($\pm 4\ \mu\varepsilon$). The next two steps of post-tensioning resulted in further changes in strain distribution (see details in Glisic 2011).

For each step of post-tensioning, the sensor AB_1 measured the highest compression, the sensor AB_3 measured the smallest compression, and the sensor AB_2 measured compression that was between the measurements of sensors AB_1 and AB_3. These results are consistent with

Equation 4.13. The results are also consistent with the assumption that the strain perturbation zone around point A is mainly introduced by transverse reaction at abutment: the results show that the magnitude and length of the perturbation zone are established after the first step of post-tensioning and kept constant through the other two steps; this is reasonable, as the first step of post-tensioning activates the dead load and generates reactions, which do not change as the other steps of post-tensioning are applied. Consequently, the strain distribution along sensors AB_1, AB_2, and AB_3 can be expressed with the following equation:

$$\varepsilon(z) = \varepsilon^{perturbed}(z) + \varepsilon^{non\text{-}perturbed}(z) \tag{4.21}$$

where z is coordinate in direction of sensors, $\varepsilon^{perturbed}(z)$ is non-linear and $\varepsilon^{non\text{-}perturbed}(z)$ is linear. According to Equation 4.16, the error inherent in gauge length is null for linear strain distribution along the sensor. Consequently, the error inherent in gauge length depends only on the error introduced in the zone of perturbed strain. Based on the above discussion, this error is constant for each step of post-tensioning, and based on elastostatic theory and Equation 4.16, it is estimated to be $-1\,\mu\varepsilon$, $-7\,\mu\varepsilon$, and $-3\,\mu\varepsilon$ respectively, for sensors AB_1, AB_2, and AB_3.

All the above error analysis considered concrete as homogeneous material, and the errors calculated in previous paragraph correspond to the error component $\delta\varepsilon_{z,C,sensor}^{GL,hom}$ from Equations 4.15a and 4.16. Sections 4.5.3 and 4.5.4 demonstrated that inclusions and discontinuities also influence the error inherent in gauge length. However, due to post-tensioning, there are no discontinuities (cracks) in the material, and the error subcomponent $\delta\varepsilon_{z,C,sensor}^{GL,discon}$ is null. The subcomponent $\delta\varepsilon_{z,C,sensor}^{GL,inclus}$ is not null, and based on Equation 4.18, it can be estimated for sensor AB_1 to be approximately $\pm0.5\,\mu\varepsilon$, $\pm0.6\,\mu\varepsilon$, and $\pm1\,\mu\varepsilon$, respectively for the first, second, and third step of the post-tensioning. For other sensors, this error is smaller as they have a longer gauge length. To simplify presentation without affecting the consistency of analysis, we can conservatively adopt the value of $\pm1\,\mu\varepsilon$ as the value of $\delta\varepsilon_{z,C,sensor}^{GL,inclus}$.

The differences between theoretical values of strains and the measurements shown in Figure 4.7 are equal to the sum of the measurement error of the monitoring system ($\pm4\,\mu\varepsilon$) and the sum of components of measurement error inherent to gauge length of the sensors. This results in the following ranges (resulting from error propagation formula): -6 to $+4\,\mu\varepsilon$ for sensor AB_1, -12 to $-2\,\mu\varepsilon$ for sensor AB_2, and -8 to $+2\,\mu\varepsilon$ for sensor AB_3.

All three virtual sensors B_1B_2, B_2B_3, and B_1B_3 were out of the perturbed zone, i.e., in the zone with linear strain distribution, and their measurement error inherent to gauge length is equal to $\delta\varepsilon_{z,C,sensor}^{GL,inclus} = \pm1\,\mu\varepsilon$ ($\delta\varepsilon_{z,C,sensor}^{GL,hom}$ is null due to linear strain distribution along the sensors). The range of differences between theoretical and measured strain is equal to the measurement error of the monitoring system combined with $\delta\varepsilon_{z,C,sensor}^{GL,inclus}$, i.e., -5 to $+5\,\mu\varepsilon$.

The ranges of differences (errors) for all real and virtual sensors are approximately confirmed in Figure 4.7. This consistency supports the theory behind the error analysis developed in this section and validates it in real-life application. Analysis presented in this subsection also illustrates how the gauge length influences the measurement and associated limits of error.

4.5.6 Recommendation for Choice of the Type of Gauge Length Depending on Scale of Monitoring

Equations 4.15 presented two components of measurement error inherent to gauge length of sensor: $\delta\varepsilon_{z,C,sensor}^{GL,hom}$ and $\delta\varepsilon_{z,C,sensor}^{GL,inhom}$; Equation 4.17 further showed that the latter has two subcomponents: $\delta\varepsilon_{z,C,sensor}^{GL,inclus}$ and $\delta\varepsilon_{z,C,sensor}^{GL,discon}$. Section 4.5.2 demonstrated that shorter gauge length provides

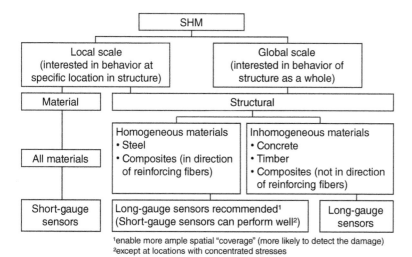

Figure 4.8 General recommendations for selection of type of gauge length of discrete sensors depending on the scale of monitoring. Source: Upgraded from Glisic (2011).

more accurate strain measurements in homogeneous material, for monitoring performed at both local (material and structural) and global scales, assuming that there are no local abrupt changes in strain distributions; however, if abrupt perturbations of strain field are present (e.g., due to presence of concentrated force), then these local abrupt perturbations might not be of interest for the purposes of monitoring at local or global structural scale, yet they might be of interest at local material scale. In the case of the former, long-gauge sensors would be a better choice, but in the case of latter, short-gauge should be chosen. Long-gauge length provides better spatial coverage of structure and consequently long-gauge sensors are more likely to perform direct detection of unusual structural behaviors (e.g., damage), regardless if the material is homogeneous or inhomogeneous. Hence the need for a trade-off between accuracy of strain measurement and sensitivity to damage detection (Equation 4.16 can be used to determine limits of error for selected gauge lengths).

Sections 4.5.3 and 4.5.4 have shown that for monitoring in inhomogeneous materials, shorter gauge length minimizes error due to variations in strain distribution, but longer gauge length minimizes error due to inhomogeneity (Equations 4.18 and 4.19). Hence, these two criteria are in contradiction, and optimization should be made using Equations 4.15–4.19. To enhance capability for direct detection, again, long-gauge sensors are preferred, and spatial coverage of structures with sensors should be taken into account as an additional trade-off for accuracy in strain measurement. Figure 4.8 summarizes the above discussion and presents general recommendations for selection of type of gauge length of discrete strain sensors depending on the scale of monitoring.

While Figure 4.8 shows monitoring at local and global scales as two separate approaches, note that global scale monitoring actually encompasses local scale monitoring due to the fact that analysis of measurements recorded by a single sensor always provides with local behavior at the point where that sensor is installed. The main difference between local- and global-scale monitoring is in data analysis approaches: the former involves individual analysis of each sensor, while the latter involves simultaneous analysis of multiple sensors. The two approaches are presented in more detail in Chapters 5 and 6, respectively.

4.5.7 Recommendation for Choice of Gauge Length for Monitoring of Beam-Like Structures at Local or Global Structural Scale

Sections 4.5.2–4.5.4 analyze in detail various aspects of error inherent to gauge length of strain sensors. Few simplified approaches for quantification of error were developed in these subsections for beam structures. More advanced analysis can be found in Glisic (2011), but recounting it here would be out of the scope of this introductory book. Nevertheless, for the purposes of completeness of presentation, the recommendations for selection of gauge length for monitoring of beam structures at local structural scale or global scale are presented below (see details in Glisic 2011). In these recommendations, L_s denotes the gauge length of the sensor, h_{CS} denotes the depth of the cross-section in which the sensor is installed, and L_{Beam} denotes the length (span) of the beam structure.

Recommendations for both homogeneous and inhomogeneous materials:

- If the sensor is installed in a cross-section close to the support of the beam (pinned or fixed at its extremity to supports or other beams), then the strain field around the sensor might be perturbed based on Saint-Venant's principle, and it is recommended to use a gauge length of sensor that is longer than the depth of the cross-section, i.e., $L_s \geq h_{CS}$; if the beam is simply supported at the observed cross-section, then $L_s \geq h_{CS}/2$.
- If the sensor is installed in a cross-section in the span with a bi-linear (broken-line) strain distribution, e.g., due to concentrated force, and the sensor is not close to the force application point, then: $L_s \leq L_{Beam}/10$.
- If the sensor is installed in a cross-section in the span with a bi-linear (broken-line) strain distribution, e.g., due to concentrated force, and the sensor is close to the force application point, then: $1.5\,h_{CS} \geq L_s \geq h_{CS}$, and this condition should be, by preference, combined with the recommendation presented in the previous point ($L_s \leq L_{Beam}/10$); if two sensors are used in cross-sections to the left and right of the force application point, then $L_s \geq h_{CS}$.
- If the sensor is installed in a cross-section in the span with a parabolic strain distribution, e.g., due to uniformly distributed force, then: $L_s \leq L_{Beam}/6$.

Recommendations for inhomogeneous materials:

- If the sensor is installed in a cross-section of material with inclusions, e.g., aggregate in concrete, then $L_s \geq 10 D_{max}$, where D_{max} is maximal nominal diameter of the inclusions.
- If the sensor is installed in a cross-section of material with discontinuities, e.g., common cracks in reinforced concrete, then $L_s \geq 10 L_c$, where L_c is distance between discontinuities (as defined in Figure 4.5).

Note that some of the recommendations presented above could be mutually contradictory, depending on properties of monitoring structure. In these cases, compromise solutions should be found that optimize the outcomes of the monitoring based on the project priorities. The above recommendations should be considered as guidelines, rather than rules, that in turn, should be implemented jointly with project specifications and the aims of monitoring, and modified, if needed, based on these specifications and aims.

4.5.8 Spatial Resolution of Distributed Sensors as an Inherent Source of Error

Note that Section 4.5 has so far considered only discrete sensors and their gauge lengths. Considerations regarding distributed strain sensors are, while similar to those regarding discrete sensors, somewhat different depending on how the sensor is coupled with structure. The notions of spatial resolution and sampling interval of distributed sensors were introduced in Section 3.6.4. The notion of spatial resolution is further developed in this subsection in the context of the inherent error it introduces in a strain measurement.

Distributed strain sensor is sensitive to strain continuously (at every point) along its length L_D. Due to the infinity of points along L_D, the strain measurement cannot be delivered at every point in a continuous manner; it is rather delivered at discrete points that are uniformly spaced along the length of the sensor. The distance L_{SI} between the points at which the measurements are delivered is called sampling interval (see Section 3.6.3). The value of strain measurement delivered at each point defined by sampling interval is not the exact value of strain at that point but rather an average value integrated over a predefined length L_{SR}, called spatial resolution (Glisic and Inaudi 2007). From that point of view, a spatial resolution of distributed sensor is equivalent to a gauge length of discrete sensors, as the measurement averages the strain along its length. Let us denote with z_i ($i = 1, 2, \ldots n$) the local coordinates of discrete points along the sensor length at which the strain measurement is delivered (note that for every $i > 0$, $z_i - z_{i-1} = L_{SI}$), then the measurement value at each point z_i is an average value of strain integrated over the length of spatial resolution L_{SR} around point z_i. Both sampling interval L_{SI} and spatial resolution L_{SR} are typically set by the user as input parameters of the monitoring system, depending on the type of the monitoring system and the aims of the project. In order to include the strain over the entire length of the sensor L_D in the measurements, the sampling interval should not be longer than a half of spatial resolution. Both sensor parameters are presented in Figure 4.9, along with a simplified illustration of the principle of distributed sensor measurement (modified from Glisic and Inaudi 2012).

Figure 4.9a shows continuous strain distribution around arbitrary point z_i, such that no abrupt strain changes are present along the part of the sensor corresponding to the full length of spatial resolution (segment $[z_i - L_{SR}/2, z_i + L_{SR}/2]$). This strain distribution is expected in homogeneous materials at points where no strain perturbation is present. The strain measurement delivered at point z_i represents the average value over the segment $[z_i - L_{SR}/2, z_i + L_{SR}/2]$ and, similar to Figures 4.2 and 4.3, the gray area in the figure indicates equivalent constant average strain along the segment. Similar to the case of discrete sensors in homogeneous materials, the difference $\delta\varepsilon_{z,z_i,sensor}^{SR,hom}$

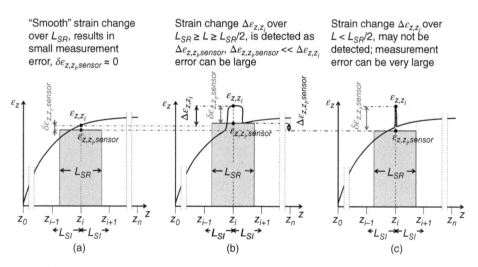

Figure 4.9 Simplified illustration of the principle of distributed sensor measurement. Source: Modified from Glisic and Inaudi (2012).

between the measured (averaged) strain $\varepsilon_{z,z_i,sensor}^{SR,hom}$ and real strain $\varepsilon_{z,z_i}^{SR,hom}$ at point z_i represents the inherent error due to spatial resolution. It is calculated using Equation 4.22, which is obtained from Equation 4.16 by appropriate substitutions.

$$\delta\varepsilon_{z,z_i,sensor}^{SR,hom} = \varepsilon_{z,z_i,sensor}^{SR,hom} - \varepsilon_{z,z_i}^{SR,hom} = \frac{1}{L_{SR}} \int_{z_i-\frac{L_{SR}}{2}}^{z_i+\frac{L_{SR}}{2}} \varepsilon_z(z)\mathrm{d}z - \varepsilon_z(z_i) \tag{4.22}$$

where superscript *SR* indicates that the error is due to spatial resolution of the sensor.

Similar to discussion about gauge length in discrete sensors, shorter spatial resolution will result in more accurate measurements in homogeneous materials.

Figure 4.9b shows a strain distribution over segment $[z_i - L_{SR}/2, z_i + L_{SR}/2]$, which features abrupt strain changes over a length shorter than the spatial resolution but not shorter than its half. For homogeneous materials, this strain distribution corresponds to locations where the structure's geometry or cross-sectional properties change or where a concentrated load is applied, similar to Figure 4.3a,b. For inhomogeneous materials, these changes can also be caused by inclusions. Such abrupt changes in strain can be detected and localized by distributed sensor measurements, but accuracy in strain measurement at these points is typically low. Let us assume that the strain distribution under normal conditions is shown in Figure 4.9a and an unusual behavior perturbs the strain distribution and modifies it into the one shown in Figure 4.9b. This perturbation will be detected but not accurately measured, i.e., $0 < |\Delta\varepsilon z, z_{i,sensor}| < < |\Delta\varepsilon z, z_i|$. Note that this might not be an issue with Rayleigh-based systems, as their spatial resolution is very small, i.e., comparable with short-gauge sensors (see Table 3.3 in Section 3.6.5), but can be an issue with Brillouin-based systems, whose spatial resolution is larger and comparable with long-gauge sensors.

To minimize the error inherent to spatial resolution in inhomogeneous materials with inclusions of maximal diameter D_{max}, the same considerations as in the case of discrete sensors can be used, i.e., the spatial resolution should be "long enough," and error can be evaluated using Equation 4.23 (obtained from Equation 4.18 by appropriate substitutions).

$$\delta^*\varepsilon_{z,z_i,sensor}^{SR,inclus} = a\left(\frac{L_{SR}}{D_{max}}\right)^{-b} \pm \delta_{fit}^* \tag{4.23}$$

where superscript *SR* indicates that the error is due to spatial resolution of the sensor.

While the error analysis of the two cases shown in Figure 4.9a,b relates well with the error analysis of discrete sensors, the same is not applicable for the cases where an abrupt strain change happens over a length smaller than half of the spatial resolution of the sensor. A simplified example of a strain distribution containing such an abrupt change is shown in Figure 4.9c. This abrupt change is typically generated by a discontinuity in an inhomogeneous material (e.g., common cracking in reinforced concrete) or damage (e.g., unusually large crack opening in concrete). Depending on the principle of decoding the encoding parameter in the reading unit, an abrupt strain change, as the one shown in Figure 4.9c, may be undetected, as it might be "invisible" for the decoding algorithm in the common mode of operation (Glisic and Inaudi 2012). Moreover, very high and localized strain changes can compromise the integrity of sensor and result in rupture of its sensing element (i.e., sensing optical fiber).

These two issues are addressed at the software and hardware levels. First, the monitoring system has to be provided with a special algorithm for the detection of abrupt strain changes occurring over lengths shorter than half of the spatial resolution (Ravet et al. 2009). Second, the design of distributed sensors and installation procedures should enable controlled strain redistribution over a length compatible with special algorithm requirements and sensor mechanical properties (Glisic and Inaudi 2012). An example of the successful real-life application of these two solutions is given in Section 1.3. Note that the first above solution might not be required for monitoring systems where spatial resolution is very small (e.g., Rayleigh-based systems) or where the encoding algorithm in common mode already has the option for detection of abrupt strain changes. However, the second solution is independent of the encoding algorithms and dependent only on the sensor design and installation procedures. These two solutions are further addressed in Section 4.6 and Chapter 7.

At this point, it is important to pay attention to some important consequences of the above discussed errors. Distributed sensor, in general, can be coupled with the structure continuously along its length L_D (bonded to the surface of the structure or embedded in concrete) or clamped on the surface of structure at discrete points $z_{clamp,i}$, and pretensioned between these points so it can measure both tension and compression.

In the case of continuous coupling, the errors inherent to spatial resolution result in unavoidable errors in strain measurements close to the endpoints of sensor (coordinates z_0 and $z_0 + L_D$), typically within approximately one-half of the spatial resolution from the endpoint into the sensor (segments $[z_0, z_0 + L_{SR}/2]$ and $[z_0 + L_D - L_{SR}/2, z_0 + L_D]$). The reason for this is that in those segments of sensor, the spatial resolution "covers" not only the part of the sensor coupled with structure but also a part of optical fiber that serves only as a connecting cable to sensor; while the strain in the former is equal to stain in the structure, the strain in the latter is null, and, as the sensor provides an average integral value of these two strains, the measurement value might be significantly different from the real strain value (up to 50% different). This "endpoint" error, which occurs at both extremities of the sensor, must be taken into account when performing the analysis of strain measurements recorded in the first and last sensor segment with lengths equal to one half of spatial resolution.

In the case of clamping installation of the sensor, the sensor is pretensioned between the clamping points, and the strain at every point of the sensor within the prestressed segment $[z_{clamp,i}, z_{clamp,i+1}]$ is theoretically constant (in reality, it might slightly vary due to variability in cross-section of the sensor). Consequently, each clamped segment of distributed sensor practically behaves as a long-gauge discrete sensor installed between the clamping points. Note that in general, the two neighboring segments (e.g., $[z_{clamp,i-1}, z_{clamp,i}]$ and $[z_{clamp,i}, z_{clamp,i+1}]$) do not have the same strain: they are usually pretensioned during installation with different values, and over time, the strain in both segments will further change as it follows the strain change in the structure. Thus, similar to "endpoint error" of a continuously bonded sensor, a "clamping error" is introduced in strain measurements close to the clamps of the sensor, typically within approximately one-half of the spatial resolution of the clamping points. For example, for clamped segment $[z_{clamp,i}, z_{clamp,i+1}]$, the error is introduced in segments $[z_{clamp,i}, z_{clamp,i} + L_{SR}/2]$ and $[z_{clamp,i+1} - L_{SR}/2, z_{clamp,i+1}]$. As a consequence, if the sensor is installed by clamping, the minimum distance between the clamps should be at least equal to two spatial resolutions of the sensor. This will allow discarding the inaccurate measurements from the two segments affected by "clamping error" and provide with at least one segment equal to spatial resolution between them where the measurements will not be affected by the error.

Another source of errors inherent to distributed sensor measurements could be related to errors in the actual length of spatial resolution and errors in the actual length of the sampling interval (and consequently the local coordinates z_i at which the strain measurement is delivered). However, both parameters depend on the optical properties of the optical components of the monitoring system (e.g., stability of the speed of the light in the optical fibers and stability of the light source, i.e., the laser), which are typically very high, resulting in negligible errors.

Example 4.5 *Error inherent to spatial resolution of distributed strain sensor*
Streicker Bridge project was introduced in Section 4.5.5, where only one aspect of SHM was presented. In this subsection, another aspect is introduced for the purposes of illustration: the error in strain measurement inherent to spatial resolution. The span P10–P11 of the south-east leg of

the bridge was instrumented with both discrete long-gauge FBG sensors (gauge length 60 cm, limits of errors ±4 με) and distributed Brillouin-based sensors (spatial resolution 1 m, limits of errors ±40 με). Both types of sensors were embedded in concrete during construction, and discrete sensors were placed at the same locations as distributed sensors, at the top and bottom of the cross-section, so that their measurements could be directly compared. Figure 4.10 shows the schematic positions of the sensors.

Figure 4.11 shows strain measurements taken by discrete and distributed sensors at location "P11up" (see Figure 4.10). Two events are of interest in this figure: (i) unusual event, i.e., thermally generated early age crack, that occurred approximately 3 days after the pouring of concrete (see also Figure 3.23), and (ii) post-tensioning of the deck of the bridge, which occurred in two stages, approximately 10 days after the pouring. Both events are indicated in Figure 4.11.

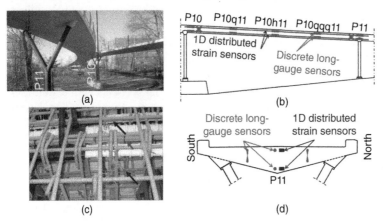

Figure 4.10 (a) View to Streicker Bridge and with part of Southeast leg P10–P11; (b) positions of discrete and distributed sensors in span P10–P11 of the Streicker Bridge, elevation view from the south side; (c) photograph taken during the installation showing discrete sensors (pointed by short arrows) and distributed sensors (pointed by long arrows); and (d) positions of sensors in the cross-section. Source: Reproduced from modified from Glisic et al. 2011 / IOP Publishing.

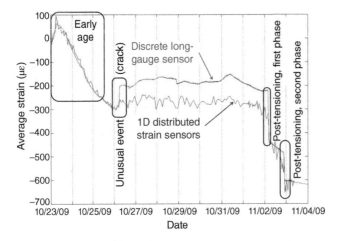

Figure 4.11 Comparison between strain measurements obtained with discrete (FBG) and distributed (BOTDA) sensors at the cross-section above the column P11 of Streicker Bridge. Source: Modified from Glisic et al. (2011).

Figure 4.11 shows good agreement in measurements of the two sensor types until the crack occurred. The crack resulted in a positive relative displacement between the endpoints of the discrete sensor, which in turn resulted in a positive increase in average strain measurement, as per Equation 4.13. However, the same crack resulted in a discontinuity that affected very short length of distributed sensor; due to the sliding of optical fiber within the sensor packaging, this scenario is similar to the one shown in Figure 4.9b (rather than the one shown in Figure 4.9c; see more details in Section 4.6) and thus, while the crack was detected, the value of strain in sensor at the location of crack is significantly lower than that of the discrete sensor. This result is in accordance with the above discussion regarding errors inherent to the spatial resolution of distributed sensors. Finally, post-tensioning closed the crack, and, as the discontinuity was removed, both sensors started measuring approximately the same value, as shown in Figure 4.11. This result is also in accordance with the above discussion regarding errors inherent to the spatial resolution of distributed sensors (only the error associated with scenario shown in Figure 4.9a is present). More details regarding cracking of the bridge deck and distributed measurements can be found in Chapters 6 and 7, and available literature (e.g., Glisic et al. 2011, Hubbell and Glisic 2013, Sigurdardottir and Glisic 2015).

Distributed sensors enable extensive spatial coverage of structures (e.g., see Figures 3.19, 3.26, and 4.10), and thus, they are particularly suited for monitoring at the integrity scale. However, given that they provide strain measurements, they can also be used for monitoring at local and global scales. For these cases, Rayleigh-based sensors can provide similar results as short-gauge sensors, while Brillouin-based sensors can provide similar results as long-gauge sensors (see Figure 4.8). Thus, integrity monitoring encompasses local- and/or global-scale monitoring. More details on integrity monitoring are given in Chapter 7.

4.6 Errors in Strain Measurement Inherent to Installation of Sensor and Sensor Packaging

4.6.1 Strain Transfer from Structure to Sensing Element of the Sensor

The accuracy of strain measurement strongly depends on the quality of the strain transfer from the structure to the sensor's sensing element. The importance of strain transfer was recognized among researchers and practitioners, and numerous analytical and experimental studies addressed this topic in the past (e.g., Ansari and Libo 1998, Calderon and Glisic 2012, Her and Huang 2016, Gerber et al. 2018, Bassil et al. 2020). The following factors influence the quality of strain transfer:

- The way the sensor is coupled with structure; this is, in general, realized by (i) gluing on the structure's surface using specialized adhesives, (ii) mechanical mounting on the structure's surface using brackets, clamps, bolts, welding, or similar, and (iii) embedding in the structure's material (e.g., concrete).
- Geometrical and mechanical properties of sensor packaging.
- Strain transfer between the sensor packaging and the sensing element.

Figure 4.12 illustrates the complexity and challenges of strain transfer using, as an example, either a discrete sensor with gauge-length L_S or a distributed sensor with spatial resolution L_{SR}, bonded to the surface of the structure. To simplify the presentation, let us assume that the structure is uniformly deformed along the z-axis. Due to uniformity of deformation, the strain distribution in material along z-axis $\varepsilon_z(z)$ is constant and denoted with $\varepsilon_{z,material}$ ($\varepsilon_z(z) = constant = \varepsilon_{z,material}$). Since the strain in material is constant, the average strain between any two points on z-axis is equal to

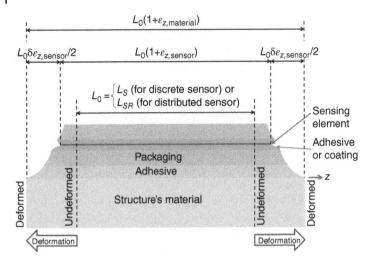

Figure 4.12 Illustrative representation of strain transfer from structure's material to sensor's sensing element for bonded sensors (not to scale). Source: Modified from Glisic (2019).

the same value $\varepsilon_{z,material}$. This is the value that the strain sensor should measure, but due to "losses" between the material and the sensing element, the sensor will measure a smaller value of strain, $\varepsilon_{z,sensor}$, and an error $\delta\varepsilon_{z,sensor}$ will be introduced.

In the example shown in the figure, the strain is transferred from the structure's surface to the sensor packaging via bonding agent (adhesive). Note that in the case of sensors embedded in material (e.g., concrete), the strain would be transferred by friction, which can be enhanced by mechanical interlocking (e.g., by "roughing" the surface of the sensor). In the case of bonding (gluing), the adhesive selected for installation must be compatible with both the construction material of the monitored structure and the packaging of the sensor. In the case of embedding, the packaging of sensor should have a high coefficient of friction with respect to the structure's material, or the geometry of the sensor packaging should enable good interlocking with the structure's material (e.g., via anchors). The strain transferred from the structure's material to the sensor "travels" through the sensor's packaging until it is transmitted to the sensing element.

Three points of strain transfer can be noticed in Figure 4.12:

1. Strain transferred from the structure to the sensor packaging suffers losses caused by:
   a. Deformation of the adhesive, for sensors bonded to the surface of the structure.
   b. Sliding of sensor within the structure's material, for embedded sensors.
2. Strain transferred from the glued surface of the packaging to the interface with the sensing element suffers losses due to deformation of the packaging.
3. Strain transferred from the sensor packaging to sensing element suffers losses due to the sliding of the sensing element at the interface.

While the points 1 and 2 above could be influenced by the user of SHM system (e.g., by selection of a sensor with specific properties of packaging or by selection of the adhesive), the point 3 cannot, as it depends only on the quality of the manufacturing of sensor. We will see in the next subsections that having an excellent strain transfer from the structure to the interface between packaging and sensing element is not always desirable; however, having an excellent (practically perfect) strain transfer at the interface is very important. Thus, the user, when selecting the sensors, must choose the manufacturer that guarantees long-term quasi-perfect strain transfer at the interface between the packaging and sensing element, even if this results in more expensive sensors. Malfunction of sensor at the interface is typically manifested as drift (sliding of the sensing element within packaging) or failure (rupture of sensing element). The former might require a long time to detect and can introduce important errors in data analysis. In both cases, the SHM system will lose performance and might fail to achieve its objectives. In addition, replacement or repair of the sensor might result in elevated costs.

Example 4.6 *Influence of quality of strain sensor components to strain transfer: polyimide- vs. acrylate-coated optical fibers*

In fiber-optic sensors, the strain sensing element is often the core or part of the core of optical fiber (see Section 3.6). Thus, at the interface, strain is transferred from packaging to the coating of optical fiber, then from the coating to the cladding, and finally from the cladding to core of the optical fiber (see Figure 3.12). The two main types of coating materials found on the optical fiber market are acrylate and polyimide. Optical fibers coated with acrylate are an order of magnitude cheaper than fibers coated with polyimide, and they are very easy to handle and modify into sensors as the coating is simply removed by mechanical stripping. Optical fibers coated with polyimide, besides the fact that they are order of magnitude more expensive, also require more expensive labor, as the safe removal of coating might require the use of strong chemicals. Consequently, sensors made of acrylate coated optical fibers are significantly less expensive than those made of polyimide coated fiber; however, the former suffer from lower quality strain transfer between coating and cladding and may feature drift in the long-term as well as inaccurate strain measurements for higher levels of strain (Glisic and Inaudi 2007). Thus, while having a lower cost, these sensors might not be a good choice, as the reliability and accuracy of strain measurements would be difficult to assess.

As stated earlier, the points 1 and 2 above are influenced by the user, and these points are discussed in the next subsections, in the context of the local, global, and integrity scales of the SHM.

4.6.2 Error Due to Installation of Sensor

A proper installation of strain sensors is crucial for high-quality strain transfer, which in turn guarantees accuracy in strain measurements. The installation must be durable over long-terms to guarantee longevity, quality of strain transfer, and reliability of strain measurements. Installation procedures are usually developed, tested, and their quality guaranteed by the manufacturer of sensors. Therefore, it is very important to strictly follow the installation procedures as provided by the manufacturer. If they have to be changed or adapted due to some project specific requirement, it is necessary to test them and confirm their performance prior to field application. Improper installation of the sensor can significantly deteriorate the quality and reliability of a strain measurement. Quantification of error due to installation depends on many parameters related to sensor, adhesive, and structural material (e.g., see Figure 4.12), and that is the reason why it is not presented here (see more details in relevant literature, e.g., Ansari and Libo 1998, Calderon and Glisic 2012, Her and Huang 2016; Gerber et al. 2018, Bassil et al. 2020). However, qualitative guidelines on how to minimize the adverse effects of sensor installation on the quality of measurement are presented below.

Figure 4.12 shows that if the sensor is installed on the surface of a structure using an adhesive, the properties of the adhesive strongly influence the quality of strain transfer. In general, to minimize the losses in strain transfer from the structure to the sensor packaging, a thin layer of adhesive is preferable to a thick layer, a uniform thickness of a layer is preferable to a non-uniform thickness, and a stiff adhesive is preferable to a soft adhesive. Needless to say, a good preparation (e.g., cleaning and smoothening) of structure's and sensor's surfaces that are in contact with the glue, as well as their material compatibility with the adhesive, are crucial to quality and longevity of the installation. A perfectly bonded sensor will ensure high-quality strain transfer and very accurate measurements, which is of high importance for SHM performed at local or global scale; however, in the event of cracking of structure, perfect strain transfer may result in damage to sensor, which is to be avoided, especially if the SHM is performed at integrity scale. This challenge is addressed in Section 4.6.4. Examples of first- and second-generation sensors bonded to the structure are shown in Figure 4.13. Examples of the third-generation sensors bonded to structure are given in Figures 3.24a (right image) and 3.24b (closeup).

Note that, due to the thickness of the adhesive and the sensor packaging, the sensing element of the sensor is typically positioned at a certain distance from the structure. As a consequence, even in the case of a perfect strain transfer from the structure to the sensing element, the strain in the former might be different from the strain in the latter, which introduces an error in sensor measurement. This is schematically shown in Figure 4.14. The potential existence of this error emphasizes the need to use thin layers of adhesive.

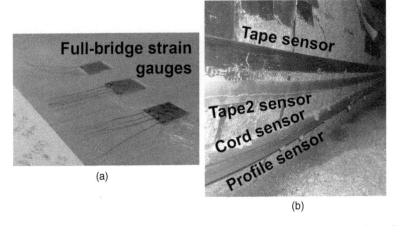

Figure 4.13 Examples of sensors installed on the surface of a structure using adhesives. (a) Full-bridge strain gauges. Source: Reproduced from Gerber et al. 2018 / MDPI / CC-BY 3.0. (b) Four types of distributed fiber-optic strain sensors. Source: Reproduced from Glisic 2009 / PRINCETON UNIVERSITY.

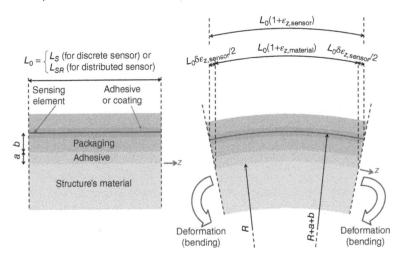

Figure 4.14 Schematic representation of the difference in strain between the sensing element and the structure introduced by the thicknesses of the adhesive and the sensor packaging; a similar error is introduced if the sensor is installed using mounting devices (e.g., brackets).

The error due to the distance of the sensing element from the structure's surface depends on the geometrical properties of the structure, the distance of the sensing element from the structure, and the strain distribution in the structure at the location of the sensor. It can frequently be evaluated using numerical modeling or structural analysis. While for most applications in civil engineering, this error can be neglected, it is always recommended to verify it by means of structural analysis or numerical modeling.

For the sensors embedded in the structure's material (e.g., concrete or soil), to improve strain transfer, a good interaction between the host material and sensor packaging should be guaranteed. This would require good adhesion of the material to the sensor packaging and a high friction coefficient between the sensor packaging and the material, which could be further enhanced by mechanical interlocking between the material and the sensor. For distributed sensors, the mechanical interlocking can be provided by creating intentionally rough texture in the sensor packaging. For discrete sensors, mechanical interlocking is provided by anchors (see Sections 3.5 and 3.6). Section 4.6.3 will show that for discrete sensors, anchors are a necessity, and their size and shape strongly influence the strain transfer. Examples of sensors embedded in the material of the monitored structure are shown in Figure 4.15.

If the sensor is installed on the surface of the structure using some accessory mounting devices such as brackets, clamps and screws, the devices have to be perfectly attached to the structure, and the same applies for the sensor packaging and device. The attachment between the device and the structure can be enabled by gluing, bolting, or welding (in the case of steel structures), and

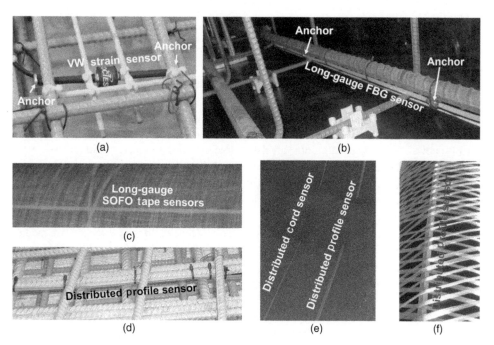

Figure 4.15 Examples of sensors installed by embedding in the structure's material; discrete sensors placed in rebar cages before pouring concrete. (a) Vibrating wire strain sensor. Source: Courtesy of Roctest, Inc. (b) Long-gauge FBG strain sensor. (c) Discrete long-gauge SOFO strain sensor embedded in composite material. Source: Courtesy of SMARTEC. (d) Distributed "Profile" strain sensor placed in rebar cage before pouring. (e) Distributed "Profile" and "Cord" strain sensors placed in soil before embedding (burying). (f) Distributed "Profile" strain sensor embedded in composite material with no curing. Source: Courtesy of SMARTEC.

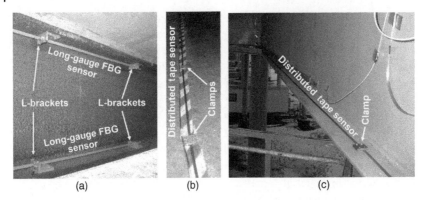

(a)   (b)   (c)

Figure 4.16 Examples of sensors installed on the surface of structure using mounting devices. (a) Long-gauge SOFO strain sensors installed using L-brackets. (b) Distributed "Tape" sensor installed on concrete structure using clamps. Source: Courtesy of SMARTEC. (c) Distributed "Tape" sensor installed on steel structure using clamps. Source: Courtesy of SMARTEC.

between the sensor and the device by mechanical interlocking, gluing, bolting, etc. Examples of sensors installed on the structure using mounting devices are shown in Figure 4.16.

Note that mounting devices usually keep the sensor's anchors at a certain distance from the structure. Thus, similar to installation by adhesive, the sensor's sensing element is positioned at a certain distance from the structure, which in turn may introduce an error in strain measurement similar to that presented in Figure 4.14.

In addition, depending on the geometrical and material properties of mounting devices and the way they are attached to the structure, additional errors might be introduced. For example, if bolted L-brackets are used to install long-gauge sensors on the surface, it is preferred to install them with the same orientation (see Figure 4.16a) rather than with the opposite orientation, as the latter can introduce multiplicative systematic errors in strain measurement and an additional error due to temperature variations. This example is schematically shown in Figure 4.17.

Figure 4.17a shows correct orientation of L-brackets. Note that in this case, the distance between the bolts d_{bolt} is equal to the gauge length of the sensor L_s. Thus, any change in distance between the bolts, Δd_{bolt}, due to deformation of the structure will induce the same change in length of strain sensor, i.e., $\Delta L_s = \Delta d_{bolt}$; consequently, the average strain in sensor ε_{sensor} would be the same as the average strain in the structure $\varepsilon_{structure}$, i.e., $\varepsilon_{sensor} = \Delta L_s/L_s = \Delta d_{bolt}/d_{bolt} = \varepsilon_{structure}$, which implies that there will be no error introduced in strain measurement. Moreover, if the L-brackets are made of material with a thermal expansion coefficient different from that of the monitored material, then the L-brackets will experience a different thermal strain than the monitored material; however, having the same orientation, this differential strain will not affect the length of the sensor, i.e., $L_{s,\Delta T} = L_s$, where $L_{s,\Delta T}$ is the length of the sensor after the thermal expansion of the L-brackets.

Figure 4.17b,c show opposite arrangements of L-brackets for the sensor of the same gauge length as the one presented in Figure 4.17a. These arrangements would be wrong, as they would introduce multiplicative error due to difference between d_{bolt} and L_s (multiplication by factor d_{bolt}/L_s), i.e., $\varepsilon_{sensor} = \varepsilon_{structure} \cdot d_{bolt}/L_s$, $\delta \varepsilon_{sensor} = \varepsilon_{structure}(d_{bolt}/L_s - 1)$, and $\delta^* \varepsilon_{sensor} = (d_{bolt}/L_s - 1)$. In addition, they would introduce a temperature-dependent error $\delta^* \varepsilon_{sensor,\Delta T} = (L_{s,\Delta T} - L_s)/L_s$.

Another error, in the case of surface mounting, can be introduced during the interpretation and analysis of data. Given that, for the surface mounting, the sensing element is distant from the structure, the two will not follow the temperature changes in the environment at the same pace; hence,

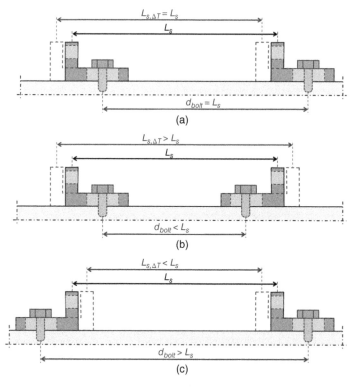

Figure 4.17 Example of error in strain measurement introduced by wrong orientation of bolted L-brackets: (a) Correct installation of L-brackets that have the same orientation, and (b) and (c) incorrect installations of L-brackets that introduce permanent multiplicative error in strain measurement, and an additional error due to temperature variations; dashed contours of L-brackets show their deformed shape due to temperature variations. Source: Modified from Glisic (2009).

at the observed time, the structure and the sensor could have different temperatures. Thus, for accurate strain measurement and data analysis, it is necessary to use a temperature sensor that is in contact with the strain sensor to perform thermal compensation for the latter (see Section 4.4) and another temperature sensor that is in contact with the surface of the structure to determine thermal strain in the structure (see Chapter 5). While the temperatures at the two locations are not excessively different (typically within the range of $\pm2°$C), if they are not individually measured and properly used in data analysis, their difference can introduce important errors in strain measurement and interpretation.

4.6.3 Error Due to Geometrical and Mechanical Properties of Sensor Packaging

Figure 4.12 points to the potential difference between strain in sensor packaging (as an intermediary of the strain transfer) and the sensing element, which results in an error in strain measurement. While Figure 4.12 shows how this error is generated for sensors installed by adhesives, the same issue is present in sensors installed using mounting devices or in those embedded in host material, as its origin is the interaction between the packaging and the sensing element. The figure shows that the thickness and stiffness of the packaging play important roles. In general, a thin and stiff packaging (e.g., "Tape" sensor in Figure 3.21a, upper image) is preferable compared to thick and soft packaging (e.g., "Profile" sensor in Figure 3.21b, upper image). However, the stiffness of the

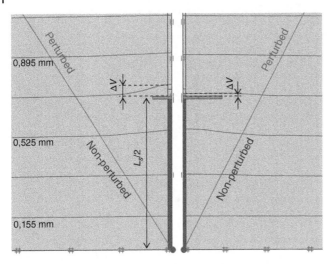

Figure 4.18 Examples of perturbation of displacement field in host material introduced by an embedded sensor with: (a) Larger anchor and (b) smaller anchor. Source: Courtesy of Prof. Pedro Calderon, Polytechnic University of Valencia.

packaging should not be excessively high because this may perturb the strain field in the material of the monitored structure, which in turn will introduce errors. To illustrate this statement, Figure 4.18 shows two examples of perturbation of the vertical displacement field (and associated strain field) due to a long-gauge sensor embedded in a host material subjected to uniaxial (vertical) compression. The main difference between the two examples is in the size of the anchors, which will be addressed further in the text. Note that due to rotational symmetry with respect to the axis of the sensor and symmetry with respect to the plane orthogonal to the center of the sensor, only one quarter of the full image is presented for each example. To ease comparison and exemplify the contrast between the two cases, the upper left quarter of the elevation of the first case is placed next to the upper right quarter of the second (due to the symmetry of both cases, they are directly comparable). Uniaxial compression applied to host material should result in a uniform vertical displacement field, yet the displacement field shown in the figure is not uniform. The main perturbation happens around the sensor, and it is largest at around location of the anchor. The difference Δv between dashed and continuous lines at the location of the anchor shows the amount of perturbation that sensors induce, and that difference divided by the gauge length of the sensor would result in an error in strain measurement induced by that difference ($\delta\varepsilon = \Delta v / L_s$). While the examples shown in the figure deal with a long-gauge sensor embedded in host material, similar anomalies can happen in the case of short-gauge and distributed sensors in the case of installation using adhesive or mounting devices for all types of sensors (short-gauge, long-gauge, and distributed).

The perturbation caused by the stiffness of the sensor embedded in the host material can be reduced by the appropriate geometry of the anchors. Figure 4.18 shows how increasing the anchor area minimizes the perturbation, which is due to engagement of a larger volume of material in deforming the sensor. This can further be improved by engaging the friction (shear) along the contact areas between the host material and the sensor, which in turn can be realized by adding indentations and increasing the roughness of sensor packaging and anchors (to improve mechanical interlocking with the host material) and using materials for sensor packaging that have a high friction coefficient in contact with the host material. More details regarding the examples shown in Figure 4.18 and the influence of geometrical and mechanical properties on the accuracy of strain measurements are given in Calderon and Glisic (2012). The main recommendations are presented below.

- For all types of sensors and sensor installation procedures: To minimize perturbation of strain field in host material of monitored structure, in general, the sensors with low axial stiffness should be used, where axial stiffness is defined as ratio between force applied to sensor and resulting deformation of sensor (typically equal to $E_s A_s$, where E_s and A_s are overall Young modulus and area of the cross-section of the sensor); by preference, a low ratio between the Young modulus of host material E_M and equivalent Young modulus of sensor E_s' should be used, where the former is defined as follows (Calderon and Glisic 2012):

$$E_s' = \frac{E_s A_s + E_M \left(A_s' - A_O\right)}{A_s'} \tag{4.24}$$

where E_s' is equivalent Young modulus of sensor, E_s is true Young modulus of sensor (overall Young modulus should be used if sensor has multiple materials in cross-section), A_s is true area of cross-section of the sensor (thus, $E_s A_s$ is axial stiffness of sensor), A_O is the area enclosed by outer perimeter of the sensor (if the cross-section of sensor is hollow, then $A_s \neq A_O$, if it is full then $A_s = A_O$), $A_s' = \pi 8^2/4 = 50.3$ mm^2 is equivalent area of cross-section of the sensor, as per Calderon and Glisic (2012), and E_M is Young modulus of host material.

Due to the high stiffness of traditional materials, such as steel and concrete, most commercial sensors satisfy the requirement of low ratio between E_s' and E_M. For soft materials such as concrete at very early age and most of soils (e.g., stiff soil such as well-compacted low-plasticity clay, or hard soil/soft rock), sensors with a ratio of $E_s'/E_M \leq 1$ or axial stiffness $E_s A_s \leq 22$ kN/m^2 are recommended (Calderon and Glisic 2012);

- Additional recommendations for sensors embedded in host material: Sensors whose anchors provide good mechanical interaction with the host material should be used; lateral surfaces of sensor packaging and the anchors should be indented and/or intentionally made rough to improve the engagement of the material in deforming the sensor; if the ratio E_s'/E_M between the equivalent moduli of the sensor and host material is not sufficiently low, then bigger anchors should be mounted on the sensors; in addition, sensors with a long-gauge length should be used (however, note that excessively long sensors can induce errors inherent to gauge length, see Section 4.5).

Finally, it is important to emphasize that in all above considerations, the overall size of the sensor was assumed to be very small in comparison with the size of the host material. If the sensor dimensions are comparable with the size of host material, especially in the case of embedded sensors, then the sensor presence can affect the safety of the structure. For example, a sensor with diameter of 10 mm embedded in a structural element whose cross-section has a depth of 40–50 mm will remove an important percentage of structural material from the cross-section, and, consequently, weaken the load-carrying capacity of the structural element at that location.

4.6.4 Error Due to Compromise Between Survivability of Sensor and Quality of Strain Transfer

Sections 4.6.1–4.6.3 demonstrate, in one way or another, the importance of the quality of strain transfer from structure to the sensing element to achieve high accuracy in strain measurement. However, transfer of an excessive localized (concentrated) strain from structure to the sensing element due to damage to structure (e.g., cracking of concrete, and permanent ground movement of soil) can damage the sensing element and consequently, partially or completely impair the functionality of the sensor. For example, a crack opening results in a relative displacement of the mouth of the crack, which in turn results in excessively high strain transferred from the mouths of the crack to the sensor. Theoretically, this strain is infinite, as before the crack opening, the distance between the mouths of the crack was zero, and after the crack opening, it became non-zero, i.e., theoretical strain is equal to the non-zero crack opening over the initial zero distance. This excessive strain, transferred to the sensing element, generates concentrated stress that can damage (break) the sensing element. The sensors where the strain is transferred to the sensing element continuously over its length (i.e., not via anchors) are the ones subjected to this type of problem, since the excessive strain due to damage to structure is then transferred to a single point or very short segment of the sensing element, which in turn generates concentrated stress in the sensing element and damages the latter. This process can be called the transfer of damage from structure to sensor. Examples of sensors subjected to damage transfer are strain gauges and distributed sensors installed by adhesives, as well as embedded distributed sensors. Sensors in which the transfer of strain to sensing element is made via anchors, such as vibrating wire sensors and most discrete fiber-optic sensors, as well as distributed sensors mounted by clamping, are not likely to suffer from damage transfer; in these sensors, the excessive strain due to damage to the structure is not transferred to the sensing element at a single point of damage but through anchors, and thus it is distributed over the entire gauge length between anchors or between clamping points, which in turn results in significantly lower strain and stress in the sensing element.

Two general approaches could be taken to mitigate the risk of damage transfer, and combinations of these approaches can be used as well, depending on project requirements. The proposed approaches target the first category of sensors described above, i.e., those where the strain is transferred to the sensing element continuously over its length. As noted above, the problem of damage transfer does not significantly affect the sensors where the strain transfer is enabled via anchors or clamps.

(1) If the sensors are installed via adhesive (see Figure 4.13), the solution is the selection of appropriate adhesive. Three approaches are possible, depending on the properties of sensor packaging.

  a. For sensors with stiff and strong packaging (e.g., distributed sensors packaged in fiber-reinforced composite tape shown in Figure 3.21a and denoted with "Tape" in Figure 4.13b), an adhesive should be chosen so that it is stiff-enough to guarantee a good strain transfer to the sensor, but its shear bonding strength should be weak-enough to allow the sensor to delaminate (unstick) from the monitored structure over a short length (e.g., 10–20 cm) in the event of local damage (e.g., cracking). This delamination allows localized strain in sensing element to redistribute over a short, finite (delaminated) length instead of being applied at a single point, which significantly lowers the maximal strain and stress in the sensing element and prevents its failure. This approach was applied and validated in real-life settings (e.g., see Section 1.3 and Glisic et al. 2007). Note that the delaminated part of the sensor, behaves as the gauge length of a discrete sensor, and thus, this approach does not significantly affect the post-cracking quality of strain measurement.

  b. For sensors with soft, thin, and weak packages (e.g., strain gauges), the approach described in Point 1a above would not be appropriate; for these sensors, the use of strong yet soft (flexible) adhesive would be more appropriate. The disadvantage of such an approach is that the soft adhesive will lower the quality of strain transfer (see Figure 4.12); however, the positive consequence of lower quality of strain transfer is that a localized strain due to damage to structure redistributes through the thickness of the adhesive, which "shields" the sensor packaging and sensing element from exposure to concentrated high strain and stress and thus prevents its failure. This approach was tested, and its feasibility was confirmed, on a third-generation 2D sensor prototype of the Sensing Sheet (Gerber et al. 2018). The adverse effects of lower strain transfer entrained by this approach (e.g., loss of accuracy in strain measurement) can be compensated by calibration of the installed sensor and adjustment of calibration constants (e.g., adjustment of gauge factor for strain gauges).

c. For sensors with a soft, thick, and weak package, (e.g., distributed sensors packaged in the polyethylene profile shown in Figure 3.21b and denoted with "Profile" in Figure 4.13b), the properties of the adhesive might be less important. If the packaging is soft- and thick-enough to redistribute concentrated strain through its own thickness, then the packaging can practically play the role of soft adhesive presented in Point 1b above. Note that soft packaging, in general, may result in lower quality strain transfer, which should be compensated through the calibration of the sensor "as installed."

(2) If the sensors are installed via embedding (see Figure 4.15), the solution is the selection of appropriate sensor packaging. Three approaches are possible, depending on the properties of the host material and the magnitude of excessive concentrated strain that can occur due to damage. As we noted in the previous text, the most important thing is to avoid the transfer of concentrated strain to the sensing element.

a. For embedded sensors, the strain transfer is mostly realized through shear adhesion and friction between the sensor packaging and host material; thus, the first approach is to select sensor packaging that would "detach" over a short length from the surrounding host material, i.e., the adhesion and friction between the sensor packaging and host material would be released over that length. This allows the sensor to locally "slide" within the host material, and enables redistribution of strain and stress in the sensor packaging and sensing element, and thus prevents the failure of the sensing element, similar to what is described in Point 1a above. If the sensor sliding happens over the controlled short length, this part of the sensor would behave as a discrete sensor with a gauge length equal to that controlled short length, and thus, this approach does not significantly affect the post-cracking quality of strain measurement. However, to ensure this, it is important to control the length of "sliding," which is not always possible.

b. The second approach aims at selecting sensor packaging that is stiff-enough for good strain transfer but soft-enough to redistribute concentrated strain through its own thickness. In this case, the thickness of the packaging also has to be relatively high (few millimeters). This approach is practically identical to the one presented in Point 1c, above, and entrain the same concerns regarding the quality of strain transfer.

c. Finally, the solution can be provided by an intentional release of bonding between the sensing element and the packaging, allowing the sensing element to "slide" within the packaging. This approach is similar to that described in Point 2a, above, except the release happens between the sensing element and packaging, as opposed to packaging and host material. The concerns regarding strain transfer are thus similar to those described in Point 2a.

Example presented in Section 4.5.5 compares the measurements of an embedded long-gauge strain sensor and a distributed sensor at the location of crack occurrence. The distributed sensor in that case was the same "Profile" sensor shown in Figures 3.21b and 4.13b, see also Figure 4.10. Profile sensor packaging is relatively thick (3 mm), and it is made of soft polyethylene, so it enables the approach described in Point 2b, above. In addition, given that polyethylene is, in general, a low-adhesion and low-friction material, it has the potential to enable the approaches presented in Points 2a and 2c, above. Figure 4.11 practically demonstrates that an embedded Profile sensor can survive the cracking of concrete by combining the approaches presented in Points 2a–c. Actually, there were in total eight crack instances in the deck of the bridge described in Section 4.5.5, and Profile sensor survived all eight of them (see more details in Glisic et al. 2011). Note that Figure 4.11 also presents measurements of an embedded discrete long-gauge sensor at the location of damage, for which the strain transfer is realized via anchors. The functionality of this type of sensor was not affected by damage transfer, and all eight sensors installed at locations of crack instances successfully detected the damage and survived.

Several approaches presented in this section involve sacrificing the quality of strain transfer to make sensors immune to damage transfer. Lower quality of strain transfer will certainly introduce errors in strain measurement. While in many cases this error can be minimized by calibrating the sensor "as installed," the conditions of such a calibration, performed in the laboratory, might not be fully reproducible in the field (e.g., the thickness of the applied adhesive is difficult to control in field conditions), and thus some error will always be present. Nevertheless, this compromise between the survivability of sensors and the decrease in quality of strain transfer is needed to optimize the use of sensors for both purposes.

Note that even if the approaches proposed in this section are implemented, very significant damage to the structure could still damage the sensors if stress levels in the sensing element caused by damage transfer reach the limits of the mechanical resistance (strength) of the sensing element. In these cases of very significant damage to structure, even the discrete sensors in which the strain transfer is made via anchors as well as distributed sensors installed by clamping can also suffer damage.

4.7 Fundamental Assessment of Reliability of Strain Measurement

4.7.1 Cleansing of Data

Collected (raw or unprocessed) monitoring data may contain some defects (i.e., undesirable inaccuracies or inconsistencies) due to various factors such as erroneously performed measurement due to malfunction of sensors (e.g., drift), environmental interferences (e.g., electro-magnetic interferences, unusual temperature or humidity changes), malfunction of SHM system components (e.g., channel switch that does not follow the schedule of measurements), outage of power or communication lines, unintentional human errors (e.g., operator who wrongly performs the measurement, incorrectly assigns sensors to channels of the switch, or accidentally deletes part of measurements), etc. As the raw data may contain defective data subsets, they should not be immediately used for interpretation and analysis, as this may result in wrong conclusions regarding structural health condition and performance. Thus, the defective data first has to be identified and corrected to ensure that only reliable and accurate data is used to perform interpretation and analysis.

In this book, the term "cleansing of data" refers to identifying and correcting inaccurate and inconsistent (defective) data. Executing the cleansing of data often requires some measurements to be deleted, some to be imputed, and some to be allocated (shifted) from one sensor to another. These actions should be performed very carefully in order not to eliminate correct data as this can adversely affect data interpretation and analysis and lead to wrong conclusions.

It is important to highlight that performing cleansing of data is not straightforward, as it often depends on several factors that are SHM-system-specific or require engineering reasoning. The three most frequent defects of data found in practice are outliers, missing data, and sensor drifts.

Examples of their identification and correction are given in the next subsections. However, these examples are all SHM-system- or project-specific, and thus not comprehensive. Nevertheless, they provide some initial guidance and logic of thinking that can be applied in some other cases.

4.7.2 Examples of Identification and Correction of Outliers

In this book, erroneous measurement values that have a pattern that sharply differs from other values measured before or after that value are referred to as "outliers." Note that in some literature, the term outlier is used for both erroneous measurement values as well as correct measurement values that indicate unusual structural behavior, which is not the definition used in this book. For example, Figures 3.23 and 4.11 show a sharp difference in measured values in encircled areas; given that these measurements were identified as correct crack detections, in this book they are not considered (erroneous) outliers.

To correctly identify outliers, it is important to examine their potential sources. Some typical (but not all) sources of outliers are:

– Malfunction of the firmware or other software components related to the decoding of the encoding parameter.
– Value measured by one sensor (e.g., labeled with "X") is wrongly attributed to another sensor (e.g., labeled with "Y") due to problems in the reading unit, switching device, or data management software.
– Malfunction of other SHM-system components.
– Unexpected change in measurement conditions.
– Third-party interference.
– Other.

One example is shown in Figure 4.19. Figure 4.19a shows two measurements that seem unusually different from all other measurements as their magnitudes are significantly different from all other measurements (see encircled). The encoding parameter in these specific sensors is deducted from interference peaks generated in reading from the signals reflected from the sensors. However, besides the main optical signal reflected from sensors, a few other reflections coming from internal components of the reading unit and cable extensions (typically connectors) will also produce interference peaks. Peak-tracking software sometimes confuses these "parasite" peaks with the main peak. Indeed, inspection of the peaks of the two measurements showed that parasite peaks were wrongly selected by the software, and after a correction was performed, i.e., correct peaks were selected, the measurement values became consistent with the other measurements, as shown in Figure 4.19b. Corrected measurements are indicated in Figure 4.19 with dashed lines.

Example shown in Figure 4.19a contains resolvable problem, which correction does not lead to removal of data (actually, the above problem is automatically resolvable by the software).

However, outliers with similar graphs can have different causes. In Section 4.4, we introduced the notion of reference reading, i.e., the reading of the sensor performed typically immediately after the sensor is installed. This reading commonly has no physical meaning, as it shows a value reflecting an offset in the sensor due to the relationship between strain and the encoding parameter and the potential deformation of the sensor itself due to installation. This notion is further developed in Chapter 5. It is important to note that due to various reasons related to manufacturing, the reference reading of different sensors is often different, and practically serves as "signature" of each sensor. For example, the reference value of reflected wavelength of FBG sensor is unique for each sensor, or the reference value of the vibration frequency is unique for each VW sensor, etc. Let us

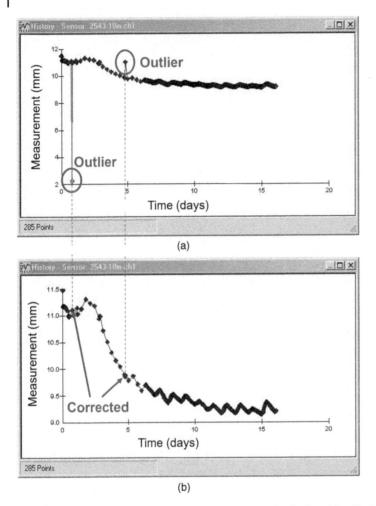

Figure 4.19 Example of outliers and their correction: (a) Outliers identified as unusual changes in measurement values; and (b) outliers are corrected and measurement values are coherent with other values (the vertical axis is automatically resized after the outliers are corrected). Source: Courtesy of SMARTEC.

assume that the channel switch that enables the reading of multiple sensors suffers malfunction and remains blocked at one channel; then, instead of reading subsequent sensors, it repeats the reading of the same sensor. Thus, the values of repeated readings, which are close to the reference value of this specific sensor, will be erroneously attributed to all other subsequent sensors. Given the difference in reference values between the sensors, this will create a similar, unusual pattern, as shown in Figure 4.19a, for each subsequent sensor. Unfortunately, in this case, the values of subsequent sensors cannot be recovered as they were not executed, and the values erroneously attributed to these sensors should be removed. Again, this has to be made carefully, after thorough analysis, to make sure that sharp changes in strain observed in graphs are not the consequence of a true unusual structural behaviors, which actually should be detected.

Both examples shown above demonstrated the need for understanding the specifics of the deployed SHM system. They also showed that outliers with similar appearances can have different causes. It is practically impossible to present all possible outlier scenarios in this book; hence, good knowledge about SHM system and conditions surrounding the sensors at time when outlier occurred are crucial for proper identification and correction of outliers.

4.7.3 Examples of Handling Missing Data

It is not unusual to have missing measurements in datasets collected in real-life settings. Typical sources of missing data are:

– Data not registered due to a temporary outage of electrical power or communication lines.
– Data not registered due to a temporary malfunction of SHM system components.
– Data registered but removed from the dataset due to lack of reliability (e.g., incorrigible outliers or data collected from drifting sensors).
– Third-party interference.
– Other.

Missing data is typically (but not only) handled as follows:

– Dataset is kept as it is, if the data interpretation and analysis are not affected by missing data.
– Missing data can be replaced (imputed) with values that are obtained using some physics-based or data-driven approach; data is typically imputed when they are needed to complement or complete data interpretation and analysis; note, however, that imputing the data introduces errors in data analysis and has to be performed carefully to avoid the potential risk of generating erroneous data.
– In the cases where imputing the data cannot be reasonably performed, the other data related to the same reading session is ignored. Note, however, that completely ignoring that data may also introduce the risk of losing some information about the monitored structure.

Let us assume that several sensors are installed in different columns of a building. If the monitoring is performed at a local structural scale, where each column is analyzed individually (see Chapter 5), the missing data related to one sensor (i.e., one column) will not affect the analysis of other columns, and thus the dataset can be kept as is.

Let us assume that two sensors (labeled "P9up" and "P9down") parallel to the centroid line are installed in a cross-section of a beam, and monitoring of the curvature at that cross-section is of interest (see Chapter 6). In this case, the measurements of both sensors are needed at every instance of time to calculate the curvature, but let us assume that the measurement of sensor P9up is missing, as shown in Figure 4.20a. Then, the missing measurement can be imputed by using some physics-based approach (e.g., analytically or numerically modeled value expected to be measured at that location at that time) or data-driven approach (e.g., average value of the two adjacent measurements or moving average, or value obtained using some machine learning method). An example of a measurement value imputed using an average value of neighboring measurements is shown in Figure 4.20b. Imputed value fits well with the pattern of other measurements and seems reasonable.

However, if there is no rational basis for estimating the imputed value, then the value of the other sensor, P9down, recorded at the time of the missing measurement from sensor P9up, can be ignored and the curvature at that time not evaluated. The ignored value is shown in Figure 4.20c. Note that while the measurement of sensor P9down is ignored at that specific time in the evaluation of curvature, it can still be used in the evaluation of local structural behavior at the location of the sensor (see Chapter 5), and thus should not be completely discarded.

Figure 4.20a shows relatively simple case of missing data (i.e., only one measurement is missing). In real-life, complete intervals of missing data can be present in datasets, as shown in Figure 4.20d,

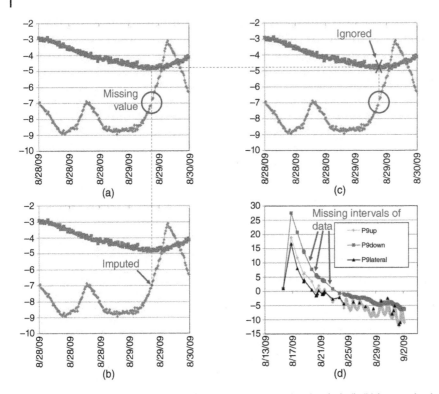

Figure 4.20 (a) Example of one missing measurement value (encircled), (b) imputed value of measurement to enable data analysis, (c) ignored value of other sensor in data analysis, as an option for handling missing data, and (d) missing intervals (shown with arrow) and more complicated patterns of missing data (graph P9lateral). Source: Modified from Glisic (2009).

with even more complicated patterns of missing data, as shown by the graph of sensor P9lateral. Values of measurements of sensor P9up hint at how the values of sensor P9lateral can be imputed, but this might be unreliable as the number of measurements of P9lateral is small, and variability of values of P9up is relatively large. In some cases where entire intervals of measurements are missing, the use of data-driven or physics-informed data-driven approaches to impute ("recover") data can be applied. Such approaches are out of the scope of this introductory book, and more detail can be found in relevant literature (e.g., see Oh et al. 2020, 2021, Pereira and Glisic 2022).

4.7.4 Examples of Evaluations of Reliability of Measurement Values

Section 4.7.2 presents outliers as potential manifestations of unreliable measurements. These outliers manifested in the form of an abrupt, unusually high, change in measurand. This abrupt change itself is a clear indicator that measurement is potentially erroneous, i.e., unreliable. However, in some cases, the loss of measurement reliability does not happen abruptly but rather slowly over long-terms. This may happen due to various causes:

- Drift in sensors due to aging of its components or manufacturing errors.
- Loss of strain transfer from structure to sensor, e.g., due to loss of performance of adhesive, and physical or chemical degradation of mounting devices (corrosion of bolts, stick-and-slip effect in clamps, etc.).

- Damage or malfunction of SHM system components (e.g., parts of the reading unit, channel switch, and sensors) due to improper installation or inadequate environmental conditions (e.g., overheating or reading unit and melting of sensor components).
- Other.

Previous subsections confirmed that ascertaining the reliability of the measurements requires expertise in the monitoring system employed, so the potential sources of loss of reliability can be understood. Note that some strain sensors require thermal compensation (see Sections 3.5 and 3.6), and for these sensors, the loss of reliability of temperature measurement can cause the loss of reliability of strain measurements too. An important challenge in identifying the loss of reliability of the long-term measurements is distinguishing the drift from slowly evolving processes that affect monitoring structures, such as rheological effects in concrete structures (creep and shrinkage), and differential settlement of foundations. Given that these processes are slow and evolve in long-terms, they may have a similar pattern as drift in sensors, and the former can be confused with the latter. Validation of measurements, i.e., ascertaining that they are reliable, can be performed via the following actions:

- Direct comparison with measurements performed using different monitoring systems with, by preference, different principles of measurement and confirmed (certified) reliability.
- Comparison with estimations obtained from analytical or numerical models.
- Comparison with estimations obtained from data-driven or hybrid physics-informed data-driven methods.
- Combined methods, involving two or more of the above actions.
- Other.

The first approach above, i.e., direct comparison with measurements performed using different monitoring systems would be ideal, but it is almost never achieved in real-life settings as it would imply the installation of two monitoring systems on the structure, which in turn would significantly increase the costs. The second approach would be more economic, yet it faces the challenges of accurate modeling of structural behaviors. These are the reasons why, in most applications, the third and fourth approaches are taken (e.g., see Abdel-Jaber and Glisic 2016, Oh et al. 2020, 2021, Pereira and Glisic 2022). Presenting these approaches in detail is out of the scope of this book; however, examples of their applications are briefly given below.

Example 4.7 *Identification of drift in long-term temperature measurements*
The Streicker Bridge project was introduced in previous sections. After several years of measurements, drift was noticed in some temperature sensors. The drift was ascertained when the values of temperature measurements from these sensors became unrealistically high or low. However, it was not clear when the drift started or how it evolved. Given that temperature measurements are necessary for the compensation of FBG sensors installed on the Streicker Bridge, it was important not only to detect the drift but also to quantify it. To perform this, a combined approach was used.

As the first step, reliable measurements of air temperature were identified at Trenton Airport, distant approximately 15.6 km from the bridge (NOAA 2013). It was assumed that the air temperature around the bridge was similar and that it influenced the temperature in the structure. However, the temperature in the structure was not expected to be identical to the air temperature due to the thermal inertia of concrete, yet it was expected to follow the overall trend. This is confirmed in Figure 4.21, where temperatures measured at the Trenton Airport and one location in the structure are plotted side-by-side.

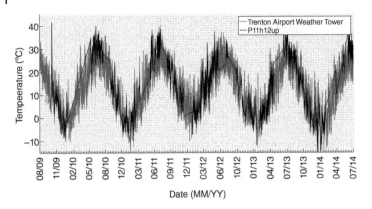

Figure 4.21 Comparison of air temperature measured at Trenton Airport and temperature in Streicker Bridge measured by sensor labeled P11h12up, installed in the middle of span P11–P12, upper in the cross-section (see Figure 4.10d) (Abdel-Jaber and Glisic 2016).

Due to the thermal inertia of concrete, the temperature in the structure at the observed time t_k was considered to be influenced not only by the air temperature at the observed time but also by the evolution of temperature over the past $n + 1$ measurement instances, i.e., at times t_k, t_{k-1}, ..., t_{k-n}. As a consequence, a weighted moving average model was used to predict temperature in the structure based on air temperature measurements; see Expression 4.25.

$$T_{C,structure}(t_k) = \sum_{i=0}^{n} w_{C,i} T_{air}(t_{k-i}) \tag{4.25}$$

where: $T_{C,structure}(t_k)$ is predicted temperature in the structure at point C at observed time t_k; $T_{air}(t_{k-i})$ are air temperatures measured at Trenton Airport at consecutive measurement times t_{k-i}, $i = 0, 1, 2, ..., n$; and $w_{C,i}$ are coefficients ("weights") that reflect the influence of air temperatures measured at times t_{k-i} to temperature in structure observed at time t_k.

Size of the model n and coefficients $w_{C,i}$ are obtained via supervised learning by using the initial two years of data, a period during which it was estimated that all sensors functioned properly, i.e., without drift. Finally, once the model is established for each location of the temperature sensor in the structure, the difference between predictions from the air temperature and the actual measurements is analyzed, and the drift is ascertained by comparing the yearly gradient of the difference with a threshold value. In this project, a threshold of 0.5°C/year is used based on engineering judgment (see Abdel-Jaber and Glisic 2016). As an illustration, Figure 4.22 shows two examples: a sensor without drift and a sensor with drift. The drift in sensors was attributed to an unintentional fabrication error. More details about the presented method are found in Abdel-Jaber and Glisic (2016).

It is important to emphasize that, in addition to evaluating drift, this work revealed an important characteristic of structures: their thermal behaviors, i.e., temperature distributions in the structures and strain responses to those temperature distributions, can be predicted with acceptable accuracy from air temperature. This was confirmed using other data-driven approaches (e.g., see Oh et al. 2021, Pereira and Glisic 2022); however, strain and temperature measurements from at least two to three years are needed to train the predictive models.

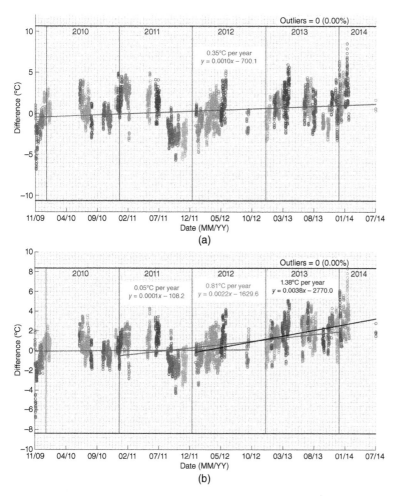

Figure 4.22 Examples of differences between predicted and measured values of: (a) Non-drifting sensor and (b) drifting sensor (Abdel-Jaber and Glisic 2016).

Example 4.8 *Ascertaining reliability of unusually high strain measurements*

The aim of this example is to present the reasoning used in a specific project to evaluate whether the strain measurements with unusually high changes, registered in that project, are reliable and thus, can be used for the evaluation of structural health conditions. Figure 3.23 shows the graphs of strain evolutions of eight sensors installed in Streicker Bridge. These sensors are labeled with P10up, P10down, P10h11up, P10h11down, P11up, P11 down, P11h12up, and P11h12down. Figure 4.10b shows the locations of cross-sections above columns P10 and P11; cross-section P10h11 is in the middle of the span P10–P11 shown in the figure, while the location of cross-section P11h12 is in the middle of the adjacent span P11–12, which is not shown in the figure (see Figure 6.3 for a more comprehensive presentation of sensors installed in Streicker Bridge). Figure 4.10d shows the "up" and "down" positions of sensors for each instrumented the cross-section (cross-section at P11 is specifically shown in the figure, but this applies to all locations).

Unusually high strain changes are encircled in Figure 3.23 and labeled as cracks. However, when these changes occurred, it was not clear whether they were an indication of unusual structural behavior or a consequence of sensor malfunction, and in the case of the former, what type of unusual structural behavior they indicates. To address these issues, first the main facts related to these measurements were identified and summarized as follows:

– Pair of sensors at locations P10h11, P11, and P11h12 exhibited unusual strain measurements on the same day, with a couple of hours of difference, respectively.
– Pair of sensors at location P10 exhibited unusual strain measurements five days later.
– In each cross-section, both sensors, "up" and "down," were simultaneously experiencing positive strain change, alluding to behavior consistent with cracking.
– No other location exhibited similar behavior during the same time frame.
– Post-tensioning performed several days later seemed to cancel this unusually high strain change (see example in Figure 4.11 for sensor P11up).
– After the post-tensioning was performed, the formworks were removed; however, there were no visible signs of cracking at these locations.

The following assumptions, related to the cause of these unusually high measurements, are analyzed and eliminated:

– Malfunction of reading unit: Each sensor had two FBGs, one for strain and the other for temperature monitoring, as shown in Figure 3.17; if the reading unit was malfunctioning, it would be unlikely, for each sensor, to have one FBG affected and the other unaffected by the malfunction; this makes it even more unlikely to have this pattern repeated eight times, and each time to have exactly strain FBG affected and temperature FBG unaffected.
– Malfunction of channel switch: As mentioned in Section 3.7.2, malfunction of switch would result in abrupt change in strain measurement value, i.e., in large "jump" between correct and incorrect measurement (similar to Figure 4.19a); that was not the case in graphs shown in Figure 3.23 – each graph is smooth with gradually increasing change that develops over approximately one to two hours.
– Malfunction of sensors: Strain-measuring FBG in the sensor is pretensioned between the anchors, to measure both elongation and shortening (see Section 3.6.3); FBG is very stable permanent change of refraction index in the core of optical fiber, and thus it does not drift; thus, the only reason for drift in the sensor would be the loss of pretension between the anchors of the optical fiber carrying the strain-measuring FBG; however, in that case, the sensor should measure negative strain change and not positive, which was the case shown in Figure 3.23.

The above analysis practically ascertained that there were no issues related to the SHM system. However, the lack of visible cracks (open or closed) on the top and bottom surfaces of the bridge deck indicated that even if the cracks were present, they were fully in the interior of each concerned cross-section. To verify this, the following actions are taken:

– Streicker Bridge, being instrumented with SHM for research and educational purposes, contained in span P10–P11 both discrete FBG sensors and distributed BOTDA sensors, see Figure 4.10; consequently, direct comparison between two different types of sensors was possible at locations P10, P10h11, and P11, and this comparison confirmed the existence of unusual strain change in both types of sensors (Glisic et al. 2011); an example is shown in Figure 4.11.
– Additional verification was performed via Ultrasonic Pulse Velocity test that confirmed slower propagation of ultrasonic waves at all four locations (Hubbell and Glisic 2013), which was consistent with cracking.

The above actions practically confirmed that sensor measurements indicate unusual structural behavior consistent with cracking. Further numerical analysis confirmed that early age cracking in the bridge deck was possible and was due to non-linear thermal gradients generated at this stage of structure's life (Hubbell and Glisic 2013). Later analysis has shown that the first three cracks, at P10h11, P11, and P11h12, occurred due to stresses built from restraint of the structure (indeterminate to degree three), while the fourth crack, at the location of joint P10, occurred later-on due to an unusually warm day. More detail about the crack analysis in this project is given in Hubbell and Glisic (2013), Abdel-Jaber and Glisic (2015), and Sigurdardottir and Glisic (2015).

More detail about the effects of non-linear thermal gradients on beam structures is given in Chapter 5. In general, this example presented the reasoning behind validation of the reliability of unusually high changes in strain measurements. While the same approach might not be possible in other projects, similar reasoning could be applied if adapted to the specifics of the deployed SHM system and monitored structure.

5

SHM at Local Scale: Interpretation of Strain Measurements and Identification of Unusual Structural Behaviors at Sensor Level

5.1 Introduction to SHM at Local Scale

SHM at local scale was briefly introduced in Section 1.2 and discussed in more details in Section 4.5. In these sections, it was noted that SHM at local scale can be performed at material or structural scale. SHM at local material scale is interested in very localized behavior at observed point. For example, in the case of reinforced concrete, which by design consists of concrete, reinforcing bars, and usual allowed cracks in tensioned zone, it might be of interest to monitor each or any of these components separately, i.e., strain in concrete, strain in rebars, or crack opening.

SHM at local structural scale is rather interested in localized behavior in a volume of material surrounding the observed point that might consist of several components that interact and create a structural entity with its own properties. For example, concrete, rebars, and allowed cracks, when acting together, make reinforced concrete that behaves differently from concrete (which is weak in tension), differently from rebars (that have a high Young's modulus), and different from cracks (that are discontinuities).

Section 4.5 has shown that for strain-based SHM at local material scale, the recommended choice would be short-gauge sensors or distributed sensors with very small spatial resolution. On the other hand, for strain-based SHM at local structural scale, the appropriate choice would be long-gauge sensors or distributed sensors with long spatial resolution. However, these recommendations are not strict, and the final choice of sensors should be made based on the project requirements. Note that for homogeneous materials, difference between local material scale and local structural scale is only in the amount of material that is covered by sensor gauge length (for discrete sensors) or spatial resolution (for distributed sensors); however, for inhomogeneous materials, this difference is significant, and the gauge lengths or spatial resolution of sensors should be carefully selected to reflect the aims of SHM.

Note that even when the primary aim of SHM is to perform monitoring at global or integrity scale, each discrete or distributed strain sensor that is installed on the structure provides measurements at specific locations, and if these locations are observed individually, the sensors also provide SHM at local (material or structural) scale at each location. Thus, interpretation of strain sensor measurements is crucial not only for SHM at local scale but also at global and integrity scales. Moreover, unusual structural behaviors often have manifestations in the form of local perturbations of strain field, and thus strain sensor measurements can be used for their detection, given that they are in their close proximity (see Chapter 4).

Hence, the aim of this chapter is to present fundamental methods for interpretation of strain measurements performed by a single sensor and for the use of measurements in identification

Introduction to Strain-Based Structural Health Monitoring of Civil Structures, First Edition. Branko Glišić.
© 2024 John Wiley & Sons Ltd. Published 2024 by John Wiley & Sons Ltd.

of unusual structural behaviors. Given that methods focus on measurements from a single sensor, they reflect SHM at local scale; whether the local SHM is at material or structural scale would depend on the type of material and gauge length or spatial resolution of sensors (see Section 4.5 and Figure 4.8). Scope is limited to beam-like structures, as they are the most frequently used in practice. To ease the presentation, the first eight sections contain topics that are fundamental for structural design and analysis and are typically covered in standard undergraduate curriculum of civil engineering. However, several examples presented in these sections demonstrate the application of these topics in strain-based SHM. The last section presents criteria for detection of unusual structural behaviors with several illustrative examples related to steel and concrete structures. Identification of cracking is not specifically addressed in this chapter, as it is presented in detail in Chapter 7.

5.2   Strain Tensor Transformations

Strain state at a point was introduced in the form of a strain tensor and mathematically represented as matrix in Section 3.2. It was noted that strain tensor consists of nine components, six of which are independent from each other. The three diagonal components of strain tensor matrix are called normal strain components, or normal strain, or simply strain (denoted as ε_x, ε_y, ε_z or ε_{xx}, ε_{yy}, ε_{zz}, respectively), and each of them is independent from all other strain components. The six non-diagonal components are called shear strain, but due to their conjugate property, only three are independent from each other (denoted as $\frac{1}{2}\gamma_{xy} = \frac{1}{2}\gamma_{yx}$, $\frac{1}{2}\gamma_{yz} = \frac{1}{2}\gamma_{zy}$, $\frac{1}{2}\gamma_{zx} = \frac{1}{2}\gamma_{xz}$ or $\varepsilon_{xy} = \varepsilon_{yx}$, $\varepsilon_{yz} = \varepsilon_{zy}$, $\varepsilon_{zx} = \varepsilon_{xz}$, respectively); see Expression 3.1. Hence, the conjugate property of shear strain results in symmetry of strain tensor matrix. Some of these components can vanish depending on dimensionality of the case: for plane strain, they reduce to four, three of which are independent (e.g., if strain components in direction x vanish, then only strain components in plane Oyz remain, i.e., ε_y, ε_z, and $\frac{1}{2}\gamma_{yz}$) and for uniaxial strain, they reduce to one (e.g., if strain components in directions x and y vanish, then only strain component ε_z remains).

While strain state is independent of the choice of coordinate system, strain components do depend on coordinate system. In Chapter 3, it was shown that a strain sensor measures only normal strain component (or, simply, strain) in the direction of sensor. If the sensor direction coincides with an axis of coordinate system (x-axis, y-axis, or z-axis), then the strain sensor measures the (normal) strain in the direction of corresponding axis (ε_x, ε_y, or ε_z). However, if the direction of strain sensor does not coincide with an axis of coordinate system, then a relationship between strain components in the given coordinate system and in direction of sensor has to be established.

Let strain tensor **E** be defined with respect to the coordinate system $Oxyz$, and unit vector **n** = $(\cos\alpha, \cos\beta, \cos\gamma)$ defines the direction of strain sensor, where α, β, and γ are the angles between vector **n** and axes x, y, and z, respectively. Then, Expression 5.1 is valid (Malvern 1969).

$$\varepsilon_n = (\mathbf{E} \cdot \mathbf{n}) \cdot \mathbf{n} = \left(\begin{bmatrix} \varepsilon_x & \frac{1}{2}\gamma_{xy} & \frac{1}{2}\gamma_{xz} \\ \frac{1}{2}\gamma_{yx} & \varepsilon_y & \frac{1}{2}\gamma_{yz} \\ \frac{1}{2}\gamma_{zx} & \frac{1}{2}\gamma_{zy} & \varepsilon_z \end{bmatrix} \begin{bmatrix} \cos\alpha \\ \cos\beta \\ \cos\gamma \end{bmatrix} \right)^{\mathrm{T}} \begin{bmatrix} \cos\alpha \\ \cos\beta \\ \cos\gamma \end{bmatrix} \tag{5.1}$$

where ε_n is the strain in direction of **n**.

Expression 5.1 shows that ε_n is a linear combination of the six independent components of strain tensor **E**. Thus, to infer the full state of strain at a point, i.e., to determine all six independent components of the strain tensor with respect to the given coordinate system $Oxyz$, it would be necessary

Figure 5.1 Schematic representation of rosette installed on the surface of monitored structure.

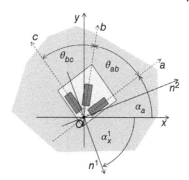

to use six strain sensors installed along six mutually different directions $\mathbf{n}_1, \mathbf{n}_2, ..., \mathbf{n}_6$, among which there are neither four co-planar (belonging to the same plane) directions nor two co-linear (belonging to the same line) directions. Assuming that angles of each direction with respect to axes of coordinate system are known, Expression 5.1 can be written for each direction; this will enable creation of six equations with six unknowns, whose solutions will result in strain tensor components.

Note that to infer the 3D strain state at a point within a material, strain sensors have to be embedded in that material. This might be possible with new concrete or composite structures, where the sensors could be embedded during the construction. However, for new steel or timber structures, this is not possible, nor is it possible for any existing structure, regardless of the material on which they were built. In these cases, sensors can only be installed on the exposed surfaces of the structure, and strain state (strain tensor) is derived using a combination of strain measurements and knowledge that normal stress component perpendicular to the structure's surface is null (see Section 5.6).

In many real-life cases, not all strain components are of interest due to absence of significant corresponding stress components. For example, in the cases of uniaxial bending of beam-like structures or membrane loading of thin-shell structures, strain components in the direction perpendicular to the centroid line of the beam or direction perpendicular to the centroid surface of shell, are frequently not of interest. In these cases, only the strain components in 2D plane might be of interest (as opposed to 3D space). For example, in the case of uniaxial bending of a beam, this plane will be the plane of bending; in the case of membrane loading of thin shells, this plane will be tangential to the centroid surface of the shell. Let's assume that coordinate system in the case where planar strain is of interest is *Oxy*, then, Expression 5.1 transforms into 5.2.

$$\varepsilon_n = \left(\begin{bmatrix} \varepsilon_x & \frac{1}{2}\gamma_{xy} \\ \frac{1}{2}\gamma_{yx} & \varepsilon_y \end{bmatrix} \begin{bmatrix} \cos\alpha \\ \sin\alpha \end{bmatrix} \right)^{\mathrm{T}} \begin{bmatrix} \cos\alpha \\ \sin\alpha \end{bmatrix} = \varepsilon_x\cos^2\alpha + \varepsilon_y\sin^2\alpha + \gamma_{xy}\sin\alpha\cos\alpha \qquad (5.2)$$

where $\gamma_{yx} = \gamma_{xy}$ due to conjugate property of shear strain and α is angle between the direction $\mathbf{n}$ and *x*-axis.

Expression 5.2 shows that for planar case, strain ε_n in a given direction $\mathbf{n}$ is linear combination of the three strain tensor components; consequently, to infer strain state at a point in a given plane, it is necessary to install three strain sensors in three different directions. This configuration of sensors is called a "rosette." Rosette made of strain gauge sensors can be found on market as a single package containing three sensors with predefined angles. An example is given in Figure 3.10b, while its schematic representation is given in Figure 5.1.

Example 5.1 *Determining strain tensor from rosette sensors measurements*
Let observe rosette sensors installed on a surface of material as shown in Figure 5.1, where $\alpha_a = 36°$ and $\theta_{ab} = \theta_{bc} = 45°$. Let assume that strain sensors within rosette measure strain

$\varepsilon_a = 20\,\mu\varepsilon$, $\varepsilon_b = 60\,\mu\varepsilon$, and $\varepsilon_c = -40\,\mu\varepsilon$, then the two-dimensional strain tensor at point O with respect to coordinate system Oxy can be calculated using Expression 5.2. Assumed limits of error in strain measurements are $\pm 1\,\mu\varepsilon$ (typical for resistive strain gauges).

$$20\,\mu\varepsilon = \varepsilon_a = \varepsilon_x \cos^2 36° + \varepsilon_y \sin^2 36° + \gamma_{xy} \sin 36° \cos 36° = 0.6545\varepsilon_x + 0.3455\varepsilon_y + 0.4755\gamma_{xy}$$

$$60\,\mu\varepsilon = \varepsilon_b = \varepsilon_x \cos^2 81° + \varepsilon_y \sin^2 81° + \gamma_{xy} \sin 81° \cos 81° = 0.0245\varepsilon_x + 0.9755\varepsilon_y + 0.1545\gamma_{xy}$$

$$-40\,\mu\varepsilon = \varepsilon_c = \varepsilon_x \cos^2 126° + \varepsilon_y \sin^2 126° + \gamma_{xy} \sin 126° \cos 126° = 0.3455\varepsilon_x + 0.6545\varepsilon_y - 0.4755\gamma_{xy}$$

The above system of equations can be rewritten in form of matrix as follows:

$$\begin{bmatrix} 20 \\ 60 \\ -40 \end{bmatrix} = \begin{bmatrix} 0.6545 & 0.3455 & 0.4755 \\ 0.0245 & 0.9755 & 0.1545 \\ 0.3455 & 0.6545 & -0.4755 \end{bmatrix} \begin{bmatrix} \varepsilon_x \\ \varepsilon_y \\ \gamma_{xy} \end{bmatrix}$$

$$\begin{bmatrix} \varepsilon_x \\ \varepsilon_y \\ \gamma_{xy} \end{bmatrix} = \begin{bmatrix} 0.6545 & 0.3455 & 0.4755 \\ 0.0245 & 0.9755 & 0.1545 \\ 0.3455 & 0.6545 & -0.4755 \end{bmatrix}^{-1} \begin{bmatrix} 20 \\ 60 \\ -40 \end{bmatrix}, \text{i.e.,} \begin{bmatrix} \varepsilon_x \\ \varepsilon_y \\ \gamma_{xy} \end{bmatrix} = \begin{bmatrix} 1.1301 & -0.9511 & 0.8210 \\ -0.1301 & 0.9511 & 0.1790 \\ 0.6421 & 0.6181 & -1.2601 \end{bmatrix} \begin{bmatrix} 20 \\ 60 \\ -40 \end{bmatrix}$$

The final matrix equation yields the following solution: $\varepsilon_x = -67.3\,\mu\varepsilon$, $\varepsilon_y = 47.3\,\mu\varepsilon$, and $\gamma_{xy} = 100.3\,\mu\varepsilon$.

Limits of error can also be calculated using the last matrix and applying error propagation formula for sum:

$$\delta\varepsilon_x = \pm(1.1301 \cdot 1 + (-0.9511) \cdot (-1) + 0.8210 \cdot 1) = \pm 2.9\,\mu\varepsilon,$$

$$\delta\varepsilon_y = \pm((-0.1301) \cdot (-1) + 0.9511 \cdot 1 + 0.1790 \cdot 1) = \pm 1.3\,\mu\varepsilon,$$

$$\delta\gamma_{xy} = \pm(0.6421 \cdot 1 + 0.6181 \cdot 1 + (-1.2601 \cdot (-1)) = \pm 2.5\,\mu\varepsilon.$$

Given that limits of error of sensor measurements are $\pm 1\,\mu\varepsilon$ and taking into account typical needed accuracy for civil engineering applications, all the final values in this example were rounded to the first decimal.

5.3 Principal Strain Components

It is well known from solid and continuum mechanics (e.g., Malvern 1969) that there exist three mutually perpendicular axes, called principal axes, such that shear strain components vanish when strain tensor is observed with respect to these axes. In that case, the matrix of strain tensor contains only diagonal (normal) strain components, called principal strain components or principal strains. Let I be a unit matrix, then principal strains can be found as roots $\varepsilon_1, \varepsilon_2$, and ε_3 ($\varepsilon_1 \geq \varepsilon_2 \geq \varepsilon_3$) of the following equation (e.g., Malvern 1969):

$$\det(\mathbf{E}^T - \varepsilon\mathbf{I}) = \begin{vmatrix} \varepsilon_x - \varepsilon & \frac{1}{2}\gamma_{yx} & \frac{1}{2}\gamma_{zx} \\ \frac{1}{2}\gamma_{xy} & \varepsilon_y - \varepsilon & \frac{1}{2}\gamma_{zy} \\ \frac{1}{2}\gamma_{xz} & \frac{1}{2}\gamma_{yz} & \varepsilon_z - \varepsilon \end{vmatrix} = 0 \tag{5.3}$$

In two-dimensional case, Expression 5.3 transforms into Expression 5.4.

$$\det\left(\mathbf{E}^T - \varepsilon\mathbf{I}\right) = \begin{vmatrix} \varepsilon_x - \varepsilon & \frac{1}{2}\gamma_{yx} \\ \frac{1}{2}\gamma_{xy} & \varepsilon_y - \varepsilon \end{vmatrix} = \varepsilon^2 - (\varepsilon_x + \varepsilon_y)\varepsilon + \varepsilon_x\varepsilon_y - \frac{1}{4}\gamma_{xy}^2 = 0$$

$$\Rightarrow \varepsilon_{1,2} = \frac{\varepsilon_x + \varepsilon_y \pm \sqrt{(\varepsilon_x - \varepsilon_y)^2 + \gamma_{xy}^2}}{2} \tag{5.4}$$

Once principal strains are determined, corresponding principal axes (or principal directions) can be determined too. Let's denote unit vectors along principal axes with $\mathbf{n}^i = (n^i_x, n^i_y, n^i_z)$, $i = 1, 2, 3$, where the axis defined with $\mathbf{n}^i$ corresponds to the principal strain ε_i. Then, for each i ($i = 1, 2, 3$) the components of unit vector $\mathbf{n}^i$ can be determined using the following two equations (e.g., Malvern 1969):

$$\left(\mathbf{E}^T - \varepsilon_i\mathbf{I}\right) \cdot \mathbf{n}^i = \begin{bmatrix} \varepsilon_x - \varepsilon_i & \frac{1}{2}\gamma_{yx} & \frac{1}{2}\gamma_{zx} \\ \frac{1}{2}\gamma_{xy} & \varepsilon_y - \varepsilon_i & \frac{1}{2}\gamma_{zy} \\ \frac{1}{2}\gamma_{xz} & \frac{1}{2}\gamma_{yz} & \varepsilon_z - \varepsilon_i \end{bmatrix}\begin{bmatrix} n^i_x \\ n^i_y \\ n^i_z \end{bmatrix} = 0 \Rightarrow n^i_y = \mathrm{f}\left(n^i_x\right) \text{ and } n^i_z = \mathrm{g}\left(n^i_x\right)$$

and

$$\left(n^i_x\right)^2 + \left(n^i_y\right)^2 + \left(n^i_z\right)^2 = 1 \Rightarrow \left(n^i_x\right)^2 + \left(\mathrm{f}\left(n^i_x\right)\right)^2 + \left(\mathrm{g}\left(n^i_x\right)\right)^2 = 1$$

yields

$$\pm n^i_x \Rightarrow \pm n^i_y, \pm n^i_z \tag{5.5}$$

Matrix equation shown in the first row of Expression 5.5 represents homogeneous system of three equations with three unknowns (components of principal direction unit vectors n^i_x, n^i_y, n^i_z); however, the determinant of this system is equal to zero (as per Expression 5.3). Thus, the system does not have only one unique trivial solution $(0,0,0)$, but rather infinite number of linearly dependent solutions. If one unknown is randomly selected (e.g., n^i_x), then the two others can be expressed as functions of that variable; $n^i_y = \mathrm{f}(n^i_x)$ and $n^i_z = \mathrm{g}(n^i_x)$. Finally, from the sum of squares of components n^i_x, n^i_y, n^i_z (which is equal to one for unit vector, in general), it is possible to calculate n^i_x. Then, the other components can be calculated using functions f and g. Note that n^i_x will have both positive and negative solution, which will define two different senses of orientation of vector $\mathbf{n}^i$, but both senses of orientation define the same principal axis. In two-dimensional case, for $\mathbf{n}^1$, Expression 5.5 transforms into (5.6).

$$\left(\mathbf{E}^T - \varepsilon_1\mathbf{I}\right) \cdot \mathbf{n}^1 = \begin{bmatrix} \varepsilon_x - \varepsilon_1 & \frac{1}{2}\gamma_{yx} \\ \frac{1}{2}\gamma_{xy} & \varepsilon_y - \varepsilon_1 \end{bmatrix}\begin{bmatrix} n^1_x \\ n^1_y \end{bmatrix} = 0 \Rightarrow \begin{array}{l} (\varepsilon_x - \varepsilon_1)n^1_x + \frac{1}{2}\gamma_{xy}n^1_y = 0 \\ \frac{1}{2}\gamma_{xy}n^1_x + (\varepsilon_y - \varepsilon_1)n^1_y = 0 \end{array}$$

$$\Rightarrow n^1_y = -2\frac{\varepsilon_x - \varepsilon_1}{\gamma_{xy}}n^1_x$$

$$\text{and } \left(n^1_x\right)^2 + \left(n^1_y\right)^2 = \left(n^1_x\right)^2 + 4\left(\frac{\varepsilon_x - \varepsilon_1}{\gamma_{xy}}n^1_x\right)^2 = 1$$

$$\Rightarrow n^1_x = \pm\frac{1}{\sqrt{1 + 4\left(\frac{\varepsilon_x - \varepsilon_1}{\gamma_{xy}}\right)^2}} \Rightarrow n^1_y = \pm\frac{2\left(\frac{\varepsilon_x - \varepsilon_1}{\gamma_{xy}}\right)}{\sqrt{1 + 4\left(\frac{\varepsilon_x - \varepsilon_1}{\gamma_{xy}}\right)^2}} \tag{5.6}$$

Example 5.2 *Determining principal axes and principal strain components from rosette measurements*

For strain measured by rosette from example 5.1, the principal strains can be calculated using Expression 5.4, as follows:

$$\varepsilon_{1,2} = \frac{-67.3+47.3\pm \sqrt{(-67.3-47.3)^2 + 100.3^2}}{2} = -10 \pm 76.1 \Rightarrow \varepsilon_1 = 66.1 \ \mu\varepsilon; \ \varepsilon_2 = -86.1 \ \mu\varepsilon$$

Application of error propagation formulae presented in Table 4.2 would be complicated in this example due to complexity of Expression 5.4. In this case, it is simpler to perform exhaustive analysis by inserting all variations of all arguments plus/minus limit of error values in Expression 5.4. This yields the following values for limits of error:

$$\delta\varepsilon_1 = \pm 2.3 \ \mu\varepsilon \ \text{for} \ \varepsilon_x = -67.3 \pm 2.9 \ \mu\varepsilon, \ \varepsilon_y = 47.3 \pm 1.3 \ \mu\varepsilon, \ \text{and} \ \gamma_{xy} = 100.5 \pm 2.5 \ \mu\varepsilon,$$

$$\delta\varepsilon_2 = \pm 3.5 \ \mu\varepsilon \ \text{for} \ \varepsilon_x = -67.3 \pm 2.9 \ \mu\varepsilon, \ \varepsilon_y = 47.3 \pm 1.3 \ \mu\varepsilon, \ \text{and} \ \gamma_{xy} = 100.5 \mp 2.5 \ \mu\varepsilon.$$

Unit vector components of the principal axis n^1 are determined by using Expression 5.6:

$$n_x^1 = \pm\frac{1}{\sqrt{1 + 4\left(\frac{-67.3 - 66.1}{100.3}\right)^2}} = \pm 0.352 \ \text{and} \ n_y^1 = \pm\frac{2\left(\frac{-67.3 - 66.1}{100.3}\right)}{\sqrt{1 + 4\left(\frac{-67.3 - 66.1}{100.3}\right)^2}} = \mp 0.936$$

Similar to principal strains, the application of error propagation formulae presented in Table 4.2 for evaluation of error limits for unit vector components of principal axis n^1 would be complicated. Again, exhaustive analysis by inserting all variations of all arguments plus/minus limit of error values in Expression 5.5 was performed, which yields the following values for limits of error:

$$\delta n_x^1 = \pm 0.020 \ \text{for} \ \varepsilon_x = -67.3 \pm 2.9 \ \mu\varepsilon, \ \varepsilon_1 = 66.1 \mp 2.3 \ \mu\varepsilon, \ \text{and} \ \gamma_{xy} = 100.5 \pm 2.5 \ \mu\varepsilon,$$

$$\delta n_y^1 = \pm 0.008 \ \text{for} \ \varepsilon_x = -67.3 \mp 2.9 \ \mu\varepsilon, \ \varepsilon_1 = 66.1 \pm 2.3 \ \mu\varepsilon, \ \text{and} \ \gamma_{xy} = 100.5 \mp 2.5 \ \mu\varepsilon.$$

More precisely, $+\delta n_x^1 = +0.020$ and $-\delta n_x^1 = -0.019$, but for practical purposes of simpler presentation, one can adopt a slightly conservative value of $\delta n_x^1 = \pm 0.020$. Similarly, $+\delta n_y^1 = +0.008$ and $-\delta n_y^1 = -0.007$, but for practical purposes of simpler presentation one can adopt a slightly conservative value of $\delta n_y^1 = \pm 0.008$.

Components n^i_x and n^i_y of the unit vector in the direction of principal axis are equal to cosines of angles α^i_x and α^i_y with respect to the x-axis and y-axis, respectively. Hence, the angle α^i_x of the first principal axis with respect to the x-axis is calculated as follows:

$$\mathbf{n}^1 = \left(\cos \alpha_x^1, \cos \alpha_y^1\right) = \left(\cos \alpha_x^1, \sin \alpha_x^1\right) = (\pm 0.352 \pm 0.020, \mp 0.936 \mp 0.008) \Rightarrow$$

$$\alpha_x^1 = -69.4° \pm 1.2° \ \text{or} \ \alpha_x^1 = 110.6° \pm 1.2°$$

Note that the difference between the two solutions shown in the equation above is 180°, which means that both solutions define the same line, i.e., the same principal axis. The second principal axis is perpendicular to the first, which means $\alpha^2_x = (-69.4° \pm 1.2°) + 90° = 20.6° \pm 1.2°$ or $\alpha^2_x = (110.6° \pm 1.2°) + 90° = 200.6° \pm 1.2°$. Both principal axes are shown in Figure 5.1 (denoted as n^1 and n^2 in the figure).

5.4   Reference Time and Reference (Zero) Measurement

Once installed onto a structure, the sensor measures strain changes in the structure that occur after its installation, i.e., the sensor cannot evaluate what strain was generated in the structure before it was installed. Thus, unless installed at the time of creation of structural material (e.g., by embedding in concrete during pouring), the sensor in general cannot measure absolute strain in the structure. Hence, the first reading of strain sensor after installation does not have a meaningful interpretation; it rather shows an arbitrary value resulting from the offset of the reading unit and potential deformation of sensor resulting from installation procedure (e.g., long-gauge fiber optic sensors are frequently slightly pretensioned during installation). However, any future reading of sensor, when made relative to the first reading, will represent the strain change that occurred in the structure between the two readings. This statement is valid for any two readings that occurred after the installation of sensor, where the time of earlier reading can serve as a reference for the later reading. To clarify this, let us define a moment in time t_{Ref} as reference time, where the earliest possible time t_{Ref} is the moment of installation of sensor on the structure. Then, at some later time t, $t > t_{Ref}$, the following equality is valid:

$$\Delta \varepsilon_n(t) = \varepsilon_n(t) - \varepsilon_n(t_{Ref}) \approx r_{n,sensor}(t) - r_{n,sensor}(t_{Ref}) = \varepsilon_{n,sensor}(t) \qquad (5.7)$$

where n is direction of strain, $\varepsilon_n(t)$ and $\varepsilon_n(t_{Ref})$ are, respectively, strain values in structure at times t and t_{Ref}, $\Delta \varepsilon_n(t)$ is change in strain between times t and t_{Ref}, $r_{n,sensor}(t)$ and $r_{n,sensor}(t_{Ref})$ are, respectively, readings of sensor at times t and t_{Ref}, and $\varepsilon_{n,sensor}(t)$ is change in strain as measured by strain sensor; note that in engineering practice, $r_{n,sensor}(t_{Ref})$ is frequently referred to as reference reading, reference measurement, zero reading, or zero measurement.

The sign of approximate equality ($\approx$) is used in Equation (5.7) only to indicate that limits of error in the measurement should also be accounted for. This is more exactly shown in Equation (5.8).

$$\Delta \varepsilon_n(t) = \varepsilon_{n,sensor}(t) \pm \delta \varepsilon_{n,sensor} \qquad (5.8)$$

where term $\pm \delta \varepsilon_{n,sensor}$ symbolically represents limits of error in strain measurement (note that positive and negative limits of error in general can have absolute values different from each other, more details about error analysis are given in Chapter 4).

To simplify presentation, the symbol delta (Δ), denoting change, is intentionally omitted next to sensor measurement $\varepsilon_{n,sensor}(t)$ in Equations 5.7 and 5.8. To further simplify presentation, this symbol will also be omitted in further text next to strain in structure $\varepsilon_n(t)$; thus, unless specified differently, the term $\varepsilon_n(t)$ should always be understood as a change in strain with respect to reference time.

Example 5.3   *Finding strain measurement from sensor readings*
The encoding parameter of (fiber-optic) FBG strain sensor is back-reflected wavelength of light λ (unit is nanometer – nm). Let's assume that an FBG sensor is installed on a metallic cantilever beam that was subjected to three steps of incremental loading. Tests were performed in the lab under controlled (constant) temperature conditions. The calibration constant of the sensor $C_\varepsilon = 0.0012$ nm/$\mu\varepsilon$, and thus, the reading of sensor at time t is calculated as $r_{n,sensor}(t) = \lambda(t)/C_\varepsilon$. Let's assume that the readings of sensors taken during the test are as presented in Table 5.1. Then, reference time t_{Ref} should be set to 8:25:11 24 Oct. 2010 and strain measurements (with respect to reference time) can be calculated using Equation 5.7. They are presented in the last column of Table 5.1.

Table 5.1 Example of determining strain measurements from sensor readings taken during an increment load test.

Load increment	Time *t* (time and date)	Value of encoding parameter λ (nm)	Sensor reading $r_{n,sensor}$ ($\mu\varepsilon$)	Strain measurement $\varepsilon_{n,sensor}$ ($\mu\varepsilon$)
No load	8:25:11 24 Oct. 2010	1550.3280	1,291,940	0
1 N	8:29:01 24 Oct. 2010	1550.3388	1,291,949	9
2 N	8:35:36 24 Oct. 2010	1550.3506	1,291,959	19
3 N	8:40:22 24 Oct. 2010	1550.3639	1,291,970	30

5.5 Strain Constituents and Total Strain

Sections 5.2–5.4 present strain components, strain tensor transformations, and strain measurements regardless of the source (origin) of strain. However, strain measurements can only be interpreted and analyzed if the source of strain is known. In this section, the most common sources of strain and resulting strain constituents are presented in order to assist in the interpretation of strain measurement and data analysis.

The total strain in construction materials commonly consists of several constituents (Muravljov 1989). The two that are present in all types of construction materials (e.g., steel, concrete, timber, and composites) are mechanical strain, which is related to stress, and thermal strain, which is caused by temperature variations. In addition, rheologic effects can cause strain in material, which commonly takes from of creep and shrinkage. Rheological strains particularly affect concrete and timber, and for these two materials, they become significant in long-term (e.g., see Neville 1975, CEB-FIP 1990, Morlier 1994). Finally, some other material-specific sources of strain can also be present. Examples are residual strain generated due to manufacturing process of steel or composite elements, endogenous, carbonation, and plastic strain in concrete at early age.

In general, all the above constituents of strain (mechanical, thermal, creep, and shrinkage) occur simultaneously. Consequently, strain sensors measure their combination, i.e., the total strain, and, in general, the individual strain constituents (i.e., contributing parts of each source of strain) cannot be simply separated. This statement is illustrated in Expression 5.9.

$$\varepsilon_{n,tot}(t) = \varepsilon_{n,\sigma}(t) + \varepsilon_{n,T}(t) + \varepsilon_{n,c}(t) + \varepsilon_{n,sh}(t) + \varepsilon_{n,other}(t)$$
$$= \varepsilon_{n,E}(t) + \varepsilon_{n,P}(t) + \varepsilon_{n,T}(t) + \varepsilon_{n,c}(t) + \varepsilon_{n,sh}(t) + \varepsilon_{n,other}(t)$$
$$\varepsilon_{n,tot}(t) = \varepsilon_{n,sensor}(t) \pm \delta\varepsilon_{n,sensor} \tag{5.9}$$

where *n* is direction of normal strain component, $\varepsilon_{n,\sigma}$ is mechanical strain, ε_E and ε_P are, respectively, elastic and plastic constituents of mechanical strain ($\varepsilon_{n,\sigma} = \varepsilon_E + \varepsilon_P$), $\varepsilon_{n,T}$ is thermal strain, $\varepsilon_{n,c}$ is creep, $\varepsilon_{n,sh}$ is a shrinkage, and $\varepsilon_{n,other}$ is combination of all other strain constituents.

To simplify presentation, subscript *n*, which denotes the direction of (normal) strain (component), will be omitted in the text whenever this omission does not alter the meaning of expressions and does not create ambiguity. Also, subscripts *tot*, σ, *E*, *P T*, *c*, *sh*, and *other* will always denote total strain, mechanical strain, elastic strain, plastic strain, thermal strain, creep, shrinkage, and other strain components, regardless of the other subscripts that may be present in the same term of the expression (e.g., $\varepsilon_{1,\sigma}$ would mean mechanical strain in the direction of principal axis n_1).

Note that the sign "plus" in Expression 5.9 could be considered both exact and symbolic, the latter meaning that in some cases a simple sum does not apply when a combination of strain constituents is present in material. However, for most applications in civil engineering, the plus sign in Expression 5.9 can be considered exact. Depending on the type of material, some strain constituents in Expression 5.9 may be absent (e.g., steel does not have a shrinkage constituent), i.e., they are equal to zero. Strain constituents found in the most frequently used construction materials are shown in Table 5.2 (Muravljov 1989, Keller 2003).

An example of total strain measurement performed on a concrete structure (Streicker Bridge) over several days following the pouring is given in Figure 4.11 (Chapter 4). In the figure, thermal strain is visibly dominant at an early age due to heat released by the hydration process (October 23–26). "Other" strain, i.e., apparent strain due to cracking, is dominant at the end of early age (October 26). For the following days (October 26–November 2), all strain constituents are slowly changing, with no visibly dominant one except thermal strain, which was slightly dominant on an unusually hot fall day, October 31. Prestressing was applied on November 2nd and 3rd, making mechanical strain dominant on these two days. Note that during the observed period, the influence of daily variations of temperature on thermal strain was minimized by the presence of formwork and cover mats that provided thermal insulation, which in turn diminished the thermal exchange between the structure and the environment. Hence, thermal strain dominance is visible only on unusually warm or cool days (e.g., October 31; see also Section 5.7 for more information about thermal strain). While during some periods presented in Figure 4.11, some strain constituents were more dominant than others, most, if not all, of them were simultaneously present at any moment of time after the hardening of concrete occurred on the first day (see the next sections for more details on individual strain constituents).

Table 5.2 Strain constituents found in most frequently used construction materials.

Source of strain	Strain constituent	Concrete	Steel	Composite	Timber
Stress[a]	Mechanical strain $\varepsilon_\sigma = \varepsilon_E + \varepsilon_P$	Yes	Yes	Yes	Yes
	Elastic strain ε_E	Yes	Yes	Yes	Yes
	Plastic strain[b] ε_P	Yes	Yes	No[c]	Yes
Temperature	Thermal strain ε_T	Yes	Yes	Yes	Yes
Rheologic effects	Creep ε_c	Yes	No[d]	No[d]	Yes
	Shrinkage ε_{sh}	Yes	No	No[e]	No[f]
Others	ε_{other}	Yes[g]	No[h]	No[h]	Yes[f]

a) Due to loads in all types of structures and, in addition, due to temperature, rheologic effects, imperfections, and differential settlements in statically indeterminate structures.
b) Plastic strain is in general considered as damage.
c) Depends on reinforcing material of composites; commonly used glass- and carbon-reinforced composites have minimal plastic deformation (if any).
d) Creep, if any, is comparable with small percentage of elastic strain, and in the most cases can be neglected.
e) Depends on matrix (resin) of composite material.
f) Swelling and shrinkage are mainly related to humidity changes and occur mostly in radial and tangential directions; however, longitudinal shrinkage can also reach important values.
g) E.g., strain at early age.
h) Practically negligible.
Source: Modified from Muravljov (1989), Keller (2003), and Glisic and Inaudi (2007).

Simplified analytical models for the estimation and analysis of the strain constituents presented in Table 5.2 are summarized in the next three subsections. The aim is to build a framework for strain analysis and illustrate its application with simplified examples. In specific projects, a more accurate analysis of strain constituents might be required, and in these cases, the use of advanced models would be needed; however, the presentation of these models exceeds the topic of this book, but they can be found in the various relevant literature. Then, within the framework of strain analysis presented in this book, the simplified analytical models presented here can be substituted with these advanced models.

5.6 Mechanical Strain and Stress

5.6.1 Scope

Mechanical strain ε_σ is directly correlated with stress (thus subscript σ) through material constitutive equations. Stress, in turn, is one of the most important indicators of structural performance and condition. Hence, identifying and measuring mechanical strain (and subsequently evaluating stress) is an important task of SHM. Once the relationship between strain and stress at a point is established, the evolution of stress can be monitored and its values compared with ultimate values (tensional and compressive strength) and design values (stress calculated based on analytical or numerical analysis of structure). Comparison with ultimate values will give an indication of whether the material is going to fail at the monitored point. Comparison with the design values will give an indication of whether the structure performs as designed. Discrepancies in comparisons might indicate unusual structural behaviors (e.g., damage or deterioration). The detected unusual behavior (if any) might have occurred at the location (point) where the mechanical strain is monitored but also at location distant from that point, as it may result in stress and strain redistributions.

As emphasized in Section 5.5, a strain sensor measures a combination of strain constituents, not exclusively mechanical strain. Thus, evaluating mechanical strain is challenging and requires determination of all other strain constituents. This section is analyzes the evolution of all these strain constituents, with particular focus on concrete and steel, as they are the most frequently used construction materials. Also, the focus is on the structural members that are the most frequently used in practice, i.e., beams-like structural members and beamlike structures (straight and curved).

5.6.2 Instantaneous Uniaxial Stress–Strain Relationship and Comparison with Ultimate Values

As a reminder, stress is a tensor consisting of nine components. Diagonal components are called normal stress components, or normal stresses, or simply stresses, while non-diagonal components are called shear stress components, or simply shear stresses. Stress tensor is symmetric due to conjugate property of shear stress, thus, six components out of nine are independent. Two notations of stress tensors are given in Expression 5.10, and more details on stress tensors can be found in literature (e.g., Malvern 1969, Brčić 1989).

$$\mathbf{\Sigma}(t) = \begin{bmatrix} \sigma_x(t) & \tau_{xy}(t) & \tau_{zx}(t) \\ \tau_{yx}(t) & \sigma_y(t) & \tau_{yz}(t) \\ \tau_{zx}(t) & \tau_{zy}(t) & \sigma_z(t) \end{bmatrix} = \begin{bmatrix} \sigma_{xx}(t) & \sigma_{xy}(t) & \sigma_{zx}(t) \\ \sigma_{yx}(t) & \sigma_{yy}(t) & \sigma_{yz}(t) \\ \sigma_{zx}(t) & \sigma_{zy}(t) & \sigma_{zz}(t) \end{bmatrix} \tag{5.10}$$

In this book, the first notation (a matrix with σ-s and τ-s) will be used to simplify presentation and emphasize the difference between normal stress components (σ-s) and shear strain components (τ-s). Note that t in Expression 5.10 represents time.

Similar to strain, the components of a stress tensor depend on choice of coordinate system, and a stress tensor has principal axes and principal values. All transformations related to change of coordinate system and determination of principal axes and principal stresses can be made by substituting components of strain tensor **E** with corresponding components of stress tensor Σ in Expressions 5.1–5.6.

The most frequently considered source of stress and mechanical strain are loads (forces); however, in statically indeterminate (hyperstatic) structures, due to over-constraints (redundancy in removal of degrees of freedom), stress and mechanical strain can also be generated by virtually any other source of strain (e.g., temperature variations, rheological effects, differential settlement of foundations, and construction imperfections). This can be understood by observing a simple example: a horizontal beam fixed at both ends with no external forces applied. In that example, a uniform increase in temperature would create compression in the beam; based on Expression 5.7, total strain is equal to zero in every cross-section (beam fixed at both ends), and in the absence of any other strain source, mechanical strain is equal to negative thermal strain; thus, negative (compressive) strain and stress are induced. Similar qualitative analysis can be carried out for other strain constituents (e.g., shrinkage of material will cause tension in the beam).

In general, mechanical strain ε_σ can have elastic and plastic sub-constituents, and both are related to stress. Elastic strain ε_E occurs at lower levels of stress, and it is reversible, i.e., it vanishes if the stress is removed. Plastic strain ε_P occurs at higher levels of stress and is irreversible, i.e., it does not vanish with the removal of stress. In typical applications, structures are designed to be in elastic regime. Thus, in most cases, plastic deformation is considered an indication of insufficient capacity or damage. Hence, it is important to identify whether the points of the structure are in an elastic or plastic state of strain.

The real behavior of materials can be complex, especially if the state of stress is multi-axial. To be able to analyze stress in structure, it is important to understand the complex stress–strain relationship, i.e., to establish constitutive equations for the observed material. To simplify the presentation, the analysis of uniaxially loaded materials is presented first. A schematic example of an instantaneous stress–strain relationship in uniaxially tensioned materials is given in Figure 5.2 (modified from Malvern 1969). While this diagram is not directly applicable to all materials, it identifies features of stress–strain relationship important for the identification of elastic state of strain. Term "instantaneous" is used in subsequent text to emphasize that only strain due to stress is considered and no other strain constituents are present (e.g., rheological or thermal strain are not present as they did not have time to develop). Note that even in the case of uniaxially applied stress, the strain state in the material is tree-axial, due to Poisson's effect (see Section 5.6.3 for more details). In Figure 5.2 and this section, only the strain component in the direction of stress is of interest, and thus, the indication of direction is omitted in subscripts denoting the strain, i.e., ε_σ represent the strain component that has the same direction as the uniaxial stress.

Figure 5.2 shows different stress and strain points of interest related to elastic state of a uniaxially loaded material: A is proportional limit or limit of linear-elastic range; B is limit of elastic range, i.e., it indicates the highest stress that results in fully reversible strain; C is practical limit of elastic range (defined by offset ε_0, typically set to 2000 $\mu\varepsilon$ for metals; line C-ε_0 is parallel to line OA, i.e., shows linear elastic recovery of strain); D indicates the ultimate stress, i.e., stress at which the material fails.

The zone between the origin of coordinate system and A represents elastic range of material, A to C represents yield range, and C to D represents the work-hardening plastic range. Note that yield

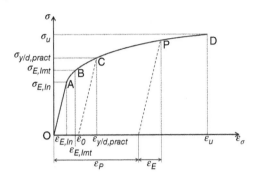

Figure 5.2 Example of instantaneous stress–strain relationship in materials under uniaxial tension. Source: Modified from Malvern (1969).

manifests as a range rather than a single point, and thus, depending on the project requirement and construction material, either of points A, B, or C can be used to define yield stress and strain. However, for the purposes of SHM and taking into account the importance of structural safety, conservative lower values of the yield range are the most frequently used as yield limits, i.e., to define uniaxial yield strain and yield stress of materials, which in turn represent important thresholds for monitored strain and stress. The strain and stress that are lower than yield strain and yield stress, respectively, indicate an acceptable state of structure, while exceeding the yield limits indicates a potentially compromised state of safety.

If the material deforms beyond the elastic limit, then it is in the plastic range. An example is illustrated by analyzing point P in Figure 5.2. Removal of stress will follow the line P-ε_P, which is parallel to the line OA. The elastic strain constituent ε_E will be fully recovered, while plastic strain constituent ε_P will permanently remain in material. Further loading will result in complex material behavior that is beyond the scope of this book (e.g., see Malvern 1969, Neville 1975, Muravljov 1989). Depending on the construction material, the plastic range can be large or small, the former indicating ductile and the latter brittle materials.

In strain-based SHM, the limit values that are important for setting thresholds that indicate the safety of the structure are yield strain ε_{yld} and ultimate strain ε_u. Given that constitutive equations for plasticity are not well established, we can illustrate these two thresholds by using the idealized constitutive relationships shown in Figure 5.3.

Figure 5.3a shows brittle linear elastic materials, that practically do not have a plastic range, such as some composite materials, some types of timber, stone, and unreinforced concrete (especially in tension), and even some alloys of steel. Figure 5.3b shows ductile linear elastic – perfectly plastic

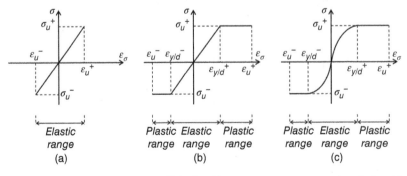

Figure 5.3 Schematic example of idealized instantaneous stress–strain relationships for uniaxially loaded materials: (a) Linear elastic, (b) linear elastic, perfectly plastic, (c) non-linear elastic, perfectly plastic.

material, such as some alloys of steel and some types of timber. Finally, Figure 5.3c shows ductile non-linear elastic – perfectly plastic material, such as reinforced concrete, and some types of timber. Note, however, that for lower levels of load, the concrete and reinforced concrete frequently behave as linear elastic materials.

For brittle materials, the positive and negative ultimate strains delimit the range of elastic strain, and if the mechanical strain in the structure exceeds any of these limits, the material fails (cracking in tension, crushing in compression). This can be described by Expression 5.11.

$$\text{(a) } \frac{\varepsilon_u^-}{\beta_\varepsilon^-} < \varepsilon_\sigma < \frac{\varepsilon_u^+}{\beta_\varepsilon^+} \Rightarrow \varepsilon_E = \varepsilon_\sigma \Rightarrow \text{material is safe, in elastic regime}$$

$$\text{(b) } \varepsilon_\sigma \le \frac{\varepsilon_u^-}{\beta_\varepsilon^-} \Rightarrow \text{material fails (crushing)}$$

$$\text{(c) } \varepsilon_\sigma \ge \frac{\varepsilon_u^+}{\beta_\varepsilon^+} \Rightarrow \text{material fails (cracking)} \tag{5.11}$$

where ε_E is elastic strain, ε_u^+ is tensional (positive) ultimate strain, and ε_u^- is compressive (negative) ultimate strain (in general $|\varepsilon_u^+| \neq |\varepsilon_u^-|$), and β_ε^+ and β_ε^- are safety factors used to account for uncertainties in the evaluation of ultimate strain values (safety factors are greater or equal to 1).

For ductile uniaxially stressed materials, positive and negative yield strains delimit the range of elastic strain, and ultimate strains delimit the range of plastic strain, as per Expression 5.12.

$$\text{(a) } \frac{\varepsilon_{yld}^-}{\beta_\varepsilon^-} < \varepsilon_\sigma < \frac{\varepsilon_{yld}^+}{\beta_\varepsilon^+} \Rightarrow \varepsilon_E = \varepsilon_\sigma \text{ and } \varepsilon_P = 0 \Rightarrow \text{material is safe in elastic regime}$$

$$\text{(b) } \frac{\varepsilon_u^-}{\beta_\varepsilon^-} < \varepsilon_\sigma \le \frac{\varepsilon_{yld}^-}{\beta_\varepsilon^-} \Rightarrow \varepsilon_E = \varepsilon_{yld}^- \text{ and } \varepsilon_P = \varepsilon_\sigma - \varepsilon_{yld}^- \Rightarrow \text{material is damaged (plastic)}$$

$$\text{(c) } \frac{\varepsilon_{yld}^+}{\beta_\varepsilon^+} \le \varepsilon_\sigma < \frac{\varepsilon_u^+}{\beta_\varepsilon^+} \Rightarrow \varepsilon_E = \varepsilon_{yld}^+ \text{ and } \varepsilon_P = \varepsilon_\sigma - \varepsilon_{yld}^+ \Rightarrow \text{material is damaged (plastic)}$$

$$\text{(d) } \varepsilon_\sigma \le \frac{\varepsilon_u^-}{\beta_\varepsilon^-} \text{ or } \varepsilon_\sigma \ge \frac{\varepsilon_u^+}{\beta_\varepsilon^+} \Rightarrow \text{material fails} \tag{5.12}$$

where ε_P is plastic strain, ε_{yld}^+ is tensional (positive) yield strain, and ε_{yld}^- is compressive (negative) yield strain (in general $|\varepsilon_{yld}^+| \neq |\varepsilon_{yld}^-|$, depending on material).

Expressions 5.12 state that if the mechanical strain in an uniaxially stressed material is by absolute value lower than the absolute value of yield, then the entire mechanical strain is elastic, but if the former exceeds the latter, then the mechanical strain consists of both elastic and plastic strain, where elastic strain is equal to yield strain. Note that the ultimate strain and yield strain are inherent properties of construction materials.

Real-life structures are frequently designed so that their mechanical strain, by absolute value, typically ranges between one quarter and one third of the elastic range and rarely reaches one half of the elastic range. This relatively low level of mechanical strain in structures is imposed by serviceability (rather than safety) requirements.

Expressions 5.11 and 5.12 do not mention mechanical strain after the material fails. Failure of the material will result in the redistribution of mechanical strain in the structure. For example, if a material is uniaxially loaded in tension and it fails, then it will break, mechanical strain will relax, and for brittle materials, it will become zero, and for ductile materials, it will become equal to plastic strain. In real-life settings, this redistribution is more complex, as it will depend on real (not idealized) material behavior, geometrical properties, and boundary conditions. On the other hand, the aim of SHM is to avoid failure, and thus, it is important to infer if the mechanical strain is approaching yield or ultimate strain over time, and this is how Expressions 5.11 and 5.12 should be considered. In the other words, Expressions 5.11(a) and 5.12(a) are the ones to be used in the analysis of structural safety of uniaxially loaded structural members.

Expressions 5.11 and 5.12 are given in terms of strain; however, the safety of structures is typically evaluated based on stresses. Therefore, expressions for an instantaneous stress–strain relationship should be established. Given that in the majority of practical applications in civil engineering, the construction materials are in the linear-elastic regime, the analysis of mechanical strain in this book focuses on elastic strain, and the occurrence of plastic strain is considered as unusual structural behavior (damage). In applications where more sophisticated stress–strain models are necessary, a process similar to the one presented in this book can be developed. However, the detailed development of those models is beyond the scope of this book.

As mentioned above, the instantaneous stress–strain relation in commonly used construction materials is considered linear elastic (unreinforced concrete, composites, and some types of steel), linear elastic – perfectly plastic (some types of steel, simplified analysis of reinforced concrete, and timber) or non-linear elastic – perfectly plastic (more sophisticated analysis of reinforced concrete and timber). This relationship is used to determine the stress in the material. For example, in a brittle, uniaxially loaded material, the stress can be determined using Expression 5.13 (Timoshenko and Goodier 1970).

$$\varepsilon_u^- < \varepsilon_\sigma < \varepsilon_u^+ \Rightarrow \varepsilon_E = \varepsilon_\sigma \Rightarrow \sigma = \varepsilon_E \cdot E = \varepsilon_\sigma \cdot E \tag{5.13}$$

where σ is uniaxial stress and E is Young's modulus.

Similar, in a ductile uniaxially loaded material, the stress can be determined using Expression 5.14 (Timoshenko and Goodier 1970).

$$
\begin{aligned}
&\text{(a)}\ \varepsilon_{yld}^- < \varepsilon_\sigma < \varepsilon_{yld}^+ \Rightarrow \varepsilon_E = \varepsilon_\sigma \Rightarrow \sigma = \varepsilon_E \cdot E = \varepsilon_\sigma \cdot E \\
&\text{(b)}\ \varepsilon_u^- < \varepsilon_\sigma \leq \varepsilon_{yld}^- \Rightarrow \varepsilon_E = \varepsilon_{yld}^- \Rightarrow \sigma = \varepsilon_E \cdot E = \varepsilon_{yld}^- \cdot E = \sigma_{yld}^- \\
&\text{(c)}\ \varepsilon_{yld}^+ \leq \varepsilon_\sigma < \varepsilon_u^+ \Rightarrow \varepsilon_E = \varepsilon_{yld}^+ \Rightarrow \sigma = \varepsilon_E \cdot E = \varepsilon_{yld}^+ \cdot E = \sigma_{yld}^+
\end{aligned}
\tag{5.14}
$$

where σ_{yld}^+ is uniaxial yield tensional stress, and σ_{yld}^- is uniaxial yield compressive stress.

Safety criterion based on stress for uniaxially loaded brittle and ductile materials are given in Expressions 5.15 and 5.16, respectively.

$$
\begin{aligned}
&\text{(a)}\ \frac{\sigma_u^-}{\beta_\sigma^-} < \sigma < \frac{\sigma_u^+}{\beta_\sigma^+} \Rightarrow \text{brittle material is safe, in elastic regime} \\
&\text{(b)}\ \sigma \leq \frac{\sigma_u^-}{\beta_\sigma^-} \Rightarrow \text{brittle material fails (crushing)} \\
&\text{(c)}\ \sigma \geq \frac{\sigma_u^+}{\beta_\sigma^+} \Rightarrow \text{brittle material fails (cracking)}
\end{aligned}
\tag{5.15}
$$

$$
\begin{aligned}
&\text{(a)}\ \frac{\sigma_{yld}^-}{\beta_\sigma^-} < \sigma < \frac{\sigma_{yld}^+}{\beta_\sigma^+} \Rightarrow \text{ductile material is safe, in elastic regime} \\
&\text{(b)}\ \sigma \leq \frac{\sigma_{yld}^-}{\beta_\sigma^-} \Rightarrow \text{ductile material is damaged or fails in compression} \\
&\text{(c)}\ \sigma \geq \frac{\sigma_{yld}^+}{\beta_\sigma^+} \Rightarrow \text{ductile material is damaged or fails in tension}
\end{aligned}
\tag{5.16}
$$

where β_σ^+ and β_σ^- are safety factors (greater or equal to 1).

Note that for idealized ductile materials, the compressive and tensional ultimate stresses are equal to the compressive and tensional yield stresses, respectively (i.e., $\sigma_{yld}^- = \sigma_u^-$ and $\sigma_{yld}^+ = \sigma_u^+$), which is not the case for most construction materials (e.g., see Figure 5.2). However, adopting this equality in SHM is justified as it provides a simplified, yet conservative, evaluation of structural safety. Exceedance of yield stress leads to plastic deformation or failure; thus, in Expression 5.15, it is written that the material is damaged (plastically deformed) or fails. Safety factors are recommended to account for errors in the estimation of ultimate stresses and errors in the estimation of stress in materials. It is important to emphasize that the schematic diagrams shown in Figure 5.3 represent idealized material behavior; stress–strain curves of real materials typically have different shapes in the plastic regime, and this should be considered if strain analysis includes that range.

In the case of idealized materials shown in Figure 5.3, Expressions 5.11 are practically equivalent to Expressions 5.15, and Expressions 5.12 to 5.16 (due to idealizations). Real materials do not have idealized behavior, and that is why determining whether the material is in an elastic regime requires both strain and stress criteria to be combined.

For brittle materials:

- If Expressions 5.11a and 5.15a are both valid, the material is in the elastic regime.
- If Expression 5.11a is valid but 5.15a is not valid, or 5.11a is not valid but 5.15a is valid, the state of the material is undetermined, but for the purposes of SHM, it can be considered that there is potential for damage or failure.
- In all other combinations of Expressions 5.11 and 5.15, the material is damaged or failed.

For ductile materials:

- If Expressions 5.12a and 5.16a are both valid, the material is in the elastic regime.
- If Expression 5.12a is valid but 5.16a is not valid, or 5.12a is not valid but 5.16a is valid, the state of the material is undetermined, but for the purposes of SHM, it can be considered that there is potential for damage or failure.
- In all other combinations of Expressions 5.12 and 5.16, the material is damaged or failed.

Similar to uniaxial ultimate strain ε_u and yield strain ε_{yld}, Young's modulus E, uniaxial yield stress σ_{yld}, and uniaxial ultimate stress σ_u, are all inherent properties of construction materials. In civil engineering applications, only uniaxial yield stress σ_{yld} and uniaxial ultimate stress σ_u are typically determined through testing. That is not the case with uniaxial ultimate strain ε_u, yield strain ε_{yld}, and Young's modulus E. To determine these parameters, in the absence of testing, the values found in the literature can be used with caution. Specifically, for the Young's modulus of concrete, empirical expressions found in the literature can be used to derive it from ultimate uniaxial compressive stress (e.g., see ACI 2008). Typical values of yield and ultimate stress and strain, along with other parameters characterizing common construction materials, are summarized in Table 5.3. Timber is not presented in the table due to large variability depending on wood species, direction (longitudinal, radial, and tangential), and humidity. These values are given for reference only; for specific materials in specific projects, they may vary and should be determined by testing.

5.6.3 Instantaneous Multidirectional Stress–Strain Relationship and Comparison with Ultimate Values

If the state of stress is multidirectional, the relationships between the stress and strain tensor components (i.e., constitutive equations) are complex. The relationship between stress and strain tensor

Table 5.3 Typical approximate values of mechanical parameters for different construction materials under static loads. Values are given for reference only – they may vary depending on material and application.

Parameter	Concrete	Steel	Composite
ε_u^+ (με)	50–100	5000–10,000	1500–3000
ε_u^- (με)	−3500	−5000 to −10,000	−1500 to −3000
ε_{yld}^+ (με)	–	2000	–
ε_{yld}^- (με)	−1350 to −2000	−2000	–
σ_u^+ (MPa)	2–6	360–550	L: 480[a] 350[b] 440[c]
σ_u^- (MPa)	20– to 70+	−360 to −550	L: 190[a] 140[b] 425[c]
σ_{yld}^+ (MPa)	–	240–450	–
σ_{yld}^- (MPa)	–	−240 to −450	–
E (GPa)	20–50	200–210	L: 76[a]220[b]45[c]T: 5.5[a]6.9[b]12[c]
v (−)	0.1–0.2	0.28–0.30	0.34[a]0.25[b]0.19[c]

L, Longitudinal; T, Transverse.
a)   Aramid fibers.
b)   High modulus carbon fibers.
c)   E-glass fibers (all composites with epoxy matrix).
Source: Modified from Glisic and Inaudi (2007).

components in an isotropic linear-elastic material (e.g., steel and concrete observed at macroscale at low levels of stress) is given by the following Expression (e.g., see Malvern 1969):

$$
\begin{bmatrix} \varepsilon_{x,E}(t) & \tfrac{1}{2}\gamma_{xy,E}(t) & \tfrac{1}{2}\gamma_{xz,E}(t) \\ \tfrac{1}{2}\gamma_{yx,E}(t) & \varepsilon_{y,E}(t) & \tfrac{1}{2}\gamma_{yz,E}(t) \\ \tfrac{1}{2}\gamma_{zx,E}(t) & \tfrac{1}{2}\gamma_{zy,E}(t) & \varepsilon_{z,E}(t) \end{bmatrix} = \begin{bmatrix} \dfrac{\sigma_x(t)-v\sigma_y(t)-v\sigma_z(t)}{E} & \dfrac{1}{2}\dfrac{\tau_{xy}(t)}{G} & \dfrac{1}{2}\dfrac{\tau_{zx}(t)}{G} \\ \dfrac{1}{2}\dfrac{\tau_{yx}(t)}{G} & \dfrac{\sigma_y(t)-v\sigma_z(t)-v\sigma_x(t)}{E} & \dfrac{1}{2}\dfrac{\tau_{yz}(t)}{G} \\ \dfrac{1}{2}\dfrac{\tau_{zx}(t)}{G} & \dfrac{1}{2}\dfrac{\tau_{zy}(t)}{G} & \dfrac{\sigma_z(t)-v\sigma_x(t)-v\sigma_y(t)}{E} \end{bmatrix}
\tag{5.17}
$$

where E is the Young's modulus, v is Poisson's ratio, and $G = \dfrac{E}{2(1+v)}$ is shear modulus of the material.

The relationship between strain and stress becomes somewhat clearer if strain and stress components are rearranged as vectors and Expression 5.17 is rewritten accordingly, as shown in Expression 5.18.

$$
\begin{bmatrix} \varepsilon_{x,E}(t) \\ \varepsilon_{y,E}(t) \\ \varepsilon_{z,E}(t) \\ \gamma_{xy,E}(t) \\ \gamma_{yz,E}(t) \\ \gamma_{zx,E}(t) \end{bmatrix} = \begin{bmatrix} \tfrac{1}{E} & -\tfrac{v}{E} & -\tfrac{v}{E} & 0 & 0 & 0 \\ -\tfrac{v}{E} & \tfrac{1}{E} & -\tfrac{v}{E} & 0 & 0 & 0 \\ -\tfrac{v}{E} & -\tfrac{v}{E} & \tfrac{1}{E} & 0 & 0 & 0 \\ 0 & 0 & 0 & \tfrac{1}{G} & 0 & 0 \\ 0 & 0 & 0 & 0 & \tfrac{1}{G} & 0 \\ 0 & 0 & 0 & 0 & 0 & \tfrac{1}{G} \end{bmatrix} \begin{bmatrix} \sigma_x(t) \\ \sigma_y(t) \\ \sigma_z(t) \\ \tau_{xy}(t) \\ \tau_{yz}(t) \\ \tau_{zx}(t) \end{bmatrix}
$$

$$
\mathbf{E}_E^v(t) = \mathbf{C}_{iso}^{-1}\mathbf{\Sigma}^v(t)
\tag{5.18}
$$

where $\mathbf{E}^v{}_E$ is a vector containing elastic strain tensor components (i.e., a strain tensor rearranged into a vector), $\mathbf{C}_{iso}{}^{-1}$ is the flexibility matrix of elastic constants for isotropic material (inverse matrix of the stiffness $\mathbf{C}_{iso}$ matrix), and $\mathbf{\Sigma}^v$ is a vector containing stress tensor components (i.e., a stress tensor rearranged into a vector).

Expression 5.19 shows the relationship between the stress tensor and the strain tensor.

$$
\begin{bmatrix}
\sigma_x(t) \\
\sigma_y(t) \\
\sigma_z(t) \\
\tau_{xy}(t) \\
\tau_{yz}(t) \\
\tau_{zx}(t)
\end{bmatrix}
=
\begin{bmatrix}
\frac{E(1-v)}{1-v-2v^2} & \frac{Ev}{1-v-2v^2} & \frac{Ev}{1-v-2v^2} & 0 & 0 & 0 \\
\frac{Ev}{1-v-2v^2} & \frac{E(1-v)}{1-v-2v^2} & \frac{Ev}{1-v-2v^2} & 0 & 0 & 0 \\
\frac{Ev}{1-v-2v^2} & \frac{Ev}{1-v-2v^2} & \frac{E(1-v)}{1-v-2v^2} & 0 & 0 & 0 \\
0 & 0 & 0 & G & 0 & 0 \\
0 & 0 & 0 & 0 & G & 0 \\
0 & 0 & 0 & 0 & 0 & G
\end{bmatrix}
\begin{bmatrix}
\varepsilon_{x,E}(t) \\
\varepsilon_{y,E}(t) \\
\varepsilon_{z,E}(t) \\
\gamma_{xy,E}(t) \\
\gamma_{yz,E}(t) \\
\gamma_{zx,E}(t)
\end{bmatrix}
$$

$$\mathbf{\Sigma}^v(t) = \mathbf{C}_{iso}\mathbf{E}_E^v(t) \tag{5.19}$$

The stiffness matrix $\mathbf{C}_{iso}$ is symmetric and contains only 12 non-zero components, all of them described by only 2 constants: E and v (G is a function of E and v); this matrix can be used only if the material is isotropic. However, if that is not the case, then either the stiffness matrix $\mathbf{C}$ has more non-zero members or it has the same number of non-zero components, but they are not described by only two constants, E and v (e.g., Malvern 1969). For anisotropic materials, in general (e.g., fiber-reinforced composites, depending on their design), all 36 components of the matrix $\mathbf{C}$ can be different from zero (in linear theory of structures, where displacements are small, 21 components are independent due to symmetry); for materials with a single plane of elastic symmetry, the matrix reduces to 21 non-zero members (in linear theory, 13 are independent due to symmetry); finally, for materials with three orthogonal planes of elastic symmetry (orthotropic materials), the matrix reduces to 12 non-zero members (in linear theory, 9 are independent due to symmetry). Matrix $\mathbf{C}_{ort}$ for orthotropic anisotropic materials has a similar appearance to matrix $\mathbf{C}_{iso}$ shown in Expression 5.17, but non-zero components are different for different directions and cannot be expressed as functions of only two constants (E and v); they are obtained by inversion of flexibility matrix $\mathbf{C}_{ort}^{-1}$ shown in Expression 5.18. An example of a material that is orthotropic but anisotropic is timber.

$$
\mathbf{C}_{ort}^{-1} =
\begin{bmatrix}
\frac{1}{E_\xi} & -\frac{v_{\xi\eta}}{E_\xi} & -\frac{v_{\xi\zeta}}{E_\xi} & 0 & 0 & 0 \\
-\frac{v_{\eta\xi}}{E_\eta} & \frac{1}{E_\eta} & -\frac{v_{\eta\zeta}}{E_\eta} & 0 & 0 & 0 \\
-\frac{v_{\zeta\xi}}{E_\zeta} & -\frac{v_{\zeta\eta}}{E_\zeta} & \frac{1}{E_\zeta} & 0 & 0 & 0 \\
0 & 0 & 0 & \frac{1}{G_{\xi\eta}} & 0 & 0 \\
0 & 0 & 0 & 0 & \frac{1}{G_{\eta\zeta}} & 0 \\
0 & 0 & 0 & 0 & 0 & \frac{1}{G_{\zeta\xi}}
\end{bmatrix}
\tag{5.20}
$$

where ξ, η, and ζ are the axes defining the planes of orthotropic symmetry in material.

Equations 5.11 and 5.12 show criteria based on strain evaluation for determining whether a uni-axially stressed material is in an elastic regime (i.e., undamaged), regardless of the type of material (concrete, steel, timber, or composites). Equations 5.15 and 5.16 define similar criteria based on stress evaluation. Equations with similar simplicity cannot be universally applied to all construction materials if they are in multi-dimensional state of stress. For ductile materials, the Von Mises criterion (e.g., Malvern 1969) can be used, with some restrictions; however, for brittle materials, this criterion does not apply, and fracture mechanics approaches should be used, but they require numerical modeling and knowledge of various material parameters that should be determined by laboratory testing (e.g., for concrete, the Concrete Damaged Plasticity [CDP] model, see Lubliner et al. 1989, Lee et al. 1998, Brune 2010, Hafezolghorani et al. 2017). In practical applications of civil engineering, for a given construction material, typically only uniaxial yield stresses σ_{yld}^+ and σ_{yld}^- (for ductile materials) and uniaxial ultimate stresses σ_u^+ and σ_u^- (for both ductile and brittle materials) are available, and hence advanced approaches for ascertaining whether the material is within the elastic range would require additional testing. While these advanced approaches are out of scope of this introductory book, two simplified methods to assess whether the material is within its elastic range or close to exceeding it are presented in this section. Note that these two methods are not fully accurate, and some restrictions apply when using them. These methods are based on strains and stresses and assume that besides uniaxial yield and ultimate stresses, yield and ultimate strains, Young's modulus, and Poisson's coefficient are also known (so that the elastic strain can be converted to stress using Expressions 5.18–5.20). As a reminder, if some of the material properties are not known, typical values or empirical formulae found in the literature can be used (with caution!) to evaluate them (see Table 5.3).

Similar to the strain tensor, it is well known from solid and continuum mechanics (e.g., Malvern 1969) that there exist three mutually perpendicular axes, called principal axes, such that shear stress components vanish when the stress tensor is observed with respect to these axes. In that case, the matrix of stress tensors contains only diagonal (normal) stress components, called principal stress components or principal stresses, denoted with σ_1, σ_2, and σ_3 ($\sigma_1 \geq \sigma_2 \geq \sigma_3$). The determination of principal stresses and principal axes follows the same formulae as in the case of the strain tensor; see Expressions 5.3–5.5 for the three-dimensional state of stress and 5.6 for the two-dimensional state of stress (i.e., symbol ϵ can be simply replaced by symbol σ in these expressions to derive expressions for principal stresses and principal axes, e.g., Malvern 1969). Note that in general, for anisotropic materials, principal stress axes are different from principal strain axes; however, for isotropic materials, they coincide.

The first simplified method to determine whether the material is out of its elastic range is applicable to brittle materials, such as (unreinforced) concrete or composites. It is based on a comparison between principal strains and stresses (principal strain and stress components) and ultimate strains and stresses (tension and compression strength) of the material, respectively. Given that the material is in a complex state of stress, a simple expression to ascertain whether the material is in an elastic regime (e.g., similar to Expressions 5.11a and 5.15a), cannot be used. However, simplified expressions can be applied to examine whether the material is out of elastic range, as follows:

(a) At least one $\epsilon_i \geq \dfrac{\epsilon_u^+}{\beta_\epsilon^+}$ and one $\sigma_i \geq \dfrac{\sigma_u^+}{\beta_\sigma^+}$ $\Rightarrow$ material fails in tension (rupture/cracking)

(b) At least one $\epsilon_i \geq \dfrac{\epsilon_u^+}{\beta_\epsilon^+}$ or one $\sigma_i \geq \dfrac{\sigma_u^+}{\beta_\sigma^+}$ $\Rightarrow$ potential failure in tension (rupture/cracking)

(c) At least one $\epsilon_i \leq \dfrac{\epsilon_u^-}{\beta_\epsilon^-}$ or one $\sigma_i \leq \dfrac{\sigma_u^-}{\beta_\sigma^-}$ $\Rightarrow$ potential failure in compression (crushing)

(d) $\dfrac{\epsilon_u^-}{\beta_\epsilon^-} < \epsilon_3 \leq \epsilon_2 \leq \epsilon_1 < \dfrac{\epsilon_u^+}{\beta_\epsilon^+}$ and $\dfrac{\sigma_u^-}{\beta_\sigma^-} < \sigma_3 \leq \sigma_2 \leq \sigma_1 < \dfrac{\sigma_u^+}{\beta_\sigma^+}$ $\Rightarrow$ potentially elastic regime

$$(5.21)$$

where ϵ_i and σ_i ($i = 1, 2, 3$) denote principal strains and stresses, respectively, and β-s denote safety factors.

The first expression, 5.21a, shows that if one principal strain component ϵ_i ($i=1,2,3$) and one principal stress component σ_i ($i=1,2,3$) both exceed the ultimate values in tension, the material will fail. Note that there is no equivalent expression for compression, as the strength of confined (three-axially loaded) materials in compression is higher than the absolute value of (uniaxial) ultimate stress.

The next two expressions, 5.21b and c, show that if either one principal strain component ε_i (i=1,2,3) or one principal stress component σ_i (i=1,2,3) exceeds ultimate value, the state of the material is undetermined (inconclusive), but there is potential that the material would fail. For SHM purposes, this indicates a critical state of material that should require further analysis to ascertain if the material is failing or not.

Finally, the last expression shows that if no principal strain or stress exceeds the ultimate values, the material is potentially in an elastic regime. However, this last statement has to be considered with a lot of caution. For example, Lemnitzer et al. (2008) performed a statistical analysis of testes on bi-axially tensioned concrete and concluded that for higher levels of tension, the concrete fails before reaching ultimate values. However, this can be accounted for, by appropriate choice of safety factors (e.g., $\beta_\sigma^+ \geq 1.5$ would guarantee an elastic regime for bi-axially tensioned concrete as per Lemnitzer et al. 2008).

Overall, if the principal strain or principal stress values are close to ultimate values, there is potential for failure, as the strain and stress in common structures, by design, rarely approach ultimate values (in tension or compression). Expressions 5.21a–c, in most of the cases, indicates that the structure's material is potentially failing and that the structure might be in distress. In the cases where Expressions 5.21b and c are valid, more sophisticated analysis should be used to ascertain whether the material is in an elastic regime or failing.

Expressions 5.21a and b pointed to criterion to identify cracking in structures with brittle materials; note that cracks in reinforced concrete are common, so crack does not necessarily represent the damage in that case.

The second simplified method to determine whether the material is out of elastic range is applicable to ductile materials, and it is based on comparison with Von Mises stress. Once principal stresses σ_1, σ_2, and σ_3 are known, Von Mises stress can be determined using Expression 5.22 (e.g., Malvern 1969).

$$\sigma_{VM} = \sqrt{\frac{(\sigma_1 - \sigma_2)^2 + (\sigma_2 - \sigma_3)^2 + (\sigma_3 - \sigma_1)^2}{2}} \tag{5.22}$$

For the cases of plane stress (e.g., uniaxial bending of beams, surface of structural elements), one principal stress is null, and this expression transforms into Expression 5.23.

$$\sigma_{VM} = \sqrt{\sigma_1^2 - \sigma_1\sigma_2 + \sigma_2^2} \tag{5.23}$$

where σ_1 and σ_2 are principal stresses in the observed plane.

It is not necessary to know principal stresses to determine Von Mises stress – it can be determined from the stress tensor established with regard to any coordinate system $Oxyz$, using the stress components as shown in Expression 5.24.

$$\sigma_{VM} = \sqrt{\frac{(\sigma_x - \sigma_y)^2 + (\sigma_y - \sigma_z)^2 + (\sigma_z - \sigma_x)^2 + 6\left(\tau_{xy}^2 + \tau_{yz}^2 + \tau_{zx}^2\right)}{2}} \tag{5.24}$$

In the case of plane stress, this expression transforms into Expression 5.25.

$$\sigma_{VM} = \sqrt{\sigma_x^2 - \sigma_x\sigma_y + \sigma_y^2 + 3\tau_{xy}^2} \tag{5.25}$$

where σ_x, σ_y, and τ_{xy} are stress components with respect to the coordinate system Oxy in the observed plane.

Once Von Mises stress is calculated, the following expressions can be used to evaluate whether the material is out of the elastic regime:

(a) At least one $\varepsilon_i \geq \dfrac{\varepsilon_{yld}^+}{\beta_\varepsilon^+}$ or one $\sigma_i \geq \dfrac{\sigma_{yld}^+}{\beta_\sigma^+}$

$\Rightarrow$ potential plastic damage or failure

(b) $\sigma_{VM} \geq \min\left(\left|\dfrac{\sigma_{yld}^+}{\beta_\sigma^+}\right|, \left|\dfrac{\sigma_{yld}^-}{\beta_\sigma^-}\right| \right)$

$\Rightarrow$ potential plastic damage or failure

(c) At least one $\varepsilon_i \geq \dfrac{\varepsilon_u^+}{\beta_\varepsilon^+}$ or one $\sigma_i \geq \dfrac{\sigma_u^+}{\beta_\sigma^+}$

$\Rightarrow$ potential failure in tension

(d) At least one $\varepsilon_i \leq \dfrac{\varepsilon_u^-}{\beta_\varepsilon^-}$ or one $\sigma_i \leq \dfrac{\sigma_u^-}{\beta_\sigma^-}$

$\Rightarrow$ potential failure in compression

(e) $\dfrac{\varepsilon_{yld}^-}{\beta_\varepsilon^-} \leq \varepsilon_3 \leq \varepsilon_2 \leq \varepsilon_1 \leq \dfrac{\varepsilon_{yld}^+}{\beta_\varepsilon^+}$ and $\sigma_{VM} \leq \min\left(\left|\dfrac{\sigma_{yld}^+}{\beta_\sigma^+}\right|, \left|\dfrac{\sigma_{yld}^-}{\beta_\sigma^-}\right| \right)$

$\Rightarrow$ potentially elastic regime   (5.26)

The first expression, 5.26a, shows that if at least one principal strain or principal stress exceeds the yield values in tension, the state of the material is undetermined (inconclusive), but there is potential that the material will be potentially in a plastic regime or failing. The second expression, 5.26b, is similar, but it uses Von Mises stress instead of principal stresses and does not require analysis of principal strain. Von Mises stress is compared with the absolute value of yield stresses, and a conservative, minimum absolute value is used if tensional and compressive yield stresses are mutually different. The next two expressions, 5.26c and d, show that if either one principal strain or one principal stress exceeds ultimate value, the state of the material is undetermined (inconclusive), but there is potential that the material would fail. For SHM purposes, all four expressions indicate a critical state of material that should require further analysis to ascertain if the material is yielding and failing or not.

Finally, the last expression shows that if no principal strain exceeds the ultimate values and Von Mises stress does not exceeds the minimum absolute value of yield stress, the material is potentially in an elastic regime. However, this last statement has to be considered with caution, as three-axial tension can result in plastic deformation at higher levels of principal stress, which, however, can be accommodated by an appropriate choice of safety factors.

Overall, if the principal strain or principal stress values are close to ultimate values, there is potential for failure, as the strain and stress in common structures, by design, rarely approach ultimate values (in tension or compression). Expressions 5.26a–d, in most of the cases, indicates that the structure's material is potentially failing and that the structure might be in distress. In the cases where Expressions 5.21b and c are valid, more sophisticated analysis should be used to ascertain whether the material is in an elastic regime or failing.

Example 5.4 *Determining principal stress components and Von Mises stress from elastic strain measurements*

Let's assume that measurements taken by the rosette in Example 5.1 are identified as elastic strain, i.e., other strain constituents (e.g., rheological and thermal strain) are not present. Let's assume that these measurements are performed on an isotropic material with a Young's modulus $E = 210\,\text{GPa}$ and a Poisson's coefficient $v = 0.3$ (e.g., steel). Finally, let's assume that there is no load applied to the surface of the material, i.e., orthogonal to the plane of sensors. In Example 5.2, the principal strain components in plane Oxy were calculated: $\varepsilon_1 = \varepsilon_{1,E} = 66.1 \pm 2.3\,\mu\varepsilon$, and $\varepsilon_2 = \varepsilon_{2,E} = -86.1 \pm 3.5\,\mu\varepsilon$. The third principal strain component is not known, but normal stress in the direction of the z-axis (orthogonal to plane Oxy) is null, i.e., $\sigma_z = 0\,\text{MPa}$. For coordinate systems defined with principal strain axes n_1, n_2, and z, shear strain components are all null. These axes are also the principal stress axes. Thus, Expression 5.15, when substituting all the known values, becomes:

$$
\begin{bmatrix} \sigma_{n_1}(t) \\ \sigma_{n_2}(t) \\ 0 \\ \tau_{n_1 n_2}(t) \\ \tau_{n_2 z}(t) \\ \tau_{z n_1}(t) \end{bmatrix} = \begin{bmatrix} 282.692 & 121.154 & 121.154 & 0 & 0 & 0 \\ 121.154 & 282.692 & 121.154 & 0 & 0 & 0 \\ 121.154 & 121.154 & 282.692 & 0 & 0 & 0 \\ 0 & 0 & 0 & 80.769 & 0 & 0 \\ 0 & 0 & 0 & 0 & 80.769 & 0 \\ 0 & 0 & 0 & 0 & 0 & 80.769 \end{bmatrix} \cdot 10^9 \begin{bmatrix} 66.1 \pm 2.5 \\ -86.1 \pm 3.5 \\ \varepsilon_{z,E}(t) \\ 0 \\ 0 \\ 0 \end{bmatrix} \cdot 10^{-6}
$$

The multiplication of the matrices above confirms that shear stress components are all null, and only three components are unknown: principal stresses in directions n^1 and n^2, and principal strain in direction z. The latter can be calculated as follows:

$$
\varepsilon_{z,E}(t) = -\frac{121,154}{282,692} \cdot (66.1 \pm 2.3) - \frac{121,154}{282,692} \cdot (-86.1 \pm 3.5) = 8.6 \mp 2.5\,\mu\varepsilon
$$

Consequently, the matrices above can be reduced and rearranged as follows:

$$
\begin{bmatrix} \sigma_{n_1}(t) \\ \sigma_{n_2}(t) \end{bmatrix} = \begin{bmatrix} 282,692 - \dfrac{121,154^2}{282,692} & 121,154 - \dfrac{121,154^2}{282,692} \\ 121,154 - \dfrac{121,154^2}{282,692} & 282,692 - \dfrac{121,154^2}{282,692} \end{bmatrix} \begin{bmatrix} 66.1 \pm 2.5 \\ -86.1 \pm 3.5 \end{bmatrix}
$$

From where, the principal stresses are calculated as follows:

$$
\sigma_{n_1}(t) = 9,293,077 \pm 782,903\,\text{Pa} = 9.3 \pm 0.8\,\text{Pa},
$$

$$
\sigma_{n_2}(t) = -15,293,077 \pm 974,931\,\text{Pa} = -15.3 \pm 1.0\,\text{Pa}.
$$

Note that $\sigma_{n_1}(t) > \sigma_z(t) > \sigma_{n_2}(t)$, and thus, $\sigma_1(t) = \sigma_{n_1}(t)$, $\sigma_2(t) = \sigma_z(t) = 0$, and $\sigma_3(t) = \sigma_{n_2}(t)$. Finally, Von Mises stress is calculated using Expression 5.16:

$$
\sigma_{VM} = \sqrt{\frac{(9.3 - 0)^2 + (0 - (-15.3)^2 + (-15.3 - 9.3)^2}{2}} = 21.5\,\text{MPa}
$$

Application of the error propagation formulae presented in Table 4.2 would be complicated in this example due to the complexity of Expression 5.16. That is the reason why an exhaustive analysis by inserting all variations of all arguments plus/minus limits of error values in Expression 5.16 is performed. This yields the following values for the limits of error:

$$\delta\sigma_{VM} = \pm 1.6 \, \text{MPa for } \sigma_1 = 9.3 \pm 0.8 \, \text{MPa and } \sigma_3 = -15.3 \mp 1.0 \, \text{MPa}.$$

5.6.4 Analytical Expressions for Stress and Strain in Beams

Sections 5.6.2 and 5.6.3 practically present simplified methods for assessing whether the structure is in a safe, elastic state of deformation. However, as stated earlier, in typical structures, the levels of elastic strain and stress are, by design, significantly lower than their ultimate values, unless the structure experiences advanced damage or deterioration. Thus, it is of interest to compare the elastic strain and stress with corresponding values obtained through structural analysis to identify discrepancies between design and reality, and identify early loss of performance, and early occurrence of unusual structural behaviors (e.g., damage or deterioration). To do this, it is necessary to understand the distribution of elastic strain and stress in structural members due to loads. Note that elastic strain and stress can also be consequences of thermal loads and rheological effects, which will be addressed in Sections 5.7 and 5.8.

The distribution of elastic strain and stress in a structural member depends on the load type (e.g., concentrated forces, distributed forces, and moments), intensity, and position, but also on the type of structural member (e.g., beam, plate, and shell), its geometrical properties (e.g., dimensions, and cross-sectional properties), its overall physical properties (e.g., isotropy, and homogeneity), its material's mechanical properties (Young's modulus, Poisson's coefficient), and its boundary conditions (i.e., structural system). Hence, the distribution of elastic strain in structural members can be very complex, and the presentation of all possible scenarios is out of the scope of this book. They represent distinct topics of study in separate branches of structural engineering such as Theory of Structures, Structural Analysis, Theory of Plate and Shells, and Solid Mechanics, and are therefore presented in detail in relevant literature (e.g., see Timoshenko and Young 1945, Billington 1965, Brčić 1989, Hibbeler 2008). Nevertheless, fundamental analytical expressions for typical beam-like structural members are presented here to facilitate an introductory comparison between the measured and designed values of elastic strain and stress and provide a workflow on how to do it.

Straight beams have one dimension (the length) that is significantly larger (at least five times larger) than the two other (cross-sectional) dimensions (width and depth), the latter two having a mutually similar order of magnitude (Đurić and Đurić-Perić 1990). In addition, curved beams have a radius of geometrical curvature significantly larger (at least five times larger) than the cross-sectional dimensions, width, and depth (Đurić and Đurić-Perić 1990).

If one principal axis of each cross-section belongs to the same plane, the beam is planar. Planar beams loaded with forces and moments acting in the beam plane are subjected to uniaxial bending, normal force, and shear force, and their stress distribution is planar (plane state of stress). However, the state of strain is not planar due to Poisson's ratio (see Expression 5.18). For planar straight beams in linear range made of isotropic material, the stress components in cross-sections at time t are related to internal forces and geometrical properties as described in Expression 5.27 (see Figure 5.4 for notation).

$$\sigma_z(t, y(t), z) = \frac{N_z(t, z)}{A(t, z)} + \frac{M_x(t, z)}{I_x(t, z)} y(t)$$

$$\tau_{zy}(t, y(t), z) = \frac{S_y(t, z) Q_x(t, y(t), z)}{I_x(t, z) w(t, y(t), z)} \tag{5.27}$$

where z-axis coincides with the centroid line of the beam and y-axis is orthogonal to z-axis in beam plane (see Figure 5.4), $N_z(t,z)$, $S_y(t,z)$, and $M_x(t,z)$ are normal force, shear force, and bending moment at cross-section with local coordinate z at time t, $A(t,z)$ is the area of the cross-section at z at time t, $I_x(t,z)$ is the moment of inertia (second moment of area) of the cross-section at z with respect to axis x at time t, $w(y,z)$ is the width of the cross-section at depth y at time t, and $Q_x(t,y,z)$ is the static moment of inertia (first moment of area) with respect to axis x of the part of the cross-section delimited with coordinates y and d_{btm} at time t, i.e., $Q_x(t, y, z) = \iint\limits_{A'(t,x,y)} y'(t) dA'(t)$ (see Figures 5.4 and 5.5).

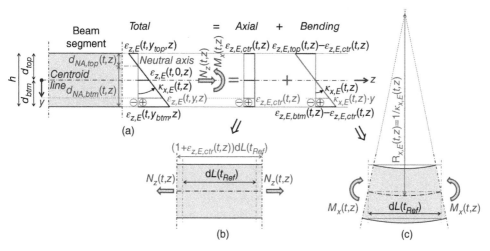

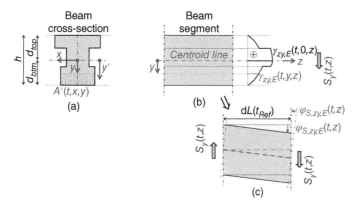

Figure 5.4 Geometrical properties of beam elevation and distribution of normal strain component in direction of z-axis under normal force $N_z(z)$ and bending moment $M_x(z)$; axes y and z define the plane of the beam, and axis x is orthogonal to the plane, i.e., it points outwards (toward the reader); gray-shaded area represents the vertical cut (elevation) of a short segment of a beam in vertical plane yz.

Figure 5.5 Geometrical properties of beam cross-section and distribution of shear strain component in the cross-sectional plane defined by axes x and y under shear force $S_y(z)$; axis z in Figure 5.5a is orthogonal to the plane, i.e., it points outwards (toward the reader).

Note that in Expression 5.27, in general, the geometrical properties of cross-section can vary along the z-axis; however, for prismatic beams (beams with invariant cross-section), they do not vary along the z-axis. Also, mechanical and geometrical properties can vary in time t, e.g., due to damage or degradation (loss of cross-section), repair or retrofitting (straightening of cross-section), the usual evolution of elastic modulus in concrete, and the common onset of cracking in reinforced concrete. It is important to emphasize that the y-axis originates from the centroid of the cross-section, and thus if the cross-sectional properties change over time, they may influence the location of the centroid, and thus the y-coordinate of an observed point will also change over time, depending on the location of the centroid.

Expression 5.27 is not applicable for shear distribution in thin-walled cross-sections, and a different expression applies in that case (e.g., see Brčić 1989 or other relevant literature). Expression 5.27.

Based on Expression 5.18, elastic strain components at time t are described by Expression 5.28. The geometrical properties of beam elevation and the distribution of the normal strain component in the direction of the z-axis are shown in Figure 5.4. The general geometrical properties of beam cross-section and distribution of shear strain components in the plane defined by axes x and y (i.e., orthogonal to axis z) are shown in Figure 5.5.

$$\varepsilon_{z,E}(t,y(t),z) = \frac{N_z(t,z)}{E(t,z)A(t,z)} + \frac{M_x(t,z)}{E(t,z)I_x(t,z)}y(t)$$

$$\varepsilon_{x,E}(t,y(t),z) = \varepsilon_{y,E}(t,y(t),z) = -v\varepsilon_{z,E}(t,y(t),z)$$

$$\gamma_{zy,E}(t,y(t),z) = \frac{S_y(t,z)Q_x(t,y(t),z)}{G(t,z)I_x(t,z)w(t,y(t),z)} \qquad (5.28)$$

Figure 5.4a and Expression 5.28 present the linear distribution of the normal elastic strain component in the direction of the z-axis across the cross-section. It also shows the independence of influences of normal (axial) force and bending moment. Figure 5.4b illustrates the influence of normal force $N_z(t,z)$ at time t at cross-section with coordinate z: extension or contraction of the infinitesimal segment of the beam dL with respect to reference time t_{Ref} (denoted with $dL(t_{Ref})$ in the figure) for the strain value at the centroid of cross-section $\varepsilon_{z,E,ctr}(t,z)$. Figure 5.4c illustrates the influence of bending moment $M_x(t,z)$ at time t at cross-section with coordinate z: curving of the infinitesimal beam segment $dL(t_{Ref})$, resulting in change of shape from straight to curved segment described by curvature $\kappa_{x,E}(t,z)$ or radius of curvature $R_{x,E}(t,z)$. Based on Expression 5.28 and the geometrical relationships shown in Figure 5.4, the normal strain component at the centroid of the cross-section with coordinate z, the curvature at the same cross-section with coordinate z, and the normal strain component at a point within the cross-section at distance y from the centroid can be expressed as follows:

$$\varepsilon_{z,E,ctr}(t,z) = \frac{N_z(t,z)}{E(t,z)A(t,z)} = \varepsilon_{z,E,btm}(t,z)\frac{d_{top}}{h} + \varepsilon_{z,E,top}(t,z)\frac{d_{btm}}{h} \qquad (5.29)$$

$$\kappa_{x,E}(t,z) = \frac{M_x(t,z)}{E(t,z)I_x(t,z)} = \frac{\varepsilon_{z,E,btm}(t,z) - \varepsilon_{z,E,top}(t,z)}{h} \qquad (5.30)$$

$$\varepsilon_{z,E}(t,y(t),z) = \varepsilon_{z,E,ctr}(t,z) + \kappa_{x,E}(t,z) \cdot y(t) \qquad (5.31)$$

where subscript *btm* indicates the bottom of cross-section, *top* indicates the top of cross-section, *ctr* indicates the centroid; d_{btm} and d_{top} are distances from centroid to bottom and top of cross-section respectively, $\varepsilon_{z,E,btm}(t,z) = \varepsilon_{z,E}(t,y_{btm},z)$ and $\varepsilon_{z,E,top}(t,z) = \varepsilon_{z,E}(t,y_{top},z)$ are strains at the bottom and top of the cross-section respectively (note that $y_{btm} = +d_{btm}$ and $y_{top} = -d_{top}$); h is the depth (height) of the cross-section ($h = d_{btm} + d_{top} = y_{btm} - y_{top}$).

Figure 5.5b and Expression 5.28 present the non-linear distribution of the shear elastic strain component in the direction of the y-axis across the cross-section. Depending on cross-sectional properties, this distribution can even be discontinuous, as shown in Figure 5.5b. Figure 5.5c illustrates the influence of shear force $S_y(t,z)$ at time t at cross-section with coordinate z: overall rotation of the beam's cross-section with respect to the centroid line $\varphi_{S,zy,E}(t,z)$, with respect to a reference time t_{Ref}. Note that in reality, the cross-section will experience deplanation due to shear strain, and this overall rotation $\varphi_{S,zy,E}(t,z)$ represents the slope of the linear approximation of this deplanation. It can be expressed as follows:

$$\varphi_{S,zy,E}(t,z) = k_S(t,z)\frac{S_y(t,z)}{G(t,z)A(t,z)} \tag{5.32}$$

where $k_S(t,z) = \frac{A(t,z)}{I_x(t,z)^2}\int_{A(t,z)}\frac{Q_x(t,y(t),z)^2}{w(t,y(t),z)^2}dA(t,z)$ is dimensionless shear coefficient (Đurić and Đurić-Perić 1990) that only depends on the geometry of the observed cross-section (e.g., $k = 1.2$ for a rectangular cross-section).

Expressions 5.29–5.32 are helpful in evaluating the design values of elastic strain and their comparison with measurements. In order to better understand these expressions, typical distributions of elastic deformation parameters $\varepsilon_{z,E,ctr}(t,z)$, $\kappa_{x,E}(t,z)$, and $\varphi_{S,zy,E}(t,z)$ in a structure made of beams with invariant cross-sections under the most common loads – concentrated forces, pure moments (couples) and distributed forces – are presented in Figure 5.6. In the figure, the beams are delimited with points a, b, c, and d, and their respective properties of cross-section, area A, moment of inertia I, and shear coefficient k_S are denoted with the delimiting points in subscript, and only a few illustrative external forces and a moment are applied: concentrated forces $F_{z,a}$ (at point a, producing normal forces in the structure) and $F_{y,f}$ (at point f, producing shear forces and bending moments), distributed force $q_{y,cd}$ (between points c and d, producing shear forces and bending moments), and pure moment (couple) $M_{x,e}$ (at point e, producing shear forces and bending moments).

For these common loads, the normal force distribution along beams is piece-wise constant or linear, the bending moment distribution is piece-wise constant, linear, or parabolic, and the shear

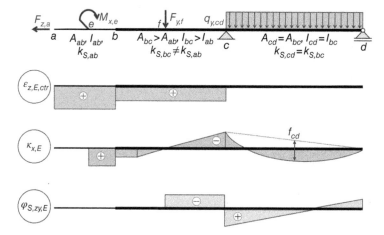

Figure 5.6 Schematic representation of the most common loads on a planar beam structure and typical resulting centroid strain, curvature, and shear angle distributions. Source: Modified from Glisic and Inaudi (2007).

force distribution is piece-wise constant or linear; consequently, the distributions of the centroid strain $\varepsilon_{z,E,ctr}$, curvature $\kappa_{x,E}$, and shear angle $\varphi_{S,zy,E}$ along the centroid line of structure are similar to those of the corresponding internal forces, as shown in Figure 5.6.

If the external forces and moments do not act in the beam's plane or the beam is not planar, then, besides the normal force $N_z(t,z)$, the beam experiences bi-axial bending moments $M_x(t,z)$ and $M_y(t,z)$, bi-axial shear forces $S_x(t,z)$ and $S_y(t,z)$, and a torsion moment $T_z(t,z)$, and its stress distribution is in general three-dimensional. However, with respect to the coordinate system $Oxyz$, only the normal stress component $\sigma_z(t,z)$ is, in general, different from zero, while two other normal stress components, $\sigma_x(t,z)$ and $\sigma_y(t,z)$ are null. Shear stress components $\tau_{zx}(t,z)$ and $\tau_{zy}(t,z)$ are, in general, different from zero, while $\tau_{xy}(t,z)$ is null. The normal and shear stress components are given in Expression 5.33.

$$\sigma_z(t,y(t),z) = \frac{N_z(t,z)}{A(t,z)} + \frac{M_x(t,z)}{I_x(t,z)}y(t) - \frac{M_y(t,z)}{I_y(t,z)}x(t)$$

$$\tau_{zy}(t,y(t),z) = \frac{S_y(t,z)Q_x(t,y(t),z)}{I_x(t,z)w(t,y(t),z)} + \frac{T_z(t,z)}{J_z(t,z)}f_{\tau zy}(y(t),z(t))$$

$$\tau_{zx}(t,x(t),z) = \frac{S_x(t,z)Q_y(t,x(t),z)}{I_y(t,z)d(t,x(t),z)} + \frac{T_z(t,z)}{J_z(t,z)}f_{\tau zx}(x(t),z(t)) \tag{5.33}$$

where $h(t,x(t),z)$ is the vertical dimension (depth) of the cross-section with coordinate z at time t and at location $x(t)$, $J_z(t,z)$ is the torsion constant (torsion moment of inertia), and functions $f_{\tau zy}(y(t),z(t))$ and $f_{\tau zx}(x(t),z(t))$ describe shear distribution in cross-sections due to torsion, in directions y and x, respectively; note that there are no general expressions to calculate $J_z(t,z)$, $f_{\tau zy}(y(t),z(t))$, and $f_{\tau zx}(x(t),z(t))$, and usually specialized tables or finite element modeling are used to determine them.

Strain components $\varepsilon_{x,E}(t,x,y,z)$, $\varepsilon_{y,E}(t,x,y,z)$, $\varepsilon_{z,E}(t,x,y,z)$, $\tau_{zx,E}(t,x,y,z)$, and $\tau_{zy,E}(t,x,y,z)$, as well as deformation parameters $\varepsilon_{z,E,ctr}(t,z)$, $\kappa_{x,E}(t,z)$, $\kappa_{y,E}(t,z)$, $\varphi_{S,zx,E}(t,z)$, and $\varphi_{S,zy,E}(t,z)$, can be expressed using equations similar to those given in Expressions 5.28–5.32 for planar beams. A deformation parameter that is not present in the planar case is the angle of twist, which can be expressed using the following expression:

$$\theta_{z,E}(t,z) = \frac{T_z(t,z)}{G(t,z)J_z(t,z)} \tag{5.34}$$

Note that Equations 5.27–5.34 apply for curvilinear planar beams too, by simply replacing coordinate z with arc-length (natural) coordinate s. Figures 5.4–5.6 apply too, with appropriate geometrical modifications. Also, note that these expressions do not apply at and around cross-sections where stress concentrations occur, i.e., where a concentrated load is applied or where an abrupt change in cross-sectional properties is present (recall the Saint-Venant principle).

5.7 Thermal Strain and Thermal Gradients

5.7.1 Thermal Strain at a Point

Temperature variations in a structure's environment result in temperature changes in the structure itself, which in turn result in dimensional changes in the structure's material. These dimensional changes are described by thermal strain, which occurs at every point in the structure's material where temperature is changed and affects all three directions, as shown in Expression 5.35.

$$\mathbf{E}_T(t) = \begin{bmatrix} \varepsilon_{\xi,T}(t) & 0 & 0 \\ 0 & \varepsilon_{\eta,T}(t) & 0 \\ 0 & 0 & \varepsilon_{\zeta,T}(t) \end{bmatrix} \tag{5.35}$$

where $\mathbf{E}_T(t)$ is tensor of thermal strain at time t, $O\xi\eta\zeta$ is coordinate system along the axes ξ, η, and ζ of thermal anisotropy of material (if any), and $\varepsilon_{n,T}(t)$ is thermal strain component (simply called "thermal strain") in direction of axis n, ($n = \xi$, η, or ζ). For thermally isotropic materials, the axes ξ, η, and ζ can be replaced by any three orthogonal axes, e.g., x, y, and z.

In typical construction materials, the relation between the thermal strain and temperature change at a point is given by the following equation (Brčić 1989):

$$\varepsilon_{n,T}(t) = \alpha_{T,n}\Delta T(t) = \alpha_{T,n}(T(t) - T(t_{\text{Ref}})) \tag{5.36}$$

where $T(t)$ is temperature at time t, $T(t_{\text{Ref}})$ is temperature at some reference time t_{Ref}, $\Delta T(t)$ is temperature change between times t and t_{Ref}, and $\alpha_{T,n}$ is thermal expansion coefficient of construction material with respect to axis n ($n = \xi$, η, or ζ).

Note that thermal strain is always relative to some reference time, and the symbol Δ should be in front of the left-hand term in Expression 5.36; however, to simplify presentation, this symbol is omitted as discussed in Section 5.4.

In an anisotropic material (e.g., fiber-reinforced composite or timber), the thermal expansion coefficient is different for different directions; in an isotropic material (e.g., steel or concrete observed at macroscale), the thermal expansion coefficient is the same for all three directions. In inhomogeneous materials, such as unreinforced, reinforced, and prestressed concrete, composites, and timber, temperature change can create internal stresses due to the thermal incompatibility of material components (e.g., cement paste, aggregate, and rebars in reinforced concrete, matrix and reinforcing fibers in composites). In these cases, thermal expansion coefficients of materials are considered at the macroscale, i.e., instead of individual thermal expansion coefficients of each material, apparent average values are used (typically "as measured" on the macroscale). While the effects of thermal incompatibility within inhomogeneous materials can be neglected for typical ranges of temperature, they should be considered for extreme temperatures.

Expression 5.36 shows, that thermal strain can be accurately and simply evaluated if the thermal expansion coefficient of the material is known and temperature changes are monitored using temperature sensors. For illustrative purposes, typical values of thermal expansion coefficients are given in Table 5.4 (Muravljov 1989, Simpson and TenWolde 1999, Ran et al. 2014).

It is important to note that the values given in Table 5.4 are only indicative and should be used with caution, as their range (variability) for all materials, except steel, is very high. In addition, in some construction materials, the thermal expansion coefficient might not necessarily be constant over time; in the case of concrete, timber, and some composites, it can depend on the temperature itself and the humidity of the material (Christensen 1991). For accurate results in a specific SHM project, the thermal expansion coefficient of construction material should be determined experimentally, either in the lab or onsite (e.g., Reilly and Glisic 2018a).

Inaccuracies in the determination of thermal strain or a complete lack of its evaluation can result in important inaccuracies in data analysis, especially in the determination of elastic strain and stress. For example, let's assume typical values of Young's modulus for steel and concrete ($E_{steel} = 210\,GPa$, $E_{concrete} = 28\,GPa$) and a temperature change of $\Delta T = 10°C$ (typical for daily variations at many geographical latitudes). Based on Table 5.4 and Expression 5.30, Expressions 5.29 and 5.36, the range of resulting thermal strain is estimated as shown in Expressions 5.37 and 5.38. Assuming that temperature was not measured and the thermal strain constituent could not

Table 5.4 Typical values of the thermal expansion coefficients of the most frequently used construction materials (in μɛ/°C).

Concrete[a]	Steel	Fiber-reinforced composite[b]	Timber (ovendry[c])
8–14	10–12	Longitudinal: −1 to 10 Transversal: 4–47	Longitudinal: 3.1–4.5 Radial: $32.4G + 9.9$ Tangential: $32.4G + 18.4$ where G is specific gravity of timber (unitless)

a) May vary over time depending on temperature and humidity.
b) Strongly depends on fiber contents and orientation.
c) In case of humid timber, an increase in temperature might trigger drying and associated shrinkage, which in turn can be higher than thermal expansion, so the final strain could be negative.

be evaluated, it would be erroneously attributed to a change in elastic strain, which, in turn, would result in an error in the evaluation of stress. The range of this error depends on Young's modulus and is presented in Expressions 5.37 and 5.38.

$$\text{Steel:} \quad \Delta T = 10°C \Rightarrow \varepsilon_{T,steel} = 100 - 120 \, \mu\varepsilon \sim \text{error in stress evaluation of } 21 - 25 \, \text{MPa} \tag{5.37}$$

$$\text{Concrete:} \quad \Delta T = 10°C \Rightarrow \varepsilon_{T,steel} = 80 - 140 \, \mu\varepsilon \sim \text{error in stress evaluation of } 2 - 4 \, \text{MPa} \tag{5.38}$$

Note that the error presented in the above equations can be several times higher if seasonal variations in temperature are taken into account. For example, typical seasonal temperature variations in Princeton, New Jersey, are approximately ranged between −15°C and +35°C ($\Delta T_{max} \approx 50°C$), which would multiply values in Expressions 5.37–5.38 five times.

Another important way temperature may influence the evaluation of elastic strain and stress is through the change of the Young's modulus. Typical construction materials will soften at higher temperatures (i.e., their Young's modulus will decrease as the temperature increases), and vice versa, stiffen at lower temperatures (i.e., their Young's modulus will increase as the temperature decreases). In common construction materials, this dependence can be neglected for a typical range of temperatures; however, in the case of extreme high or low temperatures, the influence of temperature on the Young's modulus should be considered.

5.7.2 Cross-Sectional Thermal Gradients

Thermal effects from the environment (variable and non-even sun irradiation on structure and thermal interactions between structure, air, and soil) do not affect the structure in a uniform way. In addition, due to the thermal capacity and conductivity of the construction material itself, as well as the geometrical shape of the structure (e.g., cross-section), different points in the structure experience different temperature changes over time. As a consequence, non-linear thermal gradients are frequently created in all three dimensional directions in materials, and these gradients can induce mechanical strain and stress in the structure. For beam-like structures, thermal gradients can act

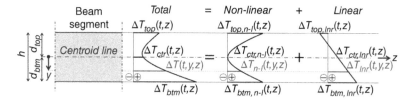

Figure 5.7 Schematic decomposition of the cross-sectional thermal gradient in beam-like structures into non-linear and linear parts.

across cross-sections (cross-sectional gradients) and along the lengths of the beams (longitudinal gradients). Longitudinal thermal gradients are usually small and non-linear. Cross-sectional gradients are typically much higher, and they can produce linear and non-linear effects, as schematically shown in Figure 5.7 for a segment of planar beam.

In Figure 5.7, a local Cartesian coordinate system is assumed as follows: z-axis is placed along the centroid line of the beam, y-axis is vertical, and x-axis is horizontal (pointing outwards from the drawing, not shown in the figure). Gray-shaded area represents the vertical cut (elevation) of a short segment of a beam in the vertical plane yz.

One simplification is made in Figure 5.7, in order to ease presentation: non-linear part is assumed to be symmetric with respect to the centroid, which, in general, is not the case. The following equalities apply for temperature changes at cross-sections with coordinates z (e.g., shown in the gray-shaded area in Figure 5.7).

$$\Delta T_{ctr,n\text{-}l}(t,z) = 0; \Delta T_{ctr,lnr}(t,z) = \Delta T_{ctr}(t,z)$$

$$\Delta T_{btm,n\text{-}l}(t,z) = \Delta T_{top,n\text{-}l}(t,z) = \frac{\Delta T_{btm}(t,z)d_{top} + \Delta T_{top}(t,z)d_{btm}}{h} - \Delta T_{ctr}(t,z)$$

$$\Delta T_{lnr}(t,y,z) = \Delta T_{ctr}(t,z) + \frac{y}{h}(\Delta T_{btm}(t,z) - \Delta T_{top}(t,z))$$

$$\Delta T_{btm,lnr}(t,z) = \Delta T_{ctr}(t,z) + \frac{d_{btm}}{h}(\Delta T_{btm}(t,z) - \Delta T_{top}(t,z))$$

$$\Delta T_{top,lnr}(t,z) = \Delta T_{ctr}(t,z) - \frac{d_{top}}{h}(\Delta T_{btm}(t,z) - \Delta T_{top}(t,z)) \tag{5.39}$$

where subscript *btm* indicates bottom of cross-section, *top* indicates top of cross-section, *ctr* indicates centroid, *n-l* indicates non-linear part of cross-sectional gradient, and *lnr* indicates the influence of the linear part of cross-sectional gradient; d_{btm} and d_{top} are distances from centroid to bottom and top of cross-section, respectively; h is depth (height) of the cross-section ($h = d_{btm} + d_{top}$).

In general, cross-sectional thermal gradients are consequences of transitional thermal effects – change in temperature due to daily variations of air temperature, sun irradiation, and vapor pressure. To understand their full effects on structure, we will analyze the non-linear part and the linear part separately, as their individual effects are different.

Let us observe the non-linear part of the gradient first (see Figure 5.7). This case, when observed alone (e.g., $\Delta T_{btm,n\text{-}l} = \Delta T_{top,n\text{-}l}$), could happen in real-life settings, e.g., due to the initiation of warming-up or cooling-down of the structure due to changes of temperature in the environment (e.g., warming-up in the morning or cooling-down in the afternoon). In the case of warming-up, the bottom and the top of the cross-sections (i.e., the skin of the cross-section, $\Delta T_{skin} = \Delta T_{btm,n\text{-}l} = \Delta T_{top,n\text{-}l}$) will warm-up first, while at the same time the interior of the cross-section will still have an initial (lower) temperature due to the thermal inertia of the material (e.g., $\Delta T_{ctr,n\text{-}l} = 0$). The bottom and the top of the cross-section would like to expand due to warming up, but their expansion will be constrained by the interior, and thus, a compression stress (and associated compressive mechanical strain) will be generated at the bottom and the top of the cross-section, while a tensional stress (and associated tensional mechanical strain) will be generated in the interior. In the case of concrete, this effect, while typically relatively small, can cause internal tensional cracks, especially at an early age when concrete has not yet gained sufficient strength. Opposite effects will be generated in the case of initiation of cooling-down of the environment – tension in the skin and compression in the interior, with potential tensional cracking on the skin in the case of concrete. This is illustrated in Figure 5.8.

The following relationships are valid for the two cases shown in Figure 5.8:

$$\Delta T_{btm,n-l}(t,z) = \Delta T_{top,n-l}(t,z) = \Delta T_{skin}(t,z)$$

$$\varepsilon_{z,T,skin}(t,z) = \alpha_{T,z}\Delta T_{skin}(t,z)$$

$$|\varepsilon_{z,tot,skin,n-l}(t,z)| < |\varepsilon_{z,T,skin,n-l}(t,z)|$$

$$|\varepsilon_{z,tot,ctr,n-l}(t,z)| > |\varepsilon_{z,T,ctr,n-l}(t,z)| = 0 \tag{5.40}$$

where z-axis coincides with the centroid line of the beam.

In Figure 5.8 and Expression 5.40, the symbol "Δ," denoting the change, is omitted next to strain for conciseness purposes. Note that the determination of the non-linear temperature distribution across the cross-section, as well as the resulting thermal and mechanical strain, is not a trivial task. Based on Figures 5.7 and 5.8, an accurate assessment would require several temperature and strain sensors installed across the cross-section of a beam. Even more sensors are required in cross-sections of large structures or in cross-sections with complex geometrical shapes. However, the installation of a large number of temperature sensors can be challenging due to their elevated cost. An option is to use less sensors and determine thermal strain and mechanical strain at non-instrumented locations by using numerical modeling. As mentioned earlier, in many practical applications, the non-linear effects are small and can be neglected when compared to the linear effects. Nevertheless, it is recommended to perform numerical analysis to confirm that this assumption holds. Again, non-linear thermal effects shown in Figure 5.8 will rarely affect the integrity

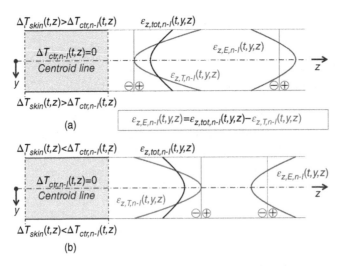

Figure 5.8 Schematic representation of examples of a non-linear thermal gradient through the cross-section of a beam and resulting total strain and strain constituents: (a) Warming-up and (b) cooling-down of the bottom and top of the cross-section.

of mature concrete or steel exposed to usual temperature changes caused by daily and seasonal changes in the environment; however, they can cause cracking in concrete at an early age, as the strength of concrete is not yet developed at that time (e.g., see Hubbell and Glisic 2013). Also, they can cause damage if the temperature changes are extreme and occur fast, so that the temperature at the skin is, by absolute value, much larger than that of the interior of concrete.

Let us now observe the linear part of the cross-sectional gradient, as per Figure 5.7. This case, when observed alone, could happen in real-life settings, e.g., when the ambient thermal conditions become stationary for an extended period of time (e.g., during the night or a cloudy day), during which non-linear effects will slowly disappear (Reilly and Glisic 2018b). For beams made of isotropic material, linear transverse gradients have twofold effects on the beam's cross-section: the first is in the axial direction (reflected through a change in strain at the centroid of the cross-section) and the second is in the transversal direction (reflected through a change in the curvature of the cross-section). If we observe a beam at rest (no loads and no boundary conditions applied) and the same linear thermal gradient is present in every cross-section of the beam, then the first effect will make the beam longer or shorter, and the second will make it curved. This is described by Expressions 5.41 through 5.43 and schematically shown in Figure 5.9. Expressions 5.41–5.43.

$$\varepsilon_{z,T,ctr}(t,z) = \alpha_{T,z}\left(\Delta T_{btm,lnr}(t,z)\frac{d_{top}}{h} + \Delta T_{top,lnr}(t,z)\frac{d_{btm}}{h}\right)$$

$$= \varepsilon_{z,T,btm,lnr}(t,z)\frac{d_{top}}{h} + \varepsilon_{z,T,top,lnr}(t,z)\frac{d_{btm}}{h} \tag{5.41}$$

$$\kappa_{x,T}(t,z) = \frac{(\varepsilon_{z,T,btm,lnr}(t,z) - \varepsilon_{z,T,top,lnr}(t,z))}{h} = \alpha_{T,z}\frac{\Delta T_{btm,lnr}(t,z) - \Delta T_{top,lnr}(t,z)}{h}$$

$$= \alpha_{T,z}\frac{\Delta T_{btm}(t,z) - \Delta T_{top}(t,z)}{h} = \frac{(\varepsilon_{z,T,btm}(t,z) - \varepsilon_{z,T,top}(t,z))}{h} \tag{5.42}$$

$$\varepsilon_{z,T,lnr}(t,y(t),z) = \varepsilon_{z,T,ctr}(t,z) + \kappa_{x,T}(t,z)\cdot y(t) \tag{5.43}$$

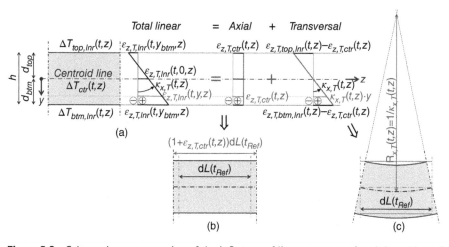

Figure 5.9 Schematic representation of the influence of linear cross-sectional thermal gradients on beam elements at rest: (a) Decomposition to axial and transversal part; (b) effect of axial; and (c) effect of transversal part.

In Figure 5.9 and Expressions 5.41–5.43, the symbol "Δ," denoting the change, is omitted next to strain for conciseness purposes. Also, subscript "*lnr*" is omitted from notations for thermal curvature "$\kappa_{x,T}$" and thermal strain at centroid "$\varepsilon_{z,T,ctr}$," as the curvature is only defined for linear gradient and the thermal strain at centroid belongs to linear gradient (see Figure 5.7); thus, the subscript would be redundant. Expressions 5.41–5.43 show relationships between thermal strain at the centroid and thermal changes and thermal strains at the bottom and top of the cross-section. These expressions are helpful in cases where temperature is measured only at the bottom and top of the cross-section but not at the centroid. Also, note the similarity and differences between Expressions 5.29–5.31 and 5.41–5.43, respectively. In particular, Expression 5.43 shows that thermal strain at the centroid produces an invariable (constant) elongation or contraction at each point of the cross-section, regardless of the point's location within the cross-section, while curvature dictates rotation of the cross-section and a linear distribution of thermal strain change across the y-axis of the cross-section. These effects are similar to the respective influences of normal force and bending moment shown in Expression 5.31. Thermal curvature and thermal strain at the centroid are mutually independent, and thus the transverse and axial thermal behaviors of a beam are also mutually independent (see Figure 5.9). Hence, they can be analyzed separately and their influences superimposed.

Linear cross-sectional thermal gradients in determinate (isostatic) beam-like structures do not generate mechanical strain and stress; however, they do generate thermally induced mechanical strain and stress in indeterminate (hyperstatic) structures, which in turn will depend on boundary conditions. That is the reason why thermal effects and their influence on total strain, elastic strain, stress, and deformation can sometimes be confusing. Example 5.5 analyzes a few specific cases with the intention of easing the understanding of the influence of linear cross-sectional thermal gradients on beam-like structures.

Example 5.5 *Illustration of influence of short-term linear cross-sectional thermal gradients to deformation, elastic strain, and stress in planar prismatic beams*

Let's observe three beams with boundary conditions as shown in Figure 5.10. The beam in Figure 5.10a is simply supported, hence determinate; the beam in Figure 5.10b is fixed at both ends, thus indeterminate; and the beam in Figure 5.10c is similar to the one in Figure 5.10b, except that the two boundaries at the right end are only partially constrained, i.e., they are equipped with axial and rotational springs. All three beams have the same material and geometrical properties, which are invariable in time and space (along the beam), and are subjected to the same linear thermal gradients that are acting in the yz-plane (similar to those shown in Figure 5.9). Given that the influence of thermal gradients in this example is observed in the short-term, the rheological constituents, creep and shrinkage, can be neglected. Assuming that no other strain constituents are present, the total strain in the beams shown in Figure 5.10 can be expressed as follows:

$$\varepsilon_{z,tot}(t,y,z) = \varepsilon_{z,E}(t,y,z) + \varepsilon_{z,T,lnr}(t,y,z) \tag{5.44}$$

and the expressions for total strain at the centroid and curvature become:

$$\varepsilon_{z,ctr,tot}(t,z) = \varepsilon_{z,E,ctr}(t,z) + \varepsilon_{z,T,ctr}(t,z)$$

$$\kappa_{x,tot}(t,z) = \kappa_{x,E}(t,z) + \kappa_{x,T}(t,z) \tag{5.45}$$

To simplify the illustration of the influence of linear cross-sectional thermal gradients on deformation, elastic strain, and stress, let's assume that there are no longitudinal thermal gradients, i.e., that the cross-sectional thermal gradient is invariable along the z-axis, i.e., $\Delta T_{btm,lnr}(t,z) = \Delta T_{top,lnr}(t)$ and $\Delta T_{top,lnr}(t,z) = \Delta T_{top,lnr}(t)$; then, the thermal strain at any coordinate y in

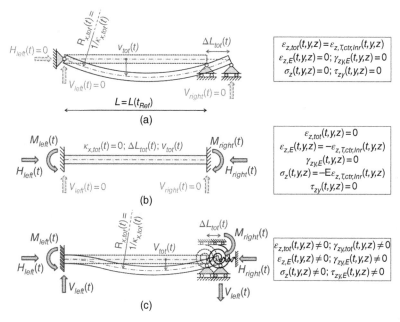

Figure 5.10 Illustration of the influence of short-term linear cross-sectional thermal gradients on beam-like structures: (a) Determinate, (b) indeterminate with full constraints, and (c) indeterminate with partial constraints.

cross-section, including the thermal strain at the centroids, as well as thermal curvature, are also invariable along the z-axis, i.e., $\varepsilon_{z,T,lnr}(t,y,z) = \varepsilon_{z,T,lnr}(t,y)$, $\varepsilon_{z,T,ctr}(t,z) = \varepsilon_{z,T,ctr}(t)$ and $\kappa_{x,T}(t,z) = \kappa_{x,T}(t)$. In order to keep the presentation of this example general, variable z is kept in all general expressions, and omitted (due to the invariability of thermal strain along the z-axis) only in some specific derivations.

The beam shown in Figure 5.10a is determinate, thus unconstrained for linear thermal influences (fully free to deform under temperature changes that result in linear thermal gradients); hence, total strain contains only thermal constituents while elastic strain is null, and the Expressions 5.44 and 5.45 transform into the following:

$$\varepsilon_{z,tot}(t,y,z) = \varepsilon_{z,T,lnr}(t,y,z); \varepsilon_{z,E}(t,y,z) = 0 \tag{5.46}$$

$$\varepsilon_{z,tot,ctr}(t,z) = \varepsilon_{z,T,ctr}(t,z); \varepsilon_{z,E,ctr}(t,z) = 0$$
$$\kappa_{x,tot}(t,z) = \kappa_{x,T}(t,z); \kappa_{x,E}(t,z) = 0 \tag{5.47}$$

Given that no elastic strain is generated, there will be no stresses, no reactions, and no internal forces generated in the structure. The change in length $\Delta L_{tot}(t)$ and deflection $v_{tot}(t,z)$ of the beam can then be determined using the following expressions:

$$\Delta L_{tot}(t) = \int_0^{L(t_{Ref})} \varepsilon_{z,tot,ctr}(t)dz = \int_0^{L(t_{Ref})} \varepsilon_{z,T,ctr,lnr}(t)dz = \varepsilon_{z,T,ctr,lnr}(t)L(t_{Ref}) = \Delta L_T(t)$$

$$v_{tot}(t,z) = \int \left(\int \kappa_{x,tot}(t)dz \right) dz + C_1 z + C_2$$
$$= \int \left(\int \kappa_{x,T}(t)dz \right) dz + C_1 z + C_2 = \frac{1}{2}\kappa_{x,T}(t)(z^2 - L(t_{Ref})z) = v_T(t,z) \tag{5.48}$$

Note that variable z is omitted from Expressions 5.48 due to the invariability of thermal strain along z-axis. In conclusion, for this specific case and for determinate structures in general, the thermal gradients result in deformation of the structure, with no stresses generated.

The beam in Figure 5.10b is indeterminate to degree three. It represents a specific case where deformation due to thermal gradients is fully prevented, i.e., the change in length $\Delta L_{tot}(t)$ and deflection $v_{tot}(t,z)$ are null and the following expressions are valid:

$$\varepsilon_{z,tot}(t,y,z) = 0; \varepsilon_{z,E}(t,y,z) = -\varepsilon_{z,T,lnr}(t,y,z) \tag{5.49}$$

$$\varepsilon_{z,tot,ctr}(t,z) = 0; \varepsilon_{z,E,ctr}(t,z) = -\varepsilon_{z,T,ctr}(t,z)$$
$$\kappa_{x,tot}(t,z) = 0; \kappa_{x,E}(t,z) = -\kappa_{x,T}(t,z) \tag{5.50}$$

As a consequence, internal forces $N_z(t,z)$ and $M_x(t,z)$ and normal stresses $\sigma(t,y,z)$, as well as horizontal reactions $H_{left}(t)$ and $H_{right}(t)$, and reactive moments $M_{left}(t)$ and $M_{right}(t)$, are generated and expressed as follows:

$$N_z(t,z) = EA\varepsilon_{z,E,ctr}(t,z) = -EA\varepsilon_{z,T,ctr}(t,z)$$
$$M_x(t,z) = EI\kappa_{x,E}(t,z) = -EI\kappa_{x,T}(t,z) \tag{5.51}$$

$$\sigma_z(t,y,z) = E\varepsilon_{z,E}(t,y,z) = -E\varepsilon_{z,T,lnr}(t,y,z) = -E(\varepsilon_{z,T,ctr}(t,z) + \kappa_{x,T}(t,z) \cdot y) \tag{5.52}$$

$$|H_{left}(t)| = |H_{right}(t)| = |EA\varepsilon_{z,E,ctr}(t)| = |-EA\varepsilon_{z,T,ctr,lnr}(t)|$$
$$|M_{left}(t)| = |M_{right}(t)| = |EI\kappa_{x,E}(t)| = |-EI\kappa_{x,T}(t)| \tag{5.53}$$

Magnitudes for reactive forces and moments are given in absolute values, as Figure 5.10 shows their correct senses of orientation. Note that bending and reactive moments depend only on thermal curvature, while normal forces and horizontal reactions depend only on thermal strain at the centroid. Given that thermal curvature and thermal strain at the centroid are mutually independent, as mentioned earlier, transverse and axial thermal behaviors of a beam are also mutually independent, and thus, the calculation of bending and reactive moments is independent from the calculation of normal and horizontal reactive forces. This is valid in general for all indeterminate beam-like structures.

In conclusion, for this specific example of indeterminate structure, the thermal gradients result in thermally generated normal stresses, internal forces, and reactions, but with no deformation of the structure (which may seem counterintuitive).

Similar to Figure 5.10b, beam shown in Figure 5.10c is also indeterminate to degree three, but it represents more general case of indeterminate structure as two out of three constraints to right are only partial, described by axial translational spring with stiffness k_t and rotational spring with stiffness k_r. In this case, the deformation of structure is not prevented, i.e., $\Delta L_{tot}(t)$ and deflection $v_{tot}(t,z)$ are not null. As a consequence, internal forces $N_z(t,z)$ and $M_x(t,z)$ and normal stresses $\sigma(t,y,z)$, as well as horizontal reactions $H_{left}(t)$ and $H_{right}(t)$, vertical reactions $V_{left}(t)$ and $V_{right}(t)$, and reactive moments $M_{left}(t)$ and $M_{right}(t)$, are generated. Derivations of expressions for all these parameters require application of structural analysis (e.g., force method) and the results are given as follows:

$$N_z(t,z) = EA\varepsilon_{z,E,ctr}(t,z) = -EA\varepsilon_{z,T,ctr}(t,z)\frac{1}{1 + \frac{EA}{k_t L(t_{Ref})}}$$

$$S_y(t) = \frac{3}{2} \frac{EI\kappa_{x,T}(t)}{L(t_{Ref})} \frac{1}{1 + \frac{k_r L(t_{Ref})}{4EI}}$$

$$M_x(t) = EI\kappa_{x,E}(t) = -EI\kappa_{x,T}(t)\left(1 + \frac{1}{2}\frac{1 - 3\frac{z}{L(t_{Ref})}}{1 + \frac{k_r L(t_{Ref})}{4EI}}\right) \tag{5.54}$$

$$\sigma_z(t,y,z) = E\varepsilon_{z,E}(t,y,z) = -E\left(\varepsilon_{z,T,ctr}(t)\frac{1}{1 + \frac{EA}{k_t L(t_{Ref})}} + \kappa_{x,T}(t)\left(1 + \frac{1}{2}\frac{1 - 3\frac{z}{L(t_{Ref})}}{1 + \frac{k_r L(t_{Ref})}{4EI}}\right)\cdot y\right) \tag{5.55}$$

$$|H_{left}(t)| = |H_{right}(t)| = |EA\varepsilon_{z,E,ctr}(t)| = \left|-EA\varepsilon_{z,T,ctr}(t)\frac{1}{1 + \frac{EA}{k_t L(t_{Ref})}}\right|$$

$$|V_{left}(t)| = |V_{right}(t)| = \left|\frac{3}{2}\frac{EI\kappa_{x,T}(t)}{L(t_{Ref})}\frac{1}{1 + \frac{k_r L(t_{Ref})}{4EI}}\right|$$

$$|M_{left}(t)| = \left|EI\kappa_{x,T}(t)\left(1 + \frac{1}{2}\frac{1}{1 + \frac{k_r L(t_{Ref})}{4EI}}\right)\right| \;;\; |M_{right}(t)| = \left|EI\kappa_{x,T}(t)\frac{1}{1 + \frac{4EI}{k_r L(t_{Ref})}}\right| \tag{5.56}$$

The expressions for total strain at the centroid and total curvature, as well as for other deformation parameters such as change in length and deflection, are the following:

$$\varepsilon_{z,tot,ctr}(t) = \varepsilon_{z,E,ctr}(t) + \varepsilon_{z,T,ctr}(t) = \varepsilon_{z,T,ctr}(t)\left(\frac{1}{1 + \frac{k_t L(t_{Ref})}{EA}}\right)$$

$$\kappa_{x,tot}(t) = \kappa_{x,E}(t,z) + \kappa_{z,T}(t,z) = -\frac{1}{2}\kappa_{x,T}(t)\frac{1 - 3\frac{z}{L(t_{Ref})}}{1 + \frac{k_r L(t_{Ref})}{4EI}}$$

$$\Delta L_{tot}(t) = \int_0^{L(t_{Ref})} \varepsilon_{z,tot,ctr}(t)dz = \varepsilon_{z,T,ctr}(t)L(t_{Ref})\frac{1}{1 + \frac{k_t L(t_{Ref})}{EA}}$$

$$v_{tot}(t,z) = \int\left(\int \kappa_{x,tot}(t)dz\right)dz + C_1 z + C_2 = \frac{1}{4}\kappa_{x,T}(t)\frac{1}{L(t_{Ref})}\frac{2z^3 - z^2 L(t_{Ref})}{1 + \frac{k_r L(t_{Ref})}{4EI}} \tag{5.57}$$

Expressions 5.54–5.57 show that in this more general case (Figure 5.10c), not only the normal stress component is generated, but also shear stress (due to shear force) is generated too. Also, total curvature as well as the normal and shear stress components are not invariable along z-axis, even if the thermal gradient is invariable along z-axis.

Now, let's assume that the stiffness of both springs is null, i.e., $k_t = 0$ and $k_r = 0$. In that case, the beam in Figure 5.10c becomes simply supported by a roller at the right end. The structure is still indeterminate (to degree one), but the indeterminacy affects only bending and sheer stresses and strains and not the axial stresses and strains. Simple substation of $k_t = 0$ and $k_r = 0$ in Expressions 5.54–5.57 shows that $N_z(t,z) = 0$, but $M_x(t,z) \neq 0$ and $S_y(t,z) \neq 0$; normal stress at centroid is null $\sigma_{z,ctr}(t,z) = 0$ and normal stress in other points in cross-section are only generated by bending moment, while shear stress is generated by shear force; horizontal reaction at left end is null, $H_{left}(t) = 0$, but vertical reactions and reactive moment at left end are not, i.e., $H_{left}(t) = 0$, $V_{left}(t) \neq 0$, $V_{right}(t) \neq 0$, and $M_{left}(t) \neq 0$; finally, the total strain at centroid is equal to thermal strain at centroid and beam is free to change length without stresses; however, total curvature is different from thermal curvature and bending strain and stresses are introduced. This last exercise clearly demonstrates the mutual independence of axial and transverse components of thermal gradients on the structural behavior of the beam. While this example deals with a beam that is practically determinate (free to expand) axially but indeterminate in transverse direction, similar conclusions about the independence of axial and transverse components could be derived if the beam was indeterminate axially but determinate in transverse direction, e.g., by considering the beam shown in Figure 5.10a with a pin support at the right end instead of a roller.

In conclusion, for this specific example that illustrates a more general case of indeterminate structure, the axial and transverse components of the cross-sectional thermal gradient (which is invariable along the z-axis), result in: thermally generated normal and shear stresses that are variable along the z-axis; all three internal forces: normal force, shear force, and bending moments (the latter two being variable along the z-axis); all three types of reactions: horizontal reactions, vertical reactions, and reactive moments; and deformation of structure: change in length, curvature, and deflection, where the latter two are variable along the z-axis. This example also emphasizes the mutual independence of the consequences of axial and transverse components of thermal gradients.

Example 5.5 presented a few illustrative cases on how linear cross-sectional thermal gradients, which are invariable along the z-axis, influence the stresses and deformations of beam-like structures. They reveal important structural behaviors that are necessary to identify when analyzing SHM strain and temperature data. The next subsection deals with the influence of longitudinal gradients on the structural behavior of beam-like structures.

5.7.3 Longitudinal Thermal Gradients

The previous section presented the influence of cross-sectional gradients on the structural behavior of beam-like structures. This section completes the analysis by considering longitudinal thermal gradients. Decompositions of cross-sectional gradients shown in Figures 5.7 and 5.8 also show that longitudinal gradients can be fully described by the change in temperature along the centroid line, $\Delta T_{ctr}(t,z)$, and the change in difference between bottom and top temperature, $\Delta T_{btm,lnr}(t,z) - \Delta T_{top,lnr}(t,z) = \Delta T_{btm}(t,z) - \Delta T_{top}(t,z)$, see Expression 5.39. These two parameters are translated into thermal strain at the centroid, $\varepsilon_{z,T,ctr}(t,z)$, and thermal curvature $\kappa_{x,T}(t,z)$, which are, in general, variable along the z-axis. Thus, the longitudinal gradient introduces variability in total strain and stress components, and internal forces along centroid line of the beam. Similar to cross-sectional gradients, the influences on structural behavior related to variability of thermal strain at centroid and thermal curvature are mutually independent, and they can be analyzed separately and then superposed. As opposed to cross-sectional gradients that affect the same cross-section in which they are present, longitudinal gradients can affect behavior in distant cross-sections, sometimes in a somewhat counterintuitive way. This is illustrated in Example 5.6.

Example 5.6 *Illustration of influence of short-term longitudinal thermal gradients on elastic strain and stress in planar prismatic beams*

Let us observe, again, the beam with boundary conditions as shown in Figure 5.10b and, to simplify presentation, let's assume that it is subjected to longitudinal thermal gradients that are constant across cross-sections but variable along the centroid line, i.e., along the z-axis. Thus, change in temperature at points in cross-section with coordinate z does not depend on coordinate y, i.e., $\Delta T_{lnr}(t,y,z) = \Delta T_{top,lnr}(t,z) = \Delta T_{btm,lnr}(t,z) = \Delta T_{ctr}(t,z) = \Delta T_{lnr}(t,z)$, where $\Delta T_{lnr}(t,z)$ depends on z but it is not necessarily linear over z. As a consequence, the thermal strain at any coordinate

z is also invariable along the y-axis, and thus, the thermal curvature is null, i.e., $\kappa_{x,T}(t,z) = 0$. Note that due to the independence of axial and transverse gradients, cases where a longitudinal gradient of curvature exists can be analyzed separately, using a similar approach as for the axial longitudinal gradient presented here. Finally, to further simplify presentation, let's assume that the longitudinal thermal gradient is increasing linearly over z, from value $\Delta T_{left}(t) = \Delta T_{lnr}(t,0) > 0$ to $\Delta T_{right}(t) = \Delta T_{lnr}(t,L(t_{Ref})) > 0$, i.e., as per the following expression:

$$\Delta T_{lnr}(t,y,z) = \Delta T_{left}(t) + \frac{\Delta T_{right}(t) - \Delta T_{left}(t)}{L(t_{Ref})} z \tag{5.58}$$

where subscripts left and right describe cross-sections at the left end of the beam ($z = 0$) and at the right end of the beam ($z = L(t_{Ref})$).

This linear function is selected for illustrative purposes only and can be replaced with any other function (e.g., inferred from SHM data) in Expressions 5.58–5.61. For the function presented in Expression 5.58, the thermal strain has the following distribution:

$$\varepsilon_{z,T,lnr}(t,y,z) = \varepsilon_{z,T,left}(t) + \frac{\varepsilon_{z,T,right}(t) - \varepsilon_{z,T,left}(t)}{L(t_{Ref})} z \tag{5.59}$$

where $\varepsilon_{z,T,left}(t) = \alpha_{T,z} \Delta T_{left}(t)$ and $\varepsilon_{z,T,right}(t) = \alpha_{T,z} \Delta T_{right}(t)$.

Since the beam is axially constrained, this longitudinal thermal gradient will generate normal forces and normal stresses in the beam, as follows (bending moments and bending stresses will be null as the thermal curvature is null):

$$N_z(t,z) = -EA \frac{\varepsilon_{z,T,right}(t) + \varepsilon_{z,T,left}(t)}{2}$$
$$\sigma_z(t,y,z) = -E \frac{\varepsilon_{z,T,right}(t) + \varepsilon_{z,T,left}(t)}{2}. \tag{5.60}$$

Elastic and total strains are then calculated using Expression 5.61.

$$\varepsilon_{z,E}(t,y,z) = -\frac{\varepsilon_{z,T,right}(t) + \varepsilon_{z,T,left}(t)}{2}$$
$$\varepsilon_{z,tot}(t,y,z) = (\varepsilon_{z,T,right}(t) - \varepsilon_{z,T,left}(t))\left(\frac{z}{L(t_{Ref})} - \frac{1}{2}\right) \tag{5.61}$$

Now, let us observe the relationship between change in temperature (which can be measured using temperature sensors) and total strain (which can be measured using strain sensors) at cross-sections at the left support ($z = 0$), in the middle ($z = L(t_{Ref})/2$), and at the right support ($z = L(t_{Ref})$). They are described by the following expressions:

At left support: $\Delta T_{left}(t,y) = \Delta T_{left}(t) > 0; \varepsilon_{z,tot,left}(t,y) = -\dfrac{\varepsilon_{z,T,right}(t) - \varepsilon_{z,T,left}(t)}{2} < 0$

In the middle: $\Delta T_{mid}(t,y) = \dfrac{\Delta T_{left}(t) + \Delta T_{right}(t)}{2} > 0; \varepsilon_{z,tot,mid}(t,y) = 0$

At right support: $\Delta T_{right}(t,y) = \Delta T_{right}(t) > 0; \varepsilon_{z,tot,right}(t,y) = \dfrac{\varepsilon_{z,T,right}(t) - \varepsilon_{z,T,left}(t)}{2} > 0$

$$\tag{5.62}$$

In Expressions 5.62, all three temperature changes are positive; yet, at the left support, the total strain is negative, in the middle, it is null, and only at the right cross-section it is positive. Expression 5.61 shows that in all cross-sections in the left half of the beam, the total strain is negative, and in all cross-sections in the right half of the beam, the total strain is positive. This demonstrates how longitudinal gradients influence distant cross-sections: despite the positivity

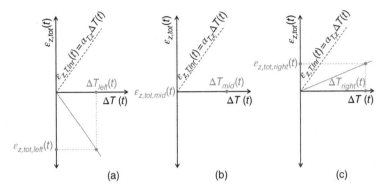

Figure 5.11 Relations between total strain and temperature change for (a) left, (b) middle, and (c) right cross-sections (as per Figure 5.10b and Expressions 5.58 and 5.62).

of temperature change in cross-sections in the left half of the beam, the total strain in these cross-sections is negative, and it is null in the middle cross-section. Negative total strain in the left half of the beam, and the null in the middle, might seem counterintuitive, yet they are consequences of the longitudinal thermal gradient and boundary conditions. This somewhat counterintuitive behavior is emphasized in Figure 5.11, where graphs between total strain and temperature change are presented for the left, middle, and right cross-sections (solid lines in the graphs). These graphs mimic hypothetical strain-temperature plots that could be obtained from SHM. The slope that corresponds to the thermal expansion coefficient is included in graphs for comparison purposes (dashed lines in the graphs).

The slopes of the graphs shown in solid lines in Figure 5.11 are equal to ratios between total strain and temperature change that are calculated as follows:

$$\text{At left support: } \frac{\varepsilon_{z,tot,left}(t,y)}{\Delta T_{left}(t,y)} = -\frac{\varepsilon_{z,T,right}(t) - \varepsilon_{z,T,left}(t)}{2\Delta T_{left}(t)} = -\frac{1}{2}\alpha_{T,z}\left(\frac{\Delta T_{right}(t)}{\Delta T_{left}(t)} - 1\right) < 0$$

$$\text{In the middle: } \frac{\varepsilon_{z,tot,mid}(t,y)}{\Delta T_{mid}(t,y)} = 0$$

$$\text{At right support: } \frac{\varepsilon_{z,tot,right}(t,y)}{\Delta T_{right}(t,y)} = \frac{\varepsilon_{z,T,right}(t) - \varepsilon_{z,T,left}(t)}{2\Delta T_{right}(t)} = \frac{1}{2}\alpha_{T,z}\left(1 - \frac{\Delta T_{left}(t)}{\Delta T_{right}(t)}\right) < \frac{1}{2}\alpha_{T,z}$$

$$(5.63)$$

Expressions 5.63 show that for indeterminate structures, the slope of graphs that represent the relation between total strain and temperature change (the ratios presented in Expression 5.63) cannot be used to determine the thermal expansion coefficient of the material. For example, for the longitudinal gradient described by Expression 5.58, Expressions 5.63 would yield a negative thermal expansion coefficient at left support, null in the middle, and a more-than-two-times underestimate at left support.

This section demonstrated the complexity of the influence of thermal gradients on the structural behavior of beam-like structures. The thermal gradient was decomposed into a cross-sectional and longitudinal component, and the former was decomposed into a non-linear and linear part. Finally, the linear cross-sectional gradient and longitudinal gradient were split into axial and transverse components that have independent influences on structural behavior. In general, in determinate structures or parts of structures, the thermal gradients do not produce stresses and internal forces, as the structure is free to expand; in indeterminate structures, they produce stresses and internal forces, and might have effects on structural behavior that appear counterintuitive.

This section used planar straight beams as illustrative examples. For non-planar beams or non-planar thermal gradients, the strain components $\varepsilon_{x,T}(t,x,y,z)$, $\varepsilon_{y,T}(t,x,y,z)$, and $\varepsilon_{z,T}(t,x,y,z)$, as well as deformation parameters $\varepsilon_{z,T,ctr}(t,z)$, $\kappa_{x,T}(t,z)$, and $\kappa_{y,T}(t,z)$, can be expressed using equations similar to those given in this section for planar beams. The transverse influences of thermal gradients in directions x and y are mutually independent and independent from axial influences, and thus, expressions in this subsection can be generalized for non-planar beams and non-planar gradients by simple algebraic extension. Similar is valid for curvilinear beams: all expressions given in this subsection apply for curvilinear planar beams too, by simply replacing coordinate z with arc-length (natural) coordinate s.

At the end of this subsection, it is important to note that, in practice, thermal strain is frequently wrongly confused with thermal compensation of strain sensors. The latter is addressed and clarified in Section 4.3. Finally, an example of cracking that occurred on a real structure, and was generated by thermal gradients, is presented in Chapter 7, Section 7.2.2.

5.8 Rheologic Strain – Creep and Shrinkage

5.8.1 Creep

Rheologic strain occurs as a consequence of time-dependent dimensional changes in construction material in the absence of load changes (Brčić 1989). The two most common types of rheologic strain are creep (and relaxation, as the opposite of creep) and shrinkage. As mentioned in Section 5.5, creep and shrinkage can result in the apparition and redistribution of stresses in indeterminate structures.

When the material is subjected to loads, it deforms, and stress and mechanical strain develop instantaneously. However, some materials keep deforming afterwards, i.e., they experience time-dependent dimensional changes even if no additional load is applied. The strain generated during that process is called creep. In some materials, creep is very small and practically negligible (e.g., steel, and fiber-reinforced composites), while in some materials it is more significant (e.g., concrete and timber, see Table 5.2). In construction materials, the creep develops slowly, and it is commonly accepted in practice (although not confirmed by science!) that for properly designed structures, it stabilizes around some final value after several years. However, if the structure is not properly designed, the creep will not stabilize and may result in damage or failure of structural members. The scientific phenomena behind the creep in concrete are not fully understood yet. To illustrate this statement, Table 5.5 presents input parameters for various advanced models of creep, used in practice in different parts of the world. The table shows that there is no unianimous agreement among researchers and practitioners on which factors influence creep.

Due to the complexity of the phenomenon, creep in materials was mostly studied under uniaxial state of stress. To simplify the presentation, let us denote the direction of uniaxial stress with z. Numerous studies found in the literature show that creep strain component $\varepsilon_{z,c}$, observed in the same direction as uniaxial stress σ_z, follows, in general, the rule, which, for ease of presentation, can be summarized in Expression 5.64. This equation was taken from the simplified CEB-FIP Model Code 1990 (MC90), i.e., from practical design recommendations based on MC90 (Fib 1999). Expression 5.64 is illustrated in Figure 5.12.

$$\varepsilon_{z,c}(t) = \varepsilon_{z,E}(t_E)K_c f_c(t - t_E) = \frac{\sigma_z}{E(t_E)}K_c f_c(t - t_E) \tag{5.64}$$

Table 5.5 Input parameters for the estimation of creep based on different codes.

	Influencing factor	ACI-209 (USA)	JSCE (Japan)	BP-KX (UK)	CEB/fip MC90	Practical design recommendations (simplified MC90)
Internal	Unit cement	X	X	X		
	Unit water		X			
	W/C		X	X		
	Unit weight	X		X		
	Compressive strength			X	X	X
	S/A			X		
	Slump	X				
	Air content	X				
	Cement type			X	X	
	V/S	X	X	X		
	Notional size	X	X		X	X
External	Curing duration	X				
	Curing method	X		X	X	
	Curing temperature			X	X	
	Relative humidity	X	X	X	X	X
	Age at loading	X	X	X	X	X

Source: Adapted from Song et al. (2001) and Glisic et al. (2013).

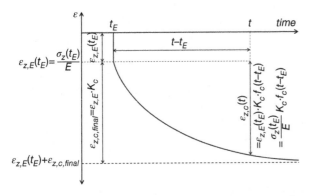

Figure 5.12 Schematic representation of creep evolution. Source: Modified from Glisic and Inaudi (2007).

where t ($t > t_E$) is time, t_E is time of loading (when instantaneous elastic strain ε_{zE} is generated), $\varepsilon_{zc}(t) = \varepsilon_{zc}(t - t_E)$ is creep developed between the time of loading t_E and observed time t, $\varepsilon_{zE}(t_E)$ is elastic strain due to stress σ_z resulting from loading applied at time t_E, $E(t_E)$ is Young's modulus of material at time t_E, K_c is a creep coefficient that depends on input parameters presented in Table 5.5, and $f_c(t - t_E)$ is a creep function that depends on input parameters presented in Table 5.5, which have the following properties: $f_c(0) = 0$ and $f_c(t) \to 1$ when $t \to \infty$, i.e., it asymptotically approaches 1 after enough time (typically 5–20 years).

Expressions 5.17–5.20 show that even a uniaxial state of stress creates a three-axial state of strain. Research has shown (e.g., Wang et al. 2016) that creep components $\varepsilon_{x,c}$ and $\varepsilon_{y,c}$ in directions x and y, perpendicular to the direction of stress z, follow the rule shown in Expression 5.65.

$$\varepsilon_{x,c}(t) = v_{zx,c}(t - t_E)\varepsilon_{z,E}(\tau_E)K_c f_c(t - t_E) = v_{zx,c}(t - t_E)\frac{\sigma_z}{E_E(t_E)}K_c f_c(t - t_E)$$

$$\varepsilon_{y,c}(t) = v_{zy,c}(t - t_E)\varepsilon_{z,E}(\tau_E)K_c f_c(t - t_E) = v_{zy,c}(t - t_E)\frac{\sigma_z}{E_E(t_E)}K_c f_c(t - t_E) \tag{5.65}$$

where $v_{zx,c}(t - t_E)$ and $v_{zy,c}(t - t_E)$ are functions defining the change of Poisson's ratios over time ("creep" of Poisson's ratio) in directions x and y, which depend on load level in the structure so that: $v_{zx,c}(t - t_E) = 0$, $v_{zx,c}(t) \to v_{zx,final}$ when $t \to \infty$, and $v_{zx,final} > v_{zx}$; and: $v_{zy,c}(t - t_E) = 0$, $v_{zy,c}(t) \to v_{zy,final}$ when $t \to \infty$, and $v_{zy,final} > v_{zy}$.

For isotropic materials, Poisson's ratios in directions x and y are equal, and the subscripts can be omitted for conciseness, i.e., $v_{zx,c}(t - t_E) = v_{zy,c}(t - t_E) = v_c(t - t_E)$. The expression shows that a uniaxial state of stress along the z-axis results in a three-axial state of elastic strain and creep, which have diagonal strain tensors, which, in turn, means that these strain and creep components are principal, and axes x, y, and z are principal as well. A similar conclusion can be carried out if uniaxial stress is applied along the x-axis or y-axis. Consequently, if the stress tensor is diagonal, the elastic strain tensor and consequently the creep strain tensor will both be diagonal (see also Expressions 5.17–5.20). Hence, in the case of a planar or spatial stress state at a point, creep at that point can be modeled using the framework shown in Expressions 5.66 and 5.67.

$$\begin{bmatrix} \sigma_x & \tau_{xy} & \tau_{xz} \\ \tau_{yx} & \sigma_y & \tau_{yz} \\ \tau_{zx} & \tau_{zy} & \sigma_z \end{bmatrix} \to \begin{bmatrix} \sigma_1 & 0 & 0 \\ 0 & \sigma_2 & 0 \\ 0 & 0 & \sigma_3 \end{bmatrix} \to \begin{bmatrix} \varepsilon_{1,c} & 0 & 0 \\ 0 & \varepsilon_{2,c} & 0 \\ 0 & 0 & \varepsilon_{3,c} \end{bmatrix} \to \begin{bmatrix} \varepsilon_{x,c} & \frac{1}{2}\gamma_{xy,c} & \frac{1}{2}\gamma_{xz,c} \\ \frac{1}{2}\gamma_{yx,c} & \varepsilon_{y,c} & \frac{1}{2}\gamma_{yz,c} \\ \frac{1}{2}\gamma_{zx,c} & \frac{1}{2}\gamma_{zy,c} & \varepsilon_{z,c} \end{bmatrix} \tag{5.66}$$

$$\varepsilon_{1,c}(t) = \frac{\sigma_1}{E_E(t_E)}K_c f_{c_1}(t - t_E) - v_{12,c}(t - t_E)\frac{\sigma_2}{E_E(t_E)}K_c f_{c_2}(t - t_E)$$
$$- v_{13,c}(t - t_E)\frac{\sigma_3}{E_E(t_E)}K_c f_{c_3}(t - t_E)$$

$$\varepsilon_{2,c}(t) = \frac{\sigma_2}{E_E(t_E)}K_c f_{c_2}(t - t_E)$$
$$- v_{23,c}(t - t_E)\frac{\sigma_3}{E_E(t_E)}K_c f_{c_3}(t - t_E) - v_{23,c}(t - t_E)\frac{\sigma_1}{E_E(t_E)}K_c f_{c_1}(t - t_E)$$

$$\varepsilon_{3,c}(t) = \frac{\sigma_3}{E_E(t_E)}K_c f_{c_3}(t - t_E) - v_{31,c}(t - t_E)\frac{\sigma_1}{E_E(t_E)}K_c f_{c_1}(t - t_E)$$
$$- v_{32,c}(t - t_E)\frac{\sigma_2}{E_E(t_E)}K_c f_{c_2}(t - t_E) \tag{5.67}$$

In Expression 5.66, the first matrix represents the stress tensor with respect to the given coordinate system $Oxyz$, the second represents the principal stress tensor (obtained from the stress tensor using the approach presented in Section 5.5), the third matrix represents the principal creep tensor, whose components are calculated using Expression 5.67, and the fourth matrix represents the creep tensor in the initial coordinate system $Oxyz$ (obtained from the stress tensor using the approach presented in Section 5.3).

Equations in Expression 5.67 are obtained by combining Expressions 5.64 and 5.65; however, it is assumed that all stress components have the same application time t_E. Note that all creep coefficients and functions in the expression have an additional subscript that indicates that they may be different for different directions as they depend on the stress magnitude.

The following properties characterize the creep in concrete under uniaxial state of stress (Fib 1999):

1. Creep exists in loaded material, and its final value is proportional to the elastic strain (if the structure is designed properly, then a final creep value will exist, i.e., the creep will stabilize in long term);

2. Creep is reversible; removing the load from the concrete will tend to remove creep (i.e., creep will tend to return to zero);

3. Among other factors, the creep also depends on maturity of the concrete at the time of application of elastic strain t_E; the more mature the concrete is at the time of loading, the less creep will be generated; typical values of the creep coefficient K_c are ranged between 1 and 3 for concrete loaded 28 days after pouring or later (mature concrete), but can be as high as 5.6 for concrete loaded 1 day after pouring (very young concrete);

4. If the load is applied to concrete in several increments, then the creep generated from each load increment may have a creep coefficient K_c and a creep function f_c that are different for different increments; the total creep at an arbitrary time t can be estimated in a simplified manner as the sum of creeps generated by each load increment separately; this statement is illustrated by Expression 5.68 for uniaxially loaded materials.

$$\varepsilon_{z,c}(t) = \sum_{i=1}^{N_{\sigma,inc}(t)} \varepsilon_{z,E,i}(t_{E,i}) K_{c,i} f_{c,i}(t - t_{E,i}) = \sum_{i=1}^{N_{\sigma,inc}(t)} \frac{\sigma_{z,i}}{E(t_{E,i})} K_{c,i} f_{c,i}(t - t_{E,i}) \tag{5.68}$$

where i in subscript indicates the stress (load) increment i, $i = 1, 2, \dots N_{\sigma,inc}(t)$, and $N_{\sigma,inc}(t)$ is total number of stress increments $\sigma_{z,i}$ applied from the beginning of loading until time t.

5. Creep of reinforced concrete is smaller than creep of non-reinforced concrete as the rebars, being made of steel, have negligible creep and a much higher Young's modulus than the concrete; heavier reinforcement would result in smaller creep of reinforced concrete and vice versa. Thus, when modeling the creep in reinforced concrete, the presence of rebars should be taken into account and the creep coefficient K_c and creep function $f_c(t)$ modified accordingly.

The consequences of creep are dimensional changes in structure. These consequences are similar to those resulting from thermal strain and thermal gradients. In isostatic (determinate) structures, creep will result in an increase in the global deformation of the structure without generating additional mechanical strain and stress. In hyper-static (indeterminate) structures, creep can cause important mechanical strain and stress redistributions. For beam-like structures, these redistributions can be described and understood using the figures and equations presented in Section 5.7. The main difference is that the processes related to creep would happen very slowly, over months and years.

As mentioned earlier in the text, steel and most fiber-reinforced composites have very small creep (their creep coefficient is in the range of a few percent) and in most practical applications, it can be neglected. Contrary to this, the long-term creep in concrete and timber structures can be equal to or several times higher than the magnitude of elastic strain, which makes its evaluation and prediction important. However, creep cannot be measured directly in real-life settings (on real structures) using strain sensors due to the presence of other strain constituents, such as mechanical strain, thermal strain, and shrinkage, and typically the following four approaches can be taken:

(1) Evaluate creep by performing laboratory tests on specimens built of the same construction material (concrete or timber) as the real structure;

(2) Use the numerical model(s) for creep proposed in the literature and evaluate creep based on these models.

(3) Apply various data-driven methods (statistical and machine learning) on the measurements taken at an early stage of a structure's life (typically two to three years) to predict the evolution of creep and calibrate numerical models from literature.

(4) Combine two or more of the approaches presented above.

The first approach can be very expensive and time-consuming as a representative number of specimens have to be created and kept in controlled lab conditions for a long period of time. Consequently, this approach is taken in practice only in the case of very important or special structures where the cost of tests is justified. In addition, due to the presence of reinforcing bars in real structures that will constrain creep, experimental data is frequently used to calibrate sophisticated numerical models that can account for the presence of rebars and evaluate constrained (by rebars) creep values. While this approach is very elaborate, time-consuming and expensive, it is also accurate as it deals with the data that are obtained from the same material as the one used to construct the structure of interest.

The second approach is simpler and less expensive, as it requires the implementation of equations (chosen from the literature) to model the creep. This model of creep can be used in analytical expressions (e.g., such as Expression 5.64) or implemented in a numerical model of the structure (i.e., in some of the professional software that was developed for the purpose). This approach is less accurate as models found in the literature are usually general and typically provide an envelope of maximal values as opposed to an accurate value for the specific construction material. Nevertheless, the obtained accuracy is frequently acceptable and justifies the use of this approach in the most cases.

The third approach is very effective as it does not require extra cost (it uses measurements from an already installed SHM system) and provides results that have acceptable accuracy; however, the major challenge is the separation of creep and shrinkage from total rheological strain. In addition, it requires a good data set containing two to three years of reliable strain and temperature measurements to train the data-driven methods, with the assumption that no unusual behaviors (e.g., damage or deterioration) occurred during the training period. For evaluation of creep, this data should be collected after major loading events such as activation of deadload (removal of the forms) or prestressing.

Given the relationship between creep and stress, in the first two approaches presented above, it is very important to evaluate stresses over time (using strain measurements or otherwise), as creep depends on stress magnitude and time of application. This means that the schedule of loadings applied to the structure (time of application and magnitude of loading) should be known for each loading increment. Still, evaluation of stresses can be challenging in indeterminate structures as the creep will slowly redistribute stresses, which in turn will affect back the creep. In the third approach, once the major load is applied, the machine learning method will "self-adjust" to stress redistributions; however, additional major loading events will perturb the algorithm and require additional training. A simplified example presented in Figure 5.13 illustrates the influence of loading schedule (time history), i.e., the time of applying the stress and stress magnitude, on the interpretation of strain measurement.

Let's assume that Figure 5.13 shows the sums of elastic strain and creep $\varepsilon_{z,E}(t_s) + \varepsilon_{z,c}(t_s)$ from two elements made of the same concrete (the same elastic and rheologic behavior). Let's assume that these two elements are loaded at two different times $t_{E,1}$ and $t_{E,2}$ ($t_{E,1} < t_{E,2}$) with different

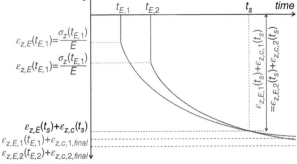

Figure 5.13 Influence of loading schedule on strain measurement in material subjected to creep. Source: Modified from Glisic and Inaudi (2007).

loads, such that the corresponding stresses $\sigma_{z,1}$ and $\sigma_{z,2}$ are generated and $\sigma_{z,1} < \sigma_{z,2}$. Given that the times of applying the stresses are different, the values of the creep coefficients $K_{c,1}$ and $K_{c,2}$ will also be different, and the values of the creep functions $f_{c,1}(t - t_{E,1})$ and $f_{c,2}(t - t_{E,2})$ as well. Note that $t_{E,1} < t_{E,2}$ implies $K_{c,1} > K_{c,2}$ (Fib 1999), i.e., the later the stress is applied to concrete, the smaller the creep coefficient will be. Hence, based on Expression 5.64, there will be a time t_s ($t_{E,1} < t_{E,2} < t_s$), when the sum of elastic strain and creep in two elements will have the same value. This demonstrates that, in general, if the time of loading is not known, then the sum of elastic strain and creep at some time t (e.g., $\varepsilon_{z,E}(t_s) + \varepsilon_{z,c}(t_s)$ at time t_s) cannot be uniquely decomposed into its constituents, elastic strain and creep. In other words, at time t_s it would be impossible to tell if stress $\sigma_{z,1}$ was applied at time $t_{E,1}$, resulting in $\varepsilon_{z,E,1}(t_s)$ and $\varepsilon_{z,c,1}(t_s)$, or stress $\sigma_{z,2}$ was applied at time $t_{E,2}$, resulting in $\varepsilon_{z,E,2}(t_s)$ and $\varepsilon_{z,c,2}(t_s)$.

Examples of the evaluation of rheological strain in concrete are given in Section 5.9.

5.8.2 Shrinkage

Shrinkage manifests as a slow, time-dependent dimensional change that occurs independently from the load. It is an intrinsic property of material, and, on a macroscale, it occurs in all three dimensions. Shrinkage in concrete is mostly caused by two processes: hydration, which causes autogenous shrinkage, and drying, i.e., the evaporation of water from concrete pores. The former develops relatively fast as it follows the dynamics of hydration, while the latter is slow as it depends on pore connectivity in concrete, the dimensions (volume) of the structural element, its exposure to the environment, and environmental conditions. Shrinkage in timber is mostly a consequence of drying. For isotropic and orthotropic materials, the shrinkage can be described as a diagonal strain tensor, as shown in Expression 5.69.

$$\mathbf{E}_{sh}(t) = \begin{bmatrix} \varepsilon_{\xi,sh}(t) & 0 & 0 \\ 0 & \varepsilon_{\eta,sh}(t) & 0 \\ 0 & 0 & \varepsilon_{\zeta,sh}(t) \end{bmatrix} \tag{5.69}$$

where $\mathbf{E}_{sh}(t)$ is shrinkage strain tensor at time t, $\varepsilon_{\xi,sh}(t)$, $\varepsilon_{\eta,sh}(t)$, and $\varepsilon_{\zeta,sh}(t)$ are shrinkage components in directions of axes of orthotropic symmetry, ξ, η, and ζ; note that specifically for isotropic material at macroscale (e.g., concrete), the three components are mutually equal (simply called "shrinkage" in the further text), and any three mutually perpendicular axes correspond to principal axes of shrinkage strain.

Shrinkage develops slowly and its final value is usually reached after several years. Similar to creep, the shrinkage in concrete can be modeled in a simplified way using the following expression (e.g., Fib 1999):

$$\varepsilon_{n,sh}(t) = \varepsilon_{sh,final} f_{sh}(t - t_0)$$ (5.70)

where $t\,(t > t_0)$ is time, t_0 is time when the shrinkage is initiated, $\varepsilon_{sh,final}$ is final shrinkage (the value at which the shrinkage stabilizes), and $f_{sh}(t - t_0)$ is shrinkage function (which, in general, depends on input parameters similar to those presented in Table 5.5), $f_{sh}(0) = 0$, $f_{sh}(t) \to 1$ when $t \to \infty$.

Expression 5.70 is illustrated in Figure 5.14.

The following properties characterize the shrinkage in concrete (Fib 1999):

1. Shrinkage is an intrinsic property of concrete;
2. The shrinkage at a time t depends on the time of shrinkage initiation t_0; although the behavior of concrete at an early age (maturity is less than 28 days) is different from mature concrete, it is commonly accepted (for practical reasons) that t_0 corresponds to the time of pouring;
3. Shrinkage is practically irreversible in real-life settings (it can be reversible only under very particular environmental conditions);
4. The final values of shrinkage in concrete are typically ranged between 200 and 530 µε (actually, between −200 and −530 µε, as shrinkage produces negative strain).

As mentioned earlier in the text, steel does not experience shrinkage; most fiber-reinforced composites also do not experience shrinkage, except in the short-term during manufacturing (shrinkage of matrix). Timber shrinks and swells due to humidity change, and the strain difference between green timber and dry timber (due to internal humidity) can range between −1000 and −2000 µε in longitudinal direction, and it is orders of magnitude larger in the tangential and radial directions.

Similar to creep and thermal gradients, the overall consequence of shrinkage in determinate structures is a change in the magnitude of the global deformation of the structure without generating additional mechanical strain and stress. In the case of indeterminate structures, shrinkage can cause important strain and stress redistributions. In both types of structures, shrinkage evolves in a gradual manner and can create non-linear shrinkage gradients, similar to non-linear thermal gradients. The consequences of these shrinkage gradients are also similar to those of thermal gradients.

Point 4 above shows that the magnitude of shrinkage in long-term can be relatively high compared to the magnitude of elastic strain. Thus, it is very important to evaluate it in the case of long-term SHM of concrete structures; however, this is not a trivial task as the issues related to the evaluation of shrinkage are similar to those of the evaluation of creep (including constraints imposed by presence of rebars). Hence, the four approaches to evaluating creep (see Section 5.8.1)

Figure 5.14 Schematic representation of shrinkage evolution. Source: Modified from Glisic and Inaudi (2007).

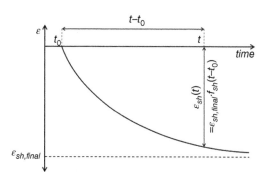

are applicable to the evaluation of shrinkage (lab tests, analytical/numerical modeling, data-driven approach, and combination of these approaches), with all associated advantages and limitations that are presented in Section 5.8.1, except the limitation associated with knowledge regarding loading schedule, as the shrinkage can be considered practically independent from loading and stresses. Examples of the evaluation of rheological strain in concrete are given in Section 5.9.

5.9 Physics-Based Interpretation of Strain Measurements and Identification of Unusual Structural Behaviors in Steel and Concrete Structures

5.9.1 Example of Physics-Based Modeling of Total Strain and Stress

The aims of the interpretation of strain measurements are to assess structural behaviors and either identify (detect, localize, and quantify) unusual structural behaviors (if any) or indicate consistency with their absence. In order to fully assess structural behaviors, the following steps are necessary:

(1) Create a full or partial model that describes structural behaviors at the location of the sensor and estimate the measures of accuracy of the model (limits of errors or uncertainty); this model can be physics-based (e.g., analytical, numerical, or design-code-based), data-driven (e.g., using statistics and machine learning on a set of SHM data to model the behavior of structure), heuristic (e.g., engineering judgment), or a combination of two or more of these; ideally, the model should be able to estimate the values or ranges of values of strain constituents, and identify possible failure modes at locations of sensors; as this chapter deals with the interpretation of strain measurement at sensor level, only local failure modes are of interest in this section: failure of the material itself (cracking or crushing) or local buckling (bowing); global failure modes will be considered in Chapter 6.

(2) Determine accuracy measures (uncertainties or limits of errors) and ascertain reliability of strain measurements; these topics are presented in Chapter 4; recall that ascertaining the reliability of the measurements requires expertise in the monitoring system employed, so the potential sources of loss of reliability can be understood; note that for the types of strain sensors that require thermal compensation, loss of reliability of the temperature sensor can cause the loss of reliability of strain measurements (i.e., reliability of total strain measurement, due to thermal compensation of the sensor, and reliability of thermal strain evaluation, due to its direct dependence on, and relationship with, temperature).

(3) Compare strain measurement or strain measurement derivatives (typically, but not limited to, elastic strain change, absolute elastic strain, stress change, and absolute stress) to various corresponding values estimated from the model, including ultimate values; the comparisons should be based on established criteria that take into account accuracy measures of both the model and the measurements.

(4) In the event that unusual structural behavior is detected, its cause, location, and magnitude should be ascertained (confirmed) by using the model.

To illustrate the application of the above-presented framework and specifically its first step, a simple physics-based model is developed in this subsection. To ease presentation, this section focuses on steel and concrete structural members and uses, for both materials, an analytical model for total strain estimations that is based, for elastic and thermal strain, on linear beam theory and, for rheologic strain in concrete, on the simplified CEB-FIP Model Code 1990 (Fib 1999). This implies that materials are considered homogeneous at macroscale. Consequently, this model can be applied to steel without restrictions regarding the gauge length of sensors, as steel is a homogeneous material. However, the use of linear theory of beams has implications in terms of validity for applications in concrete: based on considerations presented in Section 4.4, the presented model can only be used for modeling strain in sensors with an adequately long-gauge length, that are "insensitive" to material inhomogeneity (see Section 4.4). To further simplify presentation, only planar straight beams are modeled, assuming that the readers can derive corresponding expressions for non-planar and curvilinear beams based on expressions presented throughout this chapter and general theoretical approaches found in relevant literature. Hence, let us observe a beam segment instrumented with three types of strain sensors – short-gauge, long-gauge, and distributed sensors – as shown in Figure 5.15. The gauge length of the long-gauge sensor, L_s, and the spatial resolution of the distributed sensor, L_{SR}, are indicated in the figure (see also Section 4.4). Let us denote with $y_C(t)$ the position of the point C (at time t) with respect to the centroid of the cross-section with coordinate z_C. Note that the locations of sensors are chosen so that the point C is in the geometrical center of the gauge lengths of both discrete sensors (short-gauge and long-gauge), as well as in the center of the spatial resolution length of distributed sensor.

The analytical models for estimating total strain at point C in steel (combine Expressions 5.9, 5.28, and 5.36) and concrete (combine Expressions 5.9, 5.28, 5.36, 5.68, and 5.70) are given as follows:

$$\textbf{in steel: } \varepsilon_{C,z,tot,model}(t) = \varepsilon_{C,z,E,model}(t) + \varepsilon_{C,z,T,model}(t) = \frac{\sigma_{C,z,model}(t)}{E_C(t)} + \alpha_{T,C,z}(t)\Delta T_{C,model}(t)$$

(5.71)

$$\textbf{in concrete: } \varepsilon_{C,z,tot,model}(t) = \varepsilon_{C,z,E,model}(t) + \varepsilon_{C,z,T,model}(t) + \varepsilon_{C,z,c,model}(t) + \varepsilon_{C,z,sh,model}(t)$$

$$= \frac{\sigma_{C,z,model}(t)}{E_C(t)} + \alpha_{T,C,z}(t)\Delta T_{C,model}(t) + \sum_{i=1}^{N_{\sigma,inc}(t)} \frac{\sigma_{C,z,model,i}}{E_C(t_{E,i})} K_{c,i} f_{c,i}(t - t_{E,i}) + \varepsilon_{z,C,sh,final} f_{sh}(t - t_0)$$

(5.72)

where $\sigma_{C,z,model}(t) = \sum_{i=1}^{N_{\sigma,inc}(t)} \sigma_{C,z,model,i}$ and $\sigma_{C,z,model,i} = \sigma_{z,model}(t_{E,i}, y(t_{E,i}), z_C)$.

In Expressions 5.71 and 5.72, subscript C indicates the location of point C, or the cross-section containing point C. In both expressions, the stress can be given in terms of internal forces (normal force and bending moments).

$$\sigma_{C,z}(t) = \sigma_z(t, y_C(t), z_C) = \frac{N_z(t, z_C)}{A(t, z_C)} + \frac{M_x(t, z_C)}{I_x(t, z_C)} y_C(t) = \frac{N_{C,z}(t)}{A_C(t)} + \frac{M_{C,x}(t)}{I_{C,x}(t)} y_C(t).$$

(5.73)

For modern steel, Young's modulus and thermal expansion coefficient are independent of time and position in the structure (i.e., invariable in time and space); however, this might not be the case for historic steel (e.g., due to limitations in quality of steel manufacturing in the past, Young's modulus may be different at different points within the same structural member). Also, this might not be the case for concrete, modern, or historic. These are the reasons why, for both parameters, potential

Figure 5.15 Schematic representation of a beam segment instrumented with a short-gauge sensor, a long-gauge sensor, and a distributed sensor.

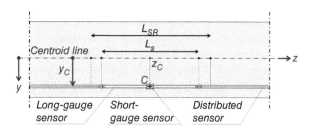

dependence on time and location is kept in Expressions 5.71 and 5.72. An advantage of the models presented in Expressions 5.71–5.73 is that they estimate absolute values of strain constituents and stress, i.e., the values that have been generated in the structure since its construction. Recall that this is not the case with sensor measurements – they provide values of strain with respect to reference time t_{Ref} (e.g., time of the first reading of installed sensors, see Section 5.4), and deliver an absolute value of total strain only if at the reference time the instrumented steel element was at rest (i.e., before construction, with no loads applied), or the instrumented concrete element was undergoing pouring (i.e., the sensors were installed before pouring, and embedded during pouring).

Note that Expressions 5.71 and 5.72 model the strain value at a point; thus, such a model is suitable for modeling the strain measured by short-gauge sensors. These expressions might be applicable with satisfactory accuracy to long-gauge and distributed sensors too, if the strain field does not vary significantly along the sensor gauge length or spatial resolution length, respectively (see Section 4.4). However, if the strain field varies significantly as per Expressions 5.71 and 5.72, then the strain measurement would be more accurately modeled if an average integral of strain along the gauge length or spatial resolution length is used. This can be expressed using the following expressions for steel and concrete:

in steel:

for long-gauge sensor:

$$\varepsilon_{C,z,tot,model}(t) = \frac{1}{L_s}\int_{z_C-\frac{L_s}{2}}^{z_C+\frac{L_s}{2}} \left(\frac{\sigma_{z,model}(t,y(t),z)}{E(t,z)} + \alpha_{T,z}(t,z)\Delta T_{model}(t,(y(t),z)) \right)dz$$

for distributed sensor:

$$\varepsilon_{C,z,tot,model}(t) = \frac{1}{L_{SR}}\int_{z_C-\frac{L_{SR}}{2}}^{z_C+\frac{L_{SR}}{2}} \left(\frac{\sigma_{z,model}(t,y(t),z)}{E(t,z)} + \alpha_{T,z}(t,z)\Delta T_{model}(t,(y(t),z)) \right)dz$$

$$(5.74)$$

in concrete:

for long-gauge sensor: $\varepsilon_{C,z,tot,model}(t)$

$$= \frac{1}{L_s}\int_{z_C-\frac{L_s}{2}}^{z_C+\frac{L_s}{2}} \left(\begin{array}{l} \dfrac{\sigma_{z,model}(t,y(t),z)}{E(t,z)} + \alpha_{T,z}(t,z)\Delta T_{model}(t,(y(t),z)) \\ + \sum_{i=1}^{N_{\sigma,inc}(t)} \dfrac{\sigma_{z,model,i}(t_{E,i},y(t_{E,i}),z)}{E(t_{E,i},z)}K_{c,i}f_{c,i}(t-t_{E,i}) + \varepsilon_{z,sh,final}(y(t),z)f_{sh}(t-t_0) \end{array} \right)dz$$

for distributed sensor: $\varepsilon_{C,z,tot,model}(t)$

$$= \frac{1}{L_{SR}}\int_{z_C-\frac{L_{SR}}{2}}^{z_C+\frac{L_{SR}}{2}} \left(\begin{array}{l} \dfrac{\sigma_{z,model}(t,y(t),z)}{E(t,z)} + \alpha_{T,z}(t,z)\Delta T_{model}(t,(y(t),z)) \\ + \sum_{i=1}^{N_{\sigma,inc}(t)} \dfrac{\sigma_{z,model}(t_{E,i},y(t_{E,i}),z)}{E(t_{E,i},z)}K_{c,i}f_{c,i}(t-t_{E,i}) + \varepsilon_{z,sh,final}(y(t),z)f_{sh}(t-t_0) \end{array} \right)dz$$

$$(5.75)$$

As shown above, modeling the strain constituents involves knowledge about loads (magnitudes and times of application to determine normal forces and bending moments), mechanical, thermal, and rheologic properties of construction material (Young's modulus, thermal expansion coefficient, creep and shrinkage coefficients and functions), and geometrical properties of the structures (dimensions and shapes of cross-sections to determine areas and moments of inertia of cross-sections, boundary conditions to determine reactions), which all introduce errors or uncertainties in the model estimates.

In general, for the analytical model presented above, typical sources of model errors or uncertainties are listed below:

(1) Appropriateness of the model (e.g., linear models cannot be used if deformations or displacements in the structure are large, or to estimate the buckling stability of the structure, etc.);

(2) Epistemic error of model (e.g., due to simplifications such as linearity, homogeneity, and isotropy, but also due to unaccounted sources of strain, if any);

(3) Constitutive law of construction material and related parameters (e.g., mechanical and physical properties of material, such as stress–strain relationship, rheologic behavior, and thermal expansion, capacity, and conductivity);

(4) Geometrical properties of the structure and structural elements (e.g., position of the centroid line in beams, areas and moments of inertia of the cross-sections, and constraints at boundaries);

(5) Sensor's exact position on the structure and its gauge length;

(6) Other.

For example, if sensor measurement is modeled using linear beam theory, then Expressions 5.71–5.75 show that error or uncertainty in determining Young's modulus E (especially in concrete, as it is frequently not directly measured and changes over time), area of cross-section A, moment of inertia I_x, and distance of sensor from the centroid y (e.g., due to construction tolerances), thermal expansion coefficient $\alpha_{T,z}$ (especially in concrete, as it can vary throughout the structure and change over time), rheological parameters $K_{c,i}$ and $\varepsilon_{z,sh,final}$, rheological functions $f_{c,i}$ and f_{sh}, but also internal forces N_z and M_x (e.g., due to boundary conditions, and unknown stiffness of joints), all contribute to the overall error or uncertainty of estimation of total strain $\varepsilon_{C,z,tot,model}$. In addition, there are epistemic errors in linear beam theory, thermal models, and rheological models for creep and shrinkage (i.e., errors due to the "goodness" of the models for applications in steel and concrete beams), which are in general unknown.

The sources of error listed above are not necessarily mutually independent and should be analyzed simultaneously rather than separately. If we denote the error of the model with $\delta\varepsilon_{C,z,tot,model}(t)$ and the limits of error of the model with $\pm\delta\varepsilon_{C,z,tot,model}$, then the following is valid:

$$\left|\delta\varepsilon_{C,z,tot,model}(t)\right| \leq \delta\varepsilon_{C,z,tot,model}$$
$$\delta\varepsilon_{C,z,tot,model}(t) = \varepsilon_{C,z,tot,model}(t) - \varepsilon_{C,z,tot,real}(t) \tag{5.76}$$
$$\varepsilon_{C,z,tot,model}(t) - \delta\varepsilon_{C,z,tot,model} \leq \varepsilon_{C,z,tot,real}(t) \leq \varepsilon_{C,z,tot,model}(t) + \delta\varepsilon_{C,z,tot,model}$$

Direct estimation of the error presented in Expressions 5.76 is not possible, as the real strain is, in general, unknown. However, assuming that the epistemic error of the model can be neglected (e.g., this is empirically proven for linear beam theory), then the error and limits of error shown in Expressions 5.76 can be evaluated by applying the error propagation formulae given in Chapter 4. Similar is valid for the evaluation of uncertainty of the model, i.e., uncertainty of model estimates of total strain $u(\varepsilon_{C,z,tot,model})$ can be evaluated by using uncertainty propagation formulae presented in Chapter 4. The limits of error of the model will be simply referred to as error of the model or model error. Beyond linear beam theory, the limits of error or uncertainty of the model can be in general determined by using propagation formulae, as long as the epistemic error of the model is negligible or known. In the further text, it will be assumed that the models presented in Expressions 5.71–5.75 are appropriate and that all parameters in the models are known (along with their measures of accuracy, i.e., limits of error or uncertainty).

5.9.2 Comparison of Strain Measurements with Corresponding Model Values

As presented in Section 5.4 and mentioned again in the previous subsection, a strain sensor measures total strain starting with reference time, t_{Ref}, and thus, unless the reference time coincides with a moment that precedes straining of material (e.g., precedes initiation of stressing in steel or coincides with pouring of concrete), the measured total strain $\varepsilon_{z,C,tot,sensor}(t)$, at observed time t, does not represent absolute total strain in material, denoted as $\varepsilon_{z,C,sensor}(t)$, but rather relative strain with respect to reference time t_{Ref}, see Expression 5.7. As mentioned above, the model provides estimates of absolute values of strain, and to avoid ambiguity in notation, let us denote the model estimate of the change in strain relative to reference time as $\Delta\varepsilon_{z,C,tot,model}(t)$. Similar notation can be implemented for the individual models of strain constituents. Then the following expressions define the model estimate of change in strain (compare with Expression 5.7).

$$\Delta\varepsilon_{C,z,tot,model}(t) = \varepsilon_{C,z,tot,model}(t) - \varepsilon_{C,z,tot,model}(t_{Ref})$$

$$\text{in steel:} \quad = \varepsilon_{C,z,E,model}(t) + \varepsilon_{C,z,T,model}(t) - \varepsilon_{C,z,E,model}(t_{Ref}) - \varepsilon_{C,z,T,model}(t_{Ref})$$

$$= \Delta\varepsilon_{C,z,E,model}(t) + \Delta\varepsilon_{C,z,T,model}(t)$$

$$\text{in concrete:} \quad = \varepsilon_{C,z,E,model}(t) + \varepsilon_{C,z,T,model}(t) + \varepsilon_{C,z,c,model}(t) + \varepsilon_{C,z,sh,model}(t)$$

$$- \varepsilon_{C,z,E,model}(t_{Ref}) - \varepsilon_{C,z,T,model}(t_{Ref}) - \varepsilon_{C,z,c,model}(t_{Ref}) - \varepsilon_{C,z,sh,model}(t_{Ref})$$

$$= \Delta\varepsilon_{C,z,E,model}(t) + \Delta\varepsilon_{C,z,T,model}(t) + \Delta\varepsilon_{C,z,c,model}(t)$$

$$+ \Delta\varepsilon_{C,z,sh,model}(t) \tag{5.77}$$

Based on Expression 5.76, the limits of error of the model estimate of change in strain $\pm\delta\Delta\varepsilon_{C,z,tot,model}$ and uncertainty of the model estimate of change in strain $u(\Delta\varepsilon_{C,z,tot,model})$ can be, respectively, expressed as follows:

$$\delta\Delta\varepsilon_{C,z,tot,model} = \delta\varepsilon_{C,z,tot,model} + \delta\varepsilon_{C,z,tot,model}(t_{Ref})$$

$$u(\Delta\varepsilon_{C,z,tot,model}) = \sqrt{u(\varepsilon_{C,z,tot,model})^2 + u(\varepsilon_{C,z,tot,model}(t_{Ref}))^2} \tag{5.78}$$

Comparison between the model and sensor measurement at time t represents the first criterion in the identification of unusual structural behaviors. It can be written as follows:

$$|\varepsilon_{C,z,tot,sensor}(t) - \Delta\varepsilon_{C,z,tot,model}(t)| \leq \delta\varepsilon_{C,z,tot,sensor} + \delta\Delta\varepsilon_{C,z,tot,model}$$

Criterion I: $\qquad\qquad\qquad\qquad\qquad\text{or}$

$$|\varepsilon_{C,z,tot,sensor}(t) - \Delta\varepsilon_{C,z,tot,model}(t)| \leq n_u\sqrt{u(\varepsilon_{C,z,tot,sensor})^2 + u(\Delta\varepsilon_{C,z,tot,model})^2}$$

$$\tag{5.79}$$

where vertical brackets denote the absolute value of the difference between the measurement and model, and the term to the right of the sign "$\leq$" represents the limits of error (upper expression) and uncertainty (lower expression) of the difference, while n_u is a multiplicator (recommended to be set between 2 and 3, i.e., $2 \leq n_u \leq 3$, see Chapter 4).

If Criterion I is fulfilled, then the strain measurement at time t does not indicate the existence of unusual structural behaviors. If Criterion I is not fulfilled, then the strain measurement at time t indicates the existence of unusual structural behavior.

Note that the above two statements are valid only if (i) the limits of error or uncertainty of both model and measurements are correctly evaluated, (ii) the sensor measurement is confirmed to be reliable, and (iii) the model is confirmed to be appropriate. If any of points (i)–(iii) is not valid, the comparison given in Criterion I is inapplicable, and points (i)–(iii) may need to be revisited and model may need to be updated or upgraded (e.g., some parameters can be updated based on measurements). Note, also, that the fulfillment of Criterion I does not necessarily mean that unusual structural behavior is not present. For example, if the limits of error or uncertainty of model or measurements are large, then Criterion I allows large discrepancies between the model and measurements, which typically occur in the case of unusual structural behavior; in that case, Criterion I (i.e., model or measurements or both) lacks sensitivity in the identification of unusual structural behaviors. Another possibility is, for example, an occurrence of unusual structural behavior that changes the strain in a direction perpendicular to the sensor, in which case the sensor might not be sensitive to that change. Thus, fulfillment of Criterion I does not necessarily guarantee the absence of an unusual structural behavior, it only indicates that an unusual structural behavior is not detected, i.e., it only indicates consistency with an usual (expected, undamaged) structural behavior.

In the cases where temperature is monitored at the location of strain sensor (which is in general highly recommended and often done), in Expressions 5.71, 5.72, 5.74, and 5.75, the term $\Delta T_{C,model}$ can be replaced with $\Delta T_{C,sensor}$ (there is no need to model temperature change, given that it is measured). In these cases, measured thermal strain is equal to the change in modeled thermal strain (i.e., $\varepsilon_{z,T,sensor} = \Delta\varepsilon_{z,T,model} = \alpha_{T,z}\Delta T_{C,sensor}$) and Criterion I is reduced to a comparison between measured and modeled elastic strain in steel or between the measured and modeled sums of elastic and rheological strain in concrete.

The next criterion (Criterion II) for assessing the condition of structure at location of sensor is based on comparison of measured elastic strain with yield strains $\varepsilon_{yld}{}^{+}$ and $\varepsilon_{yld}{}^{-}$ for ductile materials (steel, reinforced concrete) and ultimate strains $\varepsilon_{u}{}^{+}$ and $\varepsilon_{u}{}^{-}$ for brittle materials (unreinforced concrete in tension). Expressions 5.21 and 5.26 show the conditions for mechanical strain in multi-axially stressed material to be within or out of elastic range. However, using a single sensor, it is impossible to infer all components of the strain sensor. That is why the simplified approach here is based on Expressions 5.11 and 5.12, which are valid for uniaxially loaded material. Justification for the applicability of these equations is twofold: (i) frequently, planar beams are monitored with pairs of parallel sensors where one sensor is installed close to the top and the other close to the bottom of the cross-section; at these locations, shear stress is relatively small, and thus the stress state of material is approximately uniaxial; and (ii) error due to approach presented in point (i) above can be conservatively taken into account by setting relatively high safety factors when performing the comparison (see further text).

Expressions 5.11 and 5.12 are based on absolute value of mechanical strain constituent, and thus the absolute mechanical strain should be inferred from the total strain measurement. Assuming that the temperature at location of sensor is monitored, the expression for elastic strain in steel is simple to derive as it represents the difference between total strain and thermal strain. However, this is not the case with concrete due to dependence of creep on stress increments (see Expression 5.72). In order to simplify the evaluation of elastic strain in concrete, let us introduce equivalent creep coefficient $K_{c,equ}$ and equivalent creep function $f_{c,equ}$ as follows:

$$K_{c,equ} = \frac{\sum_{i=1}^{N_{\sigma,inc}(t)} \frac{\sigma_{z,model}(t_{E,i},y(t_{E,i}),z)}{E(t_{E,i},z)}K_{c,i}}{\frac{\sigma_{z,model}(t,y(t),z)}{E(t,z)}} = \frac{\sum_{i=1}^{N_{\sigma,inc}(t)} \varepsilon_{z,E,model}(t_{E,i},y(t_{E,i}),z)K_{c,i}}{\varepsilon_{z,E,model}(t,y(t),z)}$$

$$f_{c,equ}(t - t_{E,1}) = \frac{\sum_{i=1}^{N_{\sigma,inc}(t)} \frac{\sigma_{z,model}(t_{E,i},y(t_{E,i}),z)}{E(t_{E,i},z)}K_{c,i}f_{c,i}(t - t_{E,i})}{\frac{\sigma_{z,model}(t,y(t),z)}{E(t,z)}K_{c,equ}}$$

$$= \frac{\sum_{i=1}^{N_{\sigma,inc}(t)} \varepsilon_{z,E,model}(t_{E,i},y(t_{E,i}),z)K_{c,i}f_{c,i}(t - t_{E,i})}{\varepsilon_{z,E,model}(t,y(t),z)K_{c,equ}} \tag{5.80}$$

Consequently:

$$\varepsilon_{z,E,model}(t,y(t),z)K_{c,equ}f_{c,equ}(t-t_{E,1}) = \sum_{i=1}^{N_{\sigma,inc}(t)} \varepsilon_{z,E,model}(t_{E,i},y(t_{E,i}),z)K_{c,i}f_{c,i}(t-t_{E,i})$$

$$\text{i.e., } \varepsilon_{C,z,c,model}(t) = \varepsilon_{C,z,E,model}(t)K_{c,equ}f_{c,equ}(t-t_{E,1}) = \sum_{i=1}^{N_{\sigma,inc}(t)} \varepsilon_{C,z,E,model,i}K_{c,i}f_{c,i}(t-t_{E,i}). \quad (5.81)$$

Hence, the following expressions can be used for evaluation of elastic strain in steel and concrete, respectively:

in steel: $\varepsilon_{C,z,E,sensor}(t) = \varepsilon_{C,z,tot,sensor}(t) - \alpha_{T,C,z}\Delta T_{C,sensor}(t)$

in concrete:

$$\varepsilon_{C,z,E,sensor}(t) = \frac{(\varepsilon_{C,z,tot,sensor}(t) - \alpha_{T,C,z}(t)\Delta T_{C,sensor}(t) - \varepsilon_{C,z,sh,final}(f_{sh}(t-t_0) - f_{sh}(t_{Ref}-t_0)))}{1 + K_{c,equ}(f_{c,equ}(t-t_{E,1}) - f_{c,equ}(t_{Ref}-t_{E,1}))}$$

$$(5.82)$$

where $\varepsilon_{C,z,sh,final}(f_{sh}(t-t_0) - f_{sh}(t_{Ref}-t_0))$ represents model estimate of change in shrinkage $\Delta\varepsilon_{C,z,sh,model}(t)$ and $K_{c,equ}(f_{c,equ}(t-t_{E,1}) - f_{c,equ}(t_{Ref}-t_{E,1}))$ represents ratio of model estimate of change in creep and elastic strain $\Delta\varepsilon_{C,z,c,model}(t)/\varepsilon_{C,z,E,sensor}(t)$.

In Expression 5.82, total strain and temperature are directly measured by sensors. Thermal expansion coefficient has to be determined experimentally, in laboratory or from SHM data (see Chapter 6), or approximated using values given in Table 5.4. In steel, thermal expansion coefficient can be considered as constant over time, assuming normal (non-extreme) range of temperature. This is not necessarily the case with concrete, where thermal expansion coefficient can vary over the year depending on environmental conditions. For these reasons, dependence of thermal expansion coefficient on time is omitted in Expression 5.82 for steel. Creep and shrinkage coefficients and functions in Expression 5.82 cannot be directly measured, but rather determined using some of methods presented in Section 5.8.

Finally, the evaluation of absolute elastic strain at time t can be performed by using a combination of model-based estimation at time t_{Ref} and sensor-based evaluation at time t, as follows:

$$\varepsilon_{C,z,E,abs}(t) = \varepsilon_{C,z,E,sensor}(t) + \varepsilon_{C,z,E,model}(t_{Ref}) \quad (5.83)$$

Evaluation of limits of error $\pm\delta\varepsilon_{C,z,E,abs}$ or uncertainty $u(\varepsilon_{C,z,E,abs})$ in absolute elastic strain can be performed by applying corresponding propagation formulae in Expression 5.83. Criterion II, which evaluates whether the material at location of sensor is in an elastic regime, is presented in Expression 5.84.

Criterion II:
$$\frac{\varepsilon_{yld/u}^{-}}{\beta_{\varepsilon}^{-}} + \delta\varepsilon_{C,z,E,abs} < \varepsilon_{C,z,E,abs}(t) < \frac{\varepsilon_{yld/u}^{+}}{\beta_{\varepsilon}^{+}} - \delta\varepsilon_{C,z,E,abs}$$
$$\text{or} \quad (5.84)$$
$$\frac{\varepsilon_{yld/u}^{-}}{\beta_{\varepsilon}^{-}} + n_u u(\varepsilon_{C,z,E,abs}) < \varepsilon_{C,z,E,abs}(t) < \frac{\varepsilon_{yld/u}^{+}}{\beta_{\varepsilon}^{+}} - n_u u(\varepsilon_{C,z,E,abs})$$

where $\varepsilon_{yld/u}^{+}$ and $\varepsilon_{yld/u}^{-}$ denote yield or ultimate strain, depending on which one applies based on ductility or brittleness of material, and β_{ε}^{+} and β_{ε}^{-} are safety factors that account for often unknown accuracy in estimations of yield and ultimate strain.

Interpretation of outcome of Criterion II is the following: if Criterion II is fulfilled, then the strain measurement at time t does not indicate existence of unusual structural behaviors that could compromise the safety of the structure (which does not necessarily mean that unusual behavior is not present, see the comment related to Criterion I). If Criterion II is not fulfilled, then the strain measurement at time t indicates the existence of unusual structural behavior, which can compromise structural safety. All other comments given with Criterion I regarding (i) the limits of error or uncertainty of both model and measurements, (ii) the reliability of the sensor measurements, and (iii) the appropriateness of the model are valid for Criterion II and thus, they are not repeated here.

In a specific project, if the yield and ultimate strain for construction material are not accurately determined, then the conservative approximate values from Table 5.3 can be used. Recommended values for safety factors are at least 1.5–2 to account for all errors (including the error due to approach to using Expressions 5.11 and 5.12 instead of 5.21 and 5.26). This means that elastic strain that reaches 50–67% of yield strain for ductile materials or ultimate strain for brittle materials indicates unusual structural behavior. This is reasonable, as commonly, by design, the elastic strain in real structures rarely reaches 25–50% of yield or ultimate strain for ductile and brittle materials, respectively. Note that high safety factors should also account for the fact that the values in Table 5.3 are valid for a uniaxial state of stress, which is not necessarily the case for beams; due to shear, the state of stress is biaxial, except at the tops and bottoms of the cross-section. As mentioned earlier, the beams are commonly monitored with pairs of sensors parallel to the centroid line, one installed close to the top and the other close to the bottom of the cross-section, and in these cases, the shear at sensors' locations might be reasonably small and the stress state close to uniaxial.

The use of yield strain values ($\varepsilon_{yld}^- = -2000\,\mu\varepsilon$, $\varepsilon_{yld}^+ = +2000\,\mu\varepsilon$) from Table 5.3 for modern steel is straightforward, as modern steel has very stable, guaranteed properties. This might not be valid for historic steels and iron, so the values from tests or reliable literature should be used in that case.

The use of yield and ultimate strain values for concrete is not as straightforward as in the case of steel. Let us recall first that for monitoring concrete, long-gauge sensors are recommended. The yield ($\varepsilon_{yld}^- = -1300\,\mu\varepsilon$ to $-2000\,\mu\varepsilon$) and ultimate ($\varepsilon_u^+ = +50\,\mu\varepsilon$ to $+100\,\mu\varepsilon$) strain values given in Table 5.3 refer to mature (min. 28-day old concrete) unreinforced concrete. Range of compressive values of yield strain in the case of reinforced and prestressed mature concrete is the same as for unreinforced mature concrete ($\varepsilon_{yld}^- = -1300\,\mu\varepsilon$ to $-2000\,\mu\varepsilon$), as reinforcement in these cases does not significantly affect this range. Prestressed concrete, by design, is not supposed to have cracks, and cracking is considered damage; thus, the range of ultimate tensional values of unreinforced mature concrete ($\varepsilon_u^+ = +50\,\mu\varepsilon$ to $+100\,\mu\varepsilon$) can also be used in Expression 5.84 for prestressed mature concrete to detect cracking. However, reinforced concrete, by design, is supposed to have cracks, typically (but not necessarily) spaced at 20–30 cm and open for, typically, not more than 0.1–0.3 mm, and reinforcement significantly contributes to overall tensional strain. Therefore, the use of value for tensional yield strain from Table 5.3 is not justified in the case of reinforced concrete (monitored using long-gauge sensors). The tensional yield strain will vary from project to project as it depends on amount of rebar and concrete strength. It can typically range between $+350\,\mu\varepsilon$ and $+1500\,\mu\varepsilon$.

For concrete that did not reach maturity, the values for yield and ultimate strain presented above might not be valid. Model for rheological strain is not accurate during this period, and due to lower strength and Young's modulus, the yield and ultimate strain in concrete can be lower than those discussed above.

Given that Criterion I compares measurement with model (design) and Criterion II with yield or ultimate strain (damage), the former can be considered to examine the performance ("serviceability state") of material at location of sensor, and the latter the potential for damage ("ultimate state") at location of sensor. If the reference time precedes stressing in steel or coincides with pouring or concrete, then the strain sensor measures absolute strain, and in that case, if Criterion I is fulfilled, then there is no need to evaluate Criterion II. However, this is not the case if the reference time is set at some later time.

Once the unusual behavior is detected (Level I SHM) using Criteria I or II, it is automatically localized (Level II SHM), as its location is determined by the location of sensor itself. Quantification (Level III SHM) of unusual behavior is then made by evaluating the difference between the measurement and model (Criterion I) or between the measurement and yield or ultimate values (Criterion II). This quantified unusual behavior can then be inserted in the model of the structure in order to predict its current load capacity, i.e., to achieve Level IV SHM.

5.9.3 Comparison of Stress, Derived from Strain, with Corresponding Model Values

The next two criteria for identification of unusual structural behaviors are based on stress analysis. They are somewhat similar to Criteria I and II, and in some cases they coincide, but in other cases they are slightly different.

Elastic strain constituents in steel and concrete can be derived from total strain measurement as per Expression 5.82. Then, stress (more precisely, change in stress with respect to reference time) in steel and concrete at location of sensor can be evaluated from total strain measurement using the following expressions:

in steel: $\sigma_{C,z,sensor}(t) = E_C \varepsilon_{C,z,E,sensor}(t) = E_C(\varepsilon_{C,z,tot,sensor}(t) - \alpha_{T,C,z}\Delta T_{C,sensor}(t))$

in concrete: $\sigma_{C,z,sensor}(t) = E_C(t)\varepsilon_{C,z,E,sensor}(t)$

$$= E_C(t)\frac{(\varepsilon_{C,z,tot,sensor}(t) - \alpha_{T,C,z}(t)\Delta T_{C,sensor}(t) - \varepsilon_{C,z,sh,final}(f_{sh}(t - t_0) - f_{sh}(t_{Ref} - t_0)))}{1 + K_{c,equ}(f_{c,equ}(t - t_{E,1}) - f_{c,equ}(t_{Ref} - t_{E,1}))} \tag{5.85}$$

where $\sigma_{C,z,sensor}(t)$ is sensor-based evaluation of stress from short-gauge sensor measurement or estimate of average stress from a long-gauge or distributed sensor. In practice, sensor-based evaluation of stress is often referred to as stress measurement.

In steel, in addition to thermal expansion coefficient, Young's modulus can also be considered as constant over time, assuming a normal (non-extreme) range of temperature. This is not the case with concrete, where Young's modulus slowly increases over time due to the hydration process.

As mentioned earlier in text, Young's modulus E of modern steel is rather invariable in time and space, as is the thermal expansion coefficient α_T as well, which makes stress evaluation in modern steel structures relatively simple. To perform it, thus, it is necessary to determine values of E and α_T, experimentally (in laboratory or on-site, using SHM data) or by adopting the values from literature, which are given in Tables 5.3 and 5.4, respectively ($E = 200$–210 GPa and $\alpha_T = 10$–12 $\mu\varepsilon/°$C). For historic steel and iron, the value of Young's modulus can be significantly lower, and it can vary throughout the structure due to quality issues, while thermal expansion coefficient is in approximately the same range.

In the case of concrete, the evaluation of stress is not simple due to variability of Young's modulus E and the thermal expansion coefficient α_T in time and space, and complexity in evaluation of rheological strain, creep, and shrinkage. Ideally, Young's modulus E and the thermal expansion coefficient α_T should be determined experimentally, in a laboratory, or on-site, using SHM data. For concrete, standardized laboratory testing of sample cylinders or cubes at different stages is mandatory to follow the evolution of compressive strength during cure period, and verify the design value of mature concrete. Determination of Young's modulus through laboratory tests is not mandatory, but as an alternative, empirical equations for computing Young's modulus from the results of strength tests are provided in various design codes, such as ACI-318, and Euro Code. The main challenge in that case is the evaluation of accuracy in determination of Young's modulus from these expressions, due to epistemic error. Evaluation of Young's modulus from SHM data is more accurate, but it requires controlled testing of the structure (applying a known load, using an appropriate model to estimate the value of stress at location of sensor due to load, and then correlating the value of stress with measured strain). While this process seems simple, it is often challenging to realize in real-life settings (for example, it requires temporary closure of the bridge during the tests and involves provision of heavy loads, safety measures, and re-routing the traffic). Determination of thermal expansion coefficient of concrete in a laboratory is rarely performed, and its determination on-site, from SHM measurements, can also be challenging if the boundary conditions prevent free thermal expansion of monitored structural member (see Chapter 6). Finally, creep and shrinkage coefficients and functions in Expressions 5.82 and 5.85 cannot be directly measured on-site, but rather determined using some of methods presented in Section 5.8. Given numerous uncertainties related to parameters included in Expression 5.85, a relative error in determining elastic strain and stress in concrete can be high, often in the range of 20–25%.

If the limits of error $\pm \delta \Delta \sigma_{C,z,model}$ or uncertainty $u(\Delta \sigma_{C,z,model})$ of the model estimate of change in stress with respect to reference time t_{Ref} at location of sensor can be determined, then the comparison between the model stress estimate and sensor-based stress evaluation (stress measurement) at time t represents the third criterion in identification of unusual structural behaviors. It can be written as follows:

$$|\sigma_{C,z,sensor}(t) - \Delta \sigma_{C,z,model}(t)| \le \delta \sigma_{C,z,sensor} + \delta \Delta \sigma_{C,z,model}$$

Criterion III: or $\qquad\qquad$ (5.86)

$$|\sigma_{C,z,sensor}(t) - \Delta \sigma_{C,z,model}(t)| \le n_u \sqrt{u(\sigma_{C,z,sensor})^2 + u(\Delta \sigma_{C,z,model})^2}$$

where $\delta \Delta \sigma_{C,z,sensor}$ and $u(\Delta \sigma_{C,z,sensor})$ respectively represent the limits of errors and uncertainty in sensor-based evaluation of stress (obtained by applying error or uncertainty propagation formulae in Expression 5.85); vertical brackets denote the absolute value of difference between the measurement and model, and term to the right of sign "$\le$" represents the limits of error (upper expression) and uncertainty (lower expression) of the difference, while n_u is a multiplicator (recommended to be set between 2 and 3, i.e., $2 \le n_u \le 3$, see Chapter 4).

If Criterion III is fulfilled, then the sensor-based stress evaluation (or stress measurement) at time t does not indicate the existence of unusual structural behaviors. If Criterion III is not fulfilled, then the stress measurement at time t indicates the existence of unusual structural behavior. All other comments given with Criterion I regarding (i) the limits of error or uncertainty of both model and measurements, (ii) the reliability of the sensor measurements, and (iii) the appropriateness of the model, are also valid for Criterion III, and thus, they are not repeated here. Similarly, the fulfillment of Criterion III does not necessarily mean that an unusual structural behavior is not present (see the comments given for Criterion I).

Final criterion for identification of unusual structural behaviors presented in this section is based on a comparison between sensor-based evaluation of absolute stress and material yield or ultimate stress values. This comparison is based on Expressions 5.11 and 5.12 instead of Expressions 5.21 and 5.26, with the same justification as for Criterion II. Sensor-based evaluation of absolute elastic strain can be performed by either multiplying Expression 5.83 with Young's modulus or by adding the value of model-estimated stress at reference time to Expression 5.85. This results in the following expressions:

$$\sigma_{C,z,abs}(t) = E_C(t)(\varepsilon_{C,z,E,sensor}(t) + \varepsilon_{C,z,E,model}(t_{Ref})) = \sigma_{C,z,sensor}(t) + \sigma_{C,z,model}(t_{Ref})$$

in steel: $= E_C(\varepsilon_{C,z,tot,sensor}(t) - \alpha_{T,C,z}\Delta T_{C,sensor}(t)) + \sigma_{C,z,model}(t_{Ref})$

in concrete:

$$= E_C(t) \frac{(\varepsilon_{C,z,tot,sensor}(t) - \alpha_{T,C,z}(t)\Delta T_{C,sensor}(t) - \varepsilon_{C,z,sh,final}(f_{sh}(t-t_0) - f_{sh}(t_{Ref}-t_0)))}{1 + K_{c,equ}(f_{c,equ}(t-t_{E,1}) - f_{c,equ}(t_{Ref}-t_{E,1}))}$$

$$+ \sigma_{C,z,model}(t_{Ref}) \tag{5.87}$$

Evaluation of limits of error $\pm\delta\sigma_{C,z,abs}$ or uncertainty $u(\sigma_{C,z,abs})$ in absolute stress can be performed by applying corresponding propagation formulae in Expressions 5.87. Criterion IV, which evaluates whether the material at location of sensor is in a linear elastic regime, is presented in Expression 5.88.

Criterion IV:

$$\frac{\sigma^-_{yld/u/bkl}}{\beta^-_\sigma} + \delta\sigma_{C,z,abs} < \sigma_{C,z,abs}(t) < \frac{\sigma^+_{yld/u/bkl}}{\beta^+_\sigma} - \delta\sigma_{C,z,abs}$$

or

$$\frac{\sigma^-_{yld/u/bkl}}{\beta^-_\sigma} + n_u u(\sigma_{C,z,abs}) < \sigma_{C,z,abs}(t) < \frac{\sigma^+_{yld/u/bkl}}{\beta^+_\sigma} - n_u u(\sigma_{C,z,abs})$$

$$\tag{5.88}$$

where $\sigma^+_{yld/u/bkl}$ and $\sigma^-_{yld/u/bkl}$ denote yield, ultimate, or buckling stress, depending on which one applies based on ductility, brittleness, or buckling stability of material (if two or all three of these values are applicable, then the smallest absolute value should be entered in the expression), and $\beta_\sigma{}^+$ and $\beta_\sigma{}^-$ are safety factors that account for often unknown accuracy of yield, ultimate, or buckling stress.

Interpretation of outcome of Criterion IV is similar to that of Criterion II: if Criterion IV is fulfilled, then the stress measurement at time t does not indicate the existence of unusual structural behaviors that compromise the safety of the structure (which does not necessarily mean that an unusual behavior is not present; see the comment related to Criterion I). If Criterion IV is not fulfilled, then the strain measurement at time t indicates the existence of unusual structural behavior, which can compromise structural safety. All other comments given with Criterion I regarding (i) the limits of error or uncertainty of both model and measurements, (ii) the reliability of the sensor measurements, and (iii) the appropriateness of the model, are also valid for Criterion IV, and thus, they are not repeated here.

For concrete structures, the values of ultimate compressive and tensional stress (compressive and tensional strength) are defined in design phase and verified through standardized testing during construction. For steel structures, the yield stress depends on steel grade, which is defined in design phase and guaranteed by steel manufacturer (in steel, compressive and tensional yield stresses are considered equal in absolute value). If, for some reason the values of ultimate or yield stresses are not available, they have to be determined through destructive or non-destructive testing (the latter being preferable). Given their possibly wide range, conservative use of values given in Table 5.3 might lead to false positive identification of unusual structural behaviors.

Note that Criterion IV, besides yield and ultimate stress, also considers buckling stress, which corresponds to the stress at location of sensor at which either global (element level) or local (cross-section level) buckling may occur. Buckling stress is typically estimated in design phase. If, for some reason, the buckling stress is not available, appropriate non-linear models should be used to estimate it. Models based on linear beam theory cannot be used for this purpose. In concrete beams, buckling typically occurs at global lever. In steel beams, buckling can occur at local (bowing) or global level, and the smaller of the two, by absolute value, should be considered. Similar to Criteria I–III, it is recommended to set the safety factors to at least 1.5–2.

Criteria I–IV are fundamental for identifying unusual structural behaviors at sensor level, but they are not the only criteria to be applied. Other criteria can be applied at structural level based on parameters that are derived from strain measurements from multiple sensors, such as neutral axis, curvature, deflection, and force. Some of these criteria are presented in Chapter 6.

Note that Criteria I and III are practically equivalent, as are Criteria II and IV and thus, applying only one from the first set (Criteria I or III) and one from the second set (Criteria II and IV) is often sufficient for assessing whether an unusual structural behavior is present at location of sensor. Section 5.9.2 will, for the purpose of exercise, present the application of all four criteria in the case of steel structural element. Then, Section 5.9.3 will present the application of Criteria I and II to concrete structural elements.

5.9.4   Case Study on Analysis of Strain in Steel Structure

An example of strain analysis at sensor level in steel structures is presented in this subsection using the Bridge over Boonton Road, Erie-Lackawanna Railroad, and Alps Road Ramps as the case study. The bridge serves US202 and NJ23 roads close to Wayne, New Jersey, USA, and is referred to in the further text as the US202/NJ23 overpass or simply as Wayne Bridge. Structurally and functionally, Wayne Bridge represents a typical highway overpass, frequently found across the USA, that contains "twin" structures (two parallel bridges, one for each direction of traffic) consisting of a concrete deck supported by steel girders (stringers) in several simple spans. Wayne Bridge has four spans, and the focus of this book is on the second southbound span, which is skewed at its north end. The girders (stringers) are built-up sections with varying flange thicknesses. A view of the second southbound span is shown in Figure 5.16a, and a view of the girders is shown in Figure 5.16b.

In 2011, the bridge served as the case study for various testing techniques implemented by numerous international entities, universities and companies, to demonstrate effective ways to assess the performance and structural health condition of a typical US overpass. This effort is known as the

Figure 5.16   (a) View to the second southbound span of the Wayne Bridge and (b) view to steel girders (Girders #2 and #5 are labeled in the figure).

International Bridge Study (IBS), which was performed within the Long-term Bridge Performance (LTBP) Program enabled by the Federal Highway Administration (FHWA 2021). Many international teams have tested the bridge, taken data, and suggested ways of improving the bridge performance. Within this effort, Girders #2 and #5 of the second span of southbound structure were equipped with six pairs of long-gauge fiber optic sensors. Each girder was equipped with three pairs of sensors installed approximately at quarters and middles of the girder lengths, as shown in Figure 5.17a. The locations of sensors along the lengths of girders were determined based on accessibility (i.e., locations of stiffeners that obstructed installation of sensors at exact quarters and middles of spans). At each equipped cross-section, one sensor was installed at the top flange and another at the bottom flange, as shown in Figure 5.17b,c.

Sensors at Locations 2.2 and 5.1–5.3 have a gauge length of 2 m, while sensors at Locations 2.1 and 2.3 have a gauge length of 1 m. While these different gauge lengths were practical results of availability of sensors at the time of project execution, they were all in accordance with recommendations given in Section 4.4. The type of sensors used was FBG, with packaging as shown in Figure 3.14. Thus, each packaging contained one strain and one temperature FBG, where the latter was used for thermal compensation of the former. Based on specifications received from sensor manufacturer, limits of error for compensated strain measurements were estimated at ± 4 µε, and for changes in temperature at $\pm 0.2°$C. Sensors were installed on surface of flanges using L-brackets, and they were practically 17 mm distant from these steel elements (see Figure 5.17); consequently, temperature sensors were not in direct contact with structure and thus did not measure an exact temperature in steel. Based on manual measurements performed on both sensors and steel, it was determined that the difference in temperature between the two was $\pm 2°$C.

Hybrid steel–concrete nature, varying cross sections along the length of the bridge, and varying girder lengths due to the skewed shape, along with several transverse bracings, result in a structure with complex behavior. To simplify modeling, girders were observed separately as simply supported beams. Cross-section is idealized as consisting of a steel girder and concrete deck, with the width equal to half the distance between girders on both sides, as shown in Figure 5.17b. This idealization is on the conservative side of modeling as it neglects the interaction between the girders provided by concrete slab. One sensor was installed at the top and one at the bottom flange, as shown in Figure 5.17b,c. Mechanical parameters of the steel were not available, so they were taken from Table 5.3: Young's modulus of steel was assumed to be 205 ± 5 GPa and thermal expansion coefficient 11 ± 1 µε/°C. Young's modulus of concrete was estimated to 29.3 ± 5.9 GPa based on tests on cylinders, while thermal expansion coefficient was not available, and thus, to simplify presentation,

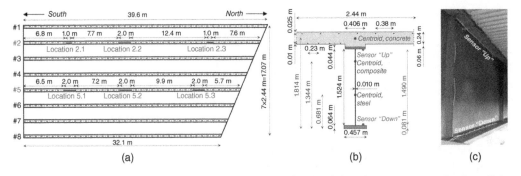

Figure 5.17 Positions of sensors in Wayne bridge: (a) Top view (each location represents pair of parallel sensors), (b) cross-sectional view (idealized cross-section at Location 5.2), and (c) photograph of installed sensors.

it was assumed to be identical to steel, i.e., $11 \pm 1\ \mu\varepsilon/^\circ C$. Note that based on Table 5.3, it would be more conservative to assume the value of thermal expansion coefficient of concrete at $10 \pm 2\ \mu\varepsilon/^\circ C$; this assumption is separately analyzed at the end of this section. This subsection focuses only on sensors installed at Location 5.2. Geometrical properties of idealized cross-section at that location were calculated from the dimensions shown in Figure 5.17b and presented in Table 5.6. Note that dimensions for steel are given to the third decimal place, reflecting millimeters, while dimensions for concrete are given to the second decimal place, reflecting centimeters. Based on specifications, the limits of error in dimensions of steel elements depend on their size, but for simplification purposes, they are adopted here to be ± 0.5 mm (± 0.0005 m). In the case of concrete, the limits of error in dimensions are commonly set to ± 1 cm (± 0.01 m). These limits of error were taken into account when calculating the properties shown in Table 5.6.

Equivalent concrete slab cross-sectional properties were calculated using Young's modulus of steel as the reference, i.e., using the following expression:

$$A_{concrete,equ} = A_{concrete} \frac{E_{concrete}}{E_{steel}}$$

$$I_{concrete,equ} = I_{concrete} \frac{E_{concrete}}{E_{steel}} \tag{5.89}$$

This transformation was necessary to enable the calculation of the theoretical location of the centroid, area, and moment of inertia of a hybrid (composite) steel–concrete cross-section. Note how large limits of error in determination of Young's modulus of concrete entrain large errors in determination of these cross-sectional properties.

Example 5.7   *Application of Criterion I to Wayne Bridge*
Measurements of strain and change in temperature with respect to November 3, 2011, which was set as reference time, are given in Tables 5.7 and 5.8. The measurements presented in tables were all taken at approximately the same time of day and with no heavy vehicles on the structure. Given that the bridge was in service at the reference time and that no extra load was added during the monitoring period (2011–2016, see Tables 5.7 and 5.8), the change in elastic strain in the structure due to load is negligible. The observed span of the bridge is simply supported, and temperature

Table 5.6   Geometrical properties of idealized cross-section at Location 5.2.

	Area A (m^2)	Moment of inertia I_x (m^4)	Distance of centroid from the bottom face of steel flange (m)
Steel girder cross-section	0.0616 ± 0.0013	0.681 ± 0.002	0.03094 ± 0.00047
Concrete slab cross-section	0.588 ± 0.027	1.814 ± 0.012	0.00286 ± 0.00037
Equivalent concrete slab cross-section	0.084 ± 0.023	1.814 ± 0.012	0.00041 ± 0.00014
Upper rebars cross-section (diameter = 16 mm)	0.00139 ± 0.00009	1.871 ± 0.012	Negligible
Lower rebars cross-section (diameter = 16 mm)	0.00089 ± 0.00006	1.719 ± 0.012	Negligible
Hybrid (composite) steel–concrete cross-section	0.148 ± 0.024	1.34 ± 0.075	0.078 ± 0.020

Table 5.7 Measurements of strain and change in temperature, and comparison of measured total strain with thermal strain of Sensor 5.2up, as per Criterion I.

Time	$\Delta T_{5.2up,sensor}$	$\delta\Delta T_{5.2up,sensor}$	$\varepsilon_{5.2up,T,sensor} =$ $\varepsilon_{5.2up,tot,model}$	$\delta\varepsilon_{5.2up,T,sensor} =$ $\delta\varepsilon_{5.2up,tot,model}$	$\varepsilon_{5.2up,tot,sensor}$	$\delta\varepsilon_{5.2up,tot,sensor}$	$\lvert\varepsilon_{5.2up,tot,sensor} -$ $\varepsilon_{5.2up,tot,model}\rvert$	$\delta\varepsilon_{5.2up,tot,sensor} +$ $\delta\varepsilon_{5.2up,tot,model}$
11/03/11	0.0	±2.2	0	±24	0	±4	0	±28
12/06/11	−0.7	±2.2	−8	±23	−8	±4	0	±27
01/26/12	−11.0	±2.2	−121	±13	−99	±4	22	±17
05/18/12	4.1	±2.2	45	±28	80	±4	35	±32
09/26/12	5.0	±2.2	55	±29	32	±4	23	±33
11/30/12	−9.2	±2.2	−101	±15	−152	±4	51	±19
02/27/13	−6.8	±2.2	−75	±17	−103	±4	28	±21
04/29/13	−1.4	±2.2	−16	±23	10	±4	26	±27
09/25/13	4.9	±2.2	54	±29	40	±4	14	±33
11/22/13	−2.3	±2.2	−25	±22	−84	±4	59	±26
02/21/14	−9.3	±2.2	−102	±15	−154	±4	52	±19
06/24/14	9.5	±2.2	105	±34	112	±4	7	±38
09/24/14	5.3	±2.2	58	±29	36	±4	22	±33
10/30/14	−2.8	±2.2	−31	±21	−75	±4	44	±25
12/01/15	−5.9	±2.2	−65	±18	−126	±4	61	±22
03/04/16	−11.1	±2.2	−123	±13	−176	±4	53	±17

Table 5.8 Measurements of strain and change in temperature, and comparison of measured total strain with thermal strain of Sensor 5.2 down, as per Criterion I.

Time	$\Delta T_{5.2down,sensor}$	$\delta\Delta T_{5.2down,sensor}$	$\epsilon_{5.2down,T,sensor} = \epsilon_{5.2down,tot,model}$	$\delta\epsilon_{5.2down,T,sensor} = \delta\epsilon_{5.2down,tot,model}$	$\epsilon_{5.2down,tot,sensor}$	$\delta\epsilon_{5.2down,tot,sensor}$	$\|\epsilon_{5.2down,tot,sensor} - \epsilon_{5.2down,tot,model}\|$	$\delta\epsilon_{5.2down,tot,sensor} + \delta\epsilon_{5.2down,tot,model}$
11/03/11	0.0	±2.2	0	±24	0	±4	0	28
12/06/11	−0.8	±2.2	−9	±23	−11	±4	2	27
01/26/12	−11.8	±2.2	−130	±12	−121	±4	9	16
05/18/12	3.6	±2.2	39	±28	32	±4	7	32
09/26/12	5.0	±2.2	55	±29	35	±4	20	33
11/30/12	−9.1	±2.2	−100	±15	−142	±4	42	19
02/27/13	−6.8	±2.2	−75	±17	−117	±4	42	21
04/29/13	−2.5	±2.2	−27	±22	−7	±4	20	26
09/25/13	5.3	±2.2	58	±29	−14	±4	72	33
11/22/13	−2.1	±2.2	−23	±22	−89	±4	66	26
02/21/14	−9.4	±2.2	−104	±15	−174	±4	70	19
06/24/14	9.6	±2.2	106	±34	36	±4	70	38
09/24/14	5.8	±2.2	64	±30	−16	±4	80	34
10/30/14	−2.3	±2.2	−25	±22	−124	±4	99	26
12/01/15	−5.6	±2.2	−62	±19	−161	±4	99	23
03/04/16	−10.9	±2.2	−120	±13	−247	±4	127	17

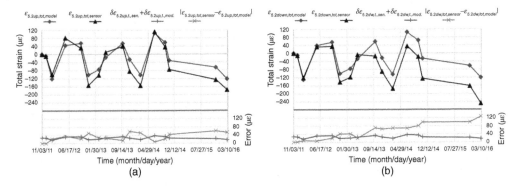

Figure 5.18 Graphs and comparisons between the measured and estimated total strain as per Criterion I: (a) For Sensor 5.2 up and (b) for Sensor 5.2 down.

changes are not supposed to produce stress and elastic strain (see Section 5.7), unless there is thermal incompatibility between the steel and concrete. To simplify the presentation, it will be assumed that this is not the case, i.e., that there is no thermal incompatibility between steel and concrete; however, this incompatibility will be discussed at the end of this section. Hence, based on model given in Expression 5.82, the total strain is estimated to be equal to thermal strain. The latter is calculated and presented in Tables 5.7 and 5.8.

Comparison between the model and the measurements as per Criterion I is performed in Tables 5.7 and 5.8 and Figure 5.18. Limits of error for thermal strain take into account both errors in temperature measurement and errors in estimation of thermal expansion coefficient. Error in temperature measurement, in turn, takes into account both error of temperature sensor and the fact that the sensor is not in direct contact with the steel. Instances in which the Criterion I was not fulfilled are shaded in Tables 5.7 and 5.8. They are also indicated in Figure 5.18 at locations where the graph with marker "x" is above the graph with marker "+".

Based on Criterion I, these instances point to unusual structural behavior. The discrepancy between the model and measurement at location of Sensor 5.2up seems to fluctuate around the limits of error, i.e., at some instances it is within and at others outside of the error limits. This indicates that unusual structural behavior is most likely triggered from time to time by some seasonal environmental effects. On the other hand, the discrepancy at location of Sensor 5.2down is present more often than in the case of Sensor 5.2up, and it increased over time starting with June 24, 2014. About that time, its magnitude started to be approximately two times higher than that of Sensor 5.2up, which can be explained by its approximately two times larger distance from the centroid of the cross-section. Possible reasons for this unusual structural behavior are discussed at the end of this section.

Example 5.8 *Application of Criterion II to Wayne Bridge*

In order to apply Criterion II, it is necessary to evaluate the absolute elastic strain at location of sensors. Based on Expression 5.82, the change in elastic strain with respect to reference time is equal to the difference between total strain and thermal strain. Both total strain and thermal strain are given in Tables 5.7 and 5.8.

Based on Expression 5.83, it is necessary to create a model to estimate the elastic strain at the reference time. The most general approach would be to perform numerical modeling and estimate the model error (e.g., using Monte Carlo simulation). For the purposes of example presented here, a simplified model based on linear beam theory is created. This simplified model has two main components: the first component addresses the loads acting on a hybrid (composite) steel–concrete beam (as shown in Figure 5.17); the second component addresses the interaction between steel and concrete due to long-term rheological strain in concrete.

The first component is practically developed in Section 5.6.4 and described in Expression 5.28. Estimations of model parameters and elastic strain due to self-weight of a hybrid beam is given in Table 5.9.

Densities of steel and concrete given in Table 5.9 are estimates based on literature. Limits of error are conservatively calculated using propagation formulae given in Section 4.3. Limits of error of elastic strain include the error due to gauge length, which is estimated to be 3 με (for a uniformly distributed load: linear weight multiplied with the square of gauge length divided by 24).

At the times the measurements presented in Tables 5.7 and 5.8 were taken, a live load due to cars on the bridge was present, in addition to self-weight (the road US202/NJ23 is very busy, and measurements were taken systematically around noon on a working day). However, no heavy vehicles were present (this was ensured by observing the magnitude of dynamic strain measurement). Hence, the values of elastic strain due to ordinary car traffic should be added to those calculated in Table 5.9. Estimation of the traffic load is made by using conservative values from literature. For example, US and European standards propose similar values: 9.4 kN/m (i.e., 0.64 kips/ft, FHWA 2015) and 9 kN/m (CEN 2010) for 3 m wide lane. For 2.44 m wide concrete deck of Girder 5, these values may be scaled and rounded up to 7.5 kN/m. Now, the beam model (Expressions 5.28) can be applied to that load, resulting in −12 με for Sensor 5.2up and 64 με for Sensor 5.2down. Given that a conservative value is used for lane load, the limits of errors are not estimated but are considered to be included in the strain estimates. Finally, the estimates of elastic strains in sensors due to lads acting on a hybrid (composite) steel–concrete beam at the reference time are calculated as simple

Table 5.9 Estimations of model parameters and elastic strain due to the self-weight of Girder 5.2.

	Concrete	Steel
Density (kg/m^3)	2400 ± 50	7850 ± 50
Gravity acceleration (m/s^3)	9.81 ± 0.02	9.81 ± 0.02
Specific weight (N/m^3)	23,544 ± 539	77,009 ± 648
Area of cross-section (m^2)	0.588 ± 0.027	0.0616 ± 0.0013
Linear weight (N/m)	13,844 ± 952	4744 ± 140
Total linear weight (kN/m)	18.6 ± 1.1	
Length of the beam (m)	29.6 ± 0.05	
Reactions (kN)	275 ± 17	
Location 5.2 (m)	16.7 ± 0.05	
Bending moment at 5.2 (kNm)	2002 ± 131	
Stiffness EI at 5.2 (kNm^2)	15,893,238 ± 4,436,705	
Elastic curvature at 5.2 (με/m)	126 ± 43	
Distance from centroid, Sensor 5.2up (m)	−0.227 ± 0.076	
Distance from centroid, Sensor 5.2down (m)	1.263 ± 0.076	
Estimated elastic strain at 5.2up (με)	−29 ± 23	
Estimated elastic strain at 5.2down (με)	159 ± 68	

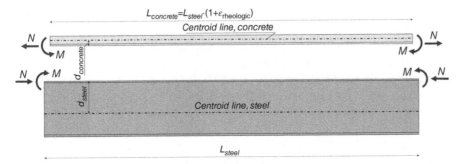

Figure 5.19 Schematic decomposition of a steel girder and a concrete slab after full development of rheological strain; both materials are loaded with force N and moment M resulting from stresses caused by interaction between materials due to rheological strain.

sum of strains due to self-weight and lane load:

$$\varepsilon_{5.2up,E,model,loads} = -41 \pm 23 \ \mu\varepsilon$$

$$\varepsilon_{5.2down,E,model,loads} = 223 \pm 68 \ \mu\varepsilon \tag{5.90}$$

A model for the second component of elastic strain, resulting from the rheologic effects in concrete slabs is developed as follows: Due to composite behavior of the beam cross-section under dead loads, the concrete part is put in compression and subjected to rheological effects. Consequently, the concrete slab becomes shorter over time, which results in an additional long-term deformation of the steel girder. The bridge was built in 1983, and thus, one can assume that creep and shrinkage were practically fully developed at the reference time (in 2011). Upon construction, both the steel girder and concrete slab had the same length; however, at the reference time, due to rheological strain, the concrete slab is shorter, as shown in Figure 5.19. In order to keep the interaction between the steel and concrete perfect, i.e., to ensure that there is no mismatch of lengths between steel and concrete at their interface, additional stresses are generated, both in steel and concrete. These stresses create resultants that can be described by the longitudinal force N and moment M applied, equal and opposite, at the extremities of the concrete and steel elements, eccentrically, at the interface between two materials, as shown in Figure 5.19.

The force N and moment M can be calculated from compatibility conditions: the lengths of the steel and concrete "fibers" at the interface have to be identical for both materials, and the curvatures in both materials have to be mutually equal. These two conditions are described by Expressions 5.91 and 5.92, respectively.

$$\left(1 + \frac{N}{E_{steel}A_{concrete,equ}} + \frac{Nd_{concrete} - M}{E_{steel}I_{concrete,equ}}d_{concrete}\right)(1 + \varepsilon_{rheologic})L_{steel}$$

$$= \left(1 - \frac{N}{E_{steel}A_{steel}} - \frac{Nd_{steel} + M}{E_{steel}I_{steel}}d_{steel}\right)L_{steel}$$

$$\Rightarrow N\left(\frac{1 + \varepsilon_{rheologic}}{A_{concrete,equ}} + \frac{1}{A_{steel}} + \frac{d_{concrete}^2(1 + \varepsilon_{rheologic})}{I_{concrete,equ}} + \frac{d_{steel}^2}{I_{steel}}\right)$$

$$- M\left(\frac{d_{concrete}(1 + \varepsilon_{rheologic})}{I_{concrete,equ}} - \frac{d_{steel}}{I_{steel}}\right) = -E_{steel}\varepsilon_{rheologic} \tag{5.91}$$

$$\frac{Nd_{concrete} - M}{E_{steel}I_{concrete,equ}} = \frac{Nd_{steel} + M}{E_{steel}I_{steel}} \quad \Rightarrow \quad N\left(\frac{d_{concrete}}{I_{concrete,equ}} - \frac{d_{steel}}{I_{steel}}\right) - M\left(\frac{1}{I_{concrete,equ}} + \frac{1}{I_{steel}}\right) = 0$$

$$(5.92)$$

To keep the model simple yet conservative, the value of creep coefficient is assumed to be 3 ($K_c = 3$) and the value for final shrinkage $\varepsilon_{sh,final} = -530$ με. Based on MC90, these values are extreme as they represent the maximum possible values (and not the values specific to the bridge case); however, the use of these extreme values keeps the evaluation in Criterion II simple and conservative, i.e., on the safe side. Using the beam model presented earlier, the elastic strain at the centroid of concrete cross-section due to loads is calculated to be −83 με, which makes the final, long-term creep at centroid of concrete equal to $\varepsilon_{c,final} = 3 \cdot (-83) = -249$ με. Thus, the final long-term rheological strain at the centroid of concrete slab is estimated to $\varepsilon_{rheological} = -530 - 249 = -779$ με. By inserting this value in Expression 5.91 and solving the system of equations described in Expressions 5.91 and 5.92, the following elastic strain is estimated at locations of sensors:

$$\varepsilon_{5.2up,E,model,rheologic} = -546 \text{ με}$$

$$\varepsilon_{5.2down,E,model,rheologic} = 62 \text{ με} \qquad (5.93)$$

Recall that these values are highly conservative, i.e., they significantly overestimate the true values, for three reasons: first, the maximal absolute value of shrinkage is used (which is not realistic, e.g., due to presence of rebars, exposure, etc.); second, the maximal value of creep coefficient is used (which is also not realistic, similar to shrinkage); and third, the final creep does not account for relaxation in concrete, which is due to horizontal force N (see Figure 5.19). For these reasons, one can assume that the limits of errors of this component of the elastic strain model are accounted for, and safety factors β_ε^+ and β_ε^- in Criterion II (see Expression 5.84) can be set to the lower end of the range, i.e., to 1.5.

The model estimates of absolute elastic strain at the locations of the two sensors at reference time are calculated as a combination of Expressions 5.90 and 5.93.

$$\varepsilon_{5.2up,E,model}(t_{Ref}) = -587 \pm 23 \text{ με}$$

$$\varepsilon_{5.2down,E,model}(t_{Ref}) = 285 \pm 68 \text{ με} \qquad (5.94)$$

Estimated values of absolute elastic strain can now be calculated by combining Expressions 5.83 and 5.94. This is performed in Table 5.10 and 5.11 for Sensors 5.2up and Sensor 5.2down, respectively.

In the absence of certified values, the yield strain values for steel can be taken from Table 5.3, as $\varepsilon_{yld}^+ = 2000$ με and $\varepsilon_{yld}^- = -2000$ με. Given that these two values are identical in absolute value, that adopted safety factors are identical for positive and negative yield strain ($\beta_\varepsilon^+ = \beta_\varepsilon^- = 1.5$), and given that the limits of errors for each time instance are identical in absolute value (i.e., they are given in plus/minus format), the upper and lower levels of thresholds given in Expression 5.84 are also identical in absolute value. To simplify comparison with the values of evaluated absolute elastic strain, these absolute values of thresholds are given in Tables 5.10 and 5.11 for Sensor 5.2up and Sensor 5.2down, respectively. The comparison shows that for both sensor locations, the absolute elastic strain is within the limits of Criterion II, and thus there is no indication of unusual structural behavior that compromises the safety of the structure.

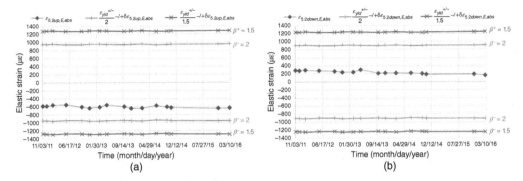

Figure 5.20 Graphs of absolute elastic strain and comparisons with yield values as per Criterion II: (a) For Sensor 5.2up and (b) for Sensor 5.2down, thresholds for safety factors of 1.5 and 2 are shown in graphs.

Figure 5.20 visually shows the evaluation of Criterion II. For illustrative purposes, two safety factor values are considered in the figures: 1.5 and 2. Figure shows no indication of unusual structural behaviors that compromises the safety of the structure for any of the two safety factors.

Given that Criterion I indicates the existence of unusual structural behaviors and Criterion II does not, one conclusion that can be drawn after the two criteria were applied is that there are unusual behaviors at the location of Sensors 5.2up and 5.2down (Criterion I), but there is no indication that these unusual behaviors compromise the safety of the structure (Criterion II).

Example 5.9 *Application of Criterion III to Wayne Bridge*
To evaluate stresses at locations of sensors and apply Criterion III as per Expression 5.86, the elastic strain at locations of sensors, determined in Tables 5.10 and 5.11, should simply be multiplied with Young's modulus of steel, as per Expression 5.85. This is performed in Tables 5.12 and 5.13. Given that the girder is simply supported and we assumed thermal compatibility between the steel and concrete (thermally generated stress is null) and the continuous presence of vehicles on the bridge (no significant stress changes due to load), the overall stress change in the girder is null, i.e., $\Delta\sigma_{5.2up,model} = \Delta\sigma_{5.2down,model} = 0$.

A comparison between the model and the measurements as per Criterion III is performed in Tables 5.12 and 5.13 (see Expression 5.86). The limits of error for the model are set to zero, given that no significant stress change is expected based on the model (this will be discussed in more detail at the end of this section). Instances in which Criterion III was not fulfilled are shaded in Tables 5.12 and 5.13.

Figure 5.21 shows the graphs of stresses at locations of Sensors 5.2up and 5.2down and the upper and lower limits of error. Instances where graphs of stresses exceed the bounds of limits of error correspond to the instances where Criterion III is not fulfilled.

Note that Criterion III confirms the findings based on Criterion I (see Example 5.7), i.e., it confirms the existence of unusual behavior at locations of Sensors 5.2up and 5.2down. This is expected, as the two criteria (Criterion I and Criterion III) are mutually equivalent – they use equivalent equations, and the only difference is multiplication with Young's modulus of steel in Criterion III.

The time history of stresses indicates the tendency of stresses to increase in absolute compressive value, and this tendency is more pronounced for location 5.2down than for location 5.2up. Possible reasons for the unusual behaviors detected at cross-section 5.2 are discussed at the end of this section.

Table 5.10 Evaluation of elastic strain and absolute elastic strain and comparison of measured absolute elastic strain of Sensor 5.2up with yield strain, as per Criterion II.

Time	$\varepsilon_{5.2up,tot,sensor}$	$\delta\varepsilon_{5.2up,tot,sensor}$	$\varepsilon_{5.2up,T,sensor}$	$\delta\varepsilon_{5.2up,T,sensor}$	$\varepsilon_{5.2up,E,sensor}$	$\delta\varepsilon_{5.2up,E,sensor}$	$\varepsilon_{5.2up,E,abs}$	$\delta\varepsilon_{5.2up,E,abs}$	$\left|\varepsilon_{yld}^{+/-}/\beta^{+/-} -/+ \delta\varepsilon_{5.2up,E,abs}\right|$
11/03/11	0	±4	0	±24	0	±28	−587	±51	1282
12/06/11	−8	±4	−8	±23	0	±27	−587	±50	1283
01/26/12	−99	±4	−121	±13	22	±17	−565	±40	1293
05/18/12	80	±4	45	±28	35	±32	−552	±55	1278
09/26/12	32	±4	55	±29	−23	±33	−610	±56	1277
11/30/12	−152	±4	−101	±15	−51	±19	−638	±42	1291
02/27/13	−103	±4	−75	±17	−28	±21	−615	±44	1289
04/29/13	10	±4	−16	±23	26	±27	−561	±50	1284
09/25/13	40	±4	54	±29	−14	±33	−601	±56	1277
11/22/13	−84	±4	−25	±22	−59	±26	−646	±49	1284
02/21/14	−154	±4	−102	±15	−52	±19	−639	±42	1291
06/24/14	112	±4	105	±34	7	±38	−580	±61	1273
09/24/14	36	±4	58	±29	−22	±33	−609	±56	1277
10/30/14	−75	±4	−31	±21	−44	±25	−631	±48	1285
12/01/15	−126	±4	−65	±18	−61	±22	−648	±45	1288
03/04/16	−176	±4	−123	±13	−53	±17	−640	±40	1293

Table 5.11 Evaluation of elastic strain and absolute elastic strain and comparison of measured absolute elastic strain of Sensor 5.2down with yield strain, as per Criterion II.

Time	$\epsilon_{5.2down,tot,sensor}$	$\delta\epsilon_{5.2down,tot,sensor}$	$\epsilon_{5.2down,T,sensor}$	$\delta\epsilon_{5.2down,T,sensor}$	$\epsilon_{5.2down,E,sensor}$	$\delta\epsilon_{5.2down,E,sensor}$	$\epsilon_{5.2down,E,abs}$	$\delta\epsilon_{5.2down,E,abs}$	$\left\lvert\epsilon_{yld}^{+/-}/\beta^{+/-}-/+\delta\epsilon_{5.2down,E,abs}\right\rvert$
11/03/11	0	±4	0	±24	0	±28	285	±96	1237
12/06/11	−11	±4	−9	±23	−2	±27	283	±95	1238
01/26/12	−121	±4	−130	±12	9	±16	294	±84	1249
05/18/12	32	±4	39	±28	−7	±32	278	±100	1234
09/26/12	35	±4	55	±29	−20	±33	265	±101	1232
11/30/12	−142	±4	−100	±15	−42	±19	243	±87	1246
02/27/13	−117	±4	−75	±17	−42	±21	243	±89	1244
04/29/13	−7	±4	−27	±22	20	±26	305	±94	1240
09/25/13	−14	±4	58	±29	−72	±33	213	±101	1232
11/22/13	−89	±4	−23	±22	−66	±26	219	±94	1239
02/21/14	−174	±4	−104	±15	−70	±19	215	±87	1247
06/24/14	36	±4	106	±34	−70	±38	215	±106	1228
09/24/14	−16	±4	64	±30	−80	±34	205	±102	1231
10/30/14	−124	±4	−25	±22	−99	±26	186	±94	1239
12/01/15	−161	±4	−62	±19	−99	±23	186	±91	1243
03/04/16	−247	±4	−120	±13	−127	±17	158	±85	1248

Table 5.12 Evaluation of stress at the location of Sensor 5.2up and comparison with the model as per Criterion III.

Time	$\varepsilon_{5.2up,E,sensor}$	$\delta\varepsilon_{5.2up,E,sensor}$	$\sigma_{5.2up,sensor}$	$\delta\sigma_{5.2up,sensor}$	$\|\sigma_{5.2up,sensor} - \sigma_{5.2up,model}\|$
11/03/11	0	±28	0	±6	0
12/06/11	0	±27	0	±6	0
01/26/12	22	±17	5	±4	5
05/18/12	35	±32	7	±7	7
09/26/12	−23	±33	−5	±7	5
11/30/12	−51	±19	−10	±4	10
02/27/13	−28	±21	−6	±4	6
04/29/13	26	±27	5	±6	5
09/25/13	−14	±33	−3	±7	3
11/22/13	−59	±26	−12	±5	12
02/21/14	−52	±19	−11	±4	11
06/24/14	7	±38	1	±8	1
09/24/14	−22	±33	−5	±7	5
10/30/14	−44	±25	−9	±5	9
12/01/15	−61	±22	−13	±4	13
03/04/16	−53	±17	−11	±3	11

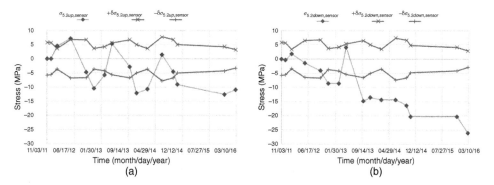

Figure 5.21 Graphs of stress changes with respect to reference time and comparisons with limits of errors as per Criterion III: (a) For Sensor 5.2up and (b) for Sensor 5.2down.

Example 5.10 *Application of Criterion IV to Wayne Bridge*

To apply Criterion IV as per Expression 5.88, first the absolute stresses at the locations of sensors have to be found. This is performed by using the model described in Example 5.8. The model estimates of absolute elastic strain at the locations of the two sensors at reference time are given in Expressions 5.94. The absolute stress estimates are obtained by simple multiplication with Young's modulus; see Expressions 5.95.

$$\sigma_{5.2up,model}(t_{\text{Ref}}) = -120 \pm 8 \text{ MPa}$$

$$\sigma_{5.2down,model}(t_{\text{Ref}}) = 58 \pm 14 \text{ MPa} \tag{5.95}$$

Table 5.13 Evaluation of stress at location of Sensor 5.2down, and comparison with the model as per Criterion III.

Time	$\varepsilon_{5.2down,E,sensor}$	$\delta\varepsilon_{5.2down,E,sensor}$	$\sigma_{5.2down,sensor}$	$\delta\sigma_{5.2down,sensor}$	$\|\sigma_{5.2down,sensor} - \sigma_{5.2down,model}\|$
11/03/11	0	28	0	±6	0
12/06/11	−2	27	0	±6	0
01/26/12	9	16	2	±3	2
05/18/12	−7	32	−1	±6	1
09/26/12	−20	33	−4	±7	4
11/30/12	−42	19	−9	±4	9
02/27/13	−42	21	−9	±4	9
04/29/13	20	26	4	±5	4
09/25/13	−72	33	−15	±6	15
11/22/13	−66	26	−14	±5	14
02/21/14	−70	19	−14	±4	14
06/24/14	−70	38	−14	±7	14
09/24/14	−80	34	−16	±7	16
10/30/14	−99	26	−20	±5	20
12/01/15	−99	23	−20	±4	20
03/04/16	−127	17	−26	±3	26

Estimated values of absolute stress can now be calculated by combining Expressions 5.87 and 5.95. In the absence of certified values, the yield stress for steel can be taken from Table 5.3, as $\sigma_{yld}^{+} = 360\,\text{MPa}\,\mu\varepsilon$ and $\varepsilon_{yld}^{-} = -360\,\text{MPa}$ (conservative value is taken). The location of Sensor 5.2up is subjected to compression; the potential local buckling stress value of steel should be determined, compared with compressive yield stress, and a smaller value should be used in Criterion IV. However, the steel flange at the location of Sensor 5.2up is stabilized by connection to the concrete slab, and local buckling is thus not expected to happen. Consequently, yield stress is used in Criterion IV. Adopted safety factors are $\beta_{\sigma}^{+} = \beta_{\sigma}^{-} = 1.5$ for reasons similar to those applied in Criterion II (see Example 5.8). Given that all arithmetical operations related to Criterion IV are similar to those of Criterion II, they are neither repeated nor presented in separate tables. However, Criterion IV is evaluated graphically in Figure 5.22. Again, two sets of safety factors are used: $\beta_{\sigma}^{+} = \beta_{\sigma}^{-} = 1.5$ and $\beta_{\sigma}^{+} = \beta_{\sigma}^{-} = 2$.

Figure 5.22 shows no indication of unusual structural behaviors that compromise the safety of the structure. Thus, Criteria I and III indicate the existence of unusual structural behaviors, and Criteria II and IV do not. The conclusion that can be drawn, after the four criteria were applied, is that there are unusual behaviors at the location of Sensors 5.2up and 5.2down (Criteria I and III), but there are no indications that these unusual behaviors compromise the safety of the structure (Criteria II and IV).

Example 5.11 *Analysis of unusual structural behavior in Wayne Bridge*
In previous examples, an unusual structural behavior was identified at cross-section 5.2, as indicated by both sensors installed at that location. A full analysis of that unusual structural behavior

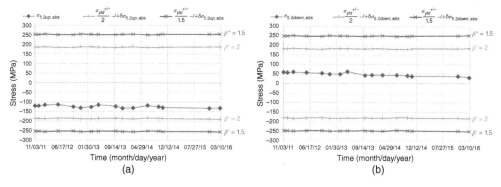

Figure 5.22 Graphs of absolute stress and comparisons with yield values as per Criterion IV: (a) For Sensor 5.2up and (b) for Sensor 5.2down; thresholds for safety factors of 1.5 and 2 are shown in graphs.

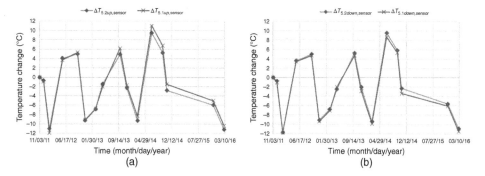

Figure 5.23 Diagrams of temperature change during monitored period, (a) for Sensors 5.1up and 5.2up, and (b) for Sensors 5.1down and 5.2down.

exceeds the scope of this example; therefore, only the most important steps in the analysis are presented.

As the first step, the reliability of sensor measurements should be verified for both strain and temperature. This is performed by comparing the measurements with measurements performed by the neighboring sensors. To simplify the presentation, only a comparison with sensors at location 5.1 is performed. Figure 5.23 shows the measurements of temperature change, and Figure 5.24 shows the measurements of elastic strain. Despite some discrepancy, which is mostly due to thermal gradients and boundary conditions, the figures show almost identical temperature change and generally consistent values of elastic strain change for both locations 5.1up and 5.2up (Figures 5.23a and 5.24a) and for locations 5.1down and 5.2down (Figures 5.23b and 5.24b). Therefore, there are no indications of sensors' malfunction, and the measurements are considered reliable.

Figure 5.24a shows that Sensors 5.1up and 5.2up measure approximately the same value of elastic strain (i.e., elastic strain change with respect to reference time). The main difference between the two diagrams is the higher sensitivity to temperature at location 5.2up, which is the most likely consequence of thermal gradients and the overall geometrical complexity of the structure (e.g., interaction with the other girders). Figure 5.24b shows that Sensors 5.1down and 5.2down also measure approximately the same values. Given that the elastic strain, i.e., elastic strain change during the monitoring period, in the two cross-sections is similar, the beam seems to be subjected to a negative bending moment and compressive normal force without shear force. This can be explained by the existence of horizontal force applied at the lower flange of the beam combined

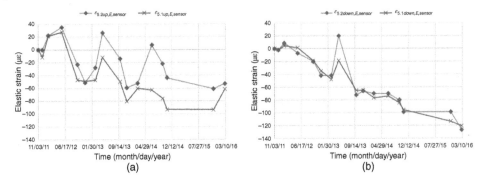

Figure 5.24 Diagrams of elastic strain change during the monitored period, (a) for Sensors 5.1up and 5.2up, and (b) for Sensors 5.1down and 5.2down.

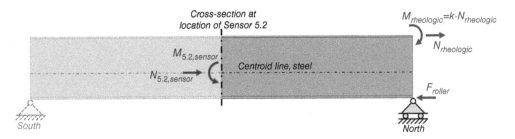

Figure 5.25 Schematic of proposed model that explains identified unusual structural behavior.

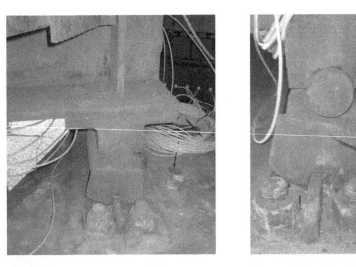

Figure 5.26 Examples of malfunctioning roller supports at north side of the second span of Wayne Bridge.

with relaxation (positive creep) in concrete slabs (due to the negative bending moment introduced by the horizontal force applied at the lower flange). This is illustrated in Figure 5.25. Given that the beam is supposed to function as simply supported, the probable reason for the horizontal force is a malfunction (blockage) of the roller support. Site inspection confirmed this assumption, as shown in Figure 5.26.

By using the elastic strain measured at locations 5.2up and 5.2down as an input in the linear beam model (Expressions 5.28) for steel cross-section only, one obtains the value of normal force $N_{5.2,sensor} \approx -3000\,kN$ (sign minus indicates compression) and the value of bending moment $M_{5.2,sensor} \approx -780\,kNm$ (sign minus indicates tensioning of the upper part of the cross-section). The model for determining the influences $N_{rheologic}$ and $M_{rheologic}$, due to rheological strain is described earlier by Expressions 5.91 and 5.92. This model shows that for given geometrical and mechanical parameters of steel and concrete, the ratio between $M_{rheologic}$ and $N_{rheologic}$ is constant, and for the Wayne Bridge, this value is $k = M_{rheologic}/N_{rheologic} \approx 0.168\,m$. From the equation of equilibrium applied for the cross-section 5.2 (see Figure 5.25), the force applied at the roller, F_{roller}, can be calculated, and it is evaluated to approximately $F_{roller} \approx 3600\,kN$.

While the blockage of the roller support explains why the force is introduced into the beam, it does not explain what causes the slow evolution of the force. The main assumption is that movements in the south embankment of the bridge are pushing the entire bridge from the south to the north, and this assumption is to be further investigated.

Note that measurements of both sensors presented in Figure 5.24b indicate the existence of an event that occurred on 29 April 2013. Measurements of sensors shown in Figure 5.24a are consistent with that event, although they are not as pronounced as those in Figure 5.24b. After the event, the elastic strain measurement continued to follow the previously established trend.

It is important to emphasize that the analysis presented in this section was limited to static strain data. In a more general case, a comprehensive analysis would need to include all possible loading scenarios. For example, for Wayne Bridge, it should include dynamic data analysis (e.g., the strain analysis for a passage of one or more heavy vehicles). Also, in a more general case, more sophisticated models of structure might need to be used. For example, for Wayne Bridge, interaction between the girders due to continuous concrete slabs, as well as skew shapes, make the bridge response complex, and to accurately describe its behavior, finite element analysis is certainly more accurate than linear beam theory (which was applied in this section to simplify presentation and illustrate methodology). However, although a small number of sensors were installed on Wayne Bridge, the measurements were taken irregularly (few times a year), and the model of the beam was drastically simplified, they were, nevertheless, sufficient to identify and diagnose unusual structural behavior and evaluate structural safety, which illustrates the great capabilities of strain-based monitoring.

5.9.5 Case Study on Analysis of Strain in Concrete Structure

An example of strain analysis at sensor level in concrete structure is presented in this subsection using the Punggol East Contract 26 (EC26) project, Singapore, as the case study. Buildings within this complex were equipped with long-term SHM by the Singapore's public housing authority, Housing and Development Board (HDB), to assure the quality and safety of new public tall buildings (Glisic and Inaudi 2007). The complex consists of six 19-story tall buildings funded on piles and supported on more than 50 columns each at ground level. In the building named "166A," 10 columns, which were considered critical based on numerical modeling, were instrumented with long-gauge fiber optic SOFO sensors. A view of the building after completion and floor plan with instrumented columns are presented in Figure 5.27 (modified from Glisic and Inaudi 2007).

The gauge length of SOFO sensors was 2 m, resolution 1 $\mu\varepsilon$ (2 μm over 2 m), and relative systematic error 0.2%. In each column, the sensor was embedded in concrete during construction and placed parallel to the centroid line of the column (see Figure 5.27c). Taking into account that SOFO sensors do not need thermal compensation (they are self-compensated to temperature, see

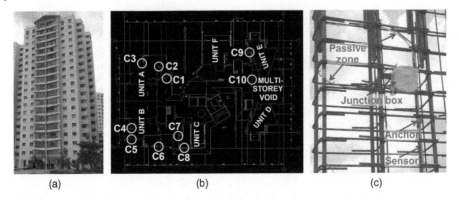

(a)   (b)   (c)

Figure 5.27 (a) View of completed building; (b) floor plan with instrumented columns C1–C10; and (c) view of a rebar cage with a SOFO sensor before the pouring of concrete. Source: Courtesy of SMARTEC.

Section 3.6.1) and that the temperature in Singapore is mildly varying over the entire year between approximately 23°C and 34°C (with extremes of 19.4°C and 35.8°C), it was decided not to install temperature sensors in order to keep the cost of SHM within the budget. Relative humidity (RH) in Singapore is also relatively stable over the year, with the morning average frequently higher than 90% and the mid-afternoon average of approximately 60% (daily mean value of 84%). The columns of the building were built of reinforced concrete. The Young's modulus of concrete was 28 GPa at 28 days and its strength was 40 MPa; for steel in reinforcing bars (rebars), the Young's modulus is 200 GPa and its strength was 160 MPa (as per information received from HDB). The column shapes, cross-sectional areas, exposures to the environment, and design loads (both dead and live loads) differ from column to column. The focus in the example presented here is on Unit A, which contains three instrumented columns; C1, C2, and C3 (see Figure 5.27b). The geometrical properties of the cross-sections of these three columns are presented in Table 5.14. More details about the project are given in Glisic and Inaudi (2007) and Glisic et al. (2013).

The columns were designed to carry predominantly axial load, and thus, the elastic strain is modeled as a ratio of load and axial stiffness (i.e., bending moment is set to zero in Expression 5.28). Table 5.14 shows the dead load, live load (actual building occupancy) at 10 years after construction, and the total load and corresponding elastic strain at 10 years after construction for each analyzed column. The relative uncertainty of the model for elastic strain was estimated to be $u^*(\varepsilon_{E,model}) = 25\%$. It was calculated taking into account construction tolerances, uncertainty in Young's modulus, uncertainty of eccentricity of load (15%), and eccentricity of sensor with respect to centroid of column.

Creep and shrinkage were modeled using a simplified approach, as per Fib (1999). Input parameters for geometrical properties (exposed perimeter and effective member size) are given in Table 5.14. The time of concrete pouring is used as a reference time, and average relative humidity (RH) used in the model is conservatively set to 75%. For each load increment i (new story built, roof, or live load increments), the values of creep coefficients $K_{c,i}$ and creep functions $f_{c,i}$ (see Expressions 5.65 and 5.68) are calculated using linear interpolation of values given in Fib (1999) (specifically, in Table 2.3 and Figure 2.7b from Fib (1999)). Values of mean final shrinkage $\varepsilon_{sh,final}$ and shrinkage function f_{sh} (see Expression 5.70) are also calculated using linear interpolation of values given in Fib (1999) (specifically, in Table 2.3 and Figure 2.7b from Fib (1999)). The creep at any observed time was estimated based on the real loading incremental schedule using

Table 5.14 Geometrical properties and strain parameters of columns C1, C2, and C3.

	Column C1	Column C2	Column C3
$A_{concrete}$ (10^3 mm^2)	300	270	420
A_{rebars} (10^3 mm^2)	2.011	2.513	8.168
$E_{concrete}$ (GPa)	28	28	28
E_{rebars} (GPa)	200	200	200
Axial stiffness $EAequivalent$ (kN)	8,802,124	8,062,600	13,393,600
Exposed perimeter (mm)[a]	2600	2400	3400
Effective member size (mm)[a]	231	225	247
Dead load (as per design, stories 1–19) (kN)	$96.4 \cdot 19 = 1830.8$	$64.8 \cdot 19 = 1231.6$	$205.8 \cdot 19 = 3909.4$
Dead load of roof (as per design) (kN)	59.7	45.2	199.7
Live load (as per design) (kN)	98.7	53.5	243.8
Total load (as per design) (kN)	1989	1330	4353
Elastic strain (as per design, both dead load and live load) $\varepsilon_{E,model}$ (µε)	−226	−165	−325
Equivalent creep coefficient (as per Expression 5.80) $K_{c,equ}$ (µε)	1.257	1.261	1.234
Final creep (as per Fib 1999) $\varepsilon_{c,final} = K_{c,equ}\varepsilon_{E,model}$ (µε)	−284	−208	−401
Final shrinkage (as per Fib 1999) $\varepsilon_{sh,final}$ (µε)	−314	−315	−312
Estimated absolute long-term total strain (based on model) $\varepsilon_{tot,model}$ (µε)	−824	−688	−1038
Relative uncertainty of the total strain in long-term u*($\varepsilon_{tot,model}$) (%)	±27	±28	±26

a) As per CEB-FIP (1990).

Expression 5.68 (see details in Glisic et al. 2013). Equivalent creep coefficients for each column were determined using Expression 5.80, and they are presented in Table 5.14. Uncertainty in estimating creep and shrinkage was set to u*($\varepsilon_{c,model}$) = 20% and u*($\varepsilon_{sh,model}$) = 35% based on Fib (1999).

The strain constituents are not mutually independent (for example, elastic strain and creep simultaneously depend on load, while creep and shrinkage simultaneously depend on relative humidity), which makes the estimation of the relative uncertainty in total strain challenging. To address this challenge, the Cauchy–Schwarz inequality for covariance was used to estimate the relative uncertainty in the total strain as follows:

$$u^*(\varepsilon_{tot,model}(t)) \leq \frac{u^*(\varepsilon_{E,model})\varepsilon_{E,model}(t) + u^*(\varepsilon_{c,model})\varepsilon_{c,model}(t) + u^*(\varepsilon_{sh,model})\varepsilon_{sh,model}(t)}{\varepsilon_{E,model}(t) + \varepsilon_{c,model}(t) + \varepsilon_{sh,model}(t)} \quad (5.96)$$

The right-hand side of Expression 5.96 is actually an overestimate of the relative uncertainty of the model, as the actual relative uncertainty is smaller due to mutual interdependencies among the strain constituents. Note that, based on Expression 5.96, the relative uncertainty of the model for total strain depends on time t, and has to be calculated for each time stamp of measurements. As an example, the relative uncertainty of the model in the long-term ($t \rightarrow \infty$) is given in the last row of Table 5.14. In general, relative uncertainties determined based on Expression 5.96 for various times ranged between 25% and 28%. These values, which are around 25%, are in agreement with the authors' personal SHM experience, as well as with other studies (e.g., Moragaspitiya 2011).

The pouring of concrete occurred on April 11, 2001, but, due to safety concerns, the reference measurement could only be made 43 days later, after the second story was built. Thus, the elastic strain generated by the first two stories, the creep generated by the first story dead load (built 15 days before the reference time), and the shrinkage were not registered. The estimations of these values based on the model presented above are given in Table 5.15 (Glisic et al. 2013).

To further keep the cost of this pioneering application within the budget, the SHM system was not centralized, and measurements were performed manually. During the construction, one measurement over all sensors was performed after each story and the roof were completed. Post-construction, the measurements were typically performed a few times a year. Temperature and live load fluctuations (tenants in and out of the building) were not directly monitored, and to assess their influence on strain, several 48-hour sessions were carried out in the years 2004–2007. A 10-year record of strain in Columns C1–C3 is presented in Figure 5.28.

The 48-hour sessions included one measurement over all sensors every hour for two consecutive days. These 48-hour sessions served as the basis for evaluation of limits of relative error of

Table 5.15 Estimation of strain constituents at the reference time (43 days after the pouring).

	Column C1	Column C2	Column C3
Elastic strain due to first two stories $\varepsilon_{E,model}$ ($t = 43$ days) ($\mu\varepsilon$)	−22	−16	−31
Creep due to the first story (15 days after completion) $\varepsilon_{c,model}$ ($t = 43$ days) ($\mu\varepsilon$)	−4	−3	−5
Shrinkage $\varepsilon_{sh,model}$ ($t = 43$ days) ($\mu\varepsilon$)	−49	−50	−48
Estimated total strain at reference time (strain not included in monitoring) $\varepsilon_{tot,model}$ ($t = 43$ days) ($\mu\varepsilon$)	−75	−69	−83

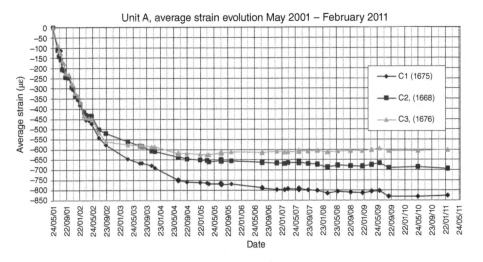

Figure 5.28 10-year record of strain monitoring in Columns C1–C3.

Table 5.16 Evaluation of relative error in strain measurement.

	Column C1	Column C2	Column C3
Range of air temperature change resulting in largest strain variation during 48-h sessions (°C)	5.0	5.0	5.0
Range of daily strain variation estimated from 48-h sessions ($\mu\varepsilon$)	10.0	10.5	8.0
Range of yearly strain variation estimated by extrapolation for expected yearly temperature range of 11°C ($\mu\varepsilon$)	22.0	23.1	17.6
Estimated yearly strain variation (± half of the value in previous row) ($\mu\varepsilon$)	±11	±12	±9
Limits of error of the monitoring system for measured strain up to 500 $\mu\varepsilon$ ($\mu\varepsilon$)	±2	±2	±2
Limits of error of the monitoring system for measured strain ranged between 500 and 1000 $\mu\varepsilon$ ($\mu\varepsilon$)	±3	±3	±3
Limits of error in (total) strain measurement $\delta\varepsilon_{tot,sensor}$ ($\mu\varepsilon$)	±14	±15	±12

total-strain measurements, which in turn was needed for data analysis and detection of unusual behaviors as per Criteria I–IV. Given that the temperature of the columns was not measured, the thermally generated strain could not be included in the model, and thus it had to be accounted for as an error. The 48-hour sessions revealed the range of daily fluctuations of total strain as shown in Table 5.16, which were then linearly extrapolated to the entire year and used to estimate the corresponding error in strain measurement (see Table 5.16). Note that the evaluations of limits of error presented in Table 5.16 are conservative, as they do not account for fluctuations in live load during the day (people out of the building for work).

To simplify the presentation and avoid repetition, only the applications of Criteria I and II to measurements from Punggol EC26, Building 166A, are given further text, while the applications of Criteria III and IV are omitted.

Example 5.12 *Application of Criterion I to Punggol EC26, Building 166A*
Strain measurements with respect to reference time, are given in Figure 5.28. Presenting all these measurements in the form of a table would not be practical, and that is the reason why only a few characteristic points in time are presented in Table 5.17. A comparison with the model is made in the same table.

In order to apply Criterion I as per Expressions 5.79, either the limits of errors in the model and monitoring system have to be combined (the first inequality in Expression 5.79) or the uncertainties of the model and monitoring system have to be combined (the second inequality in Expression 5.79). Given that in this specific project the measure of accuracy of the model is given in terms of uncertainty and the measure of accuracy of the monitoring system in terms of limits of error, a combination of the two has to be made using a somewhat hybrid method. The limits of error of the monitoring system can be considered equivalent to two or three standard deviations (standard uncertainties) of a hypothetical monitoring system with equivalent performance, whose measure of accuracy is given in terms of uncertainty. Then the uncertainty propagation formula can be applied to combine the uncertainties of the model and the equivalent monitoring system. This is performed in Table 5.18. Note that the uncertainties presented in the table change over time as they depend on the magnitude of the estimated and measured strain. In addition, the uncertainty of the model

Table 5.17 Comparison between the model and measurements as per Criterion I.

Time	Parameter	Column C1	Column C2	Column C3
7/19/2002, end of construction (464 days after pouring)	Total strain, model (με)	−479	−381	−628
	Total strain, measured (με)	−576	−519	−560
	Difference (με)	**−97**	**−138**	**68**
	Uncertainty in difference[a](με)	**±121**	**±99**	**±154**
9/25/2008, at approx. 7 years (2724 days after pouring)	Total strain, model (με)	−663	−540	−855
	Total strain, measured (με)	−809	−681	−605
	Difference (με)	**−146**	**−141**	**250**
	Uncertainty in difference[a](με)	**±172**	**±145**	**±215**
2/11/2011, at approx. 10 years (3593 days after pouring)	Total strain, model (με)	−693	−567	−891
	Total strain, measured (με)	−826	−694	−601
	Difference (με)	**−133**	**−127**	**290**
	Uncertainty in difference[a](με)	**±181**	**±153**	**±224**

a) See Table 5.18 for details.

Table 5.18 Evaluation of uncertainty of difference between the model and the measurements.

Time	Parameter	Column C1	Column C2	Column C3
7/19/2002, end of construction (464 days after pouring)	Total strain, model (με)	−479	−381	−628
	Relative uncertainty of model (%)	±28%	±29%	±27%
	Uncertainty of model (με)	±121	±99	±154
	Limits of error in measurement (με)	±14	±15	±12
	Equivalent uncertainty in measurement (με)	±14/3 = ±5	±15/3 = ±5	±12/3 = ±4
	Uncertainty in difference (see Table 4.3) (με)	**±121**	**±99**	**±154**
9/25/2008, at approx. 7 years (2724 days after pouring)	Total strain, model (με)	−663	−540	−855
	Relative uncertainty of model (%)	±28%	±29%	±27%
	Uncertainty of model (με)	±172	±145	±215
	Limits of error in measurement (με)	±14	±15	±12
	Equivalent uncertainty in measurement (με)	±14/3 = ±5	±15/3 = ±5	±12/3 = ±4
	Uncertainty in difference (see Table 4.3) (με)	**±172**	**±145**	**±215**
2/11/2011, at approx. 10 years (3593 days after pouring)	Total strain, model (με)	−693	−567	−891
	Relative uncertainty of model (%)	±28%	±29%	±27%
	Uncertainty of model (με)	±181	±153	±224
	Limits of error in measurement (με)	±14	±15	±12
	Equivalent uncertainty in measurement (με)	±14/3 = ±5	±15/3 = ±5	±12/3 = ±4
	Uncertainty in difference (see Table 4.3) (με)	**±181**	**±153**	**±224**

is dominant in Table 5.18, and the limits of error in measurement practically do not affect the uncertainty of the difference.

In this project, one standard uncertainty was used for comparison ($n_u = 1$, see Expression 5.79). The reason for this is the relatively large overall uncertainty of the model. Using two or three standard uncertainties, in this case, might be too much non-conservative. The comparison in Table 5.17 shows three instances of unusual structural behaviors. The first regards Column C2 at the end of construction; this unusual behavior did not repeat at 7 and 10 years after the pouring of concrete. The other two instances of unusual behaviors are related to Column C3 at 7 and 10 years. Repeated detection of this unusual behavior over time indicates its persistence and requires further analysis. To understand the unusual behaviors, the measurements are compared with the model in Figure 5.29. Each graph in the figure shows a 10-year record of measurements along with model estimates and their standard uncertainty thresholds for Columns C1, C2, and C3 (Figure 5.29a–c, respectively). The values of strain estimated by the model are relative to reference time (i.e., values in Table 5.15 are deducted from the values of absolute strain estimated by the model), in order to make them comparable with measurements. The points indicated with isolated markers in graphs represent the final values of the model (as per Table 5.14, values from Table 5.15 deducted).

The measurements of Column C1 were in relatively good agreement with the model during the construction, but then compressive strain increased, and from 2004 to 2008 it followed the lower threshold of standard uncertainty; after 2008 the measurements were relatively stable (no significant evolution in strain), and thus they started to approach the model. The measurements of Column C2 followed the lower threshold of standard uncertainty for almost the entire presented period, including construction. Contrary to this, the measurements of Column C3 followed the upper threshold of standard uncertainty during the construction and until 2004. After that time, the measurements show relaxation of strain in the column, and after 2005, they exceeded the upper threshold and continuously increased the discrepancy from the model over time. Two types of unusual behaviors can be identified and related conclusions can be drawn from the observations above.

First, during construction, all three sensors measure similar strain (see also Figure 5.28), despite the fact that they should experience different stress and strain based on the loading schedule and the model. A possible explanation of this behavior is the high stiffness (low deformability) of the first-floor slab, which in turn imposes approximately the same deformation on the columns. This possibility was not included in the original model used in Figures 5.28 and 5.29. However, the above behavior during construction can also be explained simply by the imperfection of the original model: during construction, the discrepancy between the measurements and the model is mostly within standard uncertainty, thus the eccentricity of the forces in columns, along with the eccentricities of sensors, could result in such a coincidence (although it is not very likely that these imperfections will result in similar values for all three columns). Finally, the observed behavior can be the result of a combination of the above two explanations. Based on available data, it is not possible to fully ascertain the reason for the presented strain measurements in columns during construction; nevertheless, given that there is mostly no exceedance of the standard uncertainty threshold, except in a few time stamps and for small values (e.g., in Column C2, at the end of construction, see the shaded field in Table 5.17, but then it was back within the threshold limits), this behavior is, in general, not alarming.

Second, while Column C2 kept the trend of following the lower standard uncertainty threshold, the two other columns exhibited a change in behavior: the strain in Column C3 started relaxing while the strain in column C1 started shifting toward the lower standard uncertainty threshold. This change in behavior started approximately in 2002 and became permanent after 2004.

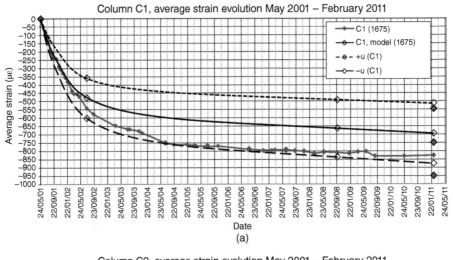

(a)

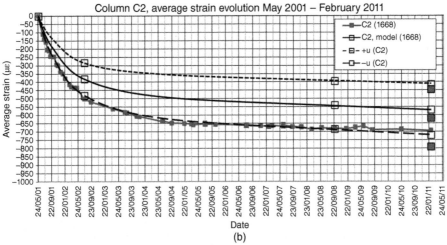

(b)

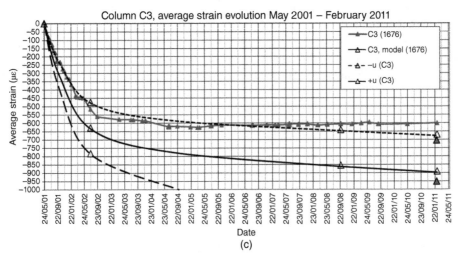

(c)

Figure 5.29 10-year record of measurements along with model estimates and their standard uncertainty thresholds; (a) for Column C1, (b) for Column C2, and (c) for Column C3.

Table 5.19 Evaluation of Criterion II.

Time	Parameter	Column C1	Column C2	Column C3
7/19/2002, end of construction (464 days after pouring)	Total strain measurement (με)	−576	−519	−560
	Thermal strain (με)	0	0	0
	Shrinkage (as per Fib 1999) (με)	−101	−102	−98
	Multiple of creep coefficient and creep function (−)	0.959	0.965	0.920
	Elastic strain (as per Expression 5.82) (με)	−243	−212	−241
	Absolute elastic strain (as per Expression 5.83) (με)	**−265**	**−228**	**−272**
	Uncertainty multiplied with $n_u = 3$	±69	±69	±68
	Upper threshold as per Expression 5.84	**−44**	**−44**	**−43**
	Lower threshold as per Expression 5.84	**−606**	**−606**	**−607**
9/25/2008, at approx. 7 years (2724 days after pouring)	Total strain measurement (με)	−809	−681	−605
	Thermal strain (με)	0	0	0
	Shrinkage (as per Fib 1999) (με)	−196	−198	−192
	Multiple of creep coefficient and creep function (−)	1.284	1.295	1.255
	Elastic strain (as per Expression 5.82) (με)	−268	−210	−183
	Absolute elastic strain (as per Expression 5.83) (με)	**−290**	**−226**	**−214**
	Uncertainty multiplied with $n_u = 3$	±98	±99	±98
	Upper threshold as per Expression 5.84	**−73**	**−74**	**−73**
	Lower threshold as per Expression 5.84	**−577**	**−576**	**−577**
2/11/2011, at approx. 10 years (3593 days after pouring)	Total strain measurement (με)	−826	−694	−601
	Thermal strain (με)	0	0	0
	Shrinkage (as per Fib 1999) (με)	−215	−216	−210
	Multiple of creep coefficient and creep function (−)	1.343	1.356	1.313
	Elastic strain (as per Expression 5.82) (με)	−261	−203	−169
	Absolute elastic strain (as per Expression 5.83) (με)	**−283**	**−219**	**−200**
	Uncertainty multiplied with $n_u = 3$	±104	±104	±103
	Upper threshold as per Expression 5.84	**−79**	**−79**	**−78**
	Lower threshold as per Expression 5.84	**−571**	**−571**	**−572**

The strain in Column C3 practically became constant, reached the upper threshold of standard uncertainty in 2005, and stayed out of threshold permanently (see shaded fields in Table 5.17). Such behavior is consistent with differential settlement of foundations: as the foundation under Column C3 settles, the strain in that column decreases; on the other hand, the load is directed toward other columns, among which is Column C1, and thus the strain in Column C1 increases (see also Glisic et al. 2013). Simplified structural analysis has shown that this differential settlement of the foundation is in the few-millimeter range, and thus does not affect the safety or serviceability of the building. That was also confirmed by applying Criterion II to the measurements, as shown in the next subsection.

Example 5.13 *Application of Criterion II to Punggol EC26, Building 166A*
To apply Criterion II, it is necessary to evaluate the absolute elastic strain at locations of sensors. The estimated elastic strain at reference time is given in Table 5.15. Expression 5.82 provides a rather complicated relationship for extracting elastic strain from the measurements. Given that creep and shrinkage were not directly measured, the values of their parameters used in Expression 5.82 are estimated using the model based on Fib (1999), described in the previous subsection. Finally, since the temperature was not measured, the value of thermal strain in Expression 5.82 is set to zero, but error limits in its estimation were conservatively set to ± 72 µε (corresponding to a temperature change of $\pm 6°C$ and using a thermal expansion coefficient of 12 µε/°C). To simplify the presentation, similar to the evaluation of Criterion I, the value of absolute elastic strain was calculated at three characteristic points in time. To set Criterion II, yielding compressive strain and ultimate tensional strain for concrete were conservatively taken from Table 5.3 as $\varepsilon_{yld}^{-} = -1350$ µε and $\varepsilon_{u}^{+} = 50$ µε. To keep the evaluation conservative, safety factors are adopted to be $\beta_{\varepsilon}^{+} = \beta_{\varepsilon}^{-} = 2$, and the multiplier of standard uncertainty is $n_{u} = 3$.

The values used in the evaluation of Criterion II are given in Table 5.19.

Table 5.19 shows that the evaluated absolute elastic strain falls well within the thresholds established by Criterion II (see Expression 5.82 for concrete), and thus there is no indication that the safety of the columns is compromised. Once again, similar to Wayne Bridge project, despite limitations in SHM approach, such as the small number of sensors installed in the columns of the building, irregular schedule of measurements, and the drastically simplified model of the columns, the results were sufficient to identify and diagnose unusual structural behavior and evaluate structural safety, which, again, illustrates the great capabilities of strain-based SHM.

6

SHM at Global Scale: Interpretation and Analysis of Strain Measurements from Multiple Sensors and Identification of Unusual Structural Behaviors at Structural Level

6.1 Introduction to SHM at Global Scale

The focus of the previous chapter was on the interpretation of single-sensor measurement and its use for the identification of unusual structural behaviors. While that approach greatly helps detect problems at the local scale, it may not be always effective in ascertaining the cause of the problem. For example, the case studies presented in Section 5.8 eventually compared the measurements from multiple sensors in order to confirm reliability of measurement and ascertain the existence and potential cause of unusual structural behaviors.

Hence, the aim of this section is to present several (but not all!) algorithms that enable assessment of condition and performance of structures based on measurements recorded by multiple sensors. Scope is limited to beam-like structures and to most frequently sought global parameters derived from strain measurements, such as, but not limited to, curvature, neutral axis, and deformed shape.

Note that Section 4.4 recommends the use of long-gauge sensors for global structural monitoring, while Chapter 5 presents relationships between strain on one side and curvature, position of centroid of stiffness, temperature changes and gradients, and internal loads on the other. These relationships will be further explored in this chapter.

6.2 Global-Scale SHM Approach

6.2.1 Implementation Guidelines

The global-scale structural health monitoring (SHM) approach for beam-like structures can be described as the three-step process (Glisic and Inaudi 2007):

– The first step is to identify the locations of interest, i.e., beam cross-sections, which will be instrumented with sensors; the selection of these locations can be justified by various reasons, such as extreme expected influences (e.g., cross-sections with maximal or minimal internal forces or deflections), expected locations of damage (e.g., sections with expected prestress loss or cracking, which might or might not coincide with extreme internal forces), improved spatial coverage of structure (e.g., enabling higher density of sensors to improve accuracy in evaluation of strain derivatives or to increase the likelihood of damage detection), and "control" cross-sections (e.g., those where the damage or significant internal forces are not expected, so their measurements can be used as reference for comparison with measurements from other instrumented cross-sections)

Introduction to Strain-Based Structural Health Monitoring of Civil Structures, First Edition. Branko Glišić.
© 2024 John Wiley & Sons Ltd. Published 2024 by John Wiley & Sons Ltd.

– The second step is to instrument the identified cross-sections with topologies of sensors that in the best way describe the expected strain distributions in these cross-sections; typical four topologies used in global-scale SHM are the following (Inaudi and Glisic, 2002, Glisic and Inaudi 2007):

  o Simple topology – used to monitor cross-sections where strain is expected to be constant across the cross-section (e.g., sections where normal forces are dominant, while the bending moments and thermal gradients along with thier resulting curvatures are very small or null).

  o Parallel topology – used to monitor cross-sections where strain is expected to vary across the cross-section (e.g., sections where bending moments or thermal gradients are significant and crate curvatures).

  o Crossed topology – used to monitor cross-sections where shear strain is significant (i.e., sections with significant shear force, e.g., close to supports).

  o Triangular topology – used to monitor relative displacement in the structure.

This second step, which determines which topologies should be used in various cells, also includes the determination of gauge lengths of sensors based on engineering experience and recommendations given in Section 4.4. Note that longitudinal beam length around the cross-section which is instrumented by a topology of long-gauge sensors is often called "cell" (Vurpillot et al. 1998, Glisic and Inaudi 2007). Given that cell has physical length, it includes all the cross-sections that are found along its length. In addition, depending on application and data analysis approach, a cell can include cross-sections that are beyond sensor's gauge length (see Figure 6.1). Examples of the four topologies are given in Figure 6.1 along with indications of various other relevant notions.

– The third step is to perform data analysis at both cross-sectional level and structural level in order to infer structural behaviors at cross-sections and of the structure as a whole.

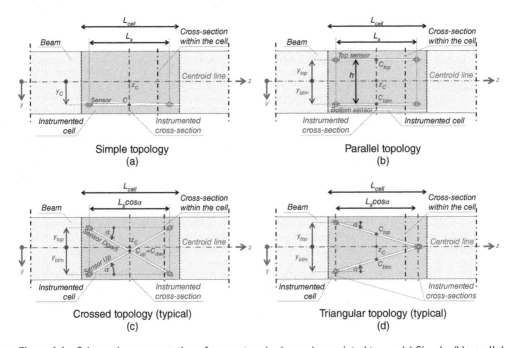

Figure 6.1 Schematic representation of sensor topologies and associated terms: (a) Simple, (b) parallel, (c) crossed, and (d) triangular topology.

All three steps make important parts in defining monitoring strategy (see Section 1.2). They are typically (but not only) performed based on structural analysis (e.g., for finding extreme influences and potential damage locations, as well as defining algorithms for data analysis), engineering experience (e.g., for typical damage locations and longitudinal spacing of sensors), and cost limitations (e.g., for selection of monitoring system and number of sensors). Depending on project aims and complexity, other factors can also influence selection of sensor locations. For example, in cases where the main objective of SHM is structural identification and assessment of structural behavior in the absence of damage, or where the expected damage scenarios are well known, the number and locations of sensors can be determined using optimization algorithms. To illustrate the three-step process behind the global-scale SHM approach, two examples are given as follows.

6.2.2 Application Example of Global-Scale SHM on a Pile Subjected to Axial Loading

This first example deals with SHM of pile subjected to axial loading: compression or tension. The schematic of instrumentation of pile is shown in Figure 6.2a. Given that the dominant internal load in pile is normal force (compressive or tensional), and that bending is expected to potentially occur only at the top of the pile due to eccentricity of the applied force, the first (top) cell was instrumented with parallel topology, while all other cells were instrumented with simple topology (based on engineering experience, it is assumed that bending "dissipates" along the pile due to its interaction with the soil). Diameter of pile was 1.2 m, and its length was approximately 35 m. Taking into account guidelines for selecting the gauge length proposed in Section 4.4 (if possible, longer than depth of cross-section and shorter than length of pile divided by six), available budget (nine sensors

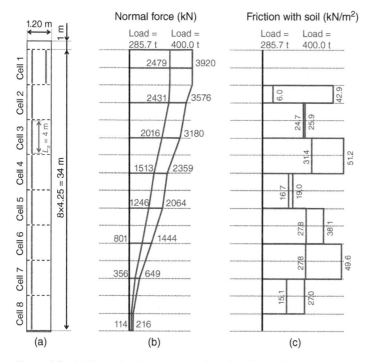

Figure 6.2 (a) Example of instrumentation of a pile subjected to axial loading with single and parallel topologies of sensors for global-scale monitoring, and examples of results at global scale obtained by "connecting" results in all cells: (b) distribution of average normal force along the pile, and (c) distribution of average frictional stress with soil along the pile. Source: Modified from Glisic et al. (2002b) and Glisic and Inaudi (2007); courtesy of SMARTEC and RouteAero.

or less per axially tested pile), and the project aim to instrument as much of a pile length as possible (to evaluate average distribution of normal force and the friction with soil with high reliability), the pile was split into eight cells of approximate length of 4.25 m, and gauge length of sensors was set to 4 m. One sensor in Cell 1 and all sensors in Cells 2–8 were aligned, so that they formed so-called "enchained" simple topology (Glisic and Inaudi 2007), see also Section 6.8.2. This enables model-based data analysis at the global level, which provides rich information regarding pile behavior, such as average normal force distribution along the pile length for different load levels, quality of interaction with soil (friction of soil along the pile), deformation of pile, and failure mode. For illustrative purposes, only the two former examples of results are presented in this section.

Data were collected during pullout test of the pile, where applied force was controlled and strain in all cells was monitored. Given that pile was buried in soil, temperature change in the pile could be neglected. In addition, entire test took only a few hours, and thus the rheological strain could also be neglected. Consequently, the measured strain in each cell consisted of mechanical strain only, elastic and plastic. Based on engineering judgment, it was assumed that friction with soil in the first cell could be neglected, and thus it was assumed that the entire elastic strain measured in the first cell was generated by the applied force. Due to symmetry, average strain from the two sensors was used to evaluate the total strain at the centroids of the cross-sections in the first cell. This is equivalent to a result expected from simple topology installed in the centroid of the cross-section (i.e, cell). Average stress in the first cell was calculated by dividing the applied force with the area of the cross-section of the pile then, the least-square linear interpolation of stress–strain relationship was used to determine modulus of elasticity of the pile (e.g., see Section 6.5 for details).

As examples of global-scale results, Figure 6.2b,c shows distribution of normal force along the pile during pullout test along and evaluated average distribution of friction between the pile and the soil, respectively (see Section 6.5 for details). More detail about this project, in general, is found in Glisic et al. (2002b) and Glisic and Inaudi (2007).

6.2.3 Application Example of Global-Scale SHM on Streicker Bridge

The second example of application of global-scale SHM is related to instrumentation of Streicker Bridge on Princeton campus (see also Sections 4.5.5 and 4.5.7). The bridge has very complex geometrical shape (see cover page of this book and Figures 4.10 and 6.3): it consists of main span and four "legs"; the main span is deck stiffened arch while each leg is curved continuous beam supported by columns. In all five components of the bridge, the deck is made of prestressed high-performance concrete (see Sigurdardottir and Glisic 2015), while arch and columns are made of weathering steel (Yoloy). Due to budget constraints and taking into account that the aim of SHM has been principally for research and education purposes, only one-half of the Main Span and the Southeast Leg (the longest of the four legs) were equipped with sensors, which were embedded in the deck during construction in 2009. Schematic of sensor locations is given in part in Figures 4.6 and 4.10 and in more detail in Figure 6.3. Due to aesthetic reasons, no sensors could be installed on the steel arch or columns. Based on recommendations given in Section 4.4 along with installation constraints (limited time available for installation), the length of sensors was set to $L_s = 60$ cm (sensors were kept to minimal recommended length, see Section 4.5.6, in order to shorten the installation time). Given that budget and selected gauge length of sensors could not guarantee full coverage of the bridge, the length of cells was simply kept the same as length of sensors, i.e., $L_{cell} = 60$ cm.

Based on loose structural analysis, the cross-sections in the middle of spans and above the columns were selected for instrumentation with parallel topology, as they were expected to experience the largest positive and negative bending moments, respectively. Being the most

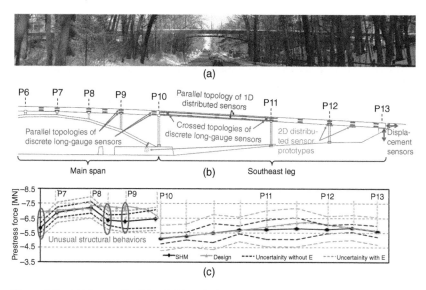

Figure 6.3 (a) View to Streicker Bridge; (b) instrumentation of the bridge with parallel and crossed topologies of sensors for global-scale monitoring; and (c) distribution of prestressing force distribution at release, as an example of global-scale results obtained by "connecting" all cross-sections. Source: Modified from Glisic (2019).

stressed, these cross-sections were considered the most likely to experience unusual structural behaviors in the case of extreme loading. Due to cost limitations and considering almost-linear distribution of bending moments in the deck between the columns of deck stiffened arch, not all middle-span sections were instrumented in the main span. The Southeast Leg was constructed approximately two months after the main span, which resulted in the creation of a continuity joint between the two parts of the structure (at P10, see Figure 6.3b). Two parallel topologies of sensors were installed at that location to assess the quality of realization of the joint, one topology at each side of the joint. In addition, in the longest span P10–P11 of the southeast leg, two topologies of sensors were installed in quarter-spans, approximately, for two reasons: first, to improve the spatial coverage of the structure with sensors, and second, to have sensors at locations where stresses are not extreme, so unusual structural behaviors are not expected to occur, and thus, these sensors can be used for comparison with sensors installed at locations where unusual behaviors are more likely to occur (in middles of spans and above columns). Thus, those were considered as "control" cross-section (i.e., cells). Finally, crossed topologies were installed at the ends of the longest span of southeast leg (P10–P11), as maximal shear forces and torsional moment (due to curved shape in horizontal plane) are expected to occur at these locations. All parallel topologies of sensors were placed parallel to centroid line and as close as possible to the vertical plane containing centroid line and vertical principal axis of cross-section. This enabled model-based data analysis at global scale, which provided rich information regarding the bridge behavior, such as determination of centroid of stiffness, prestress loss distribution along the bridge, deflection of deck, and crack detection. As an example of global-scale result, the distribution of prestressing force at release (i.e., immediately after posttensioning) is shown in Figure 6.3c. Some other results taken from the project are presented in Sections 3.7, 4.5, 4.7 and throughout this chapter and Chapter 7. More detail about this project is found in Sigurdardottir and Glisic (2015) and other author's papers.

Besides the sensors described in this example, Figure 6.3 shows several other sensors that were installed on the bridge: distributed sensor in Span P10–P11 and displacement sensors installed at abutment P13. In addition, several other sensors, which are not shown in Figure 6.3, were installed in different cross-sections in order to study various phenomena (e.g., see Figure 4.6 in Section 4.4). These sensors will be presented throughout the text as needed (see also Sigurdardottir and Glisic 2015).

The purpose of the two examples given above is to illustrate the power of global-scale monitoring. More detail on how to identify the cross-sections (cells) to be instrumented with sensor topologies, and how to evaluate parameters at local (cell) and global scale, is given throughout this chapter, while ample case studies are presented in Glisic and Inaudi (2007).

6.3   Physics-Based Interpretation of Parameters Derived from Sensor Topologies and Identification of Unusual Structural Behaviors

Sensor topologies enable evaluation of several parameters in a cell derived from sensor measurements and they can be, generally, split into two groups: (i) parameters derived from strain measurements, or their constituents, and geometrical properties of the cell, and (ii) parameters derived from stress evaluations (i.e., from evaluation of elastic strain constituents) and geometrical properties. Note that to derive strain constituents (e.g., elastic strain, thermal strain, etc.) and stresses from total strain measurement, it is also necessary to evaluate the mechanical properties of cell (e.g., Young's modulus, thermal expansion coefficient, and rheological parameters and functions).

Examples of cell parameters derived from strain measurements, or their constituents, and geometrical properties of the cell are typically the following:

– Change in length of the cell
– Average curvature in the cell
– Average position of the centroid of stiffness in the cell
– Average angle of rotation of the beam's cross-section with respect to centroid line of the cell
– Change of the slope between the extremities of the cell

Example of cell parameters derived from stress evaluations and geometrical properties of the cell are typically the following:

– Average normal force in the cell
– Average bending moment in the cell
– Average shear force in the cell
– Average prestressing force in the cell (if any)

To simplify presentation, in further text, we refer, where appropriate, to all these parameters as "measured" although they are not directly measured but rather derived from strain sensor measurements. For example, the sentence "measured average normal force" might be interchangeably used instead of "average normal force evaluated from strain sensor measurements," depending on the context, but their meaning will be assumed to be the same.

In Section 5.8, Criteria I–IV for identification of unusual structural behaviors based on measured strain and stress, and their comparison with models and ultimate values, are presented (see Expressions 5.79, 5.84, 5.86, and 5.88). Note that only Criterion I is based on total strain measurements, while Criterion II is based on elastic strain measurement and Criteria III–IV are based on stresses. Criteria I–IV can be extended to all the above-listed cell parameters, derived from strain and stress measurements, and used to identify unusual structural behaviors, as follows.

Let us denote an observed cell in a structure with i and any of the above-listed parameters (change in length, curvature, etc.) with p_{cell_i}. Then, the value of p_{cell_i} is, by definition, some function of total strains, strain constituents, or stresses at locations of sensors (these functions are derived in the further text for each parameter). For example, elastic curvature is function of elastic strain at the bottom and top of the cross-section as per Expression 5.30. Consequently, let us denote with $p^-_{cell_i,yld/u/bkl}$ and $p^+_{cell_i,yld/u/bkl}$ the values of p_{cell_i} when negative or positive yield, ultimate, or buckling values (whichever is critical) of total strain, elastic strain, or stress (based on which p_{cell_i} is evaluated) are reached in the cell.

Now let us denote the value of p_{cell_i} measured at time t (i.e., derived from sensor measurements at time t) with $p_{cell_i,sensor}(t)$ and the value of that parameter obtained by modeling (e.g., using beam theory, finite element modeling [FEM], design codes, etc.) with $p_{cell_i,model}(t)$. Given that sensors in observed cell i measure change in strain at time t with respect to reference time, the value of $p_{cell_i,sensor}(t)$ is also relative to reference time, i.e., it shows the change in the parameter with respect to reference time. However, this is not the case with $p_{cell_i,model}(t)$, which always shows absolute values (see Section 5.8). Criteria V and VI can be obtained by substituting total strain, elastic strain, and stress values (whichever applies) with p_{cell_i} in Expressions 5.79 or 5.86 (for Criterion V) and 5.84 or 5.88 (for Criterion VI). Note that substitution in Expressions 5.79 (Criterion I) and 5.86 (Criterion III) yields the same expression of Criterion V, while substitution in Expressions 5.84 (Criterion II) and 5.88 (Criterion IV) yields the same expression of Criterion VI.

Criterion V:
$$\left| p_{cell_i,sensor}(t) - \Delta p_{cell_i,model}(t) \right| \leq \delta p_{cell_i,sensor} + \delta \Delta p_{cell_i,model}$$
or
$$\left| p_{cell_i,sensor}(t) - \Delta p_{cell_i,model}(t) \right| \leq n_u \sqrt{\mathrm{u}(p_{cell_i,sensor})^2 + \mathrm{u}(\Delta p_{cell_i,model})^2}$$
(6.1)

Criterion VI:
$$\frac{p^-_{cell_i,yld/u/bkl}}{\beta^-_p} + \delta p_{cell_i,abs} < p_{cell_i,abs}(t) < \frac{p^+_{cell_i,yld/u/bkl}}{\beta^+_p} - \delta p_{cell_i,abs}$$
or
$$\frac{p^-_{cell_i,yld/u/bkl}}{\beta^-_p} + n_u \mathrm{u}(p_{cell_i,abs}) < p_{cell_i,abs}(t) < \frac{p^+_{cell_i,yld/u/bkl}}{\beta^+_p} - n_u \mathrm{u}(p_{cell_i,abs})$$
(6.2)

where, similar to Expression 5.83, absolute value of p_{cell_i} at time t is evaluated as

$$p_{cell_i,abs}(t) = p_{cell_i,sensor}(t) + p_{cell_i,model}(t_{Ref})$$
(6.3)

and β^-_p and β^+_p are safety factors that account for unknown accuracies of yield, ultimate, or buckling values of $p^-_{cell_i,yld/u/bkl}$ and $p^+_{cell_i,yld/u/bkl}$, respectively.

6.4 Defining Topology of Sensors for Instrumentation of Observed Cell

To determine which topology should be used at a specific observed cross-section and in associated cell, it is important to analyze potential sources of strain in that cross-section. Expression 5.9 shows that total strain in common materials has up to four main constituents: elastic strain, creep, shrinkage, and thermal strain. Potential presence of all these constituents should be taken into account when determining which topology of sensors should be employed in the identified cross-section.

The first equation of Expressions 5.28 provides the relationship between elastic strain constituent at a given point of observed cross-section of a plane beam and the normal force and bending moment in that cross-section. Figure 5.4 explains in more detail the influence of normal force and bending moment. Normal force in a cross-section results in uniform distribution of normal strain in that cross-section and its axial deformation (without curvature, see Figure 5.4b); thus if only normal forces are present in all cross-sections of an instrumented cell, then they will result in a simple change in length of the cell (without curving) which can be measured by single sensor, i.e., by simple topology.

Bending moment in a cross-section results in linear distribution of normal strain across that cross-section, and its curvature (without axial deformation, see Figure 5.4c); thus if only bending moments are present in all cross-sections of an instrumented cell, then they will result in curving of that cell. Expression 5.30 shows that two sensors parallel to centroid lines (one at the bottom and the other at the top of cross-section) can capture the curving of the cell, and thus parallel topology should be used in this case. If both normal force and bending moment are present in a cross-section, they result in combined axial deformation and curvature of that cross-section (see Figure 5.4a); thus, if both normal force and bending moment are present in cross-sections of an instrumented cell, then that cell will experience both change in length and curving. Expressions 5.29 and 5.30 show that this effect can be also captured by two sensors parallel to centroid line, i.e., by parallel topology.

Creep (if any) is proportional to elastic strain and thus mimics its behavior. Consequently, the above considerations regarding elastic strain, curvature, axial deformation, normal force, and bending moment also apply for creep – it is sufficient to substitute "elastic strain" with "creep" in the above considerations. More precisely, if the elastic strain is constant across the cross-section of a cell, so it is the creep, resulting in change in length of the instrumented cell (without curving), which can be captured by simple topology; if the elastic strain is linear in cross-sections of instrumented cell, so it is the creep, resulting in curving of the cell (i.e., curvature of cross-sections) that can be captured by parallel topology. In other words, the topologies determined from elastic strain analysis can be used to capture the creep.

Shrinkage does not depend on elastic strain and, ideally, it only produces axial deformation, without curvature in an observed cross-section, resulting in change in length of instrumented cell (without its curving). Thus, the topologies determined from elastic strain analysis can be used to capture the shrinkage. However, the presence of rebars in reinforced concrete influences shrinkage, and if the centroid of concrete does not coincide with centroid of rebars throughout the cell, the cell will experience curving, and parallel topology might be more suitable in that case.

Similar to Expressions 5.28, Expressions 5.41–5.43 provide the relationship between thermal strain on one hand, and temperature changes and linear thermal gradients on the other, while Figure 5.9 shows the influence of temperature changes and thermal gradients: temperature changes along the centroid line of the cell, with no transversal gradients, would result in simple change in length of the cell (axial deformation of cross-sections, see Figure 5.9b), while transversal linear gradients produce curving of the cell (curvature in cell's cross-sections, see Figure 5.9c). Thus, simple topology can be used in the former case and parallel topology in the latter.

The last equation of Expression 5.28 provides the relationship between elastic shear strain constituent at a given point of observed cross-section of a plane beam and the shear force in that cross-section. Expression 5.32 provides the relationship between slope of deplanation of the cross-section and the shear force. This slope represents an average angle of shear strain at cross-sectional level (see Figure 5.5). Thus, crossed topology of long-gauge sensors, which will "average" the influence of shear through the cross-section (and along the cell), can be used to assess the slope of deplanation.

As a reminder, in determinate structures rheological strain (creep and shrinkage), temperature changes, and linear thermal gradients do not create stresses and internal forces. Contrary, in indeterminate structures, they do create stresses and internal forces, for example, in an axially constraint beam, a uniform (across cross-section) temperature change will result in a normal force and associated stress in that cross-section, while in a beam with fixed constraints, a transversal linear thermal gradient in a cross-section will result in a bending moment and associated stress in that cross-section.

Single, parallel, and crossed topologies are directly related to internal forces due to loads, and rheological and thermal influences expected to occur in cross-sections of an instrumented cell (e.g., based on structural analysis). Assuming that elastic strain component can be extracted from strain measurements, these topologies can be used to determine the internal forces in instrumented cross-sections (e.g., by using inverse derivations of Expressions 5.28 and 5.32, see also the next sections of this chapter). Triangular topology is not related to internal forces but rather to relative displacement between two cross-sections of the beam. Due to generally small displacements in beam-like structure and limited sensitivity of triangular topology, which depends on angle of sensors and their length, triangular topology is rarely used in beam-like structures, and it is rather applied in monitoring movement of joints or opening of known cracks. That is the reason why triangular topology is not presented in detail in this chapter; however, more information on that topology is given in Glisic and Inaudi (2007). Other three individual topologies, i.e., simple, parallel, and crossed topologies, are presented in more detail in the next subsections.

6.5 Simple Topology and Evaluation of Corresponding Cell Parameters

6.5.1 Implementation of Simple Topology

As stated above, a simple topology (Inaudi and Glisic 2002, Glisic and Inaudi 2007) is used in the cells expected to have, at each point, the largest absolute value of principal strain in the direction parallel to the centroid line, and, at the same time, constant value of strain across each cross-section of the cell (note that strain can vary along the cell length, but not across cross-sections of the cell). Such a state of strain is typically found in uniaxially loaded beam-like structures such as columns, piles, and anchors, and assumes the absence of transversal thermal gradients, while longitudinal thermal gradients may exist. However, even if the cell is uniaxially loaded, it will experience three-axial state of strain due to Poisson's ratio; nevertheless, the maximal absolute value of principal normal elastic strain will be in the direction parallel to centroid line. If the strain is constant across each cross-section of the cell, then no bending occurs and the sensor can be installed parallel to centroid line at virtually any location within cross-section, i.e., not necessarily at the centroid of the cross-section (e.g., see Figures 6.1 and 6.4). Thus, the following equality is assumed to be valid:

$$\varepsilon_{cell,ctr,z,tot,sensor}(t) = \varepsilon_{C_s,z,tot,sensor}(t) \tag{6.4}$$

where $\varepsilon_{cell,ctr,z,tot,sensor}(t)$ denotes the average strain along the centroid line of the cell (centroid of stiffness).

Let us observe a cell of a vertical beam (e.g., part of column or pile) with simple topology shown in Figure 6.4. To have a constant elastic strain distribution across the cross-sections, the cell is typically loaded by constant stresses at the extremities of the cell, denoted with $\sigma_z(z_{0_{cell}})$ and $\sigma_z(z_{L_{cell}})$ as well as by "skin friction" $\tau_{sfr}(z)$ along the perimeters of cross-sections, assuming that resultant of the friction acts at the centroid of each cross-section. Other, more complex load combinations are theoretically possible but are rarely found in practice.

Given that simple topology consists of single sensor, the analysis of sensor measurement at local cell scale is the same as described in Chapter 5, and Criteria I–IV (see Chapter 5) apply directly for the identification of unusual behaviors. However, in addition to that analysis, performed at sensor

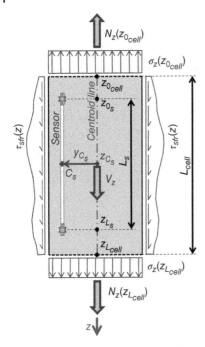

Figure 6.4 Simple topology with typical loading.

level, several other important parameters related to the observed cell can be evaluated from sensor measurements. In this section, few of them are presented: the change in length of the cell (ΔL_{cell}), axial stiffness of cell ($E_{cell}A_{cell}$), and average normal force in the cell ($N_{z,cell}$); however, depending on the type of structure and project requirements, other parameters can be evaluated as well. All these parameters can then be used to evaluate performance and safety of structure by using Criteria V and VI, developed in Section 6.3.

6.5.2 Change in Length of the Cell, $\Delta L_{cell}(t)$

Let us use the notation shown in Figure 6.4. Then, the total change in length of the cell at observed time t is evaluated by assuming that the average total strain $\varepsilon_{C_s,z,tot,sensor}(t)$, measured by sensor, can be extrapolated with sufficient accuracy over entire length of the cell, i.e., as shown in Expression 6.5.

$$\Delta L_{cell,tot,sensor}(t) = \varepsilon_{cell,ctr,z,tot,sensor}(t)L_{cell} = \varepsilon_{C_s,z,tot,sensor}(t)L_{cell} \tag{6.5}$$

The accuracy of the estimation shown in Expression 6.5 depends on the accuracy of the sensor measurement and goodness of the assumption that average strain can be extrapolated over entire length of the cell.

Similar to Expression 6.5, the change in length of a cell due to various strain constituents (elastic, thermal, creep, and shrinkage) can be determined as per Expression 6.6, assuming that these constituents can be separated from the total strain measurement.

$$\Delta L_{cell,E,sensor}(t) = \varepsilon_{cell,ctr,z,E,sensor}(t)L_{cell} = \varepsilon_{C_s,z,E,sensor}(t)L_{cell}$$

$$\Delta L_{cell,T,sensor}(t) = \varepsilon_{cell,ctr,z,T,sensor}(t)L_{cell} = \varepsilon_{C_s,z,T,sensor}(t)L_{cell}$$

$$\Delta L_{cell,c,sensor}(t) = \varepsilon_{cell,ctr,z,c,sensor}(t)L_{cell} = \varepsilon_{C_s,z,c,sensor}(t)L_{cell}$$

$$\Delta L_{cell,sh,sensor}(t) = \varepsilon_{cell,ctr,z,sh,sensor}(t)L_{cell} = \varepsilon_{C_s,z,sh,sensor}(t)L_{cell} \tag{6.6}$$

The accuracy of these expressions depends on accuracies with which the strain constituents can be evaluated from total strain measurement, along with accuracy of assumption that these values can be extrapolated over the entire length of the cell.

6.5.3 Axial Stiffness of the Cell, $E_{cell}A_{cell}(t)$

Axial stiffness of a cell can be determined from strain measurement only if elastic strain constituent can be separated from the total strain measurement and the change in normal force that is related to that strain is known or can be evaluated. For homogeneous materials (e.g., steel) the stiffness of the cell is equal to the product of the Young's modulus of the cell material and average area of the cross-section of the cell, see Expressions 6.7 and 6.8.

$$A_{cell} = \frac{1}{L_{cell}} \int_{L_{cell}} A(z)dz \tag{6.7}$$

$$E_{cell}A_{cell} = E_{cell} \frac{1}{L_{cell}} \int_{L_{cell}} A(z)dz \tag{6.8}$$

where $A(z)$ is area of the cross-section of the cell at coordinate z, assuming that in general case the cell can have variable cross-section.

If the cell has invariable cross-section, then the area of the cell (A_{cell}) is equal to the area of any cross-section of the cell, including the cross-section at the midpoint of the sensor (A_{C_s}). Thus, for homogeneous material and cell with invariable cross-section, Expressions 6.7 and 6.8 transforms into Expressions 6.9 and 6.10, respectively.

$$A_{cell} = A_{C_s} \tag{6.9}$$

$$E_{cell}A_{cell} = E_{cell}A_{C_s} \tag{6.10}$$

However, for inhomogeneous materials, such as reinforced concrete, the stiffness of the cross-section depends on stiffness of the concrete and rebars. Similarly, for prestressed concrete, the stiffness of the cross-section depends on the stiffness of the concrete, rebars, and prestressing cable. In general, in both cases, the stiffness of the cross-section is calculated as shown in Expression 6.11.

$$E_{cell}A_{cell} = \frac{1}{L_{cell}} \int_{L_{cell}} A_{concrete}(z)E_{concrete}(z)dz + E_{rebar} \int_{L_{cell}} A_{rebar}(z)dz + E_{cable} \int_{L_{cell}} A_{cable}(z)dz \tag{6.11}$$

where $A_{concrete}(z)$, $A_{rebar}(z)$, and $A_{cable}(z)$ are individual cross-sectional areas of concrete, rebars, and prestressing cables, respectively, at the cell's cross-section with coordinate z, while $E_{concrete}(z)$, E_{rebar}, and E_{cable} are, respectively, Young moduli of the three materials.

Assuming that in observed cell, the Young's modulus of concrete does not vary significantly, and assuming that the areas of cross-sections of concrete, rebars, and prestressing cables (if any) do not vary significantly over the length of cell, one can consider the cell as having invariable cross-section, and Expression 6.11 transforms into Expression 6.12.

$$E_{cell}A_{cell} = E_{concrete}A_{concrete,C_s} + E_{rebar}A_{rebar,C_s} + E_{cable}A_{cable,C_s} \tag{6.12}$$

where $A_{concrete,C_s}$, A_{rebar,C_s}, and A_{cable,C_s} are individual cross-sectional areas of concrete, rebars, and prestressing cables, respectively, at the cell's cross-section at the midpoint of the sensor.

There are two typical approaches in analyzing stiffness of inhomogeneous material, depending on the project requirements. The first approach considers the Young's modulus of concrete as the Young's modulus for entire cell, E_{cell}. This is useful when the Young's modulus of concrete can be determined independently, e.g., but testing for the purpose-made samples. In that case, Expression 6.12 transforms into Expression 6.13.

$$E_{cell}A_{cell} = E_{concrete}\left(A_{concrete,C_s} + \frac{E_{rebar}}{E_{concrete}}A_{rebar,C_s} + \frac{E_{cable}}{E_{concrete}}A_{cable,C_s}\right)$$

$$\Rightarrow$$

$$E_{cell} = E_{concrete}$$

$$A_{cell} = A_{equ,C_s} = A_{concrete,C_s} + \frac{E_{rebar}}{E_{concrete}}A_{rebar,C_s} + \frac{E_{cable}}{E_{concrete}}A_{cable,C_s} \tag{6.13}$$

where A_{equ,C_s} represents an equivalent area of cross-section at the midpoint of the sensor.

The second approach considers the geometric area of the cell's cross-section as the area for entire cell, A_{cell}. This is useful when the Young's modulus of concrete is not of interest, but composite behavior of the "mix" of materials is of interest. In that case, Expression 6.12 transforms into Expression 6.14.

$$E_{cell}A_{cell} = A_{geometric,C_s}\left(E_{concrete}\frac{A_{concrete,C_s}}{A_{geometric,C_s}} + E_{rebar}\frac{A_{rebar,C_s}}{A_{geometric,C_s}} + E_{cable}\frac{A_{cable,C_s}}{A_{geometric,C_s}}\right)$$

$$\Rightarrow$$

$$A_{cell} = A_{geometric,C_s}$$

$$E_{cell} = E_{equ,C_s} = E_{concrete}\frac{A_{concrete,C_s}}{A_{geometric,C_s}} + E_{rebar}\frac{A_{rebar,C_s}}{A_{geometric,C_s}} + E_{cable}\frac{A_{cable,C_s}}{A_{geometric,C_s}} \tag{6.14}$$

where $A_{geometric,C_s}$ represents geometrical area of cross-section at the midpoint of the sensor and E_{equ,C_s}, an equivalent Young's modulus (modulus of elasticity) of the cell; note that $A_{geometric,C_s} = A_{concrete,C_s} + A_{rebar,C_s} + A_{cable,C_s}$.

Let us assume that the change in normal force at time t, $\Delta N_{z,cell}(t)$, with respect to a time t_0 is known, as well as the change of elastic strain constituent measured by sensor, $\Delta \varepsilon_{C_s,z,E,sensor}(t)$, i.e.,

$$\Delta N_{z,cell}(t) = N_{z,cell}(t) - N_{z,cell}(t_0)$$
$$\Delta \varepsilon_{cell,ctr,z,E,sensor}(t) = \Delta \varepsilon_{C_s,z,E,sensor}(t)$$
$$= \varepsilon_{cell,ctr,z,E,sensor}(t) - \varepsilon_{cell,ctr,z,E,sensor}(t_0) = \varepsilon_{C_s,z,E,sensor}(t) - \varepsilon_{C_s,z,E,sensor}(t_0). \tag{6.15}$$

Then, the axial stiffness of the cell can be calculated using Expression 6.16.

$$E_{cell}A_{cell} = \frac{\Delta N_{z,cell}(t)}{\Delta \varepsilon_{cell,ctr,z,E,sensor}(t)} = \frac{\Delta N_{z,cell}(t)}{\Delta \varepsilon_{C_s,z,E,sensor}(t)} \tag{6.16}$$

Note that if multiple increments of normal force $\Delta N_{z,cell}(t_i)$ are known and corresponding changes in elastic strain $\Delta \varepsilon_{C_s,z,E,sensor}(t_i)$ are measured ($i = 1, 2, \ldots n$), then the axial stiffness of the cell can be evaluated by using common statistical methods such as averaging of values obtained from individual increments or linear regression (more accurate). Once the axial stiffness of the cell is evaluated, then equivalent area of cross-section or equivalent Young's modulus can be evaluated by Expressions 6.10, 6.12, 6.13, or 6.14, whichever applies.

6.5.4   Average Normal Force in the Cell, $N_{z,cell}(t)$

Let us define average normal force in the cell as the value of normal force in the cross-section at the midpoint of the sensor. Then the change in average normal force $\Delta N_{z,cell}$, at observed time t with respect to reference time t_{Ref}, is evaluated by multiplying the average elastic strain measured by sensor, $\varepsilon_{C_s,z,E,sensor}(t)$, with the stiffness of the cross-section at the midpoint of the sensor gauge length, $E_{cell}A_{cell}$, i.e., as shown in Expression 6.17.

$$\Delta N_{z,cell,sensor}(t) = \varepsilon_{cell,ctr,z,E,sensor}(t)E_{cell}A_{cell} = \varepsilon_{C_s,z,E,sensor}(t)E_{cell}A_{cell} \tag{6.17}$$

where $E_{cell}A_{cell}$ is the stiffness of the cell, which is assumed to be known or evaluated from measurements.

Average normal force in cell, $N_{z,cell}$, is then evaluated from the sensor measurement as shown in Expression 6.18 (see also Expressions 5.84 and 6.3).

$$\begin{aligned} N_{z,cell,sensor}(t) &= \varepsilon_{cell,ctr,z,E,sensor}(t)E_{cell}A_{C_s} + N_{z,cell,model}(t_{Ref}) \\ &= \varepsilon_{C_s,z,E,abs}(t)E_{cell}A_{C_s} + N_{z,cell,model}(t_{Ref}) = \Delta N_{z,cell,sensor}(t) + N_{z,cell,model}(t_{Ref}) \end{aligned} \tag{6.18}$$

Note that Expressions 6.17 and 6.18 assume that elastic strain $\varepsilon_{C_s,z,E,sensor}(t)$ can be separated from the total strain measurement $\varepsilon_{C_s,z,tot,sensor}(t)$, which might be challenging in the case of concrete structures (including reinforced and prestressed concrete structures) due to rheological effects, as discussed in Chapter 5. In addition, Expression 6.18 assumes that model for estimation of $N_{z,cell,model}(t_{Ref})$ can be accurately created.

6.5.5   Examples of Analysis of Simple Topology Sensor Measurements and Their Derivatives

To illustrate the analysis of results of measurements from simple topology, concise, averaged data of one of pull-out (axial tension) tests, as described in Section 6.2.2, are used. Pull-out force was applied by hydraulic jacks and the value of the force (expressed in tons) is given in the second column of Table 6.1. Limits of error of the applied force were estimated to ±0.05 ton (±0.490 kN), based on specifications of the hydraulic jack. Strain was measured by all sensors shown in Figure 6.2, but to keep this example concise, only the average values of the two sensors installed in Cell 1 (see Figure 6.2) are presented in the third column of Table 6.2. Assuming that Bernoulli hypothesis is valid, and considering that the sensors were symmetrically placed with respect to the centroid line, these average values can be considered as measurements of single virtual sensor installed at the centroid of the cell, i.e., as measurements of single topology with a virtual sensor installed at the centroid. Measurements were repeated four times for each load step (at 0, 2, 5, and 10 minutes from the moment of load application), but only truncated mean values are given in the table (i.e., min. and max. values were excluded from the mean), to simplify analysis in this illustrative example; the analysis that includes full set of data is more complex and informative, but it exceeds the scope of this example (for details, see Glisic et al. 2002b, Glisic and Inaudi 2007). Standard uncertainty of these values was estimated to 0.5 µε (SOFO sensors, see Chapter 3) and limits of error were set at ±1 µε (two standard uncertainties). In the analysis that follows, the geometrical cross-section of the cell was adopted as the cross-section of the cell, i.e., for pile diameter of 1.2 m, $A_{cell} = \pi \times 1.2^2/4 = 1.131 \text{ m}^2$, with error estimated to ±0.019 m² (taking into account common tolerance in concrete dimensions of 1 cm). The average axial stress in Cell 1 was modeled using Expression 6.19 and presented in the fourth column of Table 6.1.

Table 6.1 Averaged data from Cell 1 taken during the pull-out test of pile.

Load step i	Load $m_{load,i}$ (t)	Average strain $\varepsilon_{cell_1,z,E,sensor,i}$ (µε)	Average stress $\sigma_{z,cell_1,abs,i}$ (MPa)
0	0	0.0	0.000
1	28.6	7.5	0.248
2	57.1	12.6	0.495
3	85.7	18.4	0.743
4	114.3	22.9	0.991
5	142.9	29.9	1.240
6	171.4	34.9	1.487
7	200.0	44.1	1.735
8	228.6	54.5	1.983
9	257.1	63.1	2.230
10	285.7	70.5	2.478
11	314.3	113.1	2.726
12	342.9	185.1	2.974
13	371.4	330.8	3.222
14	400.0	445.6	3.470
15	400.0	459.8	3.470
16	300.0	394.3	2.602
17	200.0	310.6	1.735
18	100.0	206.3	0.867
19	0	95.1	0.000
20	0	89.7	0.000

Table 6.2 Parameters of regression functions fitting the stress–strain relationship.

Load steps	Load trend	Color code	Fitting function $\sigma(\varepsilon(µε))$ (MPa)	Coefficient of determination R^2	Young's modulus (slope of fitting function) E_{cell_1} (GPa)	Uncertainty in stress estimation using fitting function $u(\sigma(\varepsilon(µε)))$ (MPa)
1–10	Increasing	Blue or gray	$0.035195\varepsilon + 0.092664$	0.986	35.2	0.096
10–15	Increasing	Red or dark gray	$0.002358\varepsilon + 2.425870$	0.965	2.4	0.052
14–20	Decreasing	Green or light gray	$0.009463\varepsilon - 0.970867$	0.987	9.5	0.226

Note that Expression 6.19 assumes that friction between the soil and the first cell can be neglected (see Section 6.2.2). In addition, given that pile was not loaded before the test, the estimated stress was considered absolute. Based on measures of accuracy presented above, the limits of relative error for stress were estimated to ±1.8% (not derived here, to keep the presentation concise and focused).

$$\sigma_{z,cell_1,abs,i,model} = \sigma_{z,cell_1,abs,model}(t_i) = \frac{N_{z,cell_1}(t_i)}{A_{cell_1}} = \frac{F_{load}(t_i)}{A_{cell_1}} = \frac{m_{load,i}g}{A_{cell_1}} \qquad (6.19)$$

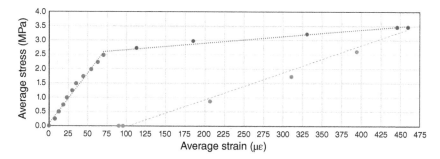

Figure 6.5 Stress–strain relationship constructed by using data from Table 6.1; colors in the figure show load progression – increase (blue and red markers, load steps 0–15) and decrease (green markers, load steps 14–20).

where $F_{load}(t_i)$ is the load at time stamp t_i (i.e., load step i), $m_{load,i}$ is applied force in terms of mass, and g is gravity acceleration, $g = 9.81$ m/s².

The test was conducted over five hours. Consequently, the measured strain change in the pile could be considered approximately equal to elastic constituent: shrinkage could be neglected as it could not significantly develop over such a short period, and thermal strain could be neglected as well (pile buried underground and thus protected from local thermal variations); the only other strain constituent that could evolve over five hours was creep, but its contribution over such a short period of time was very small and perceivable only for the highest applied load (see Figure 6.5). Thus, we assume that:

$$\varepsilon_{cell_1,z,E,sensor,i} = \varepsilon_{cell_1,z,E,sensor}(t_i) = \varepsilon_{cell_1,z,tot,sensor}(t_i) = \varepsilon_{cell_1,z,tot,sensor,i} \tag{6.20}$$

where subscript "$cell_1$" denotes average strain at the centroid of Cell 1 (see Figure 6.2).

Given the above assumptions, a stress–strain relationship in Cell 1 can be established and several conclusions regarding the pile's performance and condition from the analysis of this relationship can be carried out, such as equivalent modulus of elasticity (Young's modulus) of the cell, cracking strain, tensional strength of concrete, and load bearing capacity.

Example 6.1 *Evaluation of Equivalent Young Modulus*

The equivalent modulus of elasticity of the concrete in Cell 1 can be found by observing stress–strain relationship presented in Table 6.1. Figure 6.5 shows this relationship along with piece-wise linear regressions that fit this relationship.

Parameters of fitting functions are given in Table 6.2.

Three distinguished behaviors are noticed in the figure:

– For strain up to approximately 70.5 με (steps 0–10, colored in blue or gray in Figure 6.5), the slope of the fitting function corresponds to the Young's modulus of approximately 35.2 GPa; this value was consistent with expected (design) value.
– For strain higher than approximately 70.5 με and increasing load (steps 11–15, colored in red or dark gray in Figure 6.5), the slope of the fitting function corresponds to the Young's modulus of approximately 2.4 GPa; this very low value indicated that concrete is severely cracked and only the rebars, which were then put in high tension, were carrying almost entire load; this behavior was indicative of yielding of reinforced concrete; note that meaning of "average stress" in the cell would mean "if geometrical cross-section was intact," i.e., this is only apparent value as it considers entirety of the cross-section; true stress in rebars was much higher as the same load

was distributed over the cross-sectional area of rebars only (as opposed to entire geometrical section); finally, due to excessive cracking and yielding, one can conclude that change of the slope between blue and red points (or gray and dark gray points) in Figure 6.5 indicates failure of Cell 1 of the pile.

– For decreasing load (steps 14-20, colored in green or light gray in Figure 6.5), the Young's modulus increased; also, plastic deformation occurred, as the strain did not drop to zero after removal of the force; both these effects can be explained by imperfect closing of the cracks after the removal of the load; consequently, the accuracy of this complex post-failure behavior is difficult to evaluate; note that the measurements in load steps 14–15 were performed with the same load but at different times (their mutual difference is due to creep visible for this high load) thus they are both included in analysis of decreasing load.

Based on the above discussion, one can conclude that the Young's modulus of Cell 1 before the failure was approximately 35.2 GPa. However, if only that value is used to calculate stresses instead of fitting function, then the uncertainty in evaluation of stress is slightly higher, i.e., $u(\sigma) = 0.114$ MPa. Failure of Cell 1 occurred once the concrete tensional strength was reached, i.e., at Load step 11. This means that values related to Load step 10 define the load-bearing capacity of the pile, as well as ultimate tensional strain in concrete and tensional strength of the concrete. These values are, respectively, 285.7 tons, 70.5 ± 1 με, and 2.48 ± 0.05 MPa.

Example 6.2 *Evaluation of Normal Force*
The knowledge gathered from the analysis of Cell 1 can now be applied to examine some other parameters that can be derived from simple topologies. One example is the evaluation of normal force in each cell. Only Cell 3 is presented here for illustrative purposes, but similar approach can be applied to all other cells. The third column of Table 6.3 presents the strain measured in Cell 3 during the test, while the fourth column presents normal force evaluated by substituting the Young's modulus, evaluated in Table 6.2, into Expression 6.18, for each load step. In Cell 3, cracking occurred just before the strain reached 70 με, thus based on analysis of Cell 1 and taking into account the error of measurements of ± 1 με, it was assumed that cracking occurred at 69.5 με. Consequently, for the first 12 load steps, the value of Young's modulus used in evaluation of normal force was 35.2 GPa, for steps 13–15 it was 2.4 GPa, and for decreasing load steps 16–20, it was 9.5 GPa (see Table 6.2).

Note that accuracy (confidence) in post-failure evaluation of normal force (load steps 16–20) cannot be ascertained as the post-failure behavior is influenced by friction with soil, in addition to increase of stiffness and plastic deformation due to the imperfect closing of cracks.

At this point, it is important to note that, while assumptions presented above, along with Expressions 6.19 and 6.20, are reasonably valid based on engineering experience, it is very challenging and often impossible to assess their accuracy. This is a typical example of an unknown epistemic error in the data analysis. For example, by observing Figure 6.5, one can notice that linear regression might not be the best fit for the three parts of the graph; indeed, bi-liner regression might be better suited for the steps 1-10 in the graph ("blue or gray" part of the graph), while parabolic regression would better fit for each of the steps 11–15 and 14–20 in the graph ("red or dark gray" and "green or light gray" parts of the graph). The difference from linear regressions (which was assumed in this section for illustrative purposes) is due to small, yet visible, creep that was neglected. This is confirmed by at least two facts: (i) the measurements in load steps 14–15 and 19–20, were performed with the same load but at different times and due to creep they feature strain reading

Table 6.3 Data from Cell 3 taken during the pull-out test of pile and evaluated normal force.

Load step i	Load $m_{load,i}$ (t)	Average strain $\varepsilon_{cell_3,z,E,sensor,i}$ ($\mu\varepsilon$)	Normal force $N_{z,cell_3,i}$ (kN)
0	0	0.0	0
1	28.6	4.0	146
2	57.1	6.8	246
3	85.7	10.0	364
4	114.3	12.5	455
5	142.9	16.5	601
6	171.4	19.5	710
7	200.0	25.4	924
8	228.6	32.1	1170
9	257.1	37.6	1370
10	285.7	42.8	1557
11	314.3	47.5	1730
12	342.9	54.9	1998
13	371.4	70.0	2532
14	400.0	72.9	2540
15	400.0	120.3	2669
16	300.0	104.5	2500
17	200.0	83.6	2275
18	100.0	56.6	1985
19	0	23.4	1628
20	0	21.8	1610

clearly out of error limits of measurement system; and (ii) the measurements of all cells show the same bi-linear behavior for load steps 1–10 despite the fact that stresses in lower cells are smaller than in the first cell. For comparison purposes, observe the values shown in Figure 6.2 for Cell 3; they were calculated using bilinear stress–strain relationship, as opposed to linear relationship used in Table 6.3; note the difference in normal force in Cell 3 due to load steps 10 and 15 – this difference is a consequence of the different regression used in determining stress–strain relationship.

Example 6.3 *Evaluation of Change in Length of a Cell*

Final example of parameter derived from simple topology in this subsection is the change in length of the cells by using Expression 6.5 (see Section 6.5.2). To illustrate evaluation of this parameter, Table 6.4 presents the total strain measurement (second, fourth, and sixth column in the table) taken at characteristic load steps 10, 15, and 20, where the loads of 285.7 t (corresponding to capacity of the pile), 400 t (the highest load applied to the pile), and 0 t (the force removed from the pile) were applied, respectively. The third, fifth, and seventh columns of Table 6.4 show the change in length $\Delta L_{cell_j,tot,sensor}$ of each cell j ($j = 1, 2, \ldots 8$) for given load, evaluated using Expression 6.5.

Table 6.4 Evaluation of the change in length of Cells 1–8 and evaluation of vertical displacement of the pile.

Cell j	Load step 10 Load = 285.7 t		Load step 15 Load = 400.0 t		Load step 20 Load = 0.0 t	
	$\varepsilon_{cell_j,tot,sensor}$ (µε)	$\Delta L_{cell_j,tot,sensor}$ (mm)	$\varepsilon_{cell_j,tot,sensor}$ (µε)	$\Delta L_{cell_j,tot,sensor}$ (mm)	$\varepsilon_{cell_j,tot,sensor}$ (µε)	$\Delta L_{cell_j,tot,sensor}$ (mm)
1	70.5	0.30	459.8	1.95	89.7	0.38
2	61.8	0.26	305.8	1.30	68.3	0.29
3	42.8	0.18	120.3	0.51	21.8	0.09
4	29.7	0.13	52.9	0.22	6.7	0.03
5	24.5	0.10	43.7	0.19	5.2	0.02
6	15.8	0.07	28.1	0.12	3.4	0.01
7	7.0	0.03	12.6	0.05	1.5	0.01
8	2.2	0.01	4.2	0.02	−0.9	0.00
Displacement of top of the pile, v_{sensor} (mm)	**1.08**		**4.37**		**0.83**	
LVDT (mm)	**1.16**		**4.07**		**1.08**	

As mentioned above, the sensor gauge length was 4 m and length of each cell was 4.25 m. Limits of error in the evaluation were conservatively evaluated to approximately ±0.005 mm (limits of error of the sensor measurement was 1 µε).

In addition to evaluation of change in the length of each cell, the displacement of the top of the pile could be evaluated as well. The measured strain in the bottom cell (Cell 8) for all load cases was very low (ranged between −0.9 and 4.2 µε), meaning that the pullout force did not significantly affect the bottom cell (Cell 8), which is further confirmed by the evaluation of the normal force and friction at the bottom of the pile (see Figure 6.2). This, combined with the fact that the measured strain represents an average value along the length of the pile, leads to a reasonable assumption that the bottom of the pile did not move significantly up during the test. If this assumption holds, then the displacement of the top of the pile can be calculated by simple summation of changes in length of all cells along the pile. The result is given in the row before the last in Table 6.4, denoted with v_{sensor}. The limits of error of this estimation were conservatively calculated as $8 \times 0.005 = \pm 0.04$ mm. During the tests, vertical displacement of the top of the pile was independently measured by the two linear variable displacement transformers (LVDTs), placed symmetrically with respect to the centroid of the top cross-section. Their limits of error were estimated to ±0.25 mm. For the comparison, the average value of their measurements is given in the last row of Table 6.4 (denoted with LVDT). Comparison shows that the difference between displacement evaluated using the strain sensors and LVDTs is approximately within cumulative limits of error of the two measurement systems (±0.29 mm), confirming the validity of hypothesis that the bottom of the pile does not move or moves insignificantly.

More information about the use of simple topology for global-scale monitoring is given in the literature (e.g., see Glisic et al. 2002b, Glisic and Inaudi 2007) and Section 6.8.2.

6.6 Parallel Topology and Evaluation of Corresponding Cell Parameters

6.6.1 Implementation of Parallel Topology

A parallel topology (Inaudi and Glisic 2002, Glisic and Inaudi 2007) is used to instrument cells of beam-like structures which experience linear distribution of strain across their cross-sections. This typically happens when bending moments or transversal thermal gradients are applied to the cell. In general case, these loads can act spatially, i.e., the cell can be exposed to biaxial bending or thermal gradients acting along both principal axes of the cell's cross-sections. However, based on linear beam theory, these influences can be separately analyzed as planar and then superposed. Consequently, the scope of this section is limited to planar (uniaxial) bending case; expressions for biaxial bending analysis can be derived using approaches similar to those presented for uniaxial bending (e.g., see more detail in Glisic and Inaudi 2007).

For uniaxial bending case, parallel topology consists of two sensors with equal gauge lengths, installed parallel to the centroid line of the cell, in the plane of bending, as shown in Figure 6.6.

Given that parallel topology consists of two single sensors, the analysis of measurements of each sensor at local cell scale is the same as described in Chapter 5, and Criteria I–IV (see Chapter 5) apply directly for the identification of unusual behaviors. However, in addition to that analysis, performed at sensor level, several important parameters related to the observed cell can be evaluated from combined sensor measurements, such as (but not limited to) average curvature and change in relative rotation, average locations of neutral axis and centroid of stiffness, average strain at the centroid and the change in length of the cell, average axial and bending stiffness of a cell, and average normal force and bending moment in the cell. These parameters can then be used to evaluate performance and safety of structure by using Criteria V and VI, developed in Section 6.3.

6.6.2 Average Curvature $\kappa_{cell,x,sensor}$ and Change in Relative Rotation $\Delta\varphi_{cell,x,sensor}$

Let us use notation shown in Figure 6.6. Then, the average total curvature of the cell at observed time t is given in Expression 6.21 (see also Expressions 5.30 and 5.42)

$$\kappa_{x,cell,tot,sensor}(t) = \frac{\varepsilon_{cell,z,tot,sensor,btm}(t) - \varepsilon_{cell,z,tot,sensor,top}(t)}{h} \tag{6.21}$$

where subscripts "*btm*" and "*top*" denote sensors at the bottom and the top of cell, respectively.

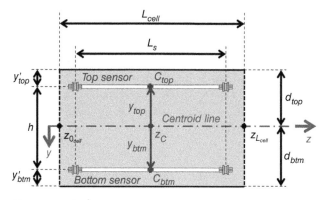

Figure 6.6 Parallel topology in uniaxial bending case.

Assuming that strain constituents can be separated from the total strain measurements (see Section 5.8), the following expressions can be derived for the curvature constituents, in general case:

$$\kappa_{x,cell,E,sensor}(t) = \frac{\varepsilon_{cell,z,E,sensor,btm}(t) - \varepsilon_{cell,z,E,sensor,top}(t)}{h} \tag{6.22}$$

$$\kappa_{x,cell,T,sensor}(t) = \frac{\varepsilon_{cell,z,T,sensor,btm}(t) - \varepsilon_{cell,z,T,sensor,top}(t)}{h} = \alpha_{T,z}\frac{\Delta T_{cell,sensor,btm}(t) - \Delta T_{cell,sensor,top}(t)}{h} \tag{6.23}$$

$$\kappa_{x,cell,c,sensor}(t) = \frac{\varepsilon_{cell,z,c,sensor,btm}(t) - \varepsilon_{cell,z,c,sensor,top}(t)}{h} \tag{6.24}$$

$$\kappa_{x,cell,sh,sensor}(t) = \frac{\varepsilon_{cell,z,sh,sensor,btm}(t) - \varepsilon_{cell,z,sh,sensor,top}(t)}{h} \tag{6.25}$$

where $\Delta T_{cell,sensor,btm}(t)$ and $\Delta T_{cell,sensor,top}(t)$ denote temperature change at time t, at the locations of bottom and top strain sensors, respectively, measured or evaluated using temperature sensors; Expression 6.23 assumes linear distribution of the thermal strain across the cross-sections of the cell (see Section 5.6 and Expression 5.42).

Note that, depending on construction material, some of the constituents might be null (e.g., curvatures due to creep and shrinkage are both null in cells of steel beams, but they might exist in cells of concrete beams).

Expression 6.25 shows the curvature due to shrinkage. This curvature will be generated in unreinforced concrete if the drying conditions at the top and the bottom of the cross-section are not the same. Curvature due to shrinkage might also be generated in reinforced and prestressed concrete even if the drying conditions at the top and the bottom of the cell are identical; this might happen if the centroid of stiffness of rebars in reinforced concrete, or the centroid of stiffness of the rebars and the prestressing cables combined, in prestressed concrete, does not coincide with the centroid of stiffness of the cell. However, in the cases where the drying conditions are similar enough at the top and the bottom of the cross-section, and the centroid of stiffness of rebars, or rebars with prestressing cable, approximately coincide with the centroid of stiffness of the cell, the curvature due to shrinkage can be neglected. If, in addition, the change in temperature is measured or evaluated at locations of sensors, the following relationship applies:

$$\kappa_{x,cell,E+c,sensor}(t) = \kappa_{x,cell,tot,sensor}(t) - \kappa_{x,cell,T,sensor}(t)$$
$$= \frac{(\varepsilon_{cell,z,tot,sensor,btm}(t) - \alpha_{T,z}\Delta T_{cell,sensor,btm}(t)) - (\varepsilon_{cell,z,tot,sensor,top}(t) - \alpha_{T,z}\Delta T_{cell,sensor,top}(t))}{h} \tag{6.26}$$

Assuming that the average total curvature $\kappa_{cell,x,tot,sensor}(t)$, measured by sensors of parallel topology, can be extrapolated with sufficient accuracy over the entire length of the cell, the change in relative rotation, $\Delta\varphi_{cell,tot,sensor}(t)$, between the first cell's cross-section (the smallest coordinate $z_{0_{cell}}$), and the last cell's cross-section (the greatest coordinate $z_{L_{cell}}$) can be evaluated using Expression 6.27 (see Figures 5.4 and 6.5).

$$\Delta\varphi_{cell,x,tot,sensor}(t) = -\kappa_{cell,x,tot,sensor}(t)L_{cell} \tag{6.27}$$

The accuracy of the estimation shown in Expression 6.27 depends on the accuracy of the sensor measurements and goodness of the assumption that average curvature can be extrapolated over the entire length of the cell.

Similar to Expression 6.27, the change in relative rotation between the extremities of a cell due to various curvature constituents (elastic, thermal, creep, and shrinkage) can be determined as per Expressions 6.28, assuming that these constitutes can be evaluated (separated) from the total strain measurements.

$$\Delta\varphi_{cell,x,E,sensor}(t) = -\kappa_{cell,x,E,sensor}(t)L_{cell}$$

$$\Delta\varphi_{cell,x,T,sensor}(t) = -\kappa_{cell,x,T,sensor}(t)L_{cell}$$

$$\Delta\varphi_{cell,x,c,sensor}(t) = -\kappa_{cell,x,c,sensor}(t)L_{cell}$$

$$\Delta\varphi_{cell,x,sh,sensor}(t) = -\kappa_{cell,x,sh,sensor}(t)L_{cell} \tag{6.28}$$

The accuracy of these expressions depends on accuracies with which the strain and curvature constituents can be evaluated from total strain measurements, along with accuracy of assumption that these values can be extrapolated over the entire length of the cell.

6.6.3 Average Locations of Neutral Axis $d_{NA,cell,btm}(t)$ and $d_{NA,cell,top}(t)$ and Centroid of Stiffness $d_{cell,btm}(t)$ and $d_{cell,top}(t)$

Neutral axis is geometrical set of points in a beam cross-section where the normal stress and corresponding normal elastic strain are null. In the case of uniaxial bending of beams, neutral axis is orthogonal to bending plane (parallel to axis of bending), and thus, it can be represented by a point at which it intersects the bending plane (see Figure 5.4a). Note that neutral axis is defined through stress and elastic strain, thus, it can only be determined if elastic strain can be separated from the total strain measurement. Assuming that this is possible, the average location of the neutral axis in a cell can be derived as follows (see Figures 5.4a and 6.6 for notation):

$$
\begin{aligned}
d_{NA,cell,btm,sensor}(t) &= \frac{\varepsilon_{cell,z,E,sensor,btm}(t) + \kappa_{x,cell,E,sensor}(t)y'_{btm}}{\kappa_{x,cell,E,sensor}(t)} \\
&= \frac{\varepsilon_{cell,z,E,sensor,btm}(t)\left(h + y'_{btm}\right) - \varepsilon_{cell,z,E,sensor,top}(t)y'_{btm}}{\varepsilon_{cell,z,E,sensor,btm}(t) - \varepsilon_{cell,z,E,sensor,top}(t)}
\end{aligned} \tag{6.29a}
$$

$$
\begin{aligned}
d_{NA,cell,top,sensor}(t) &= \frac{-\varepsilon_{cell,z,E,sensor,top}(t) + \kappa_{x,cell,E,sensor}(t)y'_{top}}{\kappa_{x,cell,E,sensor}(t)} \\
&= \frac{\varepsilon_{cell,z,E,sensor,btm}(t)y'_{top} - \varepsilon_{cell,z,E,sensor,top}(t)\left(h + y'_{top}\right)}{\varepsilon_{cell,z,E,sensor,btm}(t) - \varepsilon_{cell,z,E,sensor,top}(t)}
\end{aligned} \tag{6.29b}
$$

where $d_{NA,cell,btm,sensor}(t)$ and $d_{NA,cell,top,sensor}(t)$ are distances of the average cell neutral axis with respect to the bottom and the top of the cell, respectively (see Figure 5.4a).

Figure 5.4a shows that, in general case, the location of the neutral axis does not pass through the centroid of stiffness. However, the same figure shows that in the absence of the normal force, i.e., if only bending moment is present in a cell, then, the location of the neutral axis will pass through the centroid, i.e., in the case of planar beams it will coincide with the centroid of stiffness. Thus, Equation 6.29 can be used to evaluate the location of the centroid of stiffness, with condition that the normal force in the observed cell is null. Figure 5.4a also shows that the opposite implication is also valid, i.e., if the neutral axis coincides with the centroid, then the normal force in the cell is null. This is pointed out in Equation 6.30.

$$\text{if and only if } N_{z,cell}(t) = 0 \text{ then } d_{cell,btm}(t) = d_{NA,cell,btm}(t) \text{ and } d_{cell,top}(t) = d_{NA,cell,top}(t) \tag{6.30}$$

where $d_{cell,btm}(t)$ and $d_{cell,top}(t)$ are distances of the average cell centroid of stiffness measured from the bottom and the top of the cell, respectively (see Figures 5.4a and 6.6).

If the normal force $N_{z,cell}(t)$ in the cell is different from zero, but known (e.g., measured using load cell) along with the axial stiffness of the cell $E_{cell}A_{cell}(t)$ (e.g., calculated based on geometrical properties of the cross-section and estimated Young's modulus), then the elastic strain at centroid can be evaluated as $\varepsilon_{cell,z,E,ctr}(t) = N_{z,cell}(t)/(E_{cell}A_{cell}(t))$, and expressions for evaluation of the location of centroid can be expressed as follows:

$$d_{cell,btm,sensor}(t) = \frac{\varepsilon_{cell,z,E,sensor,btm}(t) - \varepsilon_{cell,z,E,ctr}(t) + \kappa_{x,cell,E,sensor}(t)y'_{btm}}{\kappa_{x,cell,E,sensor}(t)}$$

$$= \frac{\varepsilon_{cell,z,E,sensor,btm}(t)\left(h + y'_{btm}\right) - \varepsilon_{cell,z,E,sensor,top}(t)y'_{btm} - \varepsilon_{cell,z,E,ctr}(t)h}{\varepsilon_{cell,z,E,sensor,btm}(t) - \varepsilon_{cell,z,E,sensor,top}(t)} \tag{6.31a}$$

$$d_{cell,top,sensor}(t) = \frac{-\varepsilon_{cell,z,E,sensor,top}(t) + \varepsilon_{cell,z,E,ctr}(t) + \kappa_{x,cell,E,sensor}(t)y'_{top}}{\kappa_{x,cell,E,sensor}(t)}$$

$$= \frac{\varepsilon_{cell,z,E,sensor,btm}(t)y'_{top} - \varepsilon_{cell,z,E,sensor,top}(t)\left(h + y'_{top}\right) + \varepsilon_{cell,z,E,ctr}(t)h}{\varepsilon_{cell,z,E,sensor,btm}(t) - \varepsilon_{cell,z,E,sensor,top}(t)} \tag{6.31b}$$

Determination of location of the neutral axis at observed time requires separation of the elastic strain from total strain measurement, which has number of challenges as presented in Chapter 5. Determination of the centroid of stiffness faces the same challenges, in addition to the challenge that the normal force $N_{z,cell}(t)$ and the average stiffness $E_{cell}A_{cell}(t)$ of the cross-section must be known. However, if the beam has boundaries that are free to move in axial direction, which is often (but not always) the case for bridge decks, then normal force due to load applied in transversal direction is zero, and if a very short-term transversal load is applied, the location of the centroid of stiffness can be evaluated from measurements registered during this very short-term loading period, assuming that only elastic strain constituent is generated during this time, i.e., other strain constituents, such as thermal strain, creep, and shrinkage, could not develop due to shortness of the time during which the load is applied. Typical examples are vibration of the bridge deck after a passage of a heavy track that occurs over few seconds, or removal of the formworks after construction of the deck, where the deadload is suddenly "activated" (e.g., see Sigurdardottir and Glisic 2013). Another example is controlled load-testing of the structure (also see Sigurdardottir and Glisic 2013).

Location of the centroid of stiffness can be estimated from geometrical and mechanical properties of cell, using analytical expressions or numerical modeling, and compared with the location evaluated from measurements as described above in Expressions 6.31. It is important to note that when estimating the location from the analytical expression and numerical modeling of reinforced concrete, prestressed concrete, or composite steel–concrete cross-section, axial stiffness of each involved material has to be taken into account. This is shown in Expression 6.32 for cells with invariable cross-section.

$$d_{cell,btm,model} = \frac{E_{concrete}A_{concrete}d_{cell,btm,concrete} + E_{rebar}A_{rebar}d_{cell,btm,rebar} + E_{cable}A_{cable}d_{cell,btm,cable}}{E_{concrete}A_{concrete} + E_{rebar}A_{rebar} + E_{cable}A_{cable}}$$

$$\tag{6.32a}$$

$$d_{cell,top,model} = \frac{E_{concrete}A_{concrete}d_{cell,top,concrete} + E_{rebar}A_{rebar}d_{cell,top,rebar} + E_{cable}A_{cable}d_{cell,top,cable}}{E_{concrete}A_{concrete} + E_{rebar}A_{rebar} + E_{cable}A_{cable}}$$

$$\tag{6.32b}$$

where $d_{cell,btm,concrete}$, $d_{cell,btm,rebar}$, and $d_{cell,btm,cable}$ are distances of centroids of the areas of concrete, rebars, and prestressing cable with respect to bottom of the cross-section, respectively; $d_{cell,top,concrete}$, $d_{cell,top,rebar}$, and $d_{cell,top,cable}$ are distances of centroids of the areas of concrete, rebars, and prestressing cable with respect to top of the cross-section, respectively; and $d_{cell,btm,model}$ and $d_{cell,top,model}$ are distances of the average centroid of stiffness with respect to the bottom and the top of the cell, respectively.

Location of the centroid of stiffness can be used as a damage-sensitive feature, as, in general case, the change in its location indicates change in geometrical properties related to damage or deterioration (e.g., loss of cross-sectional area or change in moment of inertia). In materials without discontinuities, such as steel or prestressed concrete, the properties of undamaged cross-section are not expected to change in time and, consequently, the location of the centroid of stiffness is not expected to change in time, unless the damage occurs. For these materials, the use of centroid of stiffness as a damage-sensitive feature is relatively straightforward. However, in materials with discontinuities, such as reinforced concrete, geometrical properties of cross-section can change in time due to usual (normal) cracking, which does not represent the damage; thus, the use of the centroid of stiffness as a damage-sensitive feature, in this type of material should be used with caution, in order to distinguish usual (normal) changes from the unusual changes due to damage or deterioration. Detailed description of the method that uses the neutral axis and the centroid of the stiffness as the damage-sensitive feature is given in Sigurdardottir and Glisic (2013).

6.6.4 Average Strain at the Centroid $\varepsilon_{cell,ctr,sensor}(t)$ and the Change in Length of the Cell $\Delta L_{cell}(t)$

Let us keep using notation shown in Figure 6.6 and assume that the location of the centroid of stiffness of the cell is known, e.g., estimated analytically or numerically from geometrical and mechanical properties of the cross-section, or evaluated using the method shown in Section 6.6.3. Then, the average total strain at the centroid of stiffness of the cell at observed time t is given in Expression 6.33 (see also Expressions 5.29 and 5.41).

$$\varepsilon_{cell,ctr,z,tot,sensor}(t) = -\varepsilon_{cell,z,tot,sensor,btm}(t)\frac{y_{top}}{h} + \varepsilon_{cell,z,tot,sensor,top}(t,z)\frac{y_{btm}}{h} \tag{6.33}$$

Note that in Expression 6.33, y_{top} and y_{btm} are negative if above the centroid (see Figure 6.6).

Assuming that strain constituents can be separated from the total strain measurements (see Section 5.8), the following expressions can be derived for the strain constituents at the centroid of stiffness, in general case:

$$\varepsilon_{cell,ctr,z,E,sensor}(t) = -\varepsilon_{cell,z,E,sensor,btm}(t)\frac{y_{top}}{h} + \varepsilon_{cell,z,E,sensor,top}(t)\frac{y_{btm}}{h} \tag{6.34a}$$

$$\varepsilon_{cell,ctr,z,T,sensor}(t) = -\varepsilon_{cell,z,T,sensor,btm}(t)\frac{y_{top}}{h} + \varepsilon_{cell,z,T,sensor,top}(t)\frac{y_{btm}}{h}$$
$$= \alpha_{T,z}\left(-\Delta T_{cell,btm,sensor}(t)\frac{y_{top}}{h} + \Delta T_{cell,top,sensor}(t)\frac{y_{btm}}{h}\right) \tag{6.34b}$$

$$\varepsilon_{cell,ctr,z,c,sensor}(t) = -\varepsilon_{cell,z,c,sensor,btm}(t)\frac{y_{top}}{h} + \varepsilon_{cell,z,c,sensor,top}(t)\frac{y_{btm}}{h} \tag{6.34c}$$

$$\varepsilon_{cell,ctr,z,sh,sensor}(t) = -\varepsilon_{cell,z,sh,sensor,btm}(t)\frac{y_{top}}{h} + \varepsilon_{cell,z,sh,sensor,top}(t)\frac{y_{btm}}{h} \tag{6.34d}$$

Similar to Expression 6.33, in Expressions 6.34, y_{top} and y_{btm} are negative if above the centroid (see Figure 6.6). Expression 6.34b assumes linear distribution of the thermal strain across the cross-sections of the cell (see Section 5.6 and Expression 5.41).

Once the average total strain at the centroid of stiffness of the cell is evaluated (Expression 6.29), the change in length of the cell can be evaluated by using Expression 6.5. Similarly, if average strain constituents can be evaluated (Expression 6.34), the respective change in length of the cell can be evaluated by substituting Expression 6.34 in Expression 6.6.

6.6.5   Average Axial and Bending Stiffness of a Cell, $E_{cell}A_{cell}(t)$ and $E_{cell}I_{cell}(t)$

We use definition of axial stiffness of a cell $E_{cell}A_{cell}(t)$ as developed earlier in Expressions 6.7–6.14. Let us assume that the elastic strain constituent at the centroid of stiffness can be evaluated from parallel topology using Expression 6.34a. Similar to Section 6.5.4, let us assume that the change in normal force at time t, $\Delta N_{z,cell}(t)$, with respect to a time t_0 is known (e.g., measured using load cell or applied through controlled load test). Then, the change of elastic strain constituent at the centroid of stiffness, $\Delta \varepsilon_{C_s,z,E,sensor}(t)$, can be evaluated using Expression 6.15 and the axial stiffness of a cell $E_{cell}A_{cell}(t)$ using Expression 6.16 (see details in Section 6.5.3).

Similar to axial stiffness, the bending stiffness of a cell, $E_{cell}I_{cell}(t)$, can be determined from strain measurement only if elastic strain constituents can be separated from the total strain measurement so that elastic curvature can be evaluated, and the change in bending moment that is related to that curvature is known or can be evaluated or estimated. In all expressions derived in this subsection, moment of inertia is assumed to be with respect to the principal axis of the cross-section perpendicular to the plane of bending. For homogeneous materials (e.g., steel), the bending stiffness of the cell is equal to the product of the Young's modulus of the cell material and average moment of inertia of the cross-section of the cell, see Expressions 6.35 and 6.36 (compare with Expression 6.7 and 6.8, respectively).

$$I_{cell} = \frac{1}{L_{cell}} \int_{L_{cell}} I(z)\,dz \tag{6.35}$$

$$E_{cell}I_{cell} = E_{cell} \frac{1}{L_{cell}} \int_{L_{cell}} I(z)\,dz \tag{6.36}$$

where $I(z)$ is moment of inertia of the cross-section of the cell at coordinate z, assuming that in general case, the cell can have variable cross-section.

If the cell has invariable cross-section, then the moment of inertia of the cell (I_{cell}) is equal to the moment of inertia of any cross-section of the cell, including the cross-section at the midpoint of the sensor (I_{C_s}). Thus, for homogeneous material and cell with invariable cross-section, Expressions 6.35 and 6.36 transform into Expressions 6.37 and 6.38, respectively (compare with Expressions 6.9 and 6.10).

$$I_{cell} = I_{C_s} \tag{6.37}$$

$$E_{cell}I_{cell} = E_{cell}I_{C_s} \tag{6.38}$$

However, for inhomogeneous materials, such as reinforced concrete, the stiffness of the cross-section depends on stiffness of the concrete and rebars. Similarly, for prestressed concrete, the stiffness of the cross-section depends on the stiffness of the concrete, rebars, and prestressing cable. In general, in both cases, the bending stiffness of the cross-section is calculated as shown in Expression 6.39.

$$E_{cell}I_{cell} = \frac{1}{L_{cell}} \int_{L_{cell}} I_{concrete}(z)E_{concrete}(z)\mathrm{d}z + E_{rebar} \int_{L_{cell}} I_{rebar}(z)\mathrm{d}z + E_{cable} \int_{L_{cell}} I_{cable}(z)\mathrm{d}z \qquad (6.39)$$

where $I_{concrete}(z)$, $I_{rebar}(z)$, and $I_{cable}(z)$ are individual cross-sectional moments of inertia of concrete, rebars, and prestressing cables, respectively, at the cell's cross-section with coordinate z, while $E_{concrete}(z)$, E_{rebar}, and E_{cable} are, respectively, Young moduli of the three materials; above moments of inertia are calculated with respect to the principal axis (centroid) of the entire cross-section (all materials combined).

Assuming that in observed cell, the Young's modulus of concrete does not vary significantly, and assuming that the moments of inertia of cross-sections of concrete, rebars, and prestressing cables (if any) do not vary significantly over the length of cell, one can consider the cell as having invariable cross-section, and Expression 6.39 transforms into Expression 6.40.

$$E_{cell}I_{cell} = E_{concrete}I_{concrete,C_s} + E_{rebar}I_{rebar,C_s} + E_{cable}I_{cable,C_s} \qquad (6.40)$$

where $I_{concrete,C_s}$, I_{rebar,C_s}, and I_{cable,C_s} are individual moments of inertia of concrete, rebars, and prestressing cables, respectively, at the cell's cross-section at the midpoint of the sensor; similar to Expression 6.39, these moments of inertia are calculated with respect to the principal axis of the entire cross-section.

There are two typical approaches in analyzing stiffness of inhomogeneous material, depending on the project requirements. The first approach considers the Young's modulus of concrete as the Young's modulus for entire cell, E_{cell}. This is useful when the Young's modulus of concrete can be determined independently, e.g., but testing for the purpose-made samples. In that case, Expression 6.40 transforms into Expression 6.41.

$$E_{cell}I_{cell} = E_{concrete} \left(I_{concrete,C_s} + \frac{E_{rebar}}{E_{concrete}}I_{rebar,C_s} + \frac{E_{cable}}{E_{concrete}}I_{cable,C_s} \right)$$

$$\Rightarrow$$

$$E_{cell} = E_{concrete}$$

$$I_{cell} = I_{equ,C_s} = I_{concrete,C_s} + \frac{E_{rebar}}{E_{concrete}}I_{rebar,C_s} + \frac{E_{cable}}{E_{concrete}}I_{cable,C_s} \qquad (6.41)$$

where I_{equ,C_s} represents an equivalent moment of inertia of cross-section at the midpoint of the sensor.

The second approach considers the geometric moment of inertia of the cell's cross-section as the moment of inertia for entire cell, I_{cell}. This is useful when the Young's modulus of concrete is not of interest, but composite behavior of the "mix" of materials is of interest. In that case, Expression 6.39 transforms into Expression 6.42.

$$E_{cell}I_{cell} = I_{geometric,C_s} \left(E_{concrete}\frac{I_{concrete,C_s}}{I_{geometric,C_s}} + E_{rebar}\frac{I_{rebar,C_s}}{I_{geometric,C_s}} + E_{cable}\frac{I_{cable,C_s}}{I_{geometric,C_s}} \right)$$

$$\Rightarrow$$

$$I_{cell} = I_{geometric,C_s}$$

$$E_{cell} = E_{equ,C_s} = E_{concrete}\frac{I_{concrete,C_s}}{I_{geometric,C_s}} + E_{rebar}\frac{I_{rebar,C_s}}{I_{geometric,C_s}} + E_{cable}\frac{I_{cable,C_s}}{I_{geometric,C_s}} \qquad (6.42)$$

where $I_{geometric,C_s}$ represents geometrical moment of inertia of cross-section at the midpoint of the sensor with respect to the true principal axis of the entire cross-section and E_{equ,C_s}, an equivalent Young's modulus (modulus of elasticity) of the cell; note that $I_{geometric,C_s} = I_{concrete,C_s} + I_{rebar,C_s} + I_{cable,C_s}$.

Let us now assume that the change in bending moment at time t, $\Delta M_{x,cell}(t)$, with respect to a time t_0 is known, as well as the change of elastic strain constituents measured by sensors of parallel topology, $\Delta\varepsilon_{cell,z,E,sensor,btm}(t)$ and $\Delta\varepsilon_{cell,z,E,sensor,top}(t)$, i.e.,

$$\Delta M_{x,cell}(t) = M_{x,cell}(t) - M_{x,cell}(t_0)$$

$$\Delta\varepsilon_{cell,z,E,sensor,btm}(t) = \varepsilon_{cell,z,E,sensor,btm}(t) - \varepsilon_{cell,z,E,sensor,btm}(t_0)$$

$$\Delta\varepsilon_{cell,z,E,sensor,top}(t) = \varepsilon_{cell,z,E,sensor,top}(t) - \varepsilon_{cell,z,E,sensor,btm}(t_0). \tag{6.43}$$

The change of elastic curvature in cell is evaluated from sensors as follows (see Expression 6.22):

$$\Delta\kappa_{x,cell,E,sensor}(t) = \frac{\Delta\varepsilon_{cell,z,E,sensor,btm}(t) - \Delta\varepsilon_{cell,z,E,sensor,top}(t)}{h} \tag{6.44}$$

The bending stiffness of the cell can now be calculated by using Expression 6.45.

$$E_{cell}I_{cell,sensor} = \frac{\Delta M_{x,cell}(t)}{\Delta\kappa_{x,cell,E,sensor}(t)} = \frac{\Delta M_{x,cell}(t)h}{\Delta\varepsilon_{cell,z,E,sensor,btm}(t) - \Delta\varepsilon_{cell,z,E,sensor,top}(t)} \tag{6.45}$$

Note that if multiple increments of bending moments $\Delta M_{x,cell}(t_i)$ are known and corresponding changes in elastic curvature $\Delta\kappa_{x,cell,E,sensor}(t_i)$ are measured ($i = 1, 2, \ldots n$), then the bending stiffness of the cell can be evaluated by using common statistical methods such as averaging of values obtained from individual increments or linear regression (more accurate). Once the bending stiffness of the cell is evaluated, then equivalent moment of inertia of cross-section or equivalent Young's modulus can be evaluated by using Expressions 6.38, 6.40–6.42, whichever applies.

6.6.6 Average Normal Force $N_{z,cell}(t)$ and Bending Moment $M_{x,cell}(t)$ in the Cell

In Section 6.5.4, the average normal force in the cell instrumented with simple topology was defined as the value of normal force in the cross-section at the midpoint of the sensor. This definition is extended to cell instrumented with parallel topology, assuming that both sensors of parallel topology have the midpoint in the same cross-section. The change in average normal force $\Delta N_{z,cell}$, at observed time t with respect to reference time, is evaluated by multiplying the average elastic strain at centroid of cell evaluated by the sensors of parallel topology (see Expression 6.34a), $\varepsilon_{cell,ctr,z,E,sensor}(t)$, with the stiffness of the cross-section at the midpoint of the sensor gauge length, $E_{cell}A_{cell}$, as per Expression 6.17. As per Section 6.5.4, the average normal force in cell, $N_{z,cell}$, can be evaluated from the sensor measurements using Expression 6.18 (see also Expressions 5.84 and 6.3, see Section 6.5.4 for details).

Similar approach is made to evaluate bending moment in the cell $M_{x,cell}(t)$. Let us first define average bending moment in the cell as the value of the bending moment in the cross-section at the midpoint of the sensors of parallel topology. Then the change in average bending moment $\Delta M_{x,cell}$, at observed time t with respect to reference time t_{Ref}, is evaluated by multiplying the average elastic curvature measured (evaluated) by sensors of parallel topology, $\Delta\kappa_{x,cell,E,sensor}(t)$, with the bending stiffness of the cross-section at the midpoint of the sensors' gauge lengths, $E_{cell}I_{cell}$, i.e., as shown in Expression 6.46.

$$\Delta M_{x,cell,sensor}(t) = \kappa_{cell,x,E,sensor}(t)E_{cell}I_{cell} = \frac{\varepsilon_{cell,z,E,sensor,btm}(t) - \varepsilon_{cell,z,E,sensor,top}(t)}{h}E_{cell}I_{cell} \tag{6.46}$$

where $E_{cell}I_{cell}$ is the bending stiffness of the cell, which is assumed to be known or evaluated from measurements.

Average bending moment in cell, $M_{x,cell}$, is then evaluated from the sensor measurement as shown in Expression 6.47 (see also Expressions 5.84 and 6.3).

$$M_{x,cell,sensor}(t) = \kappa_{x,cell,E,sensor}(t)E_{cell}I_{cell} + M_{x,cell,model}(t_{\text{Ref}}) = \Delta M_{x,cell,sensor}(t) + M_{x,cell,model}(t_{\text{Ref}})$$

(6.47)

Note that Expressions 6.45 and 6.46 assume that elastic curvature $\kappa_{x,cell,E,sensor}(t)$ can be separated from the total curvature measurement $\kappa_{x,cell,tot,sensor}(t)$, which might be challenging in the case of concrete structures (including reinforced and prestressed concrete structure) due to rheological effects, as discussed in Chapter 5. In addition, Expression 6.46 assumes that model for estimation of $M_{x,cell,model}(t_{\text{Ref}})$ can be accurately created, which also can be challenging.

6.6.7 Examples of Analysis of Parallel Topology Sensor Measurements and Their Derivatives

To illustrate the analysis of results of measurements from parallel topology, one-month data from Southeast Leg of Streicker Bridge are used. The instrumentation of the bridge is presented in Section 6.2.3. The cross-section of the Southeast Leg can be considered as having approximately invariable cross-section (see Figure 4.10d and more details in Sigurdardottir and Glisic 2015), which simplifies the estimation of geometrical properties. The Young's modulus of concrete was estimated from the concrete compressive strength (determined using standard cylinder tests) based on ACI 318-08 (ACI 2008), and it was used as the equivalent Young's modulus for all cross-sections in each cell E_{cell} (see Sections 6.5.3 and 6.6.5). Location of the centroid of stiffness was estimated using Expression 6.32a, equivalent area of cross-section using Expression 6.13, and equivalent moment of inertia using Expression 6.41. Geometrical locations of sensors were determined manually, during installation of sensors (Sigurdardottir and Glisic 2015). Geometrical and mechanical properties of the Southeast Leg of Streicker Bridge are summarized in Table 6.5.

As an example, Figure 6.7a shows measurements of strain and temperature taken during the first month after the pouring of concrete at location P12 (see Figure 6.3 for sensor location). The uncertainty of the monitoring system was estimated to 1 με for strain and 0.1°C for temperature. Each data point in the figure is an average value of four measurements, and thus the standard error of the mean of sensor measurements was estimated to $1/\sqrt{4} = 0.5$ με for strain and $0.1/\sqrt{4} = 0.05$°C for temperature (see Section 4.2.3, also Expression 4.6 and Table 4.3). Given that sensor measurements at time t represent differences between the measurements at that time and at reference time t_{Ref}, the uncertainty of strain measurement and the uncertainty of change in temperature with respect to reference time are estimated as:

$$u(\varepsilon_{cell,z,tot,sensor}(t)) = \sqrt{0.5^2 + 0.5^2} = 0.71\,\mu\varepsilon \text{ and } u(\Delta T(t)) = \sqrt{0.05^2 + 0.05^2} = 0.071°\text{C}.$$

Two events are highlighted in the figure: prestressing of the deck, performed in two stages, approximately 10 days after the pouring, and removal of the formworks, performed 24 days after the pouring. The close-ups to measurements taken during these two events are given in Figure 6.7b,c, respectively.

The data presented in Figure 6.7 are used to demonstrate evaluation of parameters presented in Sections 6.6.1–6.6.6. To simplify presentation, the entire set of data is analyzed graphically;

Table 6.5 Geometrical and mechanical properties of cross-sections of the Southeast Leg of Streicker Bridge.

Property	Value	Limits of error
Equivalent area of cross-section A_{cell} (m^2)[a]	1.28	0.05
Equivalent moment of inertia I_{cell} (m^4)[a]	0.0300	0.0029
Location of the centroid of stiffness $d_{cell,btm,model}$ (mm)[b]	398	5
Young's modulus of concrete at five days $E_{concrete,5}$ (GPa)[c]	27	5.4
Young's modulus of concrete at 10 days $E_{concrete,10}$ (GPa)[c]	29	5.8
Young's modulus of concrete at 24 days $E_{concrete,24}$ (GPa)[c]	34	6.8
Young's modulus of concrete at 28 days $E_{concrete,28}$ (GPa)[c]	35	7.0
Young's modulus of rebars and cables E_{steel} (GPa)	205	5
Thermal expansion coefficient α_T (µε/°C)	10	0.5
Distance of bottom sensor from the bottom of cross-section y'_{btm} (mm)	100	5
Distance between sensors h (mm)	380	5

a) Venuti et al. (2021).
b) Sigurdardottir and Glisic (2013).
c) Abdel-Jaber and Glisic (2014).

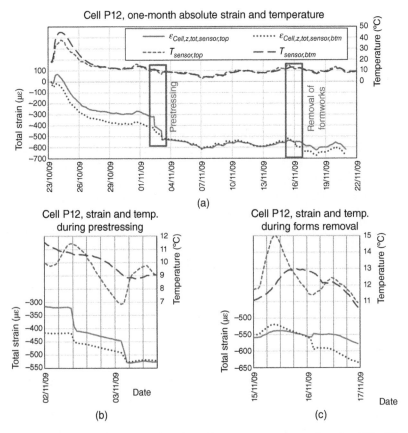

Figure 6.7 (a) Measurements of strain and temperature taken at location P12 of Streicker Bridge during the first month after the pouring of concrete, (b) close-up on measurements taken during prestressing of the deck, and (c) close-up on measurements taken during removal of the formworks.

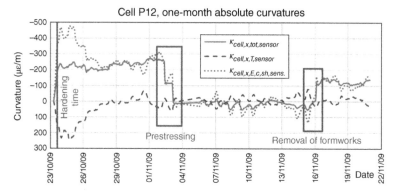

Figure 6.8 Curvature constituents evaluated at location P12 of Streicker Bridge during the first month after the pouring of concrete.

however, to emphasize the use of methods developed in Sections 6.6.1–6.6.6, the details are presented on smaller subsets, using data registered during prestressing and formworks removal.

Average total curvature $\kappa_{cell,x,tot,sensor}$ and average thermal curvature $\kappa_{cell,x,T,sensor}$, as evaluated from Expressions 6.21 and 6.23, are given in Figure 6.8.

In addition, the difference between the two curvatures, i.e., the difference $\kappa_{cell,x,E,c,sh,sensor} = \kappa_{cell,x,tot,sensor} - \kappa_{cell,x,T,sensor}$, is also presented in the figure. Note that this difference represents the sum of elastic curvature and curvatures due to creep and shrinkage, as per Expressions 5.9 and 6.21–6.25.

Example 6.4 *Evaluation of Hardening Time of Concrete*
In Figures 6.7 and 6.8, the time of pouring of concrete is used as a reference. However, few hours after the pouring, the concrete is in dormant period and measured strain reflects the internal movement of air and water, along with slow segregation (Glisic 2000). In addition, in unhardened concrete, during very early age, the thermal strain in concrete along with endogenous shrinkage does not result in generation of important elastic strain and stress (because the concrete is "liquid," i.e., viscous, see Glisic 2000). Only after the concrete is hardened some significant stresses can be generated. Thus, it is of interest to evaluate approximate hardening time of concrete.

Hardening time of concrete strongly depends on curing conditions (temperature and humidity) and, to certain extent, on concrete composition. For concrete insulated with mats and cured in outdoor conditions, so that the temperature reaches 35–45°C (which was the case of Streicker Bridge), the hardening time occurs approximately 8–24 hours after pouring of concrete (Glisic 2000). When concrete is hardened, the cross-sections of the bridge start acting as solid entities. Given that the deck of the bridge was laying on the formworks, it was not expected, during this period, to have significant sagging; in addition, due to activation of self-weight, significant hogging was not expected either. Thus, a constant, approximately null curvature was expected during the early age of concrete, after the concrete was hardened. Figure 6.8 shows that the curvature becomes constant approximately 10 hours after the pouring, which is consistent with the above estimation. Consequently, the hardening time can be evaluated as being equal to 10 hours after pouring, approximately. This time can be set as new reference. Figure 6.9a shows curvature constituents from Figure 6.8, but this time with respect to new reference time, hardening time, identified approximately 10 hours after pouring of concrete. Figure 6.9b,c shows the close-ups to the events identified earlier, i.e., prestressing and removal of the formworks, respectively.

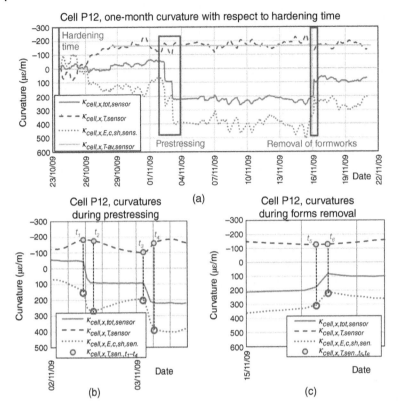

Figure 6.9 (a) Curvature constituents evaluated at location P12 of Streicker Bridge with respect to hardening time (10 hours after pouring of concrete), (b) close-up on curvature constituents evaluated during prestressing of the deck, and (c) close-up on curvature constituents evaluated during removal of the formworks.

In addition, Figure 6.9 indicates two other important observations. First, at about three days after the pouring, significant elastic curvature is generated, which means that significant bending moments and resulting stresses are generated too. This elastic curvature is the consequence of evolution of thermal gradient, which in turn is the consequence of different heating and cooling of the top and bottom parts of the cell. Note that the evolution of the thermal gradient is directly reflected through thermal curvature shown in Figure 6.9a, as the two are proportional (see Expression 6.23).

Second, the thermal gradient, reflected through thermal curvature shown in the figure, increases and becomes approximately constant approximately five days after the pouring. To emphasize this, the average value of thermal curvature ($\kappa_{cell,x,T\text{-}av,sensor} = -169.1\ \mu\varepsilon/m$) is shown in Figure 6.9a.

Figure 6.9b contains close-up on the prestressing. Four time-stamp indicators are shown in the figure: t_1 before the prestressing was applied, t_2 after the first step of prestressing was applied, t_3 before the second step of prestressing was applied, and t_4 after the second step of pressing was applied. Figure 6.9c contains close-up to measurements taken during removal of the formworks. For removal of formworks two time-stamp indicators are identified, t_5, before the removal of the forms, and t_6 after the removal of the forms.

The time indicators t_1–t_6 were used in this section to illustrate the data analysis algorithms developed in Sections 6.6.1–6.6.6. The measurements of interest are presented in Table 6.6.

Table 6.6 Absolute strain and temperature measurements recorded at characteristic times (reference time corresponds to pouring of concrete).

Description	Time	$T_{cell,sensor,top}$ (°C)	$T_{cell,sensor,btm}$ (°C)	$\varepsilon_{cell,z,tot,sensor,top}$ (με)	$\varepsilon_{cell,z,tot,sensor,btm}$ (με)
Pouring of concrete	10/23/09 7:00	18.4	18.4	0	0
Hardening time	10/23/09 17:00	31.7	37.9	58	−21
Time of reaching quasi-constant thermal gradient	10/28/09 9:00	13.4	13.4	−270	−358
t_1	11/2/09 9:00	11.4	10.7	−323	−418
t_2	11/2/09 12:00	11.0	10.6	−411	−456
t_3	11/3/09 2:00	6.8	9.2	−447	−491
t_4	11/3/09 5:00	8.7	8.8	−530	−526
t_5	11/16/09 2:00	11.4	12.8	−557	−569
t_6	11/16/09 3:00	11.4	12.7	−547	−595

Example 6.5   *Evaluation of Location of Centroid*

Calculation of curvature includes difference between the strain measured at the bottom and top of the cross-section with respect to the reference time. The uncertainty of this difference is $u(\varepsilon_{cell,z,tot,sensor,btm} - \varepsilon_{cell,z,tot,sensor,top}) = \sqrt{0.71^2 + 0.71^2} = 1.4\,\mu\varepsilon$. Similarly, the uncertainty of difference in temperature changes with respect to reference time is $u(\Delta T_{sensor,btm} - \Delta T_{sensor,top}) = \sqrt{0.071^2 + 0.071^2} = 0.14°C$.

During the removal of the formworks, only the self-weight is applied on the bridge deck, resulting in the change in bending moment and shear force distribution in the deck. The support at the abutment P13 of the bridge (see Figure 6.3) is roller, and thus the change in normal force is not expected to occur in span P12–P13 of the bridge. In addition, the bending stiffness of all columns of the bridge was estimated as negligible (Hubbell and Glisic 2013), which implies that the change in the normal force along entire southeast leg of the bridge due to the removal of formworks can be neglected. Assuming that time elapsed during the removal of the formwork, i.e., one hour approximately, was not sufficient to generate significant creep and shrinkage, these strain constituents can be neglected. Consequently, the measurements taken during the removal of formworks can be used to evaluate location of the centroid of stiffness at P12, as this event results in the change in elastic curvature only, with no change in elastic strain at the centroid of the cell, nor change in rheologic strain constituents. Hence, Expression 6.29a can be applied on difference in measurements taken during time stamps t_5 and t_6, where measurements taken at t_5 are used as a reference for the measurements taken at t_6. Table 6.7 summarizes this evaluation.

In Table 6.7, the location of the centroid of stiffness $d_{cell,btm,sensor}$ is found with respect to the bottom of the cross-section, as per Figure 6.6, using Expression 6.29a, where distance of bottom sensor from the bottom of the cell $y'_{btm} = 0.100$ m, as per Table 6.5 (see also Figure 6.6). To evaluate uncertainties in Table 6.7, the limits of error presented in Table 6.5 were assumed to be equal to two times standard uncertainty (i.e., uncertainty was assumed to be half of the limits of error). That was necessary given that measures of accuracy for sensor measurements are given in terms

Table 6.7 Evaluation of location of centroid of stiffness from measurements taken during the removal of the formworks (reference time = t_5, evaluation time = t_6).

	$\Delta T_{cell,sensor,top}$ (°C)	$\Delta T_{cell,sensor,btm}$ (°C)	$\Delta\varepsilon_{cell,z,T,sensor,top}$ (µε)	$\Delta\varepsilon_{cell,z,T,sensor,btm}$ (µε)	$\Delta\varepsilon_{cell,z,tot,sensor,top}$ (µε)	$\Delta\varepsilon_{cell,z,tot,sensor,btm}$ (µε)	$\Delta\varepsilon_{cell,z,E,sensor,top}$ (µε)	$\Delta\varepsilon_{cell,z,E,sensor,btm}$ (µε)	$\Delta\kappa_{cell,x,E,sensor}$ (µε/m)	$d_{cell,btm,sensor}$ (m)
Value	0.01	−0.07	0.1	−0.7	10.0	−25.9	9.9	−25.1	−92.3	0.373
Uncertainty	0.07	0.07	0.7	0.7	0.7	0.7	1.0	1.0	3.7	0.023

of uncertainty, while for the geometrical properties of the cross-section, they are given in terms of limits of error.

Based on Table 6.5, the location of the centroid of stiffness as estimated (modeled) from geometrical and mechanical properties of cross-section is $d_{cell,btm,model} = 398$ mm, with limits of error of 5 mm. As mentioned above, the uncertainty can be calculated as one-half of limits of error, i.e., u($d_{cell,btm,model}$) = 2.5 mm. Based on Table 6.7, the location of centroid of stiffness as evaluated from sensor measurements is $d_{cell,btm,sensor} = 0.373$ m = 373 mm, with uncertainty u($d_{cell,btm,sensor}$) = 0.023 m = 23 mm. Consequently, the absolute value of difference between evaluated (sensor) and estimated (model) location of the centroid is equal to 25 mm, i.e., $|d_{cell,btm,sensor} - d_{cell,btm,model}| = 25$ mm, while its uncertainty is equal to 23.1 mm, i.e., u($d_{cell,btm,sensor} - d_{cell,btm,model}$) = $\sqrt{23^2 + 2.5^2}$ = 23.1 mm. The absolute value of the difference is greater than one time its uncertainty, but it is smaller than two times its uncertainty, i.e.

$$25 \text{ mm} = |d_{cell,btm,sensor} - d_{cell,btm,model}| > 1 \cdot u(d_{cell,btm,sensor} - d_{cell,btm,model}) = 23.1$$

$$25 \text{ mm} = |d_{cell,btm,sensor} - d_{cell,btm,model}| < 2 \cdot u(d_{cell,btm,sensor} - d_{cell,btm,model}) = 46.2$$

Note that the first inequality above corresponds to the second expression of Criterion V (see Expression 6.1) for $n_u = 1$, and the second inequality above corresponds to the second expression of Criterion V for $n_u = 2$. One can conclude that for $n_u = 1$ Criterion V is not fulfilled, and an unusual behavior is detected, while this is not the case for $n_u = 2$. This means that (for $n_u = 1$) there is approximately 68% chance that there is unusual behavior, but that chance is less than approximately 95% ($n_u = 2$). Thus, setting $n_u = 1$ is more risk averse.

Unusual behavior, i.e., lower "measured" location of the centroid of stiffness than estimated based on analytical modeling, was found consistently for all cross-sections in Southeast Leg of Streicker Bridge (not presented here to keep presentation concise, see Sigurdardottir and Glisic 2013 for more details). This unusual behavior was explained by reduced effectiveness of the cross-section due to its geometrical properties, i.e., due to its large width-to-depth ratio (~5 times) and its quasi-triangular shape. It was shown that such cross-section behaves similar to "T" shaped cross-sections with large upper flange and short web (see more details in Sigurdardottir and Glisic 2013). Note that detected unusual behavior does not indicate the damage – it rather indicates slight inadequacy in the analytical model of the structural behavior of the bridge cross-sections. Long-term analysis of structural behavior of the bridge has shown that this model inadequacy (i.e., shift in location of the centroid of stiffness) did not significantly affect the estimation (modeling)

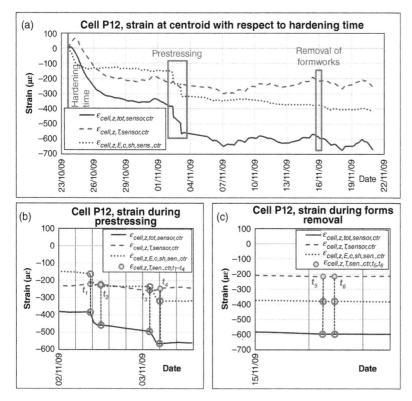

Figure 6.10 (a) Strain constituents at the centroid evaluated at location P12 of Streicker Bridge with respect to hardening time (10 hours after pouring of concrete), (b) close-up on strain constituents at the centroid evaluated during prestressing of the deck, and (c) close-up on strain constituents at the centroid evaluated during removal of the formworks.

of the performance of the bridge (e.g., see Sigurdardottir and Glisic 2015, Abdel-Jaber and Glisic 2019).

Example 6.6   *Evaluation of Normal Force (Prestressing Force)*
Once the centroid of stiffness is determined, it is possible to find the strain at the centroid using Expressions 6.34. Figure 6.10 shows one-month evolution of total strain at centroid, $\varepsilon_{cell,z,tot,sensor,ctr}$, thermal strain at centroid, $\varepsilon_{cell,z,T,sensor,ctr}$, and their difference, $\varepsilon_{cell,z,E,c,sh,sensor,ctr} = \varepsilon_{cell,z,tot,sensor,ctr} - \varepsilon_{cell,z,T,sensor,ctr}$, which corresponds to sum of elastic strain, creep, and shrinkage at the centroid.

Figure 6.10a,c confirms that removal of the forms did not change the total strain at the centroid and the difference between total strain and thermal strain, which, neglecting short-term creep and shrinkage, corresponds to elastic strain at the centroid.

Contrary, Figure 6.10a,b shows, as expected, that the prestressing did affect both the total strain at the centroid and the difference between the total strain and thermal strain at the centroid. Given that each step of prestressing was performed during short periods of time (less than three hours), creep and shrinkage can be neglected during these periods and the difference between the total strain and thermal strain can be considered as approximately equal to elastic strain. The change in

Table 6.8 Evaluation of prestressing force from measurements taken during the prestressing (reference time for the first step of prestressing $= t_1$, evaluation time $= t_2$, reference time for the second step of prestressing $= t_3$, evaluation time $= t_4$).

	$\Delta T_{cell,sensor,top}$ (°C)	$\Delta T_{cell,sensor,btm}$ (°C)	$\Delta\varepsilon_{cell,z,T,sensor,top}$ (με)	$\Delta\varepsilon_{cell,z,T,sensor,btm}$ (με)	$\Delta\varepsilon_{cell,z,tot,sensor,top}$ (με)	$\Delta\varepsilon_{cell,z,tot,sensor,btm}$ (με)	$\Delta\varepsilon_{cell,z,E,sensor,top}$ (με)	$\Delta\varepsilon_{cell,z,E,sensor,btm}$ (με)	$\Delta\varepsilon_{cell,z,E,sensor,ctr}$ (με)	$\Delta N_{z,cell,sensor}$ (MN)
First step	−0.40	−0.12	−4.0	−1.2	−88.1	−37.8	−84.1	−36.7	−70.7	−2.62
Uncertainty	0.07	0.07	0.7	0.7	0.7	0.7	1.0	1.0	5.7	0.34
Second step	1.85	−0.33	18.5	−3.3	−83.4	−35.8	−101.9	−32.6	−82.4	−3.06
Uncertainty	0.07	0.07	0.8	0.7	0.7	0.7	1.1	1.0	6.6	0.40

the elastic strain between the times t_1 and t_2 (the first step of prestressing) and t_3 and t_4 (the second step of prestressing) can be used to determine the total prestressing force. While the application of the prestressing force creates both bending moment (due to eccentricity of prestressing cables in the cross-section) and normal force, only the latter results in normal strain at the centroid of stiffness of the cross-section. Thus, Expression 6.17 can be used to evaluate the prestressing force. This evaluation is summarized in Table 6.8.

In Table 6.8, the uncertainties are evaluated using propagation formulae given in Section 4.3.1. Young's modulus of concrete at 10 days (29 GPa) was used to evaluate the change in normal force in cell P12 for each step, as per Table 6.5. Uncertainties of all parameters given in Table 6.5 are evaluated as one-half of limits of errors given in the table (as per discussion earlier in this subsection). Based on evaluations given in Table 6.8, the total prestressing force in cell P12 was evaluated (measured) to be $F_{ps,cell,sensor} = 2.62 + 3.06 = 5.68$ MN and its uncertainty $u(F_{ps,cell,sensor}) = \sqrt{0.34^2 + 0.40^2} = 0.52$ MN. Design prestress force in the cell was $F_{ps,cell,model} = 6.05$ MN, but uncertainty of its estimation was not provided, and thus it is set to zero. Note that the measured prestressing force is approximately 6% smaller than design value, and this can be considered as unusual structural behavior. However, the absolute value of the difference between the measured (sensor) and design (model) is smaller than one time its uncertainty, i.e.,

$$0.37\,\text{MN} = |F_{ps,cell,sensor} - F_{ps,cell,model}| < 1 \cdot u(F_{ps,cell,sensor} - F_{ps,cell,model}) = 0.52\,\text{MN}$$

The inequality above corresponds to the second expression of Criterion V (see Expression 6.1) for $n_u = 1$. One can conclude that for $n_u = 1$ Criterion V is fulfilled and the evaluated (measured) prestressing force can be considered within the uncertainty limits, which in turn does not ascertain (but also does not deny) the existence of unusual structural behavior.

The results presented above emphasize the difficulty in setting the thresholds for detection of unusual structural behaviors. While Criteria I–VI represent reasonable quantitative way for setting the thresholds, it remains up to engineering judgment on how to proceed in each specific project. For example, conservative judgment would be that prestress force as measured is 6% lower than design, and this indicates unusual structural behavior. Less conservative judgment would be that while the measured prestress force is lower than designed, it is not significantly smaller (i.e., it is within uncertainty of measurements), and thus there is no clear indication of presence of unusual structural behavior. In both cases, given that structures have large safety factors, 6% difference can be considered as insignificant to structural safety and performance in this specific case. Note that result presented here deals with prestressing force at single location of the bridge. By applying the analysis presented here to other locations on the bridge, it is possible to evaluate distribution of prestressing force at release (immediately after prestressing) along entire bridge, as shown in Figure 6.3b (see Abdel-Jaber and Glisic 2014, for more details).

Figure 6.10a shows other important characteristics of concrete at an early age. The difference between the total strain and thermal strain, i.e., $\varepsilon_{cell,z,E,c,sh,sensor,ctr}$ decreases rapidly until approximately two days after the pouring (October 25, 2009, 09:00), when it became approximately constant at $-135\,\mu\varepsilon$ for seven days, with minor drift of approximately $-30\,\mu\varepsilon$, until the prestressing is applied. The difference between the total strain and thermal strain consists of three strain constituents, elastic strain, creep, and shrinkage. Given that no axial compressive force was applied to the deck, the majority of the measured difference in total and thermal strain during these first two days after the pouring can be attributed to early age shrinkage. Minimal tensional elastic strain could have been generated during that period due to the friction of concrete with the formworks. Approximately constant value during the following seven days shows minimal rheological effects, which confirms that elastic strain is practically negligible (and normal force too) along with associated creep, while drying shrinkage was very small due to presence of formworks and covering mats that prevented evaporation of water from concrete. After the prestressing was introduced, the elastic strain became significant and, consequently, the creep started developing, which is visible in Figure 6.10a as downward trend of difference between total and thermal strain. The formworks and curing mats were removed approximately two weeks later, on the same day, after which drying shrinkage also started to have an important role. Based on the analysis of early age strain presented here, the change in normal force evaluated in Table 6.9 represents the absolute value of the normal force at location P12.

Example 6.7 *Evaluation of Bending Moment*

Early age thermal gradient, prestressing, and form removal, all created changes in bending moments in the bridge deck. These changes in bending moments at location P12 can be evaluated from the elastic curvature measurement by using Expression 6.46. Note that all the changes happened during the early age of concrete, and during that period the Young's modulus of concrete was still evolving (see Table 6.5). While this also influenced equivalent moment of inertia, due to low rebar-to-concrete area this influence was neglected here.

The elastic curvature due to thermal gradient evolved over the first five days then it became approximately constant (see Figures 6.8 and 6.9). Therefore, its influence on the change in bending moment should be calculated using time integration rather than Expression 6.46 (due to variability of Young's modulus over time); however, to simplify the analysis, Expression 6.46 is used assuming that the thermal gradient is applied five days after the pouring. This simplification can be justified as it is on a conservative side of evaluation (includes higher Young's modulus).

As mentioned earlier in this subsection, the prestressing force and removal of the forms are both applied in short terms and rheological curvature during these periods can be neglected; thus, for those two events, the elastic curvature can be evaluated as simple difference between the total curvature and the thermal curvature. The elastic curvature due to thermal gradient developed over five days after the pouring, and rheological effects certainly influenced the total curvature. During the first two days the shrinkage is predominant rheological effect (see Figure 6.10a), but its

Table 6.9 Evaluation of elastic curvature due to early-age thermal gradient, prestressing, and form removal, along with respective changes in bending moments.

	$\Delta\kappa_{cell,x,tot,sensor}$ ($\mu\varepsilon$/m)	$\Delta\kappa_{cell,x,T,sensor}$ ($\mu\varepsilon$/m)	$\Delta\kappa_{cell,x,E,sensor}$ ($\mu\varepsilon$/m)	E_{cell} (GPa)	I_{cell} (m^4)	$E_{cell}I_{cell}$ (MNm^2)	$\Delta M_{x,cell,sensor}$ (kNm)
Permanent bending due to early-age thermal gradient	−23.8	−169.0	145.2	27	0.0300	810	118
Uncertainty	2.6[a]	20.2[a]	20.4[a]	2.7	0.0029	113	23[a]
Prestressing, first step	132.2	7.5	124.7	29	0.0300	870	108
Uncertainty	2.7	2.6	3.8	2.9	0.0029	121	15
Prestressing, second step	125.2	−57.3	182.6	29	0.0300	870	159
Uncertainty	2.7	2.9	4.0	2.9	0.0029	121	22
Forms removal	−94.4	−2.1	−92.3	34	0.0300	1020	−94
Uncertainty	2.7	2.6	3.7	3.4	0.0029	142	14

a) To simplify presentation, evaluations of uncertainties shown in the table do not take into account variabilities of thermal and total curvature over seven-hour period, nor the averaging of strain and temperature measurements.

influence to the total curvature can be neglected as it approximately results in uniform strain across the cross-sections (i.e., curvature is approximately null). Creep developed due to elastic curvature induced by thermal gradients, but its evaluation would be complex (see Section 5.8.1); however, due to the shortness of time (five days) its value is relatively small and can be neglected too. Assumption that rheological curvature can be neglected (due to both shrinkage and creep) is justified by results shown in Figure 6.9a, where total curvature remains approximately constant between days 5 and 10, which shows that the curvature due to rheological effects is practically negligible. Hence, for all three events, the thermal gradient generation, prestressing, and removal of the forms, the elastic curvature can be evaluated as difference between total and thermal curvature. Then the changes in bending moments can be evaluated using Expression 6.46 for each event separately. Based on that expression, first, the elastic curvatures for all three events should be evaluated. For prestressing, they can be evaluated using elastic strain values given in Table 6.8. For the removal of the forms, the elastic curvature was already evaluated in Table 6.7. For elastic curvature due to thermal gradients, this is more complicated.

The thermal curvature in Figure 6.9a has two components, permanent part due to uneven heating and cooling of the top and bottom parts of the cross-sections of the deck, and variable part due to daily fluctuations of temperature; the difference in total curvature and thermal curvature, $\Delta\kappa_{cell,x,E,c,ch,sensor}$, reflects the sum of these two parts. The permanent shift in thermal curvature is evaluated as $\kappa_{cell,x,T-av,sensor} = -169.1\,\mu\varepsilon$/m, using average value of data from period October 28, 9 a.m., through November 20, 11 p.m. (see Figure 6.9a). Variable part can be evaluated using standard deviation during this period, which was $\kappa_{cell,x,T-var,sensor} = 24.4\,\mu\varepsilon$/m. During seven hours following the establishment of permanent shift in thermal curvature, i.e., over the period delimited by 9 a.m. and 4 p.m. on October 28, the average value of permanent shift was $\kappa_{cell,x,T,7d-av,sensor} = -169.0\,\mu\varepsilon$/m, which is approximately equal to the shift over entire observed period ($\kappa_{cell,x,T-av,sensor} = -169.1\,\mu\varepsilon$/m). Variability (standard deviation) during this seven-hour period was relatively low, i.e., $\kappa_{cell,x,T,7h\text{-}var,sensor} = 3.7\,\mu\varepsilon/m$, because the temperatures at the top and bottom of the cross-section were approximately constant. In addition, variability of total curvature during the same period was also low, i.e., $\kappa_{cell,x,tot,7h\text{-}var,sensor} = 2.4\,\mu\varepsilon/m$, which implies that elastic strain at observed location P12 due to temperature variation at other locations on the bridge was also relatively low, and can be neglected. Thus, this seven-hour average is used to determine elastic curvature and bending moment due to permanent shift in thermal gradient. Total average strain measured during that seven-hour period was $\kappa_{cell,x,tot,7h\text{-}av,sensor} = 23.8\,\mu\varepsilon/m$ (with variability, $\kappa_{cell,x,tot,7h\text{-}var,sensor} = 2.4\,\mu\varepsilon/m$). Now, the elastic curvature can be evaluated as the difference between seven-hour average total and thermal curvatures. Table 6.9 presents this evaluation, along with evaluations of elastic curvatures due to prestressing and form removal, and evaluations of changes in bending moments due to all three events.

Given that the bending moment is the first introduced with permanent shift due to thermal gradient, the cumulative sum of moments evaluated in Table 6.9 represents the absolute static bending moment in the observed cell. Thus, the changes in bending moments due to the variability of thermal gradients and live loads should be added to this static value. Note that the static value is not invariable over time – it will decrease in the long term due to losses in prestressing force caused by rheological effects.

Example 6.8 *Evaluation of Stress*

Assuming that rheological effects on curvature can be neglected, the maximal elastic curvature before prestressing was $\Delta\kappa_{cell,x,E,max\text{-}bp,sensor} = 201.8\,\mu\varepsilon/m$, registered on October 31 at 11:00 a.m., and maximal elastic curvature after prestressing but before removal of the formworks was $\Delta\kappa_{cell,x,E,max\text{-}ap,sensor} = 506.1\,\mu\varepsilon/m$, registered on November 15 at 8:00 a.m. (see Figure 6.9). Uncertainties in evaluations of these values were $20.1\,\mu\varepsilon/m$ and $21.0\,\mu\varepsilon/m$, respectively. Maximal elastic curvature after removal of the formworks was $\Delta\kappa_{cell,x,E,max\text{-}af,sensor} = 302.9\,\mu\varepsilon/m$, registered on November 18 at 10:00 a.m., with uncertainty of $24.8\,\mu\varepsilon/m$. Maximal (maximal positive) and minimal (maximal negative) normal stresses in cell P12 can now be evaluated using these values combined with evaluations of normal force and the first equation of Expression 5.28. However, to avoid the double counting of uncertainty in moment of inertia, Expression 5.28 is rearranged into Expression 6.48 (which is also equal to Expression 5.32 multiplied with Young's modulus E). Maximal and minimal stresses are then evaluated in Table 6.10.

$$\sigma_{z,sensor,max/min}(t) = E_{cell}\varepsilon_{cell,z,E,sensor}y_{max}/y_{min}(t) = E_{cell}(\varepsilon_{cell,z,E,sensor,ctr}(t) + \kappa_{cell,x,E,sensor}(t) \cdot y_{max/min})$$
(6.48)

where $y_{max} = d_{btm}$ and $y_{min} = -d_{top}$ (see Figures 5.4a and 6.6).

The results shown in Table 6.10 are important as they show the existence of tensional stresses at the bottom of the cross-section before prestressing and after prestressing. These stresses became finally compressive only after the form removal. However, existence of tensional stresses indicates potential for cracking, which is not desired in prestressed concrete structure. That is the reason why it is necessary to verify safety of the structure.

In order to evaluate structural safety based on results shown in Table 6.10, first the ultimate tensional and compressive stresses have to be evaluated. The concrete mix used in Streicker Bridge was based on New Jersey Department of Transportation specifications for Class A High-Performance Concrete (NJDOT 2007). Given that data presented in this subsection deals with the early age of concrete, the ultimate tensional and compressive strength were still developing over that period.

Table 6.10 Evaluation of maximal and minimal stresses before and after prestressing, and after the removal of formworks.

	$\varepsilon_{cell,z,E,sensor,ctr}$ ($\mu\varepsilon$)	$\kappa_{cell,x,E,sensor}$ ($\mu\varepsilon/m$)	$y_{max} = d_{btm}$ (m)	$y_{min} = -d_{top}$ (m)	E_{cell} (GPa)	$\sigma_{cell,z,cell,sensor}$ (kNm)
Max. stress before prestressing	0	201.8	0.373	–	27	2.03
Uncertainty	0	20.1	0.023	–	2.7	0.31
Min. stress before prestressing	0	201.8	–	−0.357	27	−1.95
Uncertainty	0	20.1	–	0.024	2.7	0.30
Max. stress after prestressing	−153.1	506.1	0.373	–	34	1.21
Uncertainty	8.7	21.0	0.023	–	3.4	0.57
Min. stress after prestressing	−153.1	506.1	–	−0.357	34	−11.35
Uncertainty	8.7	21.0	–	0.024	3.4	1.27
Max. stress after from removal	−153.1	302.9	0.373	–	27	−1.08
Uncertainty	8.7	24.8	0.023	–	2.7	0.41
Min. stress after from removal	−153.1	302.9	–	−0.357	27	−7.05
Uncertainty	8.7	24.8	–	0.024	2.7	0.81

Table 6.11 Evaluation of tensional and compressive strength of concrete in Streicker Bridge.

Time after pouring, t (day)	Tensile strength, $\sigma_u^+(t)$ (MPa)	Uncertainty, $u(\sigma_u^+(t))$ (MPa)	Compressive strength, $\sigma_u^-(t)$ (MPa)	Uncertainty, $u(\sigma_u^-(t))$ (MPa)
2	1.78	0.07	−29.3	2.45
3	1.89	0.08	−32.9	2.94
7	2.16	0.10	−42.7	4.02
28	2.30	0.07	−48.5	3.12

Compressive strength was evaluated based on standard cylinder tests carried out as per ASTM C-39 (Abdel-Jaber and Glisic 2014). Tensional strength was then evaluated based on ACI 318-02, i.e., using Expression 6.49 (ACI 2002).

$$\sigma_u^+(t) = 0.33\sqrt{f_c'(t)}$$
$$\sigma_u^-(t) = -f_c'(t) \tag{6.49}$$

where f'_c is crushing stress in the tested cylinder.

The tests were performed 2, 3, 7, and 28 days after the pouring, and at each time total of four cylinders were tested. Table 6.11 presents the average values for each time along with standard uncertainty (determined for each test separately, see details in Abdel-Jaber and Glisic 2014).

Criterion IV can be applied to the stress values found in Table 6.10. Based on the test results, the safety factor for tensional stresses can be set to $\beta_\sigma^+ = 1.1$, while for compressive tests it can be set to $\beta_\sigma^- = 1.2$ (these safety factors cover at least three standard uncertainties). For all minimal stresses evaluated in Table 6.10, Criterion IV is fulfilled by large margin, i.e., for $n_u > 3$ (see the second inequality of Expression 5.89), which means that there is no indication of existence of unusual structural behavior. For the maximal stress before prestressing, substation of stress and strength from Tables 6.10 and 6.11 in the second inequality of Expression 5.89 yields to the following expressions, taking into account that maximal stress happened approximately eight days after the pouring (thus value at seven days was used for tensile and compressive strength):

$$\frac{\sigma_u^-}{\beta_\sigma^-} + n_u u(\sigma_{cell,z,sensor,max-bp}) < \sigma_{cell,z,sensor,max-bp} < \frac{\sigma_u^+}{\beta_\sigma^-} - n_u u(\sigma_{cell,z,sensor,max-bp})$$

$$\frac{-42.7}{1.2} + n_u \cdot 0.31 < 2.03 < \frac{2.16}{1.1} - n_u \cdot 0.31$$

$$-35.6 + n_u \cdot 0.31 < 2.03 < 1.96 - n_u \cdot 0.31$$

Analysis of the last expression shows that the right inequality cannot be fulfilled for any (positive) value of n_u, which in turn indicates the existence of an unusual structural behavior with potential to result in cracking of the bridge deck. Visual inspection, as well as the sensor measurements, did not show sign of cracking, which can be explained by the fact that the true stress in cell P12 could be smaller than true strength of the bridge at that time, given that the evaluated stress in the bridge (2.03 MPa) was smaller than evaluated strength (2.16 MPa). Note, however, that such a closeness between evaluated stress and strength is dangerous, as their difference is within the limits of uncertainties of their evaluations, and there is a chance that stress may be higher than strength, which could lead to cracking. This is confirmed by cracking actually occurring at other locations of the bridge (see Hubbell and Glisic 2013, also Chapter 7 for detailed presentation)! The main cause of cracking is permanent curvature in the cell induced by thermal gradient generated at early age of concrete, which was probably not included in the design of the bridge.

Simple substitution in the above expression shows that Criterion IV is fulfilled for maximal stress after prestressing for $n_u = 1$, but not for $n_u = 2$, which indicates that structural safety is improved after the prestressing, but not yet fully achieved. Finally, the structural safety if fully guaranteed after removal of the forms, as Criterion IV was then fulfilled with $n_u > 3$.

6.7 Crossed Topology and Evaluation of Corresponding Cell Parameters

6.7.1 Implementation of Crossed Topology

Crossed topology (Inaudi and Glisic 2002, Glisic and Inaudi 2007) is used in the cells expected to experience shear strain in each cross-section. Such a state of strain is typically found in beam-like structures subjected to shear forces or torsional moments. In the general case, these loads can act spatially, i.e., the cell can be exposed to biaxial shear forces along both principal axes of the cell's cross-sections, while torsion is, by its nature, considered as spatial load. However, based on linear beam theory, these influences can be separately analyzed and then superposed. Consequently, the scope of this section is limited to planar case, which only deals with shear strain generated by shear force acting in the cell plane; expressions for biaxial shear analysis and torsion can be derived combining approaches similar to those presented here for planar shear and appropriate expressions from Solid Mechanics and Statics of Structure (e.g., see more detail in Glisic and Inaudi 2007). Note that shear strain is often accompanied by normal strain (e.g., when transversal forces introduce

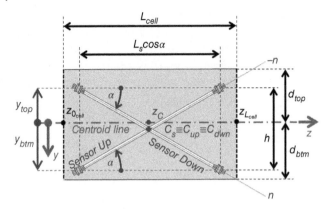

Figure 6.11 Crossed topology in planar shear case.

both shear forces and bending moments), and thus, crossed topology is often (but not necessarily) combined with parallel topology in the same cell.

Crossed topology, typically, consists of two sensors, denoted with "Sensor Down" and "Sensor Up," that have the same gauge length and are installed with, by the absolute value the same but opposite angles, as shown in Figure 6.11. The midpoints of the two sensors should coincide, so that the beginning of Sensor Down and the end of Sensor Up lay on a line parallel to centroid line (coordinate y_{top}), while the beginning of Sensor Up and the end of Sensor Down lay on another line parallel to centroid line (coordinate y_{btm}). If assumptions of linear theory of beams apply, then the deformation of cell shown in Figure 6.11, due to shear strain only, can be approximately represented as shown in Figure 5.5 (see Chapter 5). Considerations regarding determining the gauge length of sensors are given Section 6.7.2.

Given that crossed topology consists of two single sensors, the analysis of measurements of each sensor at local cell scale is the same as described in Chapter 5, and Criteria I–IV (see Chapter 5) apply directly for identification of unusual behaviors. However, in addition to that analysis, performed at sensor level, several important parameters related to the observed cell can be evaluated from combined sensor measurements, such as (but not limited to): average shear strain and change in average shear stress, average shear force and shear stiffness, and average normal strain and normal force. These parameters can then be used to evaluate the performance and safety of structure by using Criteria V and VI, developed in Section 6.3.

6.7.2 Average Shear Strain $\gamma_{cell,zy,sensor}(t)$ and Change in Average Shear Stress $\Delta\tau_{cell,zy,sensor}(t)$

Figure 6.11 shows that the extremities of both sensors are placed on lines parallel to the centroid line of the cell. Applying Expression 5.2 on both sensors (Sensor Down and Sensor Up) and finding their difference, the average strain in the cell can be evaluated using Expression 6.50.

$$\gamma_{cell,zy,E+c,sensor}(t) = \gamma_{cell,zy,tot,sensor}(t) = \frac{\varepsilon_{n,tot,sensor,down}(t) - \varepsilon_{-n,tot,sensor,up}(t)}{\sin 2\alpha} \tag{6.50}$$

where $\varepsilon_{n,tot,sensor,down}(t)$ and $\varepsilon_{-n,tot,sensor,up}(t)$ are the strain measurements of Sensor Down and Sensor Up in directions of sensors n and $-n$, respectively, and α is angle of sensors, as per Figure 6.11.

Expression 6.50 is valid only if the configuration of sensors is as shown in Figure 6.11, and it attributes evaluated shear strain to the cross-section that contains middle points of the sensors. The expression and the figure show that the crossed topology averages the shear strain, both, along the length of the cell and across the depth (height) of the cross-section. Expression 5.2 implies that sensitivity of Sensor Down and Sensor Up to shear strain is the highest for $\alpha = 45°$. In addition, Expression 6.50 implies that the influence of error in sensor measurements to evaluated shear strain is the smallest for $\alpha = 45°$. Consequently, the angle $\alpha = 45°$ provides the best performance in evaluating (measuring) the shear strain, and this angle is recommended for crossed topologies. This in turn suggests that the maximal gauge length of sensors in crossed topology should not exceed the theoretical value of the depth of cross-section divided with cos45°, i.e., multiplied with $\sqrt{2}$, and the length of the cell should be approximately equal to the depth of the cross-section. However, depending on site conditions, the true angle of sensors might be different and should be adapted to these site conditions, trying to keep it as close as possible to 45°, which in turn can lead to longer or shorter gauge length of the sensors and longer or shorter length of the cell.

If the length of the cell with crossed topology is not excessively long and temperature variations in environment are not abrupt, it is reasonable to assume that longitudinal and transversal thermal gradients are approximately constant along the depth and length of the cell; this means that their influence on each sensor is identical. For example, if transversal thermal gradient is constant along the length of the cell (e.g., temperature at bottom part of cross-section is invariable along the length of the cell and higher than temperature at the top part of the cross-section, which is also invariable along the length of the cell), then it will influence in the same way both sensors (e.g., the beginning of Sensor Down has the same lower temperature as the end of Sensor Up, and the end of Sensor Down has the same higher temperature as the beginning of the Sensor Up), and the difference in their measurements (as per Expression 6.50) will cancel out the influences of thermal gradients (see also Glisic and Inaudi 2007). This means that crossed topology is insensitive to thermal strain. In other words in materials without rheological effects, such as steel, Expression 6.50 directly yields elastic shear strain. In the case of material with rheological effects, such as concrete, the creep and shrinkage are present in long term. However, the shrinkage will affect each sensor in the same way, and thus their difference cancels out the influence of shrinkage (analysis of influence of shrinkage on crossed topology is similar to the analysis of influence of thermal gradient given above, see also Glisic and Inaudi 2007). Thus, Expression 6.50, in the case of concrete, yields the shear strain due to mechanical influence only, i.e., due to elastic shear strain and associated creep.

Assuming that the change in bending moment along the length of the cell is small, the influence of bending moment to each sensor is approximately the same and its influence on evaluation of shear strain approximately cancels out. Similar is valid for the influence of normal force (analysis of influence of bending moment and normal force on crossed topology is similar to the analysis of influence of thermal gradients given above, see also Glisic and Inaudi 2007). Thus, Expression 6.50 evaluates the shear strain only due to shear stresses in the cross-section, while the influences of normal strain to sensors apporximately cancel out.

In order to evaluate average shear stress at location of sensors, it is necessary to first evaluate elastic shear strain. For materials that do not experience rheological effects, such as steel, Expression 6.50 can be directly used. However, for materials that experience creep, such as concrete, the elastic strain should be first extracted from total strain measurements, e.g., using equivalent creep coefficient $K_{c,equ}$ and equivalent creep function $f_{c,equ}$ (see Expression 5.80) or other methods (see Section 5.8, see also Section 5.9.5), and then substituted in Expression 6.50. To summarize, Expressions 6.51

can be used to evaluate elastic shear strain.

For Steel: $\quad \gamma_{cell,zy,E,sensor}(t) = \dfrac{\varepsilon_{n,tot,sensor,down}(t) - \varepsilon_{-n,tot,sensor,up}(t)}{\sin 2\alpha}$

For Concrete: $\quad \gamma_{cell,zy,E,sensor}(t) = \dfrac{\varepsilon_{n,tot,E,down}(t) - \varepsilon_{-n,tot,E,up}(t)}{\sin 2\alpha}$

$$= \frac{\varepsilon_{n,tot,sensor,down}(t) - \varepsilon_{-n,tot,sensor,up}(t)}{\sin 2\alpha (1 + K_{c,equ} f_{c,equ}(t))} \qquad (6.51)$$

Once the elastic shear strain is evaluated, the change in average shear stress at location of crossed topology with respect to reference time can be evaluated using Expression 6.52.

$$\Delta\tau_{cell,zy,sensor}(t) = G_{cell}(t) \cdot \gamma_{cell,zy,E,sensor}(t) \qquad (6.52)$$

where $G_{cell}(t)$ is shear modulus of the construction material at time t (see also Expression 5.18).

Expression 6.52 shows that the accuracy in evaluation of change in shear stress depends on the accuracy of evaluation of elastic shear strain and accuracy in evaluation or estimation of shear modulus.

6.7.3   Average Shear Force $S_{y,cell}(t)$ and Shear Stiffness $G_{cell}A_{cell}(t)$

Based on Expression 5.29, the elastic shear strain at the top and bottom of the cross-section are null, and its distribution across the depth of the cross-section is nonlinear. Maximal value is typically found at the centroid of the cross-section. Thus, the gauge length of sensors and the position of the crossed topology within the cross-section matters when interpreting the data obtained from sensors, as the crossed topology averages the shear strain only between the points defined with y_{top} and y_{btm}. This can be expressed using the following expression (see also Expression 5.28):

$$\gamma_{cell,zy,E,sensor}(t) = \frac{1}{h}\int_{y_{top}}^{y_{btm}} \frac{S_{y,cell}(t)\,Q_{cell}(t,y(t))}{G_{cell}(t)\,I_{x,cell}(t)\,w_{cell}(t,y(t))}dy \qquad (6.53)$$

where $S_{y,cell}(t)$ is shear force in the cell at time t, $I_{cell}(t)$ is (average) moment of inertia (second moment of area) of the cell with respect to axis x at time t, $w_{cell}(y)$ is the average width of cross-section at depth y at time t, and $Q_{cell}(t,y)$ average static moment of inertia (first moment of area) with respect to axis x of the part of cross-section delimited with coordinates y and d_{btm} at time t, i.e., $Q_{cell}(t,y) = \frac{1}{l_{cell}}\int_{l_{cell}} \left(\iint_{A'(t,x,y)} y'(t)dA'(t) \right) dz$ (see Figures 5.4 and 5.5).

For cell with invariable cross-section, Expression 6.53 transforms into Expression 6.54.

$$\gamma_{cell,zy,E,sensor}(t) = \frac{S_{y,cell}(t)}{G_{cell}(t)\,A_{cell}(t)} \left(\frac{A_{cell}(t)}{I_{x,cell}(t)}\frac{1}{h}\int_{y_{top}}^{y_{btm}} \frac{Q_{cell}(t,y(t))}{w_{cell}(t,y(t))}dy \right)$$

$$= \tilde{k}_{cell,sensor}(t)\frac{S_{y,cell}(t)}{G_{cell}(t)A_{cell}(t)} \qquad (6.54)$$

where $\tilde{k}_{cell,sensor}(t) = \frac{A_{cell}(t)}{I_{x,cell}(t)}\frac{1}{h}\int_{y_{top}}^{y_{btm}} \frac{Q_{cell}(t,y(t))}{w_{cell}(t,y(t))}dy$ is dimensionless partial shear coefficient related to geometrical properties of cross-sections of the cell.

For $y_{top} \approx y_{min,cell} = -d_{top}$ and $y_{btm} \approx y_{max,cell} = d_{btm}$, the following is valid:

$$\tilde{k}_{cell,sensor}(t) \approx \frac{A_{cell}(t)}{I_{x,cell}(t)} \frac{1}{h_{cell}} \int_{y_{min,cell}}^{y_{max,cell}} \frac{Q_{cell}(t,y(t))}{w_{cell}(t,y(t)} dy = \tilde{k}_{S,cell}(t)$$

$$\varphi_{S,cell,zy,sensor}(t) = \frac{k_{S,cell}(t)}{\tilde{k}_{cell,sensor}(t)} \gamma_{cell,zy,E,sensor}(t) \approx \frac{k_{S,cell}(t)}{\tilde{k}_{S,cell}(t)} \gamma_{cell,zy,E,sensor}(t) \tag{6.55}$$

where $k_{S,cell}(t)$ is dimensionless shear coefficient of the cell and $\varphi_{S,cell,zy,sensor}(t)$ is slope of average cross-sectional deplanation in cell due to elastic shear stresses (see also Expression 5.32).

The change in average shear force $\Delta S_{y,cell}$, at observed time t with respect to reference time, is evaluated by multiplying the average elastic shear strain evaluated from crossed topology, $\gamma_{cell,zy,E,sensor}$, with the shear stiffness of the cell, $G_{cell}A_{cell}$, and dividing with partial dimensionless shear coefficient $\tilde{k}_{cell,sensor}$, i.e., as shown in Expression 6.56.

$$\Delta S_{y,cell,sensor}(t) = \gamma_{cell,zy,E,sensor}(t) \frac{G_{cell}A_{cell}(t)}{\tilde{k}_{cell,sensor}(t)} \tag{6.56}$$

where the shear stiffness of the cell, $G_{cell}A_{cell}(t)$, and the partial dimensionless shear coefficient, $\tilde{k}_{cell,sensor}(t)$, are assumed to be known – estimated using analytical or numerical modeling, or evaluated from measurements.

Average shear force in cell, $S_{y,cell}$, is then evaluated from the sensor measurement as shown in Expression 6.57 (see, for comparison, Expression 5.83).

$$S_{y,cell,sensor}(t) = \gamma_{cell,zy,E,sensor}(t) \frac{G_{cell}A_{cell}(t)}{\tilde{k}_{cell,sensor}(t)} + S_{y,cell,model}(t_{Ref})$$

$$= \Delta S_{y,cell,sensor}(t) + S_{y,cell,model}(t_{Ref}) \tag{6.57}$$

Note that Expression 6.57 assumes that that model for estimation of shear force at reference time, $S_{y,cell,model}(t_{Ref})$, can be accurately created.

Let us assume that the change in shear force at time t, $\Delta S_{y,cell}(t)$, and the value of partial dimensionless shear coefficient, $k_{cell,y_{min}y_{max}}(t)$, with respect to a time t_0 are known (estimated via analytical or numerical model, or evaluated using sensor measurements), along with the change of elastic shear strain evaluated (measured) by crossed topology sensor, $\gamma_{cell,zy,E,sensor}(t)$, i.e.,

$$\Delta S_{y,cell}(t) = S_{y,cell}(t) - S_{y,cell}(t_0)$$

$$\Delta \gamma_{cell,zy,E,sensor}(t) = \gamma_{cell,zy,E,sensor}(t) - \gamma_{cell,zy,E,sensor}(t_0). \tag{6.58}$$

Then, the shear stiffness of the cell can be evaluated using Expression 6.59.

$$G_{cell}A_{cell}(t) = \frac{\Delta S_{y,cell}(t)}{\Delta \gamma_{cell,zy,E,sensor}(t)} \tilde{k}_{cell,sensor}(t) \tag{6.59}$$

Note that if multiple increments of shear force $\Delta S_{y,cell}(t_i)$ are known and corresponding changes in elastic shear strain $\Delta \gamma_{cell,zy,E,sensor}(t_i)$ are measured ($i = 1, 2, \ldots n$), then the shear stiffness of the cell can be evaluated by using common statistical methods such as averaging of values obtained from individual increments or linear regression (more accurate). Once the shear stiffness of the cell is evaluated, then equivalent area of cross-section or equivalent shear modulus can be evaluated by exchanging G_{cell} and E_{cell} in Expressions 6.38, 6.40–6.42, whichever applies.

Note that the elastic shear strain distribution in full cross-section of a beam is different from distribution in thin-walled cross-section. Given that this book is introductory, the scope of this subsection is limited to beams with full cross-section. In other words, Expression 5.33 and expressions presented in this section do not apply for thin-walled cross-sections. However, expressions for evaluation of shear force and shear stiffness for thin-walled cross-sections can be derived based on relationship between shear force and shear stress for thin-walled cross-sections established in Solid Mechanics or Strength of Materials.

6.7.4 Average Strain $\varepsilon_{C,z,tot,sensor}(t)$ and Average Normal Force $N_{z,cell}(t)$

Based on Expression 5.2, the sum of strain in the direction of the two sensors in crossed topology transforms into the following expression:

$$\varepsilon_{C_s,n}(t) + \varepsilon_{C_s,-n}(t) = 2\left(\varepsilon_{C_s,z}(t)\cos^2\alpha + \varepsilon_{C_s,y}(t)\sin^2\alpha\right) \tag{6.60}$$

where $\varepsilon_{C_s,n}(t)$, $\varepsilon_{C_s,-n}(t)$, $\varepsilon_{C_s,z}(t)$, and $\varepsilon_{C_s,y}(t)$ are normal strains at point C_s in directions of Sensor Down, Sensor Up, z-axis, and y-axis, respectively (as per notation in Figure 6.11).

Given that, based on linear theory of beams, the stress in the direction of y-axis is negligible, the elastic strain in that direction is equal to negative product of elastic strain in the direction of z-axis multiplied with Poisson's coefficient. With this consideration, Expression 6.60 can be applied to elastic strain only and rearranged into the following expression:

$$\varepsilon_{C_s,z,E,sensor}(t) = \frac{\varepsilon_{n,E,sensor,down}(t) + \varepsilon_{-n,E,sensor,up}(t)}{2(\cos^2\alpha - v\sin^2\alpha)} \tag{6.61}$$

where v is Poisson's coefficient.

Expression 6.61 uses elastic strain constituent, and thus all previously mentioned comments regarding evaluation of elastic strain from total strain apply. In addition, the expression is only applicable if $\cos^2\alpha - v\sin^2\alpha \neq 0$, i.e.,

$$\cos^2\alpha - v\sin^2\alpha \neq 0 \Leftrightarrow \cot\alpha \neq \sqrt{v} \tag{6.62}$$

Poisson's coefficient typically ranges between 0.1 and 0.2 for concrete and 0.30–0.33 for steel, and thus, for concrete, the angle α should be out of the range 65.9°–72.5°, and for steel, it should be out of the range 60.1°–61.3°.

Expression 6.61 evaluates the elastic normal strain at center points of sensors $C_s \equiv C_{up} \equiv C_{down}$ (assuming they coincide with each other, as shown in Figure 6.11) in the direction of z-axis. In general case, the point C_s is not necessarily on a centroid line (see Figure 6.11) and evaluation of the strain given by Expression 6.61 is equivalent to measurement made by sensor with gauge length $L_s\cos\alpha$ installed at location C_s parallel to centroid line. If the point C_s lays on centroid line, then Expression 6.61 evaluates the strain at the centroid, and this evaluation can be substituted in Expressions 6.17 and 6.18 to evaluate the change in normal force $\Delta N_{z,cell}(t)$ and the normal force $N_{z,cell}(t)$, respectively.

6.7.5 Example of Analysis of Crossed Topology Sensor Measurements and Their Derivatives

To illustrate the analysis of results of measurements from crossed topology, short-term data from a test performed on a helical (also, often called spiral) cantilevered staircase of the Museum of the City of New York (MCNY) are used (see Figure 6.12a). This heritage staircase was completed in the early 1930s, and it is situated in the entrance hall (rotunda) of the museum (MCNY 2019). Each of its monolithic threads was carved out of Imperial Danby marble (MCNY 2019) and cantilevered at one extremity to the semi-cylindrical wall (Figure 6.12b,c). Elaborate rebates enable complex interactions between the threads that significantly improve structural capacity and allow for an efficient use of material, resulting in a quasi-triangular shape and minimalistic depth of

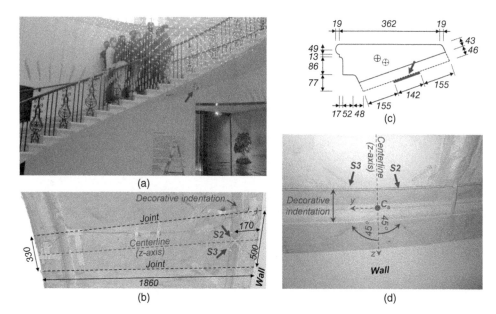

Figure 6.12 (a) View to MCNY staircase with students during the test, (b) position of sensors under the thread, source: Reproduced from Jin 2020 / PRINCETON UNIVERSITY, (c) cross-sectional properties of thread at location of crossed topology, and (d) detail showing the crossed topology; sensors S2 and S3 are pointed with red arrows; all unites are in millimeters (mm).

threads' cross-section. Helical shape, slenderness, and discrete decorative carvings both at free ends of threads and underneath the threads, render an elegant appearance to the staircase, making it a distinguished example of structural art.

In order to understand structural behavior and the flow of forces within the staircase, two threads were instrumented with total of seven Fiber Bragg-Grating (FBG) strain sensors (shown with arrows in Figure 6.12), and three load tests were performed (Jin 2020, Chen 2022, Agyarko 2023). Seven sensors were placed in various configurations at the accessible sides of the staircase threads, i.e., at their bottom cross-sectional face or free-end cross-section. In this section, only the analysis of results related to crossed topology is presented. At the time of writing this book, the data analysis was still being carried out, consequently, only partial analysis of one test (out of three) is presented here.

Crossed topology was installed at the fixed end of a selected thread to capture shear strain resulting from interaction between the threads. The sensors denoted with *S2* and *S3* were installed with an angle of 45° with respect to centerline of the exposed bottom face of the thread with their crossing point being exactly at the centerline. Gauge length of sensors of 20 cm was chosen due to the cross-sectional size of the thread and the width of decorative indentation, which had to be "bridged" by sensors, close to the fixed end of the thread. The location of sensors is chosen based on two realities: (i) shear stress and strain are expected to be greatest at fixed end and (ii) the decorative indentation weakens the involved cross-sections (reduces area for approximately 29%) and, given its proximity to fixed end, makes it the most loaded, and the most vulnerable, part of the thread. Figure 6.12c shows the cross-section at indentation (full lines) as well as cross-section without indentation (dashed lines). Respective centroids of cross-sections are also shown in the figure with full and dashed circles, with cross. Since there is a gap between the locations of centroids, and the sensors are installed distant from the actual indented cross-sections, they are capturing amplified

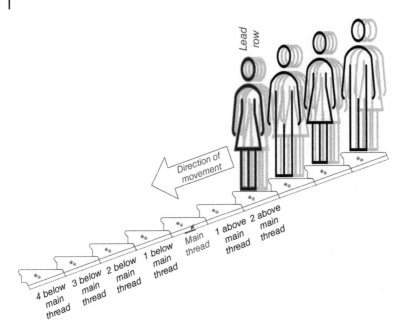

Figure 6.13 Schematic representation of load test of the staircase and notation of loading steps.

influence of torsional shear strain and normal bending strain. Global position of sensors on the thread is shown in Figure 6.12a,b, with details shown in Figure 6.12c,d. The sensors consist of bare optical fibers with FBGs that are prestressed for 2–3% and glued along length of 5 cm at each their end. The length of 5 cm was determined based on lab tests. Clear sensor gauge length between glued points was 20 cm. The glue was selected and tested prior to installation, so that it provides excellent adhesive properties, but also it can be removed after tests without leaving any trace or damage on the thread (Jin 2020).

The test presented here consisted of loading the staircase with a group of 15 people that were standing on four threads, as shown in Figure 6.13. Lead row and the subsequent two rows consisted of four persons each, while the last (uppermost) row consisted of three persons. In each row, the persons were standing distributed along the thread, shoulder to shoulder.

Reference measurement session was performed with empty staircase (denoted in the further text as "zero before"). Then, the load was applied so that the lead row was standing two threads above the thread instrumented with crossed topology, in configuration as shown in Figure 6.13, and measurement session was performed (denoted in the further text as "2 above"). In subsequent six steps of loading, the entire group of people would move down one thread, and measurement session would be performed (denoted based on the position of the lead row, i.e., "1 above," "main," "1 below," "2 below," "3 below," and "4 below"). Finally, final measurement session was performed with empty staircase (denoted in the further text as "zero after").

The entire test took less than eight minutes. That is the reason why the temperature change in and around staircase was considered as negligible, and creep in threads was considered negligible as well. Consequently, the FBG sensors were not compensated for temperature, and the measured strain was considered as purely elastic, i.e., $\varepsilon_{S2,E,sensor,down}(t) = \varepsilon_{S2,tot,sensor,down}(t)$ and $\varepsilon_{S3,E,sensor,up}(t) = \varepsilon_{S3,tot,sensor,up}(t)$. Each measurement session consisted of more than 100 readings performed at a rate of 10 readings per second, and in data analysis, an average of 100 readings was used to obtain measurement for each test. In this way, the repeatability precision of measurements

Table 6.12 Measurements of sensors S2 and S3 during the test on MCNY staircase and calculation of changes in apparent shear strain, apparent shear stress, apparent normal strain in the direction of centerline, and apparent normal stress in the direction of centerline.

Load step	Sensor S2 $\varepsilon_{S2,E,sens., down}$ (με)	Sensor S3 $\varepsilon_{S3,E,sens., up}$ (με)	Apparent shear strain $\tilde{\gamma}_{cell,zy,E,sensor}$ (με)	Apparent shear stress $\tilde{\tau}_{cell,zy,sensor}$ (MPa)	Apparent norm. strain $\tilde{\varepsilon}_{C_s,z,E,sensor}$ (με)	Apparent norm. stress $\tilde{\sigma}_{C_s,z,sensor}$ (MPa)
Zero (before)	0.0	0.0	0.0	0.00	0.0	0.00
2 above	−3.9	2.1	−6.0	−0.15	−2.3	−0.14
1 above	−4.6	2.8	−7.4	−0.18	−2.5	−0.15
Main	−4.3	2.4	−6.7	−0.16	−2.5	−0.15
1 below	−3.3	1.9	−5.2	−0.13	−1.8	−0.11
2 below	−2.0	0.8	−2.7	−0.07	−1.6	−0.10
3 below	−0.5	−0.4	−0.1	0.00	−1.1	−0.07
4 below	0.7	−1.1	1.8	0.04	−0.5	−0.03
Zero (after)	0.3	−0.2	0.5	0.01	0.2	0.01

was improved 10 times and resulted in a standard error of 0.1 με (see Section 4.2.3). Measurements resulting from each loading step are shown in Table 6.12.

Figure 6.14a shows the measured strain change in sensors S2 and S3 for each loading step. Note that the strain change is highest by absolute value for both sensors when the Lead Row is positioned on the thread "1 above"; in that case the force transmitted from that thread to the Main thread (instrumented with crossed topology) is the highest, creating the highest torsional moment and thus the highest corresponding shear strain. Another point of particular interest is the load step "4 below," when all persons are standing below Main thread: at that load step, the support of Main thread provided by thread "1 below" is partially removed due to deformation of threads below, and torsion due to self-weight of staircase is partially released resulting in inversion of trend of strain change in both sensors. Finally, once the load was removed, both sensors measure change approximately equal to zero, which confirms that potential influence of temperature and creep to measurements was minimal and justifies assumption that they can be neglected.

Note that strain values measured by sensors are not taken on the surface of monitored cross-section; due to decorative indentation, measurements were taken approximately 46 mm from the surface (see Figure 6.12c,d). That is the reason why they are not representative of the values of the strain on the surface. However, due to distance, their absolute values are greater than the values on the surface, thus they can be considered conservative. To simplify analysis in this section, we will use these values to calculate apparent shear and normal strains at Point C_s (see Figure 6.12d), as well as apparent shear and normal stresses at that point. Attribute "apparent" reflects the fact that these calculated values are not obtained from the measurements taken at the surface of the monitored cross-sections, but at 46 mm from the surface.

Example 6.9 *Evaluation of Shear Strain and Shear Stress*

Apparent shear strain was evaluated using Expression 6.50 and presented in Table 6.12 and Figure 6.14b. Symbol "~" in the table indicates that the calculated shear strain is apparent. Installation of sensors was performed with high accuracy, and error in angle α was less than 1°,

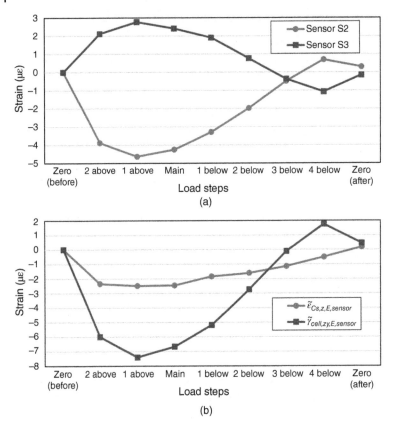

Figure 6.14 (a) Measurements of sensors S2 and S3 as functions of load steps, and (b) evaluations of apparent shear strain and apparent normal strain in the direction of centerline, as functions of load steps.

resulting in negligible error due to installation (for $\alpha = 45° \pm 1°$, $1/\sin(2\alpha) = 1$ with error smaller than -10^{-3}). Thus the uncertainty in evaluation of apparent shear strain is estimated as $u(\widetilde{\gamma}_{cell,zy,E,sensor}) = \sqrt{0.1^2 + 0.1^2} = 0.14 \, \mu\varepsilon$.

Once the apparent shear strain was evaluated, the change in apparent shear stress was evaluated using Expression 6.52 and presented in Table 6.12 (error analysis omitted for conciseness). Young's modulus of Imperial Danby Marble was estimated to $E_{cell} = 61.36 \pm 0.35$ GPa and Poisson's ration to $v_{cell} = 0.25 \pm 0.005$ (Jin 2020). Shear modulus was calculated as follows (detailed error analysis omitted for conciseness):

$$G_{cell} = \frac{E_{cell}}{2(1 + v_{cell})} = \frac{61.36}{2(1 + 0.25)} = 24.54 \pm 0.12 \, \text{GPa.}$$

Example 6.10 *Evaluation of Normal Strains and Normal Stress*

Apparent normal strain in direction of centerline was evaluated using Expression 6.61 and presented in Table 6.12 and Figure 6.14b. Symbol "~" in the table indicates that the calculated normal strain is apparent. Uncertainty of the evaluation is $u(\widetilde{\varepsilon}_{C_s,z,E,sensor}) = 0.15 \, \mu\varepsilon$ (detailed error analysis omitted for conciseness). Finally, the change in apparent normal stress in the direction of centerline is evaluated by multiplying the apparent normal strain with the Young's modulus (error analysis omitted for conciseness). Stress changes evaluated in Table 6.12 show very low values compared with the flexural strength of the marble, which was evaluated to be approximately 8.5 MPa (Jin 2020). That is the reason why a rigorous application of Criteria I–VI is not presented in this section.

The shear strain measured at location of crossed sensors was the consequence of three generalized forces: shear forces in directions of both principal cross-sectional axes and torsional moment. It was, therefore, impossible to identify shear components caused by each of these influences separately, and consequently, it was impossible to determine changes in shear forces in two principal directions as per Expression 6.56. Initial numerical modeling has shown that the shear strain at location of crossed sensors is predominantly consequence of torsional moments, and that influence of shear forces is significantly smaller. However, due to complexity of the shape of the cross-section, analytical expression that relates torsional moment with the measured shear strain is complex too, and impractical to derive, and establishment of this relationship requires the use of numerical modeling, which detailed presentation is out of the scope of this introductory book.

Figure 6.14b reveals an important fact regarding structural behavior of the staircase: as the Lead Row moves from position "2 above" through "Main" the apparent shear strain increases at "1 above" then decreases, while apparent normal strain remains approximately constant, i.e., does not indicate significant changes in bending moment; this indicates that Main thread supports the thread above, when the latter is loaded with Lead Row, mostly by torsion. This also explains, in part, extraordinary load capacity of such a slender staircase.

6.8 Examples of Global SHM Analysis

6.8.1 General Observations

Sections 6.1 and 6.2 introduced and proposed implementation steps for SHM at global scale. As an illustration, Section 6.3 presented two examples of global-scale evaluation of structural parameters, while subsequent sections presented algorithms for evaluation of these and other structural parameters at local cell scale.

For the examples shown in Section 6.3, sensor topologies and locations of sensors were selected based on structural analysis, i.e., the expected behavior of the monitored structures informed sensor placement. In the case of pile foundation, it was the knowledge that pile is predominantly loaded axially, in the case of bridge, it was the knowledge that the bridge was predominantly loaded with both bending (due to self-weight) and axial force (due to prestressing). Thus, the former was mostly equipped with simple topology, and the latter with parallel topology. Number of cells was informed by multiple criteria: locations of maximum and minimum absolute influences (max. tensional and compressive normal force, max. positive and negative bending moments), spatial resolution (density or mutual closeness of instrumented cells to achieve certain accuracy in evaluation), and accuracy measures of the monitoring system.

In the examples shown in Figures 6.2b and 6.3c, the evaluation of global behavior of structures was performed by analyzing each cell separately, using algorithms presented in Sections 6.5–6.6, and presenting them all together, without taking into account interactions between the cells. However, the evaluation shown in Figure 6.2c requires analysis of interactions between the cells: average frictional force in soil between the two midpoints of neighboring cells is evaluated as the average change in normal force in the pile between these two cells.

The aim of this section is to provide with two additional examples of global-scale SHM that take into account measurements from multiple cells in order to evaluate global structural behaviors.

These are, namely, soil–structure interaction (friction) along the pile, and deformed shape of prismatic beams. While these two examples are not exhaustive, their purpose is to illustrate the realm of possibilities when performing the data analysis and evaluation of global structural behaviors. In order to make presentations concise and avoid repetition of concepts that were presented earlier in this book, detailed, comprehensive evaluations of measures of accuracy are intentionally omitted, as well as the rigorous applications of Criteria I–VI for identification of unusual structural behaviors.

6.8.2 Evaluation of Change in Average Frictional Stress Distribution $\Delta\tau_{sfr,cell,i/i+1}$ between Pile and Soil

Evaluation of average friction between pile and soil is illustrated using example described in Sections 6.2.2 and 6.5.5. Figure 6.2a shows the division of pile in cells and sensor positions in each cell. This sequence of simple topologies placed one after another in the same structural element is referred to in literature as "enchained" simple topology (Glisic and Inaudi 2007). Diagram in Figure 6.2b shows distributions of normal force along the pile for two load instances, and the method for their determination is presented in Sections 6.5.1 and 6.5.5.

Diagram in Figure 6.2c shows distribution of average friction between the pile and soil, along the length of the pile for the pullout test described in Sections 6.2.2 and 6.5.5. This subsection presents the algorithm and assumptions behind this evaluation. Let us observe two neighboring halves of the two neighboring pile cells equipped with simple topologies, as shown in Figure 6.15. In addition, let us adopt the following assumptions:

- The neighboring cells are approximately cylindrical with approximately equal diameters $d_{cell,i/i+1} = d_{cell,i} = d_{cell,i+1}$ (see Figure 6.15).
- Friction between the pile and soil along the length of the first cell can be neglected (this assumption is taken from, and explained in, Section 6.5.5).

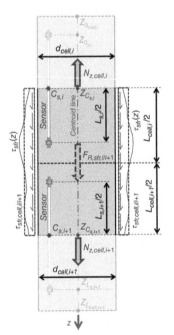

Figure 6.15 Schematic representation of forces acting on two neighboring halves of two neighboring cells of a pile instrumented with enchained simple topology.

- The soil friction $\tau_{sfr}(z)$ between the two midpoints, $z_{C_s,i}$ and $z_{C_s,i+1}$, of neighboring cells, $cell_i$ and $cell_{i+1}$ (see Figure 6.15), is approximately centrally symmetric with respect to centroid line of the pile (i.e., uniformly distributed around perimeters of cells' cross-sections); consequence of this assumption is that change in resultant frictional force, $\Delta F_{R,sfr,i/i+1}$, with respect to some reference time t_{Ref}, generated between the two observed midpoints of cells, acts along centroid line of the pile (see Figure 6.15), and the following expression is valid:

$$\Delta F_{R,sfr,i/i+1}(t) = \int_{A_{S,i/i+1}} \Delta\tau_{sfr}(z,t)\mathrm{d}A = \pi d_{cell,i/i+1} \int_{z_{C_s,i}}^{z_{C_s,i+1}} \Delta\tau_{sfr}(z,t)\mathrm{d}z$$

$$= \Delta\tau_{sfr,cell,i/i+1}(t)\frac{\pi d_{cell,i/i+1}(L_{cell,i} + L_{cell,i+1})}{2} \tag{6.63}$$

where $A_{S,i/i+1}$ is the lateral surface of the two halves of cells i and $i+1$ (i.e., the surface in contact with soil), $\Delta\tau_{sfr}(z,t)$ is change in friction between soil and pile at location z with respect to reference time t_{Ref}, and $\Delta\tau_{sfr,cell,i/i+1}(t)$ is change in average frictional stress between soil and pile over observed halves of cells i and $i+1$ (with respect to reference time t_{Ref}).
- Self-weight of the pile between the two midpoints of observed cells ($z_{C_s,i}$ and $z_{C_s,i+1}$) can be neglected.

Considering the above assumptions, and given that the change in normal force $\Delta N_{z,cell,i}(t)$ can be evaluated from strain measurements using Expression 6.17 (see Sections 6.5.4 and 6.5.5), the change in resultant frictional force, $\Delta F_{R,sfr,i/i+1}(t)$, can be evaluated from equations of equilibrium, as follows (see Figure 6.15):

$$\Delta F_{R,sfr,i+1,sensor}(t) = \Delta N_{z,cell,i,sensor}(t) - \Delta N_{z,cell,i+1,sensor}(t) \tag{6.64}$$

where subscript "*sensor*" indicates evaluation from strain sensor measurements.

Finally, by combining Expressions 6.63 and 6.64, the change in average frictional stress, $\Delta\tau_{sfr,cell,i/i+1,sensor}(t)$, along the length of every pair of neighboring halves of cells i and $i+1$, can be evaluated from strain measurements as follows:

$$\Delta\tau_{sfr,cell,i/i+1,sensor}(t) = 2\frac{\Delta N_{z,cell,i,sensor}(t) - \Delta N_{z,cell,i+1,sensor}(t)}{\pi d_{cell,i/i+1}(L_{cell,i} + L_{cell,i+1})} \tag{6.65}$$

The expression above is applicable directly for $i \geq 2$. However, given the assumption that friction in the first cell can be neglected, the length belonging to cell 1 should not be considered as developing friction stresses, i.e., for $i = 1$, the value $L_{cell,1} = 0$ should be inserted in Expression 6.65.

Thus, for distribution of change in normal force along the length of the pile (with respect to reference time corresponding to zero load applied to pile), which is evaluated from strain sensor measurements as shown in Figure 6.2b, the distribution of the change in average frictional stress $\Delta\tau_{sfr,cell,i/i+1}$ along the length of pile is given in Figure 6.2c; the values shown in the figure were calculated using Expression 6.65.

Note that two unusual behaviors can be identified in Figure 6.2c. First, the friction between Cells 2 and 3 in Figure 6.2c decreases from 25.9 kN/m^2 to 24.7 kN/m^2 after load increased from 285.7 tons to 400 tons, which indicates that the soil–pile interaction in that layer of soil practically met frictional capacity at or below 285.7 tons, and any further increase of load results in sliding of pile through that layer of soil. Second, the friction between Cells 4 and 5 only slightly increased with increase of load, i.e., from 16.7 kN/m^2 to 19.0 kN/m^2, which indicates that the soil–pile interaction in that layer of soil is close to frictional capacity or exceeded it.

In summary, this example illustrates how the global structural behavior of a structural element (pile) can be evaluated by linking derivatives of strain measurements from different cells of the observed structural element.

6.8.3 Evaluation of Deformed Shape and Deflection of a Prismatic Beam Subjected to Bending

Strain in a beam-like structure is directly correlated to its deformed shape and deflection. If the hypotheses of linear theory apply, axial deformation (e.g., due to normal force) and bending (e.g., due to bending moments) can be analyzed separately. Example of analysis of axial deformation is given in Section 6.5.5, for pile foundation. This section deals with deformed shape and deflection of beams subjected to bending.

Deformed shape and deflection, although similar, are actually different, and often confused, due to their similarity. Deformed shape of a beam is consequence of strain only, and is described by the change in shape of its centroid line after the generation of the strain (i.e., after the structure is deformed due to load, temperature changes, etc.). Given that it depends only on strain, which is internal property of structure, the deformed shape is also internal property of structure. Deflection or displacement diagram is consequence of both deformed shape and global, rigid-body displacement of the beam. Thus, it depends on both strain and boundary conditions of the beam. This is illustrated in Figure 6.16, assuming small deformation in linear theory of beams.

In Figure 6.16, deformed shape of the beam is depicted with light gray contour lines, and described with change in position of centroid line, $w(z)$, assuming that the beam did not move as a rigid body. Deflection (i.e., diagram of orthogonal displacements) is depicted with black contour lines, and described with change of position of centroid line, $v(z)$, taking into account its movement as a rigid body (e.g., translation for value $v_{0_{Beam}} = v(z_{0_{Beam}})$ and rotation for value ψ_{Beam}). The following relationships apply:

$$\frac{d^2 w(z,t)}{dz^2} = \frac{d^2 v(z,t)}{dz^2} = -\kappa(z,t)$$

$$\frac{dw(z,t)}{dz} = \varphi(z,t)$$

$$\frac{dv(z,t)}{dz} = \phi(z,t) = \varphi(z,t) + \psi_{Beam}(t) \tag{6.66}$$

where $v(z,t)$ denotes deflection at time t (with respect to some reference time t_{Ref}), $w(z,t)$ denotes deformed shape at time t, $\kappa(z,t)$ denotes curvature at time t, $\Phi(z,t)$ denotes rotation of cross-section

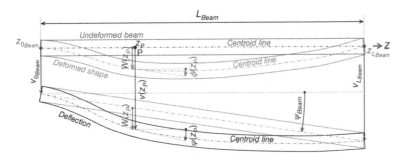

Figure 6.16 Illustrative example of difference between deformed shape and deflection (displacement) diagram (assumption of small deformation applies).

in deformed shape at time t, $\phi(z,t)$ denotes rotation of cross-section in deflection at time t, and $\psi_{Beam}(t)$ denotes rigid-body rotation of the beam.

While differential equation for deflection and deformed shape are identical (second derivative equals to negative curvature), their solutions depend on boundary conditions, and thus they are not identical in general (they might be identical in some special cases). These boundary conditions are the following:

Boundary conditions for deformed shape: $\quad w(z_{0_{Beam}}, t) = 0$ and $w(z_{L_{Beam}}, t) = 0$

Boundary conditions for deflection: $\quad v(z_{0_{Beam}}, t) = v_{0_{Beam}}(t)$ and $v(z_{0_{Beam}}) = v_{L_{Beam}}(t)$

$$(6.67)$$

where $v_{0_{Beam}}(t)$ and $v_{L_{Beam}}(t)$ are displacements of the extremities of the beam (see Figure 6.16).

Displacements of the extremities of the beam are correlated to rotation of the beam ψ_{Beam} as follows (assuming that hypothesis of small displacements applies), see Figure 6.16:

$$\psi_{Beam}(t) = \frac{v_{L_{Beam}}(t) - v_{0_{Beam}}(t)}{L_{Beam}} \tag{6.68}$$

In addition, for any point P with coordinate z_P, the following relationships are valid, see Figure 6.16:

$$\psi_{Beam}(t) = \frac{\left(v(z_P, t) - w\left(z_P, t\right)\right) - v_{0_{Beam}}(t)}{z_P - z_{0_{Beam}}} = \frac{v_{L_{Beam}}(t) - (v_P(t) - w_P(t))}{z_{L_{Beam}} - z_P} \tag{6.69}$$

Expressions 6.68 and 6.69, as well as Expressions 6.66, and equivalent equations derived using their combinations, show that in order to set boundary conditions for the deflection of a beam, it is sufficient to determine a set of two parameters that describe rigid-body movement:

- Displacements of two points (e.g., at extremities, but also any two points along the beam), or
- Displacement of one point (i.e., any point along the beam) and rotation of any cross-section of the beam ϕ (i.e., at any point along the beam).
  or
- Displacement of one point (i.e., any point along the beam) and rotation of the beam ψ_{Beam},

Note that all three sets of parameters above are external, independent from strain in the structure, and thus they cannot be evaluated using strain sensors. To evaluate these parameters, it is necessary to use sensors that measure displacements, such as LVDTs, fiber-optic (FO) displacement sensors, vibrating wire (VW) displacement sensors, sensors based on Global Positioning System – GPS, digital image correlation – DIC, etc., combined, if necessary, with sensors that measure rotation, i.e., tilt-meters or inclinometers based on various principles such as FO or VW.

As opposed to deflection, deformed shape depends only on strain and thus, it can be evaluated using strain sensors. Let us observe a beam instrumented with a set of parallel topologies as shown in Figure 6.17. Expression 6.21 can be used to evaluate total curvature in each cell, while Expressions 6.22–6.25 can be used to evaluate curvature constituents (elastic, thermal, creep, and shrinkage constituent, respectively). Once the curvatures are evaluated in each cell, i.e., the curvature distribution along the beam is known, then deformed shape can be evaluated by double integration of Expression 6.66 and setting boundary conditions for deformed shape given in Expression 6.67. This is summarized in Expression 6.70.

$$w(z, t) = -\int \left(\int \kappa(z, t)dz\right) dz + C_1(t)z + C_0(t) \tag{6.70}$$

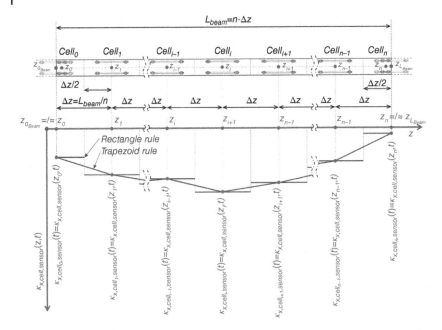

Figure 6.17 Schematic representation of beam or beam segment equipped with uniformly distributed (equidistant) parallel topologies, distribution of curvature along the beam measured by each topology, and approximations of distribution using rectangle and trapezoid rules.

where $C_1(t)$ and $C_0(t)$ are constants of integration that fulfil boundary conditions set in Expression 6.67.

There are multiple numerical integration methods that can be used to solve Expression 6.70 (e.g., polynomial interpolation of entire set of points, see Vurpillot et al. 1998). Only the two most common methods, rectangle rule and trapezoid rule, are briefly summarized here, along with error resulting from double-integration using these two methods. Rectangle rule and trapezoid rules represent subset of Newton–Cotes formulae for numerical integration using polynomial degrees 0 and 1, respectively (e.g., see Bradie 2006).

Let us assume that monitored beam of length L_{Beam} is instrumented with $n+1$ parallel topologies, equally spaced at distance $\Delta z = L_{Beam}/n$, as shown in Figure 6.17. Regarding extremities of the beam, three possible scenarios are considered:

1. The beam is individual, i.e., does not extend beyond points with coordinates $z_{0_{Beam}}$ and $z_{L_{Beam}}$; in this case $Cell_0$ and $Cell_n$ should be equipped with sensors having sufficiently short gauge length, so their midpoints are close to beam extremities, i.e., $z_0 \approx z_{0_{Beam}}$ and $z_n \approx z_{L_{Beam}}$; this is necessary in order to evaluate curvature at beam extremities.

2. The beam is not individual, i.e., it extends beyond points with coordinates $z_{0_{Beam}}$ and $z_{L_{Beam}}$, on both ends (e.g., beam is part of continuous girder with support at one or both points with coordinates $z_{0_{Beam}}$ and $z_{L_{Beam}}$, see for example beam P11–P12 in Figure 6.3); in this case $Cell_0$ and $Cell_n$ could be equipped with sensors having the same gauge length as the sensors in other cells, and in this case $z_0 = z_{0_{Beam}}$ and $z_n = z_{L_{Beam}}$.

3. The beam ends at one extremity but extends at the other (e.g., see beam P12–13 in Figure 6.3). In this case, Point 1 above applies for ending extremity and Point 2 to extending extremity.

Note that, in general, while the gauge lengths of sensors belonging to the same cell should be equal, the gauge lengths of sensors in different cells can be different; important is that cell midpoints are uniformly distributed along the length of the beam.

Lower graph in Figure 6.17 shows distribution of curvature measurements at time t, $\kappa_{x,cell_i,sensor}(t) = \kappa_{x,cell,sensor}(z_i, t)$. Type of curvature constituent (elastic, thermal, creep, or shrinkage) is omitted in the graph and in the expressions in the following text, as they apply to total curvature and each constituent individually (see Expressions 6.21–6.25). Deformed shape with values at cell midpoints, z_i, can be calculated using Expressions 6.71 and 6.72, for rectangle rule and trapezoid rule, respectively.

Rectangle rule:

$$
\begin{aligned}
w_R(z_i, t) = w_{cell_i,R}(t) = &-\frac{L_{Beam}^2}{n^2} \sum_{j=1}^{i} \kappa_{x,cell,sensor}(z_{j-1}, t)(i-j+1) \\
&+ \frac{iL_{Beam}^2}{n^3} \sum_{j=1}^{n} \kappa_{x,cell,sensor}(z_{j-1}, t)(n-j+1) \\
= &-\frac{L_{Beam}^2}{n^2} \sum_{j=1}^{i} \kappa_{x,cell_{j-1},sensor}(t)(i-j+1) \\
&+ \frac{iL_{Beam}^2}{n^3} \sum_{j=1}^{n} \kappa_{x,cell_{j-1},sensor}(t)(n-j+1)
\end{aligned}
\tag{6.71}
$$

where $i \in \{1, 2, \ldots n\}$ and $w_R(z_0, t) = 0$.

Trapezoid rule:

$$
\begin{aligned}
w_T(z_i, t) = w_{cell_i,T}(t) = \\
&-\frac{L_{Beam}^2}{n^2} \sum_{j=1}^{i} \kappa_{x,cell,sensor}(z_{j-1}, t)(i-j+1) + \frac{iL_{Beam}^2}{n^3} \sum_{j=1}^{n} \kappa_{x,cell,sensor}(z_{j-1}, t)(n-j+1) \\
&-\left[\kappa_{x,cell,sensor}(z_i, t) - \frac{i}{n}\kappa_{x,cell,sensor}(z_n, t) - \left(1 - \frac{i}{n}\right)\kappa_{x,cell,sensor}(z_0, t) \right] \\
=&-\frac{L_{Beam}^2}{n^2} \sum_{j=1}^{i} \kappa_{x,cell_{j-1},sensor}(t)(i-j+1) + \frac{iL_{Beam}^2}{n^3} \sum_{j=1}^{n} \kappa_{x,cell_{j-1},sensor}(t)(n-j+1) \\
&-\left[\kappa_{x,cell_i,sensor}(t) - \frac{i}{n}\kappa_{x,cell_n,sensor}(t) - \left(1 - \frac{i}{n}\right)\kappa_{x,cell_0,sensor}(t) \right]
\end{aligned}
\tag{6.72}
$$

where $i \in \{1, 2, \ldots n\}$ and $w_T(z_0, t) = 0$.

It is important to highlight that the integration methods presented in Expressions 6.71 and 6.72 do not consider any particular cause of deformation, i.e., they are independent of the loading. The only requirement for them to be applicable is the equidistant distribution of cells with parallel topologies.

The accuracy of integration methods depends not only on the number of cells along the beam but also on the strain distribution, i.e., on loading. Typical loading on a beam structure can be split into three elementary cases as follows:

Case 1: Linear distribution of curvature along the beam; in this case, no load is applied along the beam, except at extremities, resulting in linear distribution of curvature along the beam (e.g., moments and forces applied at extremities will result in linear bending moment distribution along the beam, which in turns results in linear curvature distribution along the beam with invariable cross-sectional properties, see Figure 6.18).

Case 2: Parabolic distribution of curvature along the beam; in this case, a uniformly distributed load is applied, resulting in a parabolic distribution of curvature; the loads applied at beam extremities (as in Case 1 above) may or may not be present (see Figure 6.18).

Case 3: Bilinear (broken line) distribution of curvature along the beam; a concentrated transversal force is applied on a beam, at a point between the extremities, resulting in bilinear (broken line) distribution of curvature, with change in slope at a point of application of the force; the loads applied at beam extremities (as in Case 1 above) may or may not be present. Two sub-cases will be considered: (a) point of application of the force is on sensors, and (b) point of application of the force is between sensors (see Figure 6.18).

It was demonstrated in a published work that an uncorrected error in estimation of deformed shape at point z_i at time t, $\delta w(z_i,t)$, due to integration using rectangle and trapezoid rules can be evaluated using Expression 6.73 and Table 6.13 (Sigurdardottir et al. 2017).

$$\delta w_{cell_i,uncorrected,R/T}(t) = \delta w_{uncorrected,R/T}(z_i, t)$$

$$= C_{\delta,R/T} \kappa_{x,cell,sensor}(z_i, t) \frac{L^2_{Beam}}{n^2} = C_{\delta,R/T} \kappa_{x,cell,sensor}(z_i, t) \Delta z^2$$

$$= C_{\delta,R/T} \kappa_{x,cell_i,sensor}(t) \frac{L^2_{Beam}}{n^2} = C_{\delta,R/T} \kappa_{x,cell_i,sensor}(t) \Delta z^2 \tag{6.73}$$

where C_δ represents error constant that depends on load case as shown in Figure 6.18 and Table 6.13, and where subscript R/T denotes that expression applies for both rectangle and trapezoid rule, when appropriate error constant is chosen. i.e., $C_{\delta,R}$ for rectangular rule and $C_{\delta,T}$ for trapezoidal rule.

The reason why the error above is considered as uncorrected is explained later in the text.

Expression 6.73 contains the term for curvature, $\kappa_{x,cell,sensor}(z_i,t)$, which is evaluated using strain measurements and is subject to errors due to accuracy of monitoring system and tolerances in sensor positioning due to installation (see Section 6.6.7); consequently, the total error in evaluation of deformed shape should account for error in evaluation of curvature by applying error propagation formula in Expression 6.73.

Values of error constant C_δ presented in Table 6.13 are calculated based on linear theory of beams; thus, they contain epistemic model error that depends on how good the linear theory of beams is in estimating deformed shape. While the estimation of epistemic model error of linear theory could not be found in the literature, general agreement is that it is negligible (assuming that hypotheses of linear theory are applicable to the monitored structure).

Table 6.13 Error constant C_δ for different elementary load cases and different integration methods; subscript "*R*" denotes rectangle rule, subscript "*T*" denotes trapezoid rule.

Distribution of curvature	Integration using rectangle rule	Integration using trapezoid rule
Case 1: Linear	$C_{\delta,R,Case\ 1} = 0$	$C_{\delta,T,Case\ 1} = 0$
Case 2: Parabolic	$C_{\delta,R,Case\ 2} = 2/24$	$C_{\delta,T,Case\ 2} = -4/24$
Case 3a: Bilinear (broken line), force on sensors	$C_{\delta,R,Case\ 3a} = 4/24$	$C_{\delta,T,Case\ 3a} = -2/24$
Case 3b: Bilinear (broken line), force between sensors	$C_{\delta,R,Case\ 3b} = 1/24$	$C_{\delta,T,Case\ 3b} = -5/24$

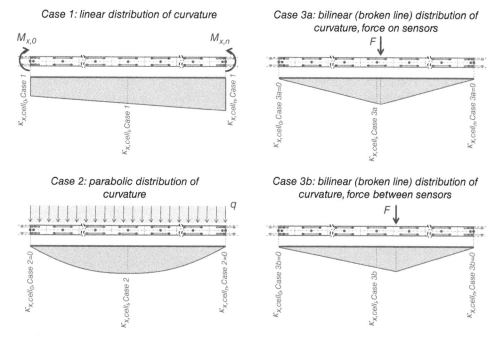

Figure 6.18 Elementary load cases and resulting curvature distributions; Case 1: moments at extremities of the beam, Case 2: uniformly distributed load along the beam, and Case 3: concentrated force, with subcases 3a: the force acts on sensors, and 3b: the force acts between sensors.

Expression 6.73 shows that the uncorrected error in evaluation of deformed shape at midpoint of *Cell i* due to double integration has the following important properties:

A. Uncorrected error depends on type of load, i.e., error constant $C_{\delta,R/T}$, which can be accurately determined based on Table 6.13; error constant $C_{\delta,R/T}$ is positive for rectangle rule and negative for trapezoid rule, which implies that double integration with rectangle rule provides upper limit (overestimate) and trapezoid rule provides lower limit (underestimate) for evaluated deformed shape.

B. Simplicity of expression enables calculation of error in the case of combined load scenarios by applying error propagation formulae, which can be symbolically written as:

$$\delta w_{cell_i,uncorrected,R/T,Cases\,1+2+3a+3b}(t) = \sum_{[j]\in\{1,2,3a,3b\}} \delta w_{cell_i,uncorrected,R/T,Case[j]}(t) =$$

$$= \sum_{[j]\in\{1,2,3a,3b\}} C_{\delta,R/T,Case[j]}\,\kappa_{x,cell_i,sensor,Case[j]}(t)\frac{L_{Beam}^2}{n^2}$$

$$= \sum_{[j]\in\{1,2,3a,3b\}} C_{\delta,R/T,Case[j]}\,\kappa_{x,cell_i,sensor,Case[j]}(t)\Delta z^2 \quad (6.74)$$

where subscript R/T denotes that expression applies for both rectangle and trapezoid rule.

C. Uncorrected error is proportional to curvature evaluated from the pair of parallel sensors installed in *Cell i*, i.e., larger measured curvature implies larger error in evaluation of value of deformed shape at that location, which is, in turn, dependent on the location of the observed *Cell i* within the beam and on the type of applied load.

D. Uncorrected error is proportional to the square of the distance between midpoints of neighboring cells Δz (i.e., inversely proportional to the square of number of spaces between the cells n); this means that decrease in distance between the midpoints (i.e., increase in number of instrumented cells $n + 1$) would result in quadratic decrease in error in evaluation of deformed shape at every location along the beam.

Point C above highlights that the uncorrected error at an observed cell depends on the curvature at that cell, and this applies to the first and last cell, too. However, in these two cells, by definition, the value of deformed shape is set to zero. This means that if an error is allowed at these points, this error will further affect the errors in other points. Thus, to correct the errors in other points, a corrective error term has to be added to uncorrected error, and this term is obtained by imposing corrected error equal to zero to cells at end points, which in turn is made by linear shift (rotation in linear theory beams). In practice, this is obtained by applying the following expression:

$$\delta w_{cell_i,corrective,R/T}(t) = -\frac{\delta w_{cell_n,uncorrected,R/T}(t) \cdot z_i + \delta w_{cell_0,uncorrected,R/T}(t) \cdot (L_{Beam} - z_i)}{L_{Beam}} \quad (6.75)$$

Expression 6.75 shows that the absolute value of corrective error is smaller than the greater of the two absolute values of uncorrected errors at beam extremities, i.e., $|\delta w_{cell,i,corrective,R/T}| \leq \max(|\delta w_{cell,0,uncorrected,R/T}|, |\delta w_{cell,n,uncorrected,R/T}|)$. The corrected error can now be calculated by summing Expressions 6.73 and 6.75, i.e.:

$$\delta w_{cell_i,R/T}(t) = \delta w_{cell_i,uncorrected,R/T}(t) + \delta w_{cell_i,corrective,R/T}(t) \quad (6.76)$$

Point D above implies that the sensor network, i.e., the minimum number of instrumented cells, $n + 1$, can be designed based on a set (i.e., desired or preferred) accuracy δw_{set} in evaluation of the deformed shape, as shown in Expression 6.77. The set accuracy depends on monitoring project specifics, and is typically estimated (i.e., set) based on project requirements.

$$\text{If } \delta^* \varepsilon w_{cell,corrective,R/T}(t) \neq 0 : n_{R/T} + 1 \geq L_{Beam} \sqrt{2 \frac{C_{\delta,R/T} \kappa_{x,estimated,max}}{\delta w_{set,R/T}}} + 1$$

$$\text{If } \delta^* \varepsilon w_{cell,corrective,R/T}(t) = 0 : n_{R/T} + 1 \geq L_{Beam} \sqrt{\frac{C_{\delta,R/T} \kappa_{x,estimated,max}}{\delta w_{set,R/T}}} + 1 \quad (6.77)$$

where $\kappa_{x,estimated,max}$ is the maximal estimated curvature along the beam (e.g., estimated using numerical or analytical models).

Note that estimation of $\kappa_{x,estimated,max}$ includes curvatures at extremities of the beam and takes into account Expression 6.76. Expression 6.77 should be considered approximative, yet conservative, as it assumes that maximal absolute value of curvature occurs in both right-hand terms of Expression 6.76, thus number 2 in the first Expression 6.77; however, if, based on boundary conditions, the curvature at extremities of the beam are know to be null, then corrective error is also null and the second Expression 6.77 applies.

In the cases where estimation of maximal curvature along the beam is not practical, various other approaches can be used to assess the error in evaluation of deformed shape, and one of them is summarized here, for three common cases: distributed load and concentrated force (parabolic and bi-linear distribution of curvatures, Cases 2, 3a, and 3b above, see Figure 6.18). For all three cases, the curvatures at the extremities of the beam are expected to be null, and consequently, corrective error is null too. The approach is based on an analytical solution for the cells in the beam with the largest curvature and hance, based on Expression 6.73 (or Expression 6.76 by setting corrective error to zero), the largest error in evaluation of deformed shape.

For Case 2, max. curvature is in the middle of the beam (see Figure 6.18). Commonly, the cell in the middle of the beam is equipped with sensors, to capture max. curvature, and let us denote this cell with i, and coordinate of its midpoint with z_i. Then the following is valid based on linear theory of beams (derivation is omitted for conciseness):

$$w_{cell_i,Case\,2}(t) = \frac{5qL_{Beam}^4}{384EI} = \frac{qL_{Beam}^2}{8EI}\frac{5L_{Beam}^2}{48} = \frac{5}{48}\kappa_{x,cell_i,Case\,2}(t)L_{Beam}^2 \tag{6.78}$$

By division of left- and right-hand sides of Expression 6.73 with left- and right-hand sides of Expression 6.78, we obtain relative error in evaluation of deformed shape at midpoint of Cell i:

$$\delta^* w_{cell_i,R/T,Case\,2}(t) = \frac{\delta w_{cell_i,sensor,Case\,2}(t)}{w_{cell_i,Case\,2}(t)} = \frac{C_{\delta,R/T,Case\,2}}{\frac{5}{48}}(1 + \left|\delta^* \kappa_{x,cell_i,sensor,Case\,2}(t)\right|)\frac{1}{n_{R/T,Case\,2}^2},$$

$$\tag{6.79}$$

where $\left|\delta^* \kappa_{x,cell_i,sensor,Case\,2}(t)\right|$ represents absolute value of relative error in measurement of curvature in *Cell i* at time t. From here we obtain minimal number of cells to be instrumented by substituting relative error in deformed shape at *Cell i* with the set (desired, preferred) relative error $\delta^* w_{set}$ and relative error in curvature measurement with limits of relative error $\left|\delta^* \kappa_{x,cell_i,sensor}\right|$, i,e, by substituting Expression 6.79 into second Expression 6.77, as follows.

$$n_{R,Case\,2} + 1 \geq \sqrt{\frac{\left|C_{\delta,R,Case\,2}\right|}{\frac{5}{48}} \cdot \frac{1 + \left|\delta^* \kappa_{x,cell_i,sensor}\right|}{\delta^* w_{set}}} + 1 = \sqrt{\frac{4}{5} \cdot \frac{1 + \left|\delta^* \kappa_{x,cell_i,sensor}\right|}{\delta^* w_{set}}} + 1$$

$$n_{T,Case\,2} + 1 \geq \sqrt{\frac{\left|C_{\delta,T,Case\,2}\right|}{\frac{5}{48}} \cdot \frac{1 + \left|\delta^* \kappa_{x,cell_i,sensor}\right|}{\delta^* w_{set}}} + 1 = \sqrt{\frac{8}{5} \cdot \frac{1 + \left|\delta^* \kappa_{x,cell_i,sensor}\right|}{\delta^* w_{set}}} + 1. \tag{6.80}$$

For Cases 3a and 3b, max. curvature is at the location of applied force z_F (see Figure 6.18). For Case 3a, the cell with sensors is at that location. For Case 3b, the curvatures in the two closest cells instrumented with sensors (left and right of the force, see Figure 6.18) are smaller than the maximal curvature in the beam (at force location), thus, based on Expression 6.73, the error in deformed shape at both locations will be smaller than at location of the force, and the latter can be used as the bound for the former. The following is valid for both Cases 3a and 3b (derivation is omitted for conciseness):

$$w_{Case\,3a,3b}(z_F, t) = \frac{F(t)(L_{Beam} - z_F)^2 z_F^2}{3EIL_{Beam}} = \frac{F(t)(L_{Beam} - z_F)z_F}{3EIL_{Beam}}(L_{Beam} - z_F)z_F$$

$$= \frac{1}{3}\kappa_{x,Case\,3a,3b}(z_F, t)(L_{Beam} - z_F)z_F \tag{6.81}$$

As mentioned above, for Case 3a, the sensors are installed in the cell at location of force; let us denote this cell with i and coordinate of its midpoint with z_i. Note that curvature at that cell will be smaller or equal than the curvature at location of force, z_F (they will be equal if $z_i = z_F$).

For Case 3b, the two cells closest to force application point will experience the two largest curvatures that can be measured by sensors. Let us denote grater of the two with i and coordinate of its midpoint with z_i. Note that curvature at that cell will be smaller than the curvature at location of force, z_F.

Similar to Case 2, by division of left- and right-hand sides of Expression 6.73 with left- and right-hand sides of Expression 6.81, we obtain relative error in evaluation of deformed shape at midpoint of *Cell i* (*Cell i* is defined above).

$$
\begin{aligned}
\delta^* w_{cell_i, R/T, Case\,3a,3b}(t) &= \frac{\delta w_{cell_i, sensor, Case\,3a,3b}(t)}{w_{cell_i, Case\,3a,3b}(t)} \\
&\geq \frac{C_{\delta, R/T, Case\,3}\kappa_{x, sensor, Case\,3a,3b}(z_i, t)L_{Beam}^2}{\frac{1}{3}\kappa_{x, Case\,3a,3b}(z_F, t)(L_{Beam} - z_F)z_F} \frac{1}{n_{R/T, Case\,3a,3b}^2}.
\end{aligned}
\tag{6.82}
$$

Taking into account that $\kappa_{x, Case\,3a,3b}(z_i, t)$ is equal to either $\kappa_{x, Case\,3a,3b}(z_F, t) \cdot z_i / z_F$ or $\kappa_{x, Case\,3a,3b}(z_F, t) \cdot (L_{Beam} - z_i)/(L_{Beam} - z_F)$, whichever is larger, which in turn is smaller than $\kappa_{x, Case\,3a,3b}(z_F, t)$, by substitution in Expression 6.82, then substituting into second Expression 6.77, and by rearranging the terms, we obtain Expression 6.83 for evaluation of number of cells needed to achieve the set relative error.

$$
n_{R/T, Case\,3a,3b} + 1 \geq \sqrt{\left|\frac{C_{\delta, R/T, Case\,3a,3b}}{\frac{1}{3}}\right| \cdot \frac{1 + \left|\delta^* \kappa_{x, cell_i, sensor}\right|}{\delta^* w_{set}} \frac{L_{Beam}^2}{z_F(L_{Beam} - z_F)} + 1}
\tag{6.83}
$$

The above Expression 6.83 can now be written for each case and each type of integration as shown in Expression 6.84.

$$
n_{R, Case\,3a} + 1 \geq \sqrt{\frac{1}{2} \cdot \frac{1 + \left|\delta^* \kappa_{x, cell_i, sensor}\right|}{\delta^* w_{set}} \frac{L_{Beam}^2}{z_F(L_{Beam} - z_F)} + 1}
$$

$$
n_{T, Case\,3a} + 1 \geq \sqrt{\frac{1}{4} \cdot \frac{1 + \left|\delta^* \kappa_{x, cell_i, sensor}\right|}{\delta^* w_{set}} \frac{L_{Beam}^2}{z_F(L_{Beam} - z_F)} + 1}
$$

$$
n_{R, Case\,3b} + 1 \geq \sqrt{\frac{1}{8} \cdot \frac{1 + \left|\delta^* \kappa_{x, cell_i, sensor}\right|}{\delta^* w_{set}} \frac{L_{Beam}^2}{z_F(L_{Beam} - z_F)} + 1}
$$

$$
n_{T, Case\,3b} + 1 \geq \sqrt{\frac{5}{8} \cdot \frac{1 + \left|\delta^* \kappa_{x, cell_i, sensor}\right|}{\delta^* w_{set}} \frac{L_{Beam}^2}{z_F(L_{Beam} - z_F)} + 1}.
\tag{6.84}
$$

To complete this subsection with an example, deformed shape of Streicker Bridge is determined after removal of formworks (see more detail about Streicker Bridge in Sections 6.2.3, and 6.6.7 and also Sections 4.5.5 and 4.5.7). To keep the demonstration concise and focused, only the deformed shape of span P10–P11 is analyzed, and only the error due to double integration of curvature is presented, while the analysis of other sources of error is omitted. Figures 4.10 and 6.3 show the sensor locations in the observed span P10–P11. The length between the midpoints of sensors installed in Cells P10 and P11 is $L_{Beam} = 18.20$ m. In total, five parallel topologies of sensors are installed approximately equidistantly along the span, i.e., $n = 5 - 1 = 4$, with mutual distance of approximately $L_{Beam}/n = \Delta z = 4.55$ m. The square of this term, occurring in Expressions 6.71–6.73, is calculated as $L_{Beam}^2/n^2 = \Delta z^2 = 20.7025$.

Table 6.14 summarizes the curvature measured in the span P10–P11 using pairs of parallel sensors along with deformed shapes evaluated using rectangle and trapezoid rules, as per Expressions 6.71 and 6.72. Figure 4.10 and insert in Figure 6.19a show the naming and locations of cells with parallel sensors (P10, P10q11, P10h11, P10qqq11, and P11). Curvature change in each cell, due to the removal of formworks during the construction of the bridge, was evaluated using

Table 6.14 Curvature distribution in span P10–P11 of Streicker Bridge due to removal of formworks, evaluation of deformed shape using rectangle and trapezoid rules, and estimation of errors due to double integration.

Location	Curvature change $\kappa_{cell,x,tot,sensor}$ ($\mu\varepsilon$/mm)	Deformed shape, rectangle rule $w_{cell,R}$ (mm)	Error in rectangle rule $\delta w_{cell,R}$ (mm)	Deformed shape, trapezoid rule $w_{cell,T}$ (mm)	Error in trapezoid rule $\delta w_{cell,T}$ (mm)
P10	−0.12	0.00	0.00	0.00	0.00
P10q11	0.14	4.81	0.54	3.18	−1.09
P10h11	0.21	6.73	0.76	4.45	−1.52
P10qqq11	0.09	4.30	0.65	2.35	−1.29
P11	−0.34	0.00	0.00	0.00	0.00

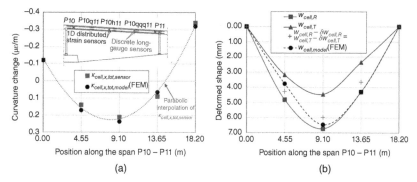

Figure 6.19 (a) Distribution of curvature measurements and curvature estimated using FEM, and (b) deformed shapes evaluated using rectangle and trapezoid rules, their difference with error due to double integration, and comparison with deformed shape estimated using FEM.

Expression 6.21. More precisely, campaign of measurements performed over all sensors just before the formworks were removed is used as the reference. Then, another campaign of measurements over all sensors performed immediately after the removal of formworks is used to evaluate curvature changes and deformed shape as the consequence of the removal and activation of self-weight.

Given that the cross-section of the bridge is approximately invariable along the length of span P10–P11, the activation of self-weight of the deck was expected to create parabolic distribution of curvature change along the span, which was confirmed by fitting the parabola to the curvature measurements as shown in Figure 6.19a (coefficient of determination $R^2 > 0.99$). For each of double-integration methods, the difference between the deformed shape and the error calculated using Expression 6.76 is presented as well; for approximately parabolic distribution of curvature, the error constants corresponding to Case 2 were used, i.e., for integration with rectangle rule $C_{\delta,R,Case\,2} = 2/24$ and for integration with trapezoid rule $C_{\delta,T,Case\,2} = -4/24$ (see Table 6.13). Given the parabolic distribution of the curvature, the differences between the deformed shape and the error are identical for the two methods (i.e., $w_{cell,R} - \delta w_{cell,R} = w_{cell,T} - \delta w_{cell,T}$), which is not necessarily the case for non-parabolic distributions. To simplify presentation, the observed differences are, for both methods, denoted with the same symbol, "+," in Figure 6.19b.

To evaluate the accuracy of the methods, the results of double integration are compared with the values estimated using FEM. The model was calibrated, based on curvature measurements and the quality of its fit to the measurements is shown in Figure 6.19a. Then deformed shape of observed span is shown in Figure 6.19b. As expected, the deformed shape estimated using FEM is found between the deformed shapes evaluated based on rectangle (overestimate) and trapezoid (under-estimate) rules. It was, for some cells, within the error margin of one method, and for some cells it was within the error margin of the other method. This is expected, as the limits of error shown in Figure 6.19b are rather non-conservative, given that they don't include the errors related to the accuracy of monitoring system, measurements of geometry of the bridge, ascertaining of positions of sensors, and the accuracy of FEM. As stated earlier, these errors were not accounted for, to simplify the presentation. These unaccounted errors provide explanation for the spatial variability of closeness of model to evaluations obtained using one or the other integration method. Nevertheless, in overall, the model is in good agreement with double-integration evaluations, even when only the error inherent to double integration is considered.

7

SHM at Integrity Scale: Interpretation and Analysis of Sensor Measurements at Location of Damage

7.1 Introduction to SHM at Integrity Scale

In this book, the term "integrity" is used to describe the quality of being whole and complete, or the state of being unimpaired. Hence, SHM at the integrity scale, or simply integrity monitoring, aims at identifying a lack of integrity, i.e., identifying the existence (presence) of an unusual behavior that led to a breach of integrity. A typical example of such an unusual behavior is cracking in concrete (unreinforced, reinforced, or prestressed) or steel structural elements, and the scope of this chapter will be limited to that form of unusual structural behavior. This chapter presents basic approaches to integrity monitoring; more advanced approaches can be found in Glisic and Inaudi (2007).

Several types of strain sensors can be used for integrity monitoring: discrete short-gauge, discrete long-gauge, and distributed 1D and 2D sensors (see Chapter 3 and Figure 3.26). Given that SHM at the integrity scale uses strain sensors, it also encompasses certain aspects of SHM at both local and global scales, i.e., it may provide data for the analysis presented in Chapters 5 and 6; however, depending on the type of sensor used, it may also have some distinct differences.

Section 3.7 presents in detail two principles applicable in crack detection: direct and indirect detection, along with examples of crack detection. To avoid repetition, only the main highlights from that section are presented here, but it is strongly suggested that readers refer to that section before reading this chapter.

As a reminder, indirect detection is based on the assumption that sensors are not necessarily placed in the proximity of locations of occurrence of cracks, in which case the latter may or may not affect the strain field at locations of sensors, and in many cases, sensors installed only 50 mm from the crack tip would not be likely to capture a change in strain field, i.e., to detect the crack (see also Figure 3.22). In some cases, sophisticated algorithms for damage detection can be used to analyze data from multiple sensors and identify cracking; however, their applicability in real-life settings is often challenged by changes in environmental conditions that induce strain changes that, in turn, can confuse the algorithms. In addition, in some cases, even sophisticated algorithms could not detect the cracking – for example in a determinate structure, a potential cracking would not affect the structural system, flow of forces, and strain field at the distances from cracking that are approximately equal to, or greater than the depth of the cross-section; thus, strain sensors installed at these locations would not experience any change in strain due to crack occurrence, and consequently, would not be able to detect the crack. In summary, the indirect detection approach might not be always successful in crack detection, i.e., might not be reliable in crack detection.

In the case of the direct detection, the main assumption is that the sensors are placed at or in proximity to the locations of potential crack occurrences. In that case, cracking will create a strain

Introduction to Strain-Based Structural Health Monitoring of Civil Structures, First Edition. Branko Glišić.
© 2024 John Wiley & Sons Ltd. Published 2024 by John Wiley & Sons Ltd.

field anomaly that is detectable by the sensors in form of an unusually high change in the measured strain (positive or negative), i.e., by simple thresholding (e.g., see Figure 3.23). Even a small crack opening would result in an unusually high strain change – positive if the sensor is installed on the crack (i.e., "crosses" crack opening) – and negative if the sensor is installed very close to the crack, but not on the crack (i.e., does not "cross" crack opening). This results in a very high sensitivity to crack occurrence, and a simple thresholding algorithm for crack detection can be used, which makes the direct detection approach reliable even in variable on-site conditions.

The above "simple thresholding" actually means applying Criterion I (see Expression 5.80) by setting the reference time "shortly" before the observed measurement time, with the assumption that cracks occur during a short period of time. This will be further discussed on a case-by-case basis throughout this chapter.

An issue that can arise with direct detection, in the cases where the sensor is in contact with the crack, is that the crack opening can damage the sensor, as it generates, theoretically, infinite strain in the sensor. To explain this statement in layman's words, let us observe the simplified definition of strain, i.e., $\varepsilon = \Delta L/L$, for some initial length L. In the case of crack opening, let us observe two opposite points of the crack mouth; the distance between them is $\Delta L > 0$ (i.e., equal to the crack opening); but $L = 0$ as the distance between the two observed points is null (they practically coincide with each other) if the crack is closed, i.e., if the material is put back in its original state before the cracking. To avoid damaging of sensors due to crack opening, it is necessary to make compromise between survivability of sensor and quality of strain transfer, which is described in detail in Section 4.6.4. The main idea is to choose the means of sensor installation or the types of sensor packaging that would allow for that compromise. More details are given in Section 4.6.4, but some parts are repeated in this chapter to ease the presentation.

The main challenge of the direct detection approach occurs in the situations where the exact location of damage is unknown or difficult to ascertain. In these cases, spatial coverage provided by sensors plays an important role, as shown in Figure 3.27 (see Section 3.7 for more details).

The focus of this chapter is on the principles of the implementation of a direct detection approach using short-gauge, long-gauge, and 1D distributed strain sensors, which are all commercially available. The main advantages and challenges are illustrated through real applications. Examples of applications of 2D distributed sensors, which are not yet fully commercially available, are given in Section 3.7.

To simplify the presentation, rigorous error analysis is not performed in the examples shown in this chapter; nevertheless, in real-life projects, rigorous error analysis should be performed, as presented in many examples in Chapters 4 through 6.

7.2 Crack Identification: Detection, Localization, and Quantification at Discrete-Sensor Level

7.2.1 Crack Detection and Localization Using Soft-Packaged Short-Gauge Strain Sensors Installed Via Adhesive

The main detection principles and implementation challenges for the use of soft-packaged short-gauge sensors installed via adhesive for integrity monitoring are presented in this subsection. As stated in Section 7.1, reliable integrity monitoring is based on direct detection; therefore, we assume that a short-gauge sensor is installed at or close to the position of the crack occurrence. If the sensor is installed at the location of crack occurrence, an unusually high positive strain change measured by the sensor would indicate the existence of the crack, as the latter would put the sensor in tension. Alternatively, the high (theoretically infinite) strain would damage the sensor, and in that case, the malfunction of the sensor would serve as an indication of crack occurrence. Contrary to this, if the sensor is installed close enough, but not at the crack location, the crack opening will result in strain relaxation at the location of the sensor, and the sensor will measure an unusually high negative strain change. To support the above statements, let us observe the results of a fatigue cycling test performed on a standard steel plate with dimensions of 25.4 × 25.4 cm, as shown in Figure 7.1a. The plate was equipped with holes for mounting on the machine that introduces the cycling load, and the notch, whose purpose was to initiate fatigue cracking in a controlled and predictive manner (see dashed line in Figure 7.1). The plate was instrumented with a dense arrangement of full-bridge strain gauges (see Figures 3.10c and 4.13a) with external dimensions of 11 × 14 mm and gauge lengths of approximately 5 mm, as shown in Figure 7.1a. A set of sensors was installed along the projected (estimated) crack propagation path, so they were expected to be in direct contact with the crack when it occurs and propagates; another set of sensors was installed about 10.5 mm from the projected crack propagation path, and thus, they were not expected to be in direct contact with the crack yet to be in its close proximity (see Figure 7.1a for details). Other sets of sensors were installed too, but their analysis is out of the scope of this section (the tests were performed in the frame of the development of the 2D Sensing Sheet, an example of a third-generation strain sensor; see Section 3.7.2 and Yao and Glisic 2015a). The measurements of two sensors, one from each group, are presented here for illustrative purposes: sensor C4, which was in direct contact with the crack, and sensor D5,

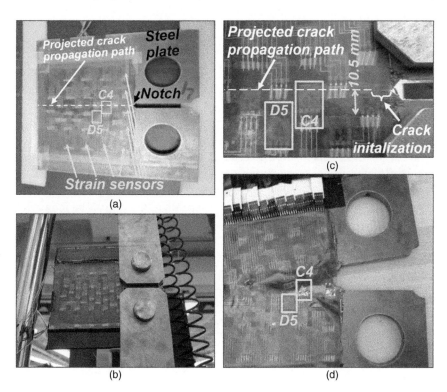

Figure 7.1 (a) View of the plate with sensors and projected crack propagation path; (b) plate mounted in the loading machine; (c) detail of the area around the notch after initialization of the crack; and (d) plate with a visibly open crack after completion of the test. Source: Reproduced from Yao and Glisic 2015a / MDPI /CC-BY 4.0.

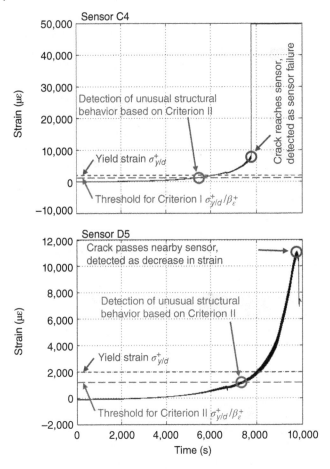

Figure 7.2 Strain measured by sensors C4 and D5 during the tests.

which was in proximity of the crack. The plate was installed in the loading machine using pins (see Figure 7.1b) and cyclically loaded to create fatigue, which initiated crack (see Figure 7.1c). As the cycling test continued, the crack propagated and eventually reached sensor C4 and then passed nearby sensor D5 (see Figure 7.1d). More detail regarding the test can be found in Yao and Glisic (2015a).

The results of the test are summarized in Figure 7.2, and they serve as the basis for explaining the crack detection principle using short-gauge sensors in this specific case.

All sensors in the test were installed using a very thin layer of epoxy adhesive with high bonding quality, which was supposed to guarantee a very high level of strain transfer between the plate and sensors (see Section 4.6.4). Both graphs in Figure 7.2 show the change in strain over time since the beginning of the tests. Given that the sensors were installed on plates before any load was applied, the measured strain is equal to the absolute strain. Yield strain and threshold for Criterion II (using $\beta_\varepsilon^+ = 1.25$, see Expression 5.84) are indicated with dashed lines. Both graphs show exponential growth in strain, and in both cases, the threshold for Criterion II was exceeded, which was the first indication of unusual structural behavior. This exceedance actually serves as a precursor of an incoming (propagating) crack, and in real-life settings, it would be sufficient to trigger intervention on the structure.

However, the precursor described above would exist in real-life settings only if similar conditions were present on the structure, i.e., in the case of cycling loading. In general case, the crack may occur suddenly, without precursor, and in that case, Figure 7.2 indicates two possible scenarios.

The first scenario, when there is a direct contact between sensor and crack, is illustrated with the graph corresponding to Sensor C4 (upper graph in Figure 7.2). When the crack tip reaches the sensor, damage to the sensor occurs due to a high level of strain transfer between the plate and the sensor (guaranteed with the selected epoxy adhesive). While the crack is successfully detected and localized (the location of the sensor indicates the location of the crack), the drawback of this detection is that the sensor is damaged, and it becomes impossible to quantify the size of the crack opening.

The second scenario, when there is no direct contact between the sensor and crack, is illustrated with the graph corresponding to Sensor D5 (lower graph in Figure 7.2). When the crack tip reaches the sensor proximity and passes nearby the sensors, the strain starts to decrease significantly, and this significant decrease (relaxation) is an indicator of the existence of a crack. The decrease of approximately $4000\,\mu\varepsilon$ is registered within approximately 110 seconds, without change of loading (i.e., cycling loading continued), and this sets conditions to apply thresholding by using Criterion I. By setting reference time $t_{Ref} = 9750$ seconds, for measurement made at 9860 seconds, we obtain $\varepsilon_{C4,y,tot,sensor}(9860 \text{ seconds}) \approx -4000\,\mu\varepsilon$ and $\Delta\varepsilon_{C4,y,tot,model}(9860 \text{ seconds}) \geq 0\,\mu\varepsilon$ (a specific non-linear model for this test was not created; however, no change in loading implies that the estimated strain should follow the non-negative trend with cycling). Based on the specifications of the strain gauge, the limits of error of measurement were $\pm 2\,\mu\varepsilon$. The error of model could not be estimated accurately (model was not created), but given that loading did not change, we can assume that the limits of error for the model are "small enough" and, based on engineering judgment, can be set to $\pm 100\,\mu\varepsilon$, i.e., the limits of error of the relative estimates of the model are $\pm 100\,\mu\varepsilon$. Then application of Criterion I yields the following inequality:

$$|\varepsilon_{C4,y,tot,sensor}(9860 \text{ seconds}) - \Delta\varepsilon_{C4,y,tot,model}(9860 \text{ seconds})|$$
$$\geq 4000\,\mu\varepsilon \geq \delta\varepsilon_{C4,y,tot,sensor} + \delta\Delta\varepsilon_{C4,y,tot,model} \approx 202\,\mu\varepsilon \qquad (7.1)$$

which shows that Criterion I is not fulfilled (see Expression 5.79) and a crack is detected.

Note that inequality in Expression 7.1 would be fulfilled even if the error of the model was 10 times greater (which would be unrealistic), which makes the direct detection principle very robust.

The advantage of this scenario, compared with the first scenario, where the sensor is in direct contact with a crack, is that the sensor is not damaged by the crack. In this case, the crack is successfully detected and localized (location of sensor indicates location of crack), however, quantification of the size of crack opening using sensor measurement is complicated, if not impossible, as the strain relaxation measured by the sensor depends on the complex relationships between loads, non-linear state of material, and structure's geometry and boundary conditions.

While in the presented test, the size of the crack opening was impossible to assess from the measurements, it was possible to monitor the crack propagation. Sensor C4 reached the threshold for Criterion II after approximately 5500 seconds, while that was the case for sensor D5 approximately 1000 seconds later. Similar, the crack tip reached sensor C4 after approximately 7800 seconds, while that was the case for sensor D5 approximately 1700 seconds later. This indicates that the set of densely spaced strain measurements can indicate crack propagation, i.e., the size of the crack extent, which is very useful information. This, in turn, further justifies the development of the third-generation strain sensors (see Section 3.7.2). As an example, four snapshots of the results of the evaluation of the crack extent (progression) using a dense arrangement of sensors installed on the plate (see Figure 7.1a) are shown in Figure 7.3. The unrealistically high strain shown in the

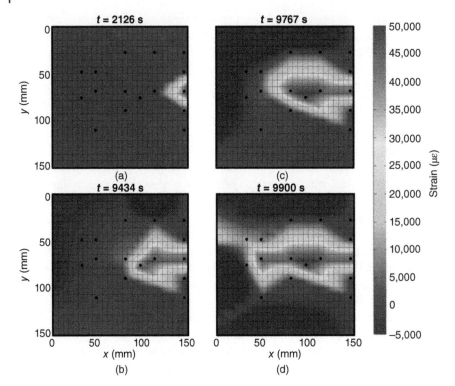

Figure 7.3 Monitoring of crack extent at (a) 2126 seconds, (b) 9434 seconds, (c) 9767 seconds, and (d) 9900 seconds; time counted after the beginning of the test. Source: Modified from Yao and Glisic (2015a).

figure around the crack is a consequence of approximative calculations (see Yao and Glisic 2015a, for more details).

One way to address the challenge of sensor survival when in direct contact with a crack is to use a bonding agent (adhesive) that is soft and applied in a relatively thick layer, thus, providing a less good strain transfer. This approach is in detail explained in Section 4.6.4, Point 1(b), and thus, it is not repeated here. We only highlight that while this approach can improve the survivability of the sensor, it decreases sensitivity and accuracy in strain measurement, which have to be taken into account when analyzing the data (see more details in Gerber et al. 2018).

Note that solving the challenge of survivability of the short-gauge strain sensors as described above would not necessarily lead to a simpler evaluation of the size of the crack opening. The later remains challenging as the measurement of the sensor will be distorted by the stress concentrations induced by the crack. In addition, the angle of the crack with respect to the direction of the sensor would play an important role (e.g., see the detailed study in Tung et al. 2014).

This subsection considered only soft discrete short-gauge strain sensors that are installed using adhesive, i.e., resistive strain sensors (strain gauges); similar considerations apply for other types of short-gauge sensors with soft packaging installed via adhesive (e.g., some types of FBG strain sensors; see upper group of sensors in Figure 3.18a). Short-gauge sensors with stiff packaging (e.g., some types of FBG strain sensors; see lower group of sensors in Figure 3.18a) may show better robustness in contact with cracks, but they can also experience delamination from the structure when the crack opens (see more detail in Section 4.6.4).

Expression 7.1 shows how Criterion I can be applied to detect cracks; nevertheless, this expression uses a very specific reference time and bases detection on a single measurement, which might be suitable for a lab test where damage does not need to be immediately detected. Generalization and more formal presentation of crack detection approach are given in the next section, which deals with discrete sensors installed via clamping or embedding.

7.2.2 Crack Detection and Localization Using Short- and Long-Gauge Strain Sensors Installed Via Clamping or Embedding

The main detection principles and implementation challenges for the use of short- and long-gauge strain sensors, installed via clamping or embedding, for integrity monitoring are presented in this subsection. Note that the commercially available short-gauge sensors that can be installed via clamping or embedding are, in principle, sensors based on vibrating wires (see Figures 3.7 and 4.15a) and Fabry–Perot interferometry (see Figures 3.26, bottom left). Soft-packaged short-gauge sensors, i.e., strain gauges, are described in previous subsection and thus not considered in this subsection. Short-gauge sensors based on Fiber Bragg-Gratings (FBG) are not specifically considered in this section; there is a large variety of packagings for these sensors that determine their survivability if in contact with cracks (see Figure 3.18a). While the short-gauge FBG sensors are not specifically considered in this subsection, their applicability can be inferred from this subsection and Section 7.2.1, depending on their packaging, the material of the monitored structure, and the method of installation. Long-gauge sensors included in consideration in this section are those with anchors (see Figures 4.15b and 4.16a); sensors without anchors, such as the one shown in Figure 4.15c, are not specifically considered, but their applicability can be inferred from this subsection and Section 7.2.1, depending on their packaging, the material of the monitored structure, and the method of installation.

In this subsection, Expression 7.1, presented in the previous subsection, is generalized and expanded into a simplified algorithm so that it is applicable for any type of strain sensor. Comments related to integrity monitoring based on direct and indirect detection are the same as for short-gauge sensors presented in the previous subsection; therefore, we assume that a short- or long-gauge strain sensor installed via clamping or embedding is located at or close to the location of crack occurrence. Again, the method is presented based on an example, and sections and subsections relevant to strain sensors, measurement errors induced by installation, and criteria for detection of unusual structural behaviors (Chapters 3–6) are evoked.

Streicker Bridge project was introduced in Sections 4.5.5 and 4.5.8, and presented in more detail in Sections 6.2.3, 6.6.7, and 6.8.3. Long-gauge strain sensors based on FBGs (see Figure 3.17 and middle sensor in Figure 3.18b) were embedded in concrete by attaching them to rebars using plastic ties prior to concrete pouring (shown with red arrows in Figure 4.10c, similar to those shown in Figure 4.15b). Examples of typical expected early age strain evolutions at locations of all sensors are given in Figure 6.7a. However, multiple sensors measured different behaviors, as shown in Figure 3.23 (see also Figure 4.11), which indicated the existence of unusual structural behaviors. More specifically, unusual structural behaviors, identified as cracking (Hubbell and Glisic 2013), were detected as the "jumps" in graphs of strain evolutions. To ease the presentation and analysis and complement the measurements shown in Figure 3.23, temperature change and strain measurements extending to the time of prestressing are shown in Figure 7.4 for cell (location) P10h11, and these graphs are used to develop and present the algorithm for crack detection. The same approach is applicable for all other locations. Note that the cross-section P10h11 contained three embedded parallel sensors: the usual bottom and top sensors (see Figure 4.10d), but also a "lateral" sensor, installed approximately 1 m south of and 2 cm above the top sensor, as shown in Figure 7.4a. Measurements were taken every 15 minutes.

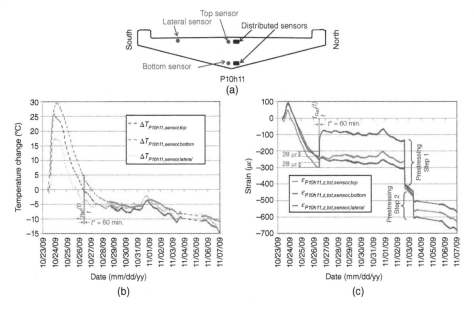

Figure 7.4 (a) Positions of sensors in a cross-section of cell P10h11; (b) evolution of change in temperature; and (c) evolution of total strain, from the pouring of concrete to the time of prestressing.

The following patterns characterize the graphs in Figures 3.23 and 7.4 (where applicable):

– Significant strain changes ("jumps") occur in pairs, i.e., both the bottom and the top sensor installed in the same cell (the same cross-section) registered the jump at the same time.
– Once a jump occurred, it was not short-term reversible, i.e., the strain was permanently shifted.
– When the jump occurred, the bottom sensor systematically measured a higher strain change than the top sensor.
– For each location, the inversion of the trend of strain evolution ("jump"), i.e., tension, has three phases:

  (i) The first phase consists of a slow inversion of trend, which, depending on location, took from less than an hour to several hours. During this phase, tension was generated at the locations of sensors;

  (ii) The second phase consists of one or two unusually high strain changes that took between 15 and 30 minutes to develop, depending on location; during this phase, the crack occurred and opened;

  (iii) The third phase consisted of the slow stabilization of crack opening and again took from less than an hour to several hours, depending on the location of the sensor and environmental influences (mostly temperature).

On one hand, knowledge of the typical usual early age behavior of concrete (e.g., see Figure 6.7a) seems sufficient to conclude, based on engineering judgment, that the behavior shown in Figures 3.23 and 7.4 (i.e., jumps in graphs) is consistent with the existence of unusual behaviors. On the other hand, appropriate implementation of Criterion I results in the quantitative detection of unusual structural behaviors. However, to apply Criterion I, it is necessary to create a model of concrete behavior at an early age, which in turn is very complex due to endogenous shrinkage and other rheological effects. Nevertheless, assuming that during short periods of time, e.g., 15–60 minutes, all rheological strain can be neglected, the only two strain components, which change would be non-negligible, are elastic and thermal strain. Considering that the concrete was still in the form-works, i.e., activation of self-weight load was not complete during early age, and considering that the structure is indeterminate, temperature changes and thermal gradients would be practically the only acting loading, which would result in both elastic and thermal strain.

Consequently, to detect cracking at time t, it is sufficient to apply Criteria I on measurement performed at that time using as a reference time, $t_{\text{Ref}}(t)$, the time that occurred t^* minutes before the observed time t (e.g., if $t^* \approx 60$ minutes, then $t_{\text{Ref}}(t) \approx t - t^* \approx t - 60$ minutes). In general case, let us assume that for each measurement performed at time t, the reference time is set to $t_{\text{Ref}}(t) \approx t - t^*$ and the measurements of an observed sensor, relative to the reference time, are described as follows:

$$\varepsilon_{cell,z,tot,sensor}(t, t_{\text{Ref}}(t)) = \varepsilon_{cell,z,tot,sensor}(t) - \varepsilon_{cell,z,tot,sensor}(t_{\text{Ref}}) = \varepsilon_{cell,z,tot,sensor}(t) - \varepsilon_{cell,z,tot,sensor}(t - t^*)$$
$$(7.2)$$

For the example shown in Figures 3.23 and 7.4, subscript "cell" should be simply replaced with "P10h11" and subscript "bottom" or "top" added to describe which sensor in the cell is analyzed.

Regarding the model, the total strain change between the two readings is expected in absolute value to be in the order of magnitude, or smaller than, the thermal strain due to the highest temperature change in the observed cell. Thus, the model estimating the total strain change between the time t and $t_{\text{Ref}}(t)$ can be expressed as follows:

$$\left| \varepsilon_{cell,z,tot,model}\left(t, t_{\text{Ref}}(t)\right) \right| = \left| \varepsilon_{cell,z,tot,model}(t) - \varepsilon_{cell,z,tot,model}(t_{\text{Ref}}) \right|$$
$$\lesssim \alpha_T \max_{cell}(|T_{model}(t) - T_{model}(t_{\text{Ref}}(t))|) =\sim \alpha_T \max_{cell}(|\Delta T_{model}\left(t, t_{\text{Ref}}(t)\right)|)$$
$$(7.3)$$

where α_T is thermal expansion coefficient of concrete and symbol "~" denotes approximate value.

The typical maximal temperature change described on the right-hand side of inequality in Expression 7.3 depends on the properties of concrete and the geographical region and should be assessed on a case-by-case basis. For example, the thermal expansion coefficient of concrete is $\alpha_T = 10 \pm 2\ \mu\varepsilon/°C$, and for the geographical area of central New Jersey, USA (i.e., the geographical area of Streicker Bridge), typical temperature change occurring in the bridge deck is lower than $\Delta T_{max,typ.} < 1$–$2°C/h$; then, by setting $t^* \approx 60$ minutes, i.e., $t_{\text{Ref}}(t) \approx t - t^* \approx t - 60$ minutes, one obtains $\max|\Delta T_{model}(t,t_{\text{Ref}}(t))| \leq 2°C$, and consequently $|\varepsilon_{cell,z,tot,model}| \leq 20\ \mu\varepsilon$, with the limits of error in model $\delta\Delta\varepsilon_{cell,z,tot,model} \leq 4\ \mu\varepsilon$ (excluding the epistemic error of the model). For the example of measurements shown in Figures 3.23 and 7.4, the limits of error in strain measurement were evaluated to $\delta\varepsilon_{cell,z,tot,sensor} = \pm 4\ \mu\varepsilon$ (see Section 4.5.5). The first expression of Criterion I, based on limits of error (see Expression 5.79), can now be rewritten as follows:

$$|\varepsilon_{cell,z,tot,sensor}\left(t, t_{\text{Ref}}(t)\right) - \Delta\varepsilon_{cell,z,tot,model}\left(t, t_{\text{Ref}}(t)\right)| \leq \delta\varepsilon_{cell,z,tot,sensor} + \delta\Delta\varepsilon_{cell,z,tot,model}$$
$$\Rightarrow$$
$$|\varepsilon_{cell,z,tot,sensor}\left(t, t_{\text{Ref}}(t)\right) - (\pm 20\ \mu\varepsilon)| \leq \delta\varepsilon_{cell,z,tot,sensor} + \delta\Delta\varepsilon_{cell,z,tot,model} = 4\ \mu\varepsilon + 4\ \mu\varepsilon$$
$$\Rightarrow$$
$$|\varepsilon_{cell,z,tot,sensor}\left(t, t_{\text{Ref}}(t)\right)| \leq 28\ \mu\varepsilon \qquad (7.4)$$

The second expression of Criterion I, based on uncertainties (see Expression 5.79), can be rewritten using a similar approach.

Thus, Criterion I was applied in the form of simple thresholding given in the last inequality of Expression 7.4. This expression can be applied in a continuous manner to every measurement of each sensor, i.e., at every time t when measurement is performed. If the inequality of the expression is not fulfilled, the crack is detected. Note that the right-hand value in Expression 7.4 depends on loading, specific geographical area, and time t^*, and thus, it has to be derived on a case-by-case basis. In the case presented above, there was no loading applied, but if the loading is applied, it has to be included in the model, i.e., in Expression 7.3. For example, in the case of pull-out test of pile presented in Section 6.5.5, the model should include the change in load as per Table 6.1. In that example, the cracking occurred at Load Step 11 (314.3 tons), when strain exceeded 70.5 µε and made a jump of 42.6 µε, which was approximately four times greater than expected (expected approximately 10 µε; see data in Table 6.1; see also Figure 6.5).

To increase the chances of crack detection in real-life settings, several values of t^* (e.g., $t_1^* = 15$ minutes, $t_2^* = 30$ minutes, $t_3^* = 60$ minutes, $t_4^* = 90$ minutes etc.) can be used to define multiple thresholds based on Expressions 7.4 and apply them simultaneously and continuously. This means the algorithm for damage detection should use multiple predefined values of t_i^*, and for every time of measurement t, apply Expression 7.4 for each of t_i^*.

Due to the simplification of the model, too short intervals defined by t_i^* can lead to false positive detections, and too long intervals defined by t_i^* can lead to false negative detections; therefore, the values to be used for t_i^* should be fine-tuned on-site during the initial stage of SHM to avoid these false detections.

Every sensor that detects the crack occurrence also points to the location of the crack occurrence, which is identical to the location of the sensor. However, a long-gauge sensor can have considerable length (from 0.1 m to 10+ m), and thus, based on the algorithm described above, the crack location can only be identified as being within the gauge length of the sensor, but its exact location within the gauge length of the sensor cannot be inferred. Moreover, due to the long-gauge length, it is also impossible to know whether only one crack opened within the gauge length of the sensor or multiple cracks opened simultaneously. If the cracks are visible on the surface of the structure, then their exact location and number can be inferred by visual inspection, which in turn is triggered by sensor detection of crack occurrence. However, if the cracks are not visible on the surface of the structure, inference of their exact position and number would require the deployment of non-destructive evaluation tools, typically ultrasonic pulse velocity probing.

Important feature of sensors installed by clamping or embedding is that the strain induced by crack opening is distributed between the clamps or anchor points of the sensors, i.e., along the gauge lengths of the sensors, which significantly reduces its magnitude and makes sensors very likely to survive the crack opening and to continue functioning properly. Thus, they can be used for crack quantification, as described in the next subsection.

7.2.3 Crack Quantification Using Short- and Long-Gauge Strain Sensors Installed Via Clamping or Embedding

This subsection considers the same type of sensors that are considered in Section 7.2.2.

Crack quantification requires the evaluation of the moments of crack initiation and crack stabilization, as per phases I and III, described in Section 7.2.2. These phases are indicated in Figure 7.5, where the evolution of strain change is graphed with respect to reference time $t_{Ref} = 2.95$ days (2 days, 22 hours, and 45 minutes) after the pouring of concrete. Initiation of cracking, i.e., the reverse in strain sign, starts approximately 30 minutes later, i.e., 2 days, 23 hours, and 15 minutes,

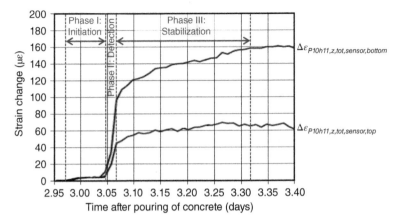

Figure 7.5 Strain change evolution measured by sensors in cell P10h11 2 days, 22 hours, and 45 minutes after the pouring of concrete, and the identification of phases of crack opening.

and finishes 1 hour and 45 minutes later, i.e., 3 days, 1 hour, and 0 minutes after the pouring of concrete. Phase II starts at the end of Phase I and takes 30 minutes (two larger jumps); note that the detection of cracks, based on Expression 7.4, happens at the end of Phase II, i.e., some 2 hours and 15 minutes after Phase I begun. Finally, Phase III takes approximately 6 hours, i.e., it finishes 3 days, 7 hours, and 30 minutes after the pouring of concrete.

The crack opening (or crack openings, if multiple cracks are present along the gauge length of sensors; see Section 7.2.1) generates strain change in the sensor, which is equal to the addition of strain changes in non-cracked parts of concrete (along the gauge lengths of the sensors) and of the crack opening (or openings). Given that the influence of the latter is significantly larger than the former, the former can be neglected, and the crack opening (or the sum of crack openings) can be evaluated using Expression 7.5.

$$\Delta_{C,crack,tot,sensor}(t) \approx (\varepsilon_{C,z,tot,sensor}(t) - \varepsilon_{C,z,tot,model}(t_{crack,initiation}))L_s \qquad (7.5)$$

where $\Delta_{C,crack,tot,sensor}(t)$ denotes crack opening or sum of crack openings at time t at location of gauge length of sensor with midpoint C, $t_{crack,initiation}$ is time of crack initiation (beginning of Phase I), and L_s is gauge length of the sensor.

Expression 7.5 is not perfectly accurate (it depends on the strain distribution along the gauge length of the sensor, which is, in general, difficult to evaluate), but it is sufficiently accurate for applications in civil engineering.

For example, based on Expression 7.5, and taking into account that the gauge length of sensors was $L_s = 0.6$ m, the crack opening (or sum of crack openings) at the location of the bottom sensor at the end of Phase III was $\Delta_{P10h11,crack,tot,sensor,bottom} \approx 162$ µε $\cdot$ 0.6 m = 0.10 mm, and at the location of the top sensor it was $\Delta_{P10h11,crack,tot,sensor,bottom} \approx 67$ µε $\cdot$ 0.6 m = 0.04 mm.

Besides the size of the crack opening (or sum of crack openings), it is important to evaluate the extent of the cracking. In Section 7.2.2, it was shown that having a dense arrangement of sensors greatly helps to identify the depth of penetration of cracks within material, see Figure 7.3. In the specific case analyzed here, there were three sensors placed in the cell P10h11 of the Streicker bridge, as shown in Figure 7.4a; however, based on Figure 7.4c, only the sensors installed at the bottom and top of the cross-section detected cracking, while the sensor installed laterally did not. The latter did show some small reverse in the sign corresponding to Phase I, but never experienced Phase II, and thus did not register cracking. This also means that cracking only happened in the central area of the cross-section, surrounding the bottom and top sensors, and did not extend laterally to the location of lateral sensor. This also shows that the lateral sensor did not robustly detect the cracking occurring less than 1 m from it, which emphasizes the importance and demonstrates significantly higher reliability of direct detection over indirect detection of damage.

7.2.4 Crack Diagnosis and Prognosis Using Short- and Long-Gauge Strain Sensors Installed Via Clamping or Embedding

Similar to Section 7.2.3, this subsection considers the same type of sensors that are considered in Section 7.2.2.

Once the cracks are detected, it is important to ascertain the reason for their occurrence and their consequence for structural performance. Ascertaining a plausible reason for the crack occurrence is considered as diagnosis in this book, while a change in structural behavior due to cracking is considered as prognosis.

A crack occurs if the tensional stress in the structure exceeds the tensional strength of the material. Note that the strength of a material is not necessarily invariable over time – it can vary depending on the age of the material, its loading history, or degradation processes. The stress is correlated to elastic strain, and the latter typically occurs in real-life settings due to (but not limited to) one of the following reasons:

- Loading
- Thermal actions
- Rheological effects
- Differential settlement of foundations
- Construction imperfections
- Fatigue
- Degradation processes (e.g., corrosion, alkali–silicate reaction, etc.)
- Combination of two or more from above.

To perform a diagnosis, it is necessary to analyze all potential sources of stress, to analyze measurements at each cracked location, but also to analyze measurements and structural system of the entire structure holistically. To perform the diagnosis related to cracking described in Section 7.2.3 and shown in Figures 3.23 and 7.4, the summary of crack initiation times and crack openings at the end of Phase III is presented in Table 7.1. Cracks are listed in order of occurrence.

Table 7.1 shows that all cracks initiated during the early age of concrete, while the bridge was still supported with formworks, covered with mats that prevented rapid evaporation of water (desiccation), and continuously wetted on the surface. This fact narrows down potential sources of cracking to thermal actions and early age shrinkage (rheologic effect).

Table 7.1 Summary of crack initiation times and crack openings at the end of Phase III.

No.	Cell	Crack initiation time (after pouring) (days hours minutes)	Crack opening at the end of Phase III, bottom sensor (mm)	Crack opening at the end of Phase III, top sensor (mm)
1	P10h11	02 d 23 h 15 min	0.10	0.04
2	P11	03 d 01 h 45 min	0.07	0.06
3	P11h12	03 d 04 h 15 min	0.09	0.02
4	P10	07 d 20 h 00 min	0.11	0.04

Figure 7.4b shows that hydration temperature did not develop uniformly across the cell P10h11, which means that in the cooling phase, after hardening time, significant thermal gradients were generated along both principal axes of the cross-sections of the bridge. The temperature at the bottom was higher than the temperature at the top at hardening time, which resulted in negative thermal curvature during cooling (bottoms of the cross-sections were thermally contracting more than the tops). Given that columns prevented negative thermal curving of the deck, as a consequence, a positive elastic curvature was generated along the deck of the bridge. This phenomenon is described in more detail in Section 6.6.7 and Figures 6.8 and 6.9. The vertical thermal gradient resulted in elastic curvature (similar to that shown in Figure 6.9), which generated stresses in the cell, with tension at the bottom and compression at the top.

Cell P10h11 was instrumented with a lateral sensor (south of the axis of symmetry of the cross-section, see Figure 7.4a) that showed significantly lower hydration temperature (see Figure 7.4b), due to thinner depth of cross-sections at its location, which not only produced less hydration heat but also enabled faster exchange of temperature between the concrete and the air at that location. It is therefore reasonable to assume that the other, north-lateral side of the cell, i.e., the one without a lateral sensor (see Figure 7.4a), experienced similar thermal conditions. Thus, there was a significant difference in thermal contraction due to overall cooling between the north and south extremities of the cross-sections on the one hand and the central area of the cross-sections on the other. The north and south extremities had to cool less and thermally deform less, which then put the central area of the cross-sections in tension. This effect is presented in more detail in Section 5.7.2 and Figure 5.8a.

In addition, overall cooling of the deck after hardening time resulted in its contraction, which was in part prevented by columns, which in turn resulted in tension over the entire observed cell. Early age shrinkage had the same effect, so it added to tensional stresses in the observed cell.

The combination of the three thermal effects described above provides an explanation of how tensional stresses could be generated in all identified cracked areas. In addition, we have to take into account that three to eight days after the pouring of concrete (approximate time span of crack occurrences, see Table 7.1), the concrete did not develop full strength, which means it was significantly weaker than mature concrete. This fact, combined with tension generated by thermal gradients and cooling and early age shrinkage, could provide an explanation for crack occurrence. Indeed, this hypothesis was tested as plausible by creating a numerical model that takes into account the evolving strength of young concrete, vertical thermal gradients, overall thermal contraction, and uniform early age shrinkage. Horizontal thermal gradients were omitted due to complexity. The first important conclusion was that the very slender columns of Streicker Bridge had very small bending stiffness and thus provided very little restraint for the deck to contract under the overall cooling and early age shrinkage (see Hubbell and Glisic 2013).

While the tensional stresses due to these effects, estimated at 0.1 MPa, increased the vulnerability to cracking, they were very low and thus not the main cause of cracking.

The results of numerical modeling are shown in Figure 7.6 (see Hubbell and Glisic 2013). The cracking strength is estimated using three different approaches: ACI 318-02 (ACI 2002), CEB-FIB Model Code MC1990 (CEB-FIP 1990), and Oluokun (Oluokun 1991), and the stresses in the deck using two boundary conditions: fixed connection at P10 and pinned connection at P10.

The results confirmed that the vertical thermal gradient played an important role in cracking occurrence. Figure 7.6 shows that stresses in the deck approached the cracking strength of the concrete approximately three days after the pouring, i.e., at the approximate time when the cracking at locations P10h11, P11, and P11h12 happened. Note that there is a discrepancy between the model and measurements, which can be explained by the simplification and epistemic errors of

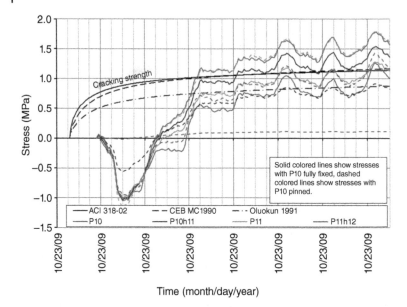

Figure 7.6 Estimated stresses on the deck of Streicker Bridge due to vertical thermal gradient measured onsite and comparison with estimated cracking strength.

the numerical model as well as by errors in estimating the Young modulus and cracking strength of concrete (based on empirical formulae applied to standard cylinder compressive tests and the size effect). Moreover, the influence of the horizontal nonlinear thermal gradient that added to tensional stresses in the central area of the cross-section was not taken into account.

The results of the modeling were considered acceptable, as the aim was to infer the reason for cracking rather than accurate modeling at each cracked location. Note that negative vertical thermal gradients create positive bending moments in the deck, which tension the bottom of the cross-section; indeed, the crack openings are systematically greater at the bottoms of the cracked cross-sections, when compared with the openings at the tops of the cracked cross-sections (see Table 7.1), which is consistent with the hypothesis that vertical gradients are the principal cause of cracking.

There is another interesting fact that is consistent with the hypothesis that vertical gradients are the principal cause of cracking. The southeast leg of Streicker Bridge is statically indeterminate to degree three. The first crack was initiated at location P10h11, the second at location P11, and the third at location P11h12. All three cracking occurrences happened on the same day, within five hours interval, but then cracking propagation was stopped, i.e., did not occur for the next four days approximately (see Table 7.1). Given that cracking creates partial hinges at locations of their occurrences, the static system after the three cracking occurrences can be considered as becoming determinate, which means that structure would not generate further stresses due to vertical thermal gradients, and that may explain why the cracking did not occur in other instrumented cross-sections.

Here we also highlight the influence of the horizontal non-linear thermal gradient, whose distribution is similar to that shown in Figure 5.8a: it compressed the lateral areas of the cross-sections and tensioned the central areas of the cross-sections; this explains why cracking did not happen at the location of the lateral sensor but also explains why it did happen at the location of the top sensor.

The cracking in cell P10 occurred four days, approximately, after the cracking in the three other cells. Based on measurements, this cracking happened on an unusually warm day (see the peak in temperature in Figure 7.4b on October 31, 2009). Thus, the reason for this cracking is rather related to non-linear thermal gradients in both vertical and horizontal directions due to temperature change (see Figure 5.8a and Section 5.7.2), which added to the tensional stresses already generated in previous days due to vertical thermal gradients.

The prognosis on how the cracking affects the structural behavior of the bridge is performed by a holistic analysis of the relationship between loading and measured strain, numerical modeling, and the results of the diagnosis. As mentioned earlier, the deck of the southeast leg of Streicker Bridge was prestressed by performing post-tensioning in two steps, approximately on the tenth and eleventh days after the pouring of concrete (see Figures 6.7a and 7.4c). Thus, it was of interest to evaluate to what extent prestressing was affected by the cracking, and whether the cracks were closed by prestressing. Detailed analysis based on numerical modeling confirmed both, that cracking did not adversely affect prestressing (see Abdel-Jaber and Glisic, 2014) and that cracks are practically closed by prestressing (see Abdel-Jaber and Glisic, 2015). Figure 6.3c shows that, at the time of post-tensioning (or release of the prestressing), the distribution of prestressing force evaluated from measurements along the southeast leg of Streicker Bridge is within the uncertainty limits of the model. In addition, Figure 7.7 shows that the crack openings at centroids of cross-sections (i.e., at centroid lines of cells), as evaluated from measurements, are practically negligible and well within the limit of 0.3 mm, set for the maximum allowable crack opening for prestressed concrete structures in humid environments (Nilson 1987).

A detailed presentation on the evaluation of prestress force and crack closure exceeds the scope of this introductory book and can be found in Abdel-Jaber and Glisic (2014, 2015), respectively. Here we only emphasize that for both evaluations, the strain and crack openings at the centroid of the cross-section were computed and used in the analysis to minimize the influence of thermal gradients and bending moments (e.g., see Section 6.6.7 for the evaluation of prestressing force at a non-cracked location).

The above analysis confirmed that the prestress force is correctly transmitted to the deck and that the cracks are practically closed at each instrumented location. However, the above analysis was performed individually for each cell without taking into account the entirety of the bridge deck, and the question remained whether some alteration was present at the global structural level. To answer

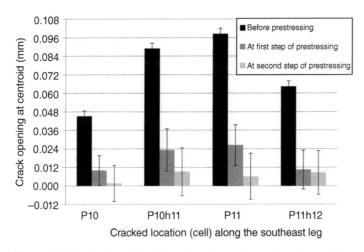

Figure 7.7 Evaluation of the sizes of crack openings at locations of the centroids of cracked cells.
Source: Modified from Abdel-Jaber and Glisic (2015).

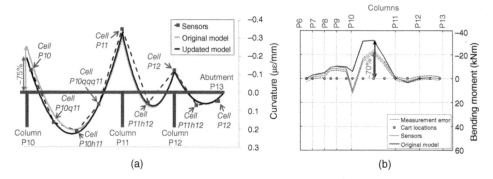

Figure 7.8 (a) Curvature distributions due to activation of self-weight based on measurements, the original model, and the updated model of the bridge deck, and (b) influence lines for the bending moment at the location of cell P10 based on the original model and measurements. Source: Modified from Sigurdardottir and Glisic (2015).

this question, it was necessary to perform an analysis of the structural behavior of the entire deck. For this purpose, curvatures and bending moments evaluated from measurements were compared with the finite element model during the two controlled load applications. The first was the activation of self-weight after the removal of the formwork, while the second was the load test of the bridge. The finite element model was created using SAP2000 and it assumed, initially, that the structure was intact, i.e., that no unusual structural behaviors were present.

The activation of self-weight is described in detail in Sections 6.6.7 and 6.8.3 (see also Figures 6.7–6.10 and 6.19). The curvature distribution along the southeast leg due to the activation of self-weight was evaluated based on sensor measurements and compared with the original (initial) finite element model, as shown in Figure 7.8a. While the model and measurements were mostly in good agreement, a rather large discrepancy was observed at the locations between cells P10 and P10q11. The model was updated to match the measurements, and comparison between the model updated based on the measurements, and the original model, showed decrease in curvature at the cross-section above the column P10 to approximately a 75% of the original model value (note that this cross-section was not directly monitored as sensors in cell P10 were placed about 1 m from that cross-section).

The load test was performed on March 11, 2011, and four golf carts with known weights were used to load the bridge. The carts were placed at 13 locations along the bridge, one after another, and for each location, the measurements were taken from all sensors while the golf carts were still at that location. After one location was loaded and measurements for that location recorded, the golf carts were moved to the next location and measurement were taken from all sensors at that location. This was repeated for all 13 locations. In that way, influence lines for bending moments were created based on measurements and compared with influence lines estimated using the original model of the bridge deck. While for most of the locations there was an excellent agreement between the model and measurements, a decrease of approximately 70% was noticed at location P10, as shown in Figure 7.8b. This result is consistent with the result of self-weight activation (Figure 7.8a) and confirms that there is a permanent alteration of structural behavior around the cross-section above column P10 in the form of reduced stiffness. More details about the analysis of the results of both self-weight activation and load tests can be found in Sigurdardottir and Glisic (2015).

The detected unusual behavior can be explained as the consequence of cracking, but also by the fact that the cross-section above column P10 represents the joint between the two stages of pouring of concrete and between prestressing cables: the main span and the other three legs of the bridge were all poured in August 2009, while the southeast leg was poured in October 2009. The joint was designed to guarantee continuity between the two parts of the structure (the main span and the southeast leg), but this type of connection is in general difficult to realize in real-life settings, and the combination of a partial lack of continuity in the joint combined with cracking is the most likely explanation for the detected unusual behavior.

The updated finite element model confirmed that the detected unusual behavior does not reduce the overall capacity of the bridge and that the bridge is safe for its users. Note that this final evaluation actually reflects Level IV SHM as per the definition given in Chapter 1.

The final comment in this subsection is related to direct sensing: Figure 7.8a shows that the original and updated models are different only at the locations of cells P10 and P10q11, while they are almost identical in all other cells. This is consistent with and emphasizes the statement that sensors distant from the damage are not likely to detect it.

7.3 Distributed Sensing for Integrity Monitoring

7.3.1 Brief Overview on 1D Distributed Sensing

Currently available one-dimensional (1D) distributed sensors are based on fiber-optic technologies. Their spatial configuration, functional principles (Brillouin and Rayleigh scattering), and appearances are presented in Section 3.6.4 (see Figures 3.19–3.21), and their best performances in Section 3.6.5 (see Table 3.3). In addition, important characteristics of distributed sensing, such as spatial resolution and sampling interval are presented in Section 4.5.8 (see Figure 4.9). Examples of sensors installed by bonding and embedding are given in Figures 4.13b and Figures 4.15d–f, respectively. Examples of distributed sensors installed by clamping are given in Figures 4.16b,c. The most appealing aspect of distributed sensors is their spatial continuity, which makes them ideal for direct sensing (see Section 3.7) and makes them advantageous over discrete sensors from that point of view (see Section 3.8 and Figure 3.27). The aspects of packaging, strain transfer, and survivability when in contact with cracking are presented in Section 4.6.

7.3.2 Sensor Packaging, Installation, Survivability of Sensor, and Thermal Compensation

Basic notions of sensor packaging, installation, survivability of the sensor when in contact with cracking, and thermal compensation, are presented in previous chapters of this book. In order to emphasize their features important for the selection and implementation of distributed sensing technologies (see Section 3.7 and Table 3.3), in this subsection we focus on the aspects of sensor packaging, installation, and thermal compensation that are specific to distributed sensing.

Packaging of distributed sensors is presented in detail in Section 4.6 and shown in Figures 3.21a,b, 4.13b, and 4.15d–f, and here we only summarize important trade-offs related to distributed sensor packaging. As stated in Section 4.6, packaging has a double role: to protect the sensing element, i.e., the optical fiber, during handling and installation in typically harsh real-life settings, and to guarantee strain transfer from the monitored structure to the sensor. There are multiple trade-offs that have to be considered when the packaging of the sensor is selected. In general, the following performances are intertwined, and improving one of them necessarily means worsening some of the others:

– Mechanical and chemical protection of sensing element (optical fiber)
– Quality of strain transfer (accuracy of measurement)

- Survivability of sensing element when in direct contact with the cracking (see Section 4.6.4)
- Spatial range of the sensor (see Table 3.3)
- Ease of installation (by bonding, embedding, or clamping)

For example, to guarantee excellent strain transfer (and excellent accuracy of strain measurement), the packaging must have excellent mechanical "gripping" to the optical fiber, that allows no sliding between the fiber and the packaging (e.g., "Tape" sensor shown in Figure 4.13b). In addition, its thickness should be small and its overall stiffness low (see Section 4.6.4), so it follows the deformation of the structure. However, realizing an excellent gripping might create micro-bending in the optical fiber along its entire length, which generates high cumulative optical losses and results in a reduction of the spatial range of the sensor. In addition, the small thickness and low stiffness may result in the delicacy of the sensor, so the handling and installation become tiresome. Finally, such a sensor, if embedded in the host material, might experience damage if in direct contact with cracking due to excellent strain transfer.

On the other hand, the sensor with less good gripping on the optical fiber (e.g., "Profile" sensor shown in Figures 4.13b, and 4.15d–f) and higher thickness has worse strain transfer quality (and thus worse accuracy in strain measurement), but a larger spatial range, better survivability in contact with cracking (see Section 4.6.4), and is easier to handle and install than "Tape" sensor.

Bare optical fiber can also be used as a distributed sensor. While it would have the best strain transfer and the longest spatial range, all other properties will be worse, and it will be very impractical to handle it and install it in real-life settings, especially for long-term monitoring.

The installation of distributed sensors is particularly challenging due to the large size of the structure to be instrumented. That is the reason why, typically, to instrument large areas of structures, it is recommended to use multiple shorter sensors (sensor segments) that are then connected onsite, by connectors or by splicing, into a single sensing length. This is schematically shown in Figure 7.9.

For example, to instrument 300-m-long structural element, it might be more practical to install three segments of 100 m instead of a single, 300-m long, sensor. Nevertheless, an excessive segmentation comes with the complications: every connector or splice has to be protected, which requires additional accessories (connection boxes, connectors or splice protectors, etc.) and their installation, which incurs additional costs but also introduces optical losses that affect (shorten) the spatial range of the sensing length; in addition, every connection requires overlap of two adjacent segments of sensor over the length of the spatial resolution to guarantee spatial continuity of measurements

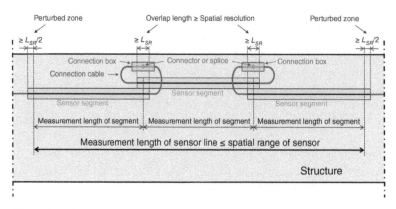

Figure 7.9 Schematic representation of a single sensing line divided into sensor segments and their deployment characteristics, which include overlap lengths, perturbed zones at the ends of the sensing line, connections, and associated accessories.

with the same accuracy (see Sections 4.5.8; see also Figure 7.9). Finally, it is important to highlight that there are perturbed zones at the ends of the sensing line, caused by the transition from sensor to connecting cable. These zones are affected by the spatial resolution of the sensor (see Section 4.5.8), and they are not shorter than half of the spatial resolution. To avoid misinterpretation of results, it is recommended to exclude these zones from measurement length, if possible, i.e., to install the sensor segments longer than the needed measurement length for at least half of the spatial resolution. Note that, considering perturbed zones and overlap lengths, the physical length of the sensor segment should be longer than the measurement length of the sensor for at least one length of spatial resolution (and, by preference, more than one length; see Figure 7.9).

Each type of installation of a distributed sensor, by bonding, embedding, or clamping, has its own challenges. Many details are given in Section 4.6; in this subsection, only some specific challenges related to distributed sensing are emphasized.

Installation by bonding requires smoothening and thorough cleaning of the surface over which the sensor is to be bonded, which can be very tiresome. Any sharp geometrical feature (e.g., stiffener, joint, etc.) on the structure has to be adapted, so no distortion or severe bending of the sensor occurs during installation (distortion or severe bending can introduce losses and lower the accuracy of strain measurement). Note that proper bonding on steel structures would require removal of protective paint (if any), which is often not permitted due to concerns related to corrosion. In these cases, the sensor can be glued over the paint, but this would worsen the strain transfer from the structure to the sensor, as the strain transfer goes from the structure to the paint, to the bonding agent, and to the sensor. In the case of bonding, the survival of the sensor at the location of cracking (see Section 4.6.4) is guaranteed either by sliding of the optical fiber within the packaging (e.g., "Profile" sensor, which has less good gripping on optical fiber) or by limited delamination of the sensor at that location (e.g., "Tape" sensor that has excellent gripping on optical fiber). In the latter case, the bonding agent (glue) has to be carefully selected so that it fails at interface with the sensor or the structure, and allows delamination of sensor before the cracking penetrates into the sensor packaging (e.g., see Ravet et al. 2009).

Challenges related to installation by embedding are related to guaranteeing the desired geometrical shape and position of sensors. To enable this, the most practical is to install the sensor along rebars (e.g., using plastic ties, see Figure 4.15d). However, at some locations, the installation might require the use of accessory holders, e.g., to place the sensor at locations where no rebars are installed or to avoid severe bending at locations where the sensor crosses perpendicular rebars. Particular attention should be made to protect ingress and egress points of sensors, as the sensor is prone to damage at these points, and is irreparable if damaged there. The survival of the sensor at the location of cracking is guaranteed either by sliding of the optical fiber within the packaging, or by sliding between the packaging and host material, or by both combined (e.g., "Profile" sensor, which has less good gripping on optical fiber but also less good adhesion with host material, as its packaging is made of polyethylene, see also Section 4.6.4).

The main challenge when installing a distributed sensor by clamping is to guarantee uniform or close to uniform prestressing of the sensor during the installation. Prestressing is necessary to enable the sensor to measure both tension and compression. Uniform or close to uniform prestressing is desirable to avoid sharp changes in strain in the sensor at the locations of clamps; this affects the sensing signal and may introduce inaccuracy in measurements. The recommended minimal distance between the clamps is at least equal to two spatial resolutions, so that the sensor middle point between two clamps is measured without interference from strain changes in the sensor occurring at the clamps. Given that the sensor is clamped and assuming that the strain in the sensor between the two clamps is uniform, the measurement at the middle point is sufficient to get the average value of strain between clamped points. An additional challenge is related to the geometry of the sensor: space between the two adjacent clamping points must be clear, and the angle at the clamp between two adjacent sensor parts must be small to avoid distortion or severe bending of the sensor. A distributed sensor installed by clamping behaves as a continuous chain of long-gauge sensors rather than a spatially continuous sensor. Thus, all considerations related to long-gauge sensors apply. This includes the consideration that a crack occurring at location of sensor will not damage the sensor, regardless of the type of packaging (see Section 3.2).

Distributed sensors need thermal compensation, and for this purpose, a distributed temperature sensor has to be installed, by preference, next to the distributed strain sensor. The need for thermal compensation incurs costs related to the temperature sensor and accessories, and their installation. Alternatively, both strain and temperature sensing fibers can be placed in the same packaging, as in the case of "Profile" sensor, which contains two strain fibers embedded in polyethylene packaging and two temperature fibers placed in a loose tube, which is then embedded in the packaging. Details about thermal compensation are given in Section 4.4, and only specifics related to distributed sensing are summarized here.

A distributed temperature sensor consists of packaging and a "loose" tube that contains one or more strain-free optical fiber (or fibers). The fiber (or fibers) can freely move within the loose tube, and any deformation of the packaging (and the loose tube) should not introduce strain in the fibers. To enable and guarantee this free movement, the diameter of the loose tube has to be sufficient to accommodate an extra length of the temperature-sensing optical fiber (or fibers) so that expansion or contraction of the packaging and loose tube (e.g., due to deformation of the structure or temperature changes) do not entrain strain in the optical fiber (or fibers). Thermal compensation, in general, affects accuracy in strain measurements (see Section 4.4), but in the case of distributed sensors, this can be more severe than in the case of discrete sensors, especially when the temperature trends change from increasing to decreasing and vice versa. There are two reasons for this: first, the temperature sensor cannot be installed at exactly the same location as the strain sensor, i.e., it is usually installed at some distance from it; consequently, the temperature at location of temperature sensor is not necessarily the same as the temperature at location of strain sensor; and second, the thermal conductivity of temperature sensor packaging, including the air in the loose tube, is not necessarily the same as thermal conductivity of strain sensor, and thus, temperature changes might not be introduced at the same pace in both sensors. Note that the difference in temperature at the observed point between the strain and temperature sensor of only 1°C will introduce an error in strain measurement at the observed point in the order of approximately 10 με.

The best accuracy of thermally compensated strain measurements is achieved when the temperature sensor and strain sensor are interconnected and red simultaneously. In addition, to complete the process of thermal compensation, it is necessary to map the coordinates of the strain and the temperature sensor; this process is described in the next subsection.

7.3.3 Sampling Interval and Coordinates Mapping

In this subsection, we review the definition of sampling interval and introduce coordinate mapping as an important aspects specific to the implementation of distributed sensing technologies.

Due to its spatial continuity, at each instance in time, a distributed sensor provides strain distribution along the sensor. Theoretically, this includes the infinity of points. To effectuate measurements and make them presentable, distributed sensor measures and present results at discreet points along the length of sensor. Distance between the two successive such points is called sampling interval (see also Section 4.5.8). For example, if the length of distributed sensor installed on the structure is 1 km, and sampling interval is set to 10 cm, at each instance of time the distributed sensor will measure and present the strain at 10,000 points, i.e., every 10 cm along 1 km. Each of these points has a local coordinate measured along the sensor from the reading unit (in mathematical terms, if the distributed sensor is considered a curve in 3D space, the local coordinate is identical to the arc-length coordinate of that curve with the origin set at the connection with the reading unit). In the other words, the point with local coordinate s is found at the distance s measured from the reading unit along the length of the sensor. To interpret the measurement of a distributed sensor, it is necessary to associate each point of the sensor with local coordinate s, with the point on the structure with coordinates (x, y, z) at which the sensor point with local coordinate s is installed, and with which the measurement is associated. The process of association of local coordinates of sensor with the coordinates of points in the structure is called coordinate mapping. This process is often accomplished by heating, as follows:

After the installation, a set of n characteristic points P_i $(i = 1, 2, \ldots n)$ on the sensor are identified and marked, and their coordinates on the structure (x_i, y_i, z_i) are ascertained (i.e., measured on-site). Note that at that stage, the local coordinates of points P_i are not known, but will be found through the process described below. These characteristic points are chosen so that the mapping of sensor points between any two successive points P_i and P_{i+1} can be performed using simple geometrical analysis, i.e., for any point Q on the sensor with local coordinate s_Q and with coordinates on structure (x_Q, y_Q, z_Q), there are points P_i, with local coordinate s_i mapped to structural coordinates (x_i, y_i, z_i), and P_{i+1}, with local coordinate s_{i+1} mapped to structural coordinates $(x_{i+1}, y_{i+1}, z_{i+1})$, and function F_i, such that: $s_i \leq s_Q \leq s_{i+1}$ and $(x_Q, y_Q, z_Q) = F_i[(x_i, y_i, z_i), (x_{i+1}, y_{i+1}, z_{i+1})]$. Function F_i is often s simple linear function, and for every segment $P_i - P_{i+1}$ it is determined from a simple geometrical analysis of the sensor location on the structure.

Once the points P_i are identified and marked, their coordinates (x_i, y_i, z_i) on the structure ascertained, and the functions F_i established, the baseline measurement is performed. Then points P_i are heated (e.g., using a heating gun) and measurements performed, one by one, and each measurement is compared with the baseline measurement. When a point P_i, with the known coordinates on the structure (x_i, y_i, z_i) is heated, the strain and temperature at that location will be changed, and the measurement diagram, which is always displayed in local sensor coordinates, will show a peak at the heated location, i.e., at location with the local coordinate s_i that corresponds to the heated point P_i. In that way, the local coordinate s_i is determined and mapped to coordinates on the structure (x_i, y_i, z_i). This process is repeated for all characteristic points P_i, and their mapping is then completed. Now, for any point Q with local coordinate s_Q, first the points P_i and P_{i+1} are identified, so that $s_i \leq s_Q \leq s_{i+1}$, and then the mapping is performed using the function F_i. This process has to be meticulously and rigorously conducted, as any error in coordinate mapping will result in the errors in data analysis and the errors in localization of damage. An example of coordinate mapping, that includes a temperature and strain sensor is described in Figure 7.10.

In Figure 7.10, the points of interest (denoted with P_1, P_2, P_3, and P_4) are at the extremities of each strain sensor segment (denoted with subscripts $s1$, $s2$, and $s3$ on strain sensor) and their temperature counterparts (denoted with subscript "T" on temperature sensors). Assuming that all sensors are parallel to the z-axis of the structure, coordinates x and y are the same for all points of the sensors and the structure, so they are not considered in further text to simplify presentation. Local coordinates s of these points of interest are mapped with corresponding coordinates z on structure (e.g., by heating) as follows (mapping is denoted with "$\sim$"):

$$z(P_1) \sim s(P_{1,s1}) \sim s(P_{1,T})$$

$$z(P_2) \sim s(P_{2,s1}) \sim s(P_{2,s2}) \sim s(P_{2,T})$$

$$z(P_3) \sim s(P_{3,s2}) \sim s(P_{3,s3}) \sim s(P_{3,T})$$

$$z(P_4) \sim s(P_{4,s3}) \sim s(P_{4,T})$$

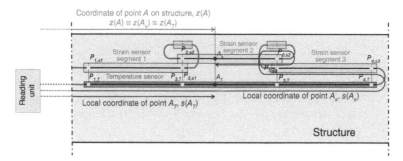

Figure 7.10 Example of coordinates mapping where temperature and strain sensors are interconnected.

Then, the mapping functions are created; for example, mapping function for point A, found on strain sensor segment 2 (see Figure 7.10), is created as follows:

$$F_{2,s}(s) = z(P_3) - (s - s(P_{3,s2})), \text{ i.e., } z(A) = z(A_s) = z(P_3) - (s(A_s) - s(P_{3,s2}))$$

$$F_{2,T}(s) = z(P_2) + (s - s(P_{2,T})), \text{ i.e., } z(A) = z(A_T) = z(P_2) + (s(A_T) - s(P_{2,T}))$$

Mapping functions for strain sensor segments 1 and 3 can be created using similar reasoning:

$$F_{1,s}(s) = z(P_2) - (s - s(P_{2,s1})) \text{ and } F_{1,T}(s) = z(P_1) + (s - s(P_{1,T}))$$

and

$$F_{3,s}(s) = z(P_4) - (s - s(P_{4,s3})) \text{ and } F_{3,T}(s) = z(P_3) + (s - s(P_{3,T})).$$

As stated in the previous subsection, the accuracy of thermally compensated strain measurements is highest if the temperature sensor and strain sensor are interconnected and red simultaneously. If they are red one after another, there is a possibility that the temperature changes between the strain and temperature readings, which would then affect the thermal compensation and introduce an error. Given that the optical fibers of a temperature sensor are strain-free, they have minimal optical losses, and thus interconnecting strain and temperature sensors in a single sensing length would not significantly affect the spatial range of the strain sensor. Typically, a temperature sensor is connected to the reading unit, and a strain sensor to the temperature sensor, as shown in Figure 7.10. It is important to notice that both sensors cover the same length, but the local coordinate s on the temperature sensor rises with the coordinate z of the structure, while the local coordinate s on the strain sensor decreases with the coordinate z of the structure. Thus, even if the temperature and strain sensors are installed parallel to the axis z and at almost the same location, the functions F_i for the two sensors are different, as shown above and in Figure 7.10. This clearly emphasizes the importance of coordinate mapping.

7.3.4 Spatial Resolution and Crack Identification

The notion of spatial resolution for distributed sensors is introduced and presented in detail in Section 4.5.8. The main take aways are:

- Spatial resolution of a distributed sensor has a meaning similar to the gauge length of discrete sensors, i.e., it represents the length over which the strain is "integrated" and its average value computed.
- Strain changes occurring over the lengths smaller than spatial resolution, yet greater than its half, will be detected, but their measurement might be inaccurate.

– Strain changes occurring over lengths smaller than one half of the spatial resolution might not be detected, and if detected, their measurement might be inaccurate.

For sensors installed via clamping, it is sufficient to place the clamping points to a distance larger than the spatial resolution, and the challenges presented in the last two points above will be avoided. As mentioned in Section 7.3.2, a distributed sensor installed in that way will behave as a chain of long-gauge sensors, and the only challenge is that if the distance between the clamping points is too large, then sensitivity to crack opening will be small due to large averaging length (for example, for a distance of 10 m, a crack opening of 0.1 mm will generate a strain change of 10 με, which might be in the range of error due to thermal compensation and thus undetectable).

For distributed sensors installed via bonding or embedding, the cracking will introduce a large strain change in the sensor (at the location of crack occurrence), but this will happen over a very short length, and, based on the last point above, there is a risk of not detecting it (see also Section 4.5.8). Experience from experiments has shown that this is not a challenge for Rayleigh-based distributed sensing (e.g., see Broth and Hoult 2020), as its spatial resolution is very small (under 10 mm, see Table 3.3).

However, this might be a challenge for Brillouin-based sensing, as its spatial resolution is typically in the order of magnitude of meters (see Table 3.3), and the integration of strain along the spatial resolution is only minimally affected by the cracking strain, which develops over a very short length. To address this challenge, a special algorithm has to be implemented at the optical signal level so that both the average strain along the gauge length and the peak strain at the cracking location can be detected (e.g., see Ravet et al. 2009). In that case, the diagram of distributed measurements contains two graphs: the first, always present, is the graph of distributed average strain measurements along the distributed sensor; and the second, the graph of cracking, is present only at the location of cracking (if any). Figure 1.10, in Chapter 1, shows the results of measurements of a sensor installed on a specimen subjected to controlled crack openings of 0.7 mm, 3.4 mm, and 5.18 mm. The graph of average strain is shown in continuous lines for each crack opening, while the graph of cracking is denoted by markers (circles). The spatial resolution of the sensor was 1 m. For the first two crack opening values (0.7 mm and 3.4 mm), the graph of average strain did not detect the change in strain that would indicate cracking, but the cracking graph did. In the case of the last crack opening value (5.18 mm), both graphs indicated the crack location; in this particular case, the graph of average strain could detect the crack as the crack opening was very large and affected integration of the strain along the length of spatial resolution despite the short length over which it occurred. Note the magnitude of strain in the graph of cracking – it is in the order of magnitude of several thousand microstrains, which is very high and close to the dynamic range of fiber-optic sensors ($\sim$1.0% = 10,000 με), i.e., to the survivability limit of the sensor. Note also that the cracking graph, indicated with markers (circles), has an accuracy of localization of the crack within plus/minus length of the spatial resolution (0.5 m).

Depending on the method of installation (bonding, embedding, or clamping) and the physical principle behind the distributed sensing technology (Rayleigh or Brillouin), the crack detection is performed either using the algorithm presented in Section 7.2.2 (based on the approach developed in Section 7.2.1) or by using the above-mentioned special algorithm implemented at the optical signal level (to create a cracking graph), or both.

The crack detection algorithms have to be applied at every point at which the strain measurement is performed, as defined by the sampling interval of the monitoring system. Given that the length of the sampling interval is usually set to be smaller than the length of the spatial resolution, the crack will be detected at multiple points that are all within the length of the spatial resolution (see Figure 1.10). This also provides robustness (redundancy) in crack detection.

The localization of cracks in the structure is performed in two steps. First, once the crack is detected by the sensor (i.e., the algorithms based on sensor measurements), its location is automatically associated with local sensor coordinates, as the crack detection algorithms are run over all the points along the sensor defined by the sampling interval. In other words, the local coordinates of each point at which the crack detection algorithms detected the crack indicate locations on the sensor where the crack was detected. Then, the coordinate mapping function is used (see Section 7.3.3) to translate the local sensor coordinate into corresponding coordinates on the structure, i.e., to identify the location of the cracking on the structure.

While crack detection and localization using distributed sensors based on the principle of direct detection are very reliable, the quantification of the crack opening is challenging for sensors installed by bonding or embedding. The average strain values measured by distributed sensors installed by bonding or embedding are noised by the strain concentrations at the location of the crack opening and redistribution of the strain in order to make the sensor survive the crack opening (sliding of optical fiber within the packaging or limited delamination of the sensor, etc., see Sections 4.5.8 and 7.3.2). These effects introduce errors in measurement that cannot be simply determined, and thus the crack opening cannot be accurately quantified. As an example, the cracking graph in Figure 1.10 shows smaller values for crack opening of 5.18 mm than the values for crack opening of 3.4 mm, which is not reflecting reality. Also, Figure 4.11 shows that the average strain value measured by the distributed sensor at the location of cracking is smaller than the value measured by the discrete long-gauge sensor; note that once the crack is closed by prestressing, both sensors measure the same value of the strain again.

Hence, in the case of a distributed sensor installed by bonding or embedding, the quantification of crack opening cannot be accurately performed based on sensor measurements, and it has to be performed using some other means, if possible. Nevertheless, for a distributed sensor installed by clamping, given that it will behave as a chain of long-gauge sensors, quantification can be performed as per Section 7.2.3. Diagnosis and prognosis can then be performed using approaches similar to those presented in Subsection 7.2.4.

7.3.5 Implementation Examples

To illustrate the implementation of integrity monitoring, two brief examples are presented in this subsection. Details about Streicker Bridge are presented in several subsections of this book (e.g., Sections 4.5.5, 4.5.8, 6.2.3, 6.6.7, 6.8.3, and 7.2.2). Figure 4.10b,d show the schematic position of the distributed "Profile" sensor, which was embedded in the deck of the longest span of the southeast leg of the bridge. The sensor was attached to rebars using plastic ties, as shown in Figure 4.10c, before the concrete was poured. The monitoring system was based on stimulated Brillouin scattering, more precisely Brillouin Optical Time Domain Analysis (BOTDA, see Section 3.6.4). The spatial resolution was set to 1 m and the sampling interval to 10 cm. The measurements were taken throughout early age and prestressing, and an example of measurements is shown in Figure 4.11 for location P11up. For comparison, the same figure shows measurements taken by a long-gauge FBG sensor installed at the same location. As presented in Section 7.2, cracking occurred at that location (see also Figure 3.23) and was detected by a long-gauge FBG sensor. Figure 4.11 shows that the distributed sensor also detected the crack as a high positive change in strain (jump in the diagram), although the magnitude of the jump was visibly smaller than the one observed by the long-gauge FBG sensor. The reasons for this are the sliding of the distributed sensor packaging within the concrete and sliding of the optical fiber within the sensor packaging, which both enabled strain redistribution in optical fiber around the location of the crack occurrence and,

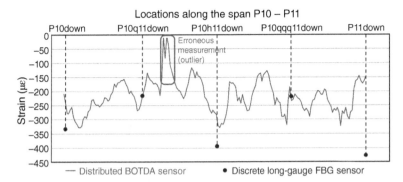

Figure 7.11 Strain measurements taken with distributed BOTDA and discrete long-gauge FBG sensors after prestressing of Streicker Bridge; reference time was set just before prestressing started.

consequently, the survival of the sensor (see Section 4.5.8 and above Sections 7.3.2 and 7.3.4). As expected, the measurements of the two sensors went back in agreement after the crack was closed by prestressing (see Sections 7.2.3–7.2.4).

Figure 4.11 represents one way of displaying distributed measurements: it shows the time history of strain for one observed location. Figure 7.11 represents another way of displaying distributed measurements: it shows strain change for all locations along the distributed sensor length at one observed moment in time. More precisely, Figure 7.11 shows the change in strain immediately after prestressing is applied, with respect to time just before the prestressing started. For comparison, measurements performed using long-gauge FBG sensors are displayed as well. The figure shows that at the three cracked locations (P10down, P10h11down, and P10h11down), there is a discrepancy between the measurements of the two systems, which was expected based on the above analysis and the analysis presented in Section 7.3.4. At non-cracked locations (P10q11down and P10qqq11down), the measurements from the two different sensor types were in agreement, which was also expected as there was no reason for these measurements to disagree (there was no cracking at these locations); this confirmed the good performance of both systems. More details about the comparison between the two systems are found in Glisic et al. 2011.

The graph of distributed strain measurements in Figure 7.11 shows anomalous behavior at approximately one-third of the distance between locations P10q11 and P10h11: the measured strain change due to prestressing is close to zero at that location. Based on engineering experience, this anomalous behavior is attributed to erroneous measurements (outliers), most likely generated by sharp bending of the sensor intersecting a cross rebar. Approximately 10 measurement points are involved, spatially corresponding to the length of one spatial resolution of the monitoring system. All these measurements were excluded from the data analysis. This anomaly emphasizes the importance of proper installation of sensor, as per the guidelines provided in Section 7.3.2.

Another example of the implementation of distributed sensing is given in Section 1.3. In that case, a total of approximately 5 km of distributed sensor of type "Tape" was surface mounted by bonding on the steel girders (see Figure 1.7) of the Gota Bridge in Gothenburg, Sweden (see Figure 1.5). As opposed to Streicker Bridge example, where sensors were embedded in concrete and crack detection did not require the use of a special algorithm (see Section 7.3.4), for Gota Bridge, the special algorithm was used as shown in the test results in Figure 1.10. More detail about this project can be found in Glisic et al. (2007), Enckell et al. (2011), and Glisic and Inaudi (2012). The distributed monitoring system installed on Gota Bridge in 2008 helped extend its lifespan until 2021, when a new Hisingen Bridge was inaugurated, and Gota Bridge closed for traffic and decommissioned after 82 years of service.

8

Closing Remarks and Future Perspectives

Strain-based SHM was conceived approximately one century ago, with the birth of the first strain sensors and their first applications in large real-life structures: dams and tunnels. Since then, the research, applications, and commercialization of strain sensing technologies have flourished, resulting in three generations of strain sensors and widespread implementations that encompass virtually any type of structure and any type of construction material. Thanks to its long tradition, mature technology, intuitive understanding of data, close relation to structural analysis, and affordable cost, strain-based SHM is nowadays considered one of the most viable Level IV SHM approaches and is among the most deployed SHM techniques in real-life settings. This, however, does not mean that the story of strain-based SHM is finished; on the contrary, it continues with great perspectives.

Given present-day vital societal challenges related to aging infrastructure but also climate change, accelerated urban growth, and depletion of resources, strain-based SHM certainly has great potential to help address these challenges and, in combination with other SHM techniques, enable a safe, resilient, sustainable, and livable built environment. Consequently, fundamental and applied research on strain-based SHM continues, and this happens principally in three distinct yet interwoven areas: sensing technologies, data analytics, and implementation policies.

First, as mentioned in Chapter 3, the fourth generation of strain sensors, aiming at 2D and 3D strain sensing technologies, is under development and exploratory applications. These sensing technologies are based on a variety of contact-based and non-contact-based (contactless) sensing principles (see Chapter 3), and their applications will greatly augment the applicability and improve the performance of strain-based SHM, especially at the integrity scale but also at local and global structural scales.

Second, the rapid development of informatics technologies and data analytics based on Machine Learning and Artificial Intelligence incited prolific research on new, powerful methods for the analysis and interpretation of large SHM data sets. These methods can be purely data-driven, or hybrid, physics-and-data-driven, or physic-informed data-driven, and they can enable not only identification of unusual structural behaviors that could not be detected using basic physics-based approaches (e.g., model-based approaches, such as those presented in this introductory book), but they can also provide improved modeling of structural behavior, either by creating surrogate data-driven models or by optimal identification of parameters for physics-based models. These models can then be used for diagnosis and prognosis purposes. While already at present-day, a substantial amount of research and exploratory applications of these new methods can be found in the literature, research will certainly continue in order to develop, identify, and validate in real-life settings the methodologies that are the most effective for various aspects of strain-based SHM

Introduction to Strain-Based Structural Health Monitoring of Civil Structures, First Edition. Branko Glišić.
© 2024 John Wiley & Sons Ltd. Published 2024 by John Wiley & Sons Ltd.

(e.g., concrete vs. steel structures, short-term vs. long-term SHM, contact-based vs. contactless sensing, static vs. dynamic monitoring, and discrete vs. distributed sensing).

Third, while strain-based SHM technologies are mature and data analytics is at an advanced stage, their widespread applications in real-life settings are hampered by a lack of policies and implementation guidelines or codes. This challenge is not specific to strain-based SHM; this is a general SHM challenge, independent of the type of SHM. The main cause is the lack of relevant policies and guidelines for the implementation of SHM. SHM is currently mostly implemented on a case-by-case basis, typically on structures with recognized deficiencies or on some important ("signature") structures. However, SHM is not implemented systematically, as it is not recommended or mandated by the design or maintenance codes nor enforced by relevant agencies (e.g., as they are visual inspections of bridges). To a large extent, the reason why the policies or guidelines are not created is the lack of means to ascertain the value of the information provided by SHM and, consequently, to perform a cost-benefit analysis of the implementation of SHM. The solutions for these challenges are currently being explored, and there is a growing body of research that is ongoing in that area.

In summary, strain-based SHM is a well-established approach, yet it also represents a vibrant multidisciplinary area of both fundamental and applied research. This book presents a comprehensive introduction to strain-sensing technologies, basic methods for implementation of strain-based SHM, and model-based strain data analytics, related principally to short-term and long-term static monitoring of structures built of the two most common materials, steel and concrete (other materials are sporadically represented). The author's hope is that this book will be used by students, teachers, researchers, practitioners, and policymakers, as a departure point in their endeavors to understand and implement strain-based SHM in a variety of structures, or to advance it and raise its performance and applicability to the next level.

References

Abdel-Jaber, H. and Glisic, B. (2014). A method for on-site determination of prestressing forces using long-gauge fiber optic strain sensors, *Smart Materials and Structures*, 23(7), art. no. 075004 (16pp).

Abdel-Jaber, H. and Glisic, B. (2015). Analysis of the status of pre-release cracks in prestressed concrete structures using long-gauge sensors, *Smart Materials and Structures*, 24, art. no. 025038 (12pp).

Abdel-Jaber, H. and Glisic, B. (2016). Systematic method for the validation of long-term temperature measurements, *Smart Materials and Structures*, 25, art. no. 125025 (12pp).

Abdel-Jaber, H. and Glisic, B. (2019). Monitoring of long-term prestress losses in prestressed concrete structures using fiber optic sensors, *Structural Health Monitoring*, 18(2), 254–269. doi: 10.1177/1475921717751870.

ACI (2002). *Building Code Requirements for Structural Concrete (ACI 318-02) and Commentary*, American Concrete Institute, Farmington Hills, MI.

ACI (2008). *Building Code Requirements for Structural Concrete (ACI Standard) ACI 318-08*, American Concrete Institute, Farmington Hills, MI.

Adler, A.J. (1971). Wireless strain and temperature measurement with radio telemetry - Development of miniature strain and temperature telemetry transmitters enables measurement of physical variables in industrial applications where wires are difficult or impossible to use, *Experimental Mechanics*, 11(8), 378–384.

Agyarko, T. (2023). *Understanding the Structural Behavior of Spiral Cantilevered Staircases A Case Study of the Museum of the City of New York Spiral Staircase*. Senior Thesis, Department of Civil and Environmental Engineering, Princeton University, Princeton, NJ, USA.

AISC (2005). *Steel Construction Manual*, 13th edn, American Institute of Steel Construction, Chicago, IL.

Aktan, H.M. (1986). Pseudo-dynamic testing of structures, *Journal of Engineering Mechanics*, 112(2), 183–197.

Altounyan, P.F.R. (1981). Measurement of stresses, strain and temperature in concrete in shafts and insets, *Developments in Geotechnical Engineering*, 32(C), 154–159.

Andersen, J.E. and Fustinoni, M. (2006). *Structural Health Monitoring Systems*, COWI-Futurec, L&S S.r.l.Servizi Grafici, Milano, Italy.

Ansari, F. (editor) (2005). *Sensing Issues in Civil Structural Health Monitoring*, Springer.

Ansari, F. and Libo, Y. (1998). Mechanics of bond and interface shear transfer in optical fiber sensors, *Journal of engineering mechanics*, 124 (4), pp. 385–394.

Asawa, C.K., Yao, S.K., Stearns, R.C., Mota, N.L. and Downs, J.W. (1982). High-sensitivity fibre-optic strain sensors for measuring structural distortion, *Electronics Letters*, 18(9), 362–364.

Introduction to Strain-Based Structural Health Monitoring of Civil Structures, First Edition. Branko Glišić.
© 2024 John Wiley & Sons Ltd. Published 2024 by John Wiley & Sons Ltd.

ASCE (2011). *Failure to Act: The economic impact of current Investment Trends in surface Transportation Infrastructure*, Report by Economic Development Research Group, Inc, American Society of Civil Engineers, Reston, VA, USA.

Aygun, L.E., Kumar, V., Weaver, C., Gerber, M.J., Wagner, S., Verma, N., Glisic, B. and Sturm, J.C. (2020). Large-area resistive strain sensing sheet for structural health monitoring, *Sensors*, 20, 1386.

Bado, M.F., Casas, J.R. and Barrias, A. (2018). Performance of Rayleigh-based distributed optical fiber sensors bonded to reinforcing bars in bending, *Sensors (Switzerland)*, 18(9), 3125.

Balageas, D., Fritzen, C.P. and Güemes, A. (editors) (2006). *Structural Health Monitoring*, John Wiley & Sons, Inc.

Bao, X., Webb, D.J. and Jackson, D.A. (1993), Characteristics of Brillouin gain based distributed temperature sensors, *Electronics Letters*, 29(17), 1543–1544.

Bao, Y., Hoehler, M.S., Smith, C.M., Bundy, M. and Chen, G. (2017). Temperature measurement and damage detection in concrete beams exposed to fire using PPP-BOTDA based fiber optic sensors, *Smart Materials and Structures*, 26(10), 105034.

Bassil, A., Chapeleau, X., Leduc, D. and Abraham, O. (2020). Concrete crack monitoring using a novel strain transfer model for distributed fiber optics sensors, *Sensors*, 20(8), 2220.

Bernard, O. (2000). *Comportement à long terme des éléments de structures formés de bétons d'âges différents*. Ph.D. Thesis No 2283, EPFL, Lausanne, Switzerland.

Billington, D. (1965). *The Thin Shell Concrete Structures*, McGraw-Hill, New York, USA.

Bradie, B. (2006). *A Friendly Introduction to Numerical Analysis—With C and Materials on Website*, Pearson international ed. Prentice Hall, Person Upper Saddle River, NJ.

Brčić, V. (1989). *Otpornost Materijala (Strength of Materials)*, 6th edn, Gradevinska knjiga, Belgrade, Serbia.

Bridges, M.C., Gladwin, M.T., Greenwood-Smith, R. and Muirhead, K. (1976). Monitoring of Stress, Strain and Displacement in and around a Vertical Pillar at Mount Isa Mine, *National Conference Publication - Institution of Engineers, Australia* 76 /4, pp. 44–49.

Brönnimann, R., Nellen, Ph.M., Anderegg, P. and Sennhauser, U. (1998). Packaging of fiber sensors for civil engineering applications, MRS Proceedings, 531. doi: 10.1557/PROC-531-327.

Broth, Z.E. and Hoult, N.A. (2020). Field monitoring of RC-structures under dynamic loading using distributed fiber-optic sensors, *Journal of Performance of Constructed Facilities*, 34 (4), 04020070.

Brune, P.F. (2010). *The Mechanics of Imperial Roman Concrete and the Structural Design of Vaulted Monuments*. Ph.D. Thesis, University of Rochester, Rochester, NY.

Burdet, O.L. (1993). *Load testing and monitoring of Swiss bridges,* Comité Européen du béton, safety and performance concepts, bulletin d'information nr. 219, Lausanne, Switzerland.

Calderon, P. and Glisic, B. (2012). Influence of mechanical and geometrical properties of embedded long-gauge strain sensors on the accuracy of strain measurement, *Measurement Science and Technology*, 23 (6), 065604 (15pp).

Cappello, C., Zonta, D., Laasri, H.A., Glisic, B. and Wang, M. (2018). Calibration of elasto-magnetic sensors on in-service cable-stayed bridges for stress monitoring, *Sensors*, 18(2), 466.

Carlson, R.W. (1939). *Development and Analysis of a Device for Measuring Compressive Stress in Concrete*. Ph.D. Thesis, Massachusetts Institute of Technology (MIT), Boston, MA, USA.

Catbas, N. (2013). Case Studies on the Structural Identification of Bridges. In: Çatbaş F.N., Kijewski-Correa T. and Aktan A.E. (editors), *Structural Identification of Constructed Systems: Approaches, Methods, and Technologies for Effective Practice of St-Id*, American Society of Civil Engineers (ASCE), Reston, VA, pp. 169–226.

CEB-FIP (1990). *CEB-FIP Model Code*, fib, Lausanne, Switzerland.

CEN (2010). *Eurocode 1: Actions on Structures - Part 2: Traffic Loads on Bridges*, European Committee for Standardization, Brussels, Belgium.

Chen, J. (2022). *Structural Analysis of the Museum of the City of New York (MCNY) Staircase*. Senior Thesis, Department of Civil and Environmental Engineering, Princeton University, Princeton, NJ, USA.

Choi, W., Kim, B.R., Lee, H.M., Kim, Y. and Park, H.S. (2013). A deformed shape monitoring method for building structures based on a 2D laser scanner, *Sensors*, 13, 6746–6758.

Choquet, P., Juneau, F. and Bessette, J. (2000). New generation of Fabry-Perot fiber optic sensors for monitoring of structures, *Proceedings of the SPIE*, 3986, 418–426.

Christensen, R.M. (1991). *Mechanics of Composite Materials*, Krieger Publishing Company, Malabar, FL.

Chung, W., Kim, S., Kim, N.-S. and Lee, H. (2008). Deflection estimation of a full scale prestressed concrete girder using long-gauge fiber optic sensors, *Construction and Building Materials*, 22, 394–401.

Cielo, P.G. (1979). Fiber optic hydrophone: Improved strain configuration and environmental noise protection, *Applied Optics*, 18(17), 2933–2937.

Coyne, A. (1938). Quelques résultats d'auscultation sonore sur les ouvrages en béton, béton armé ou métal, *Annales ITBTP*, July and August.

Culshaw, B., Davies, D.E.N. and Kingsley, S.A. (1981). Multimode optical fiber sensors, *Advances in Ceramics*, 2, 515–528.

Culverhouse, D., Farahi, F., Pannell, C.N. and Jackson, D.A. (1989). Potential of stimulated Brillouin scattering as sensing mechanism for distributed temperature sensors, *Electronics Letters*, 25(14), 913–915.

Dadpay, C., Sivakumar, N.R. and Mrad, N. (2008). Strain distribution and sensitivity in fiber Bragg grating sensors, *Proceedings of the SPIE*, 7099, 70992E.

Davidenkoff, N. (1934). The vibrating wire method of measuring deformations, *Proceedings ASTM*, 34(2), 847–860.

De Roeck, G. (2003). The state-of-the-art of damage detection by vibration monitoring: The SIMCES experience, *Journal of Structural Control*, 10(2), 127–134.

DEED (2009). *Economic Impacts of the I-35W Bridge Collapse*, Minnesota Department of Employment and Economic Development.

Dunphy, J.R., Meltz, G. and Elkow, R.M. (1986). Distributed strain sensing with a twin-core fiber optic sensors, *Instrumentation in the Aerospace Industry*, 32, 145–149.

Đurić, M. and Đurić-Perić, O. (1990). *Statics of Structures (Statika Konstrukcija)*, 4th edn, Gradevinska Knjiga, Belgrade, Serbia.

Enckell, M., Glisic, B., Myrvoll, F. and Bergstrand, B. (2011). Evaluation of a large-scale bridge strain, temperature and crack monitoring with distributed fibre optic sensors, *Journal of Civil Structural Health Monitoring*, 1(1–2), 37–46.

Fan, L., Bao, Y. and Chen, G. (2018). Feasibility of distributed fiber optic sensor for corrosion monitoring of steel bars in reinforced concrete, *Sensors (Switzerland)*, 18(11), 3722.

Fang, C. and Yunfei, W. (2012). A new system for measuring bridge deflections based on laser imaging process, *Applied Mechanics and Materials*, 143–144, 211–215.

Farar, C.R. (2009). *Introduction in Structural Health Monitoring Using Statistical Pattern Recognition*. A 3 day short course, la-dynamics, Palo Alto.

Farrar, C.R. and Worden, K. (2007). An introduction to structural health monitoring, *Philosophical Transactions of the Royal Society A*, 365, 303–315.

Feng, X., Wu, W., Meng, D., Ansari, F. and Zhou, J. (2017). Distributed monitoring method for upheaval buckling in subsea pipelines with Brillouin optical time-domain analysis sensors, *Advances in Structural Engineering*, 20(2), 180–190.

FHWA (2015). *Load and Resistance Factor Design (LRFD) for Highway Bridge Superstructures*, Reference Manual for NHI Course No. 130081, 130081A, and 130081B, Federal Highway Administration, Washington DC, USA.

FHWA (2021). *Long-Term Bridge Performance*, Federal Highway Administration, Washington DC, USA, https://highways.dot.gov/research/long-term-infrastructure-performance/ltbp/long-term-bridge-performance (accessed on December 25, 2021).

Fib (1999). *Practical Design of Structural Concrete*, Fib Commission on Practical Design, SETO, London.

Figurski, M., Galuszkiewicz, M. and Wrona, M. (2007). A bridge deflection monitoring with GPS, Artificial Satellites, 42(4), 229–238.

Frangopol, D.M. and Liu, M. (2006). *Life-Cycle Cost and Performance of Civil Structures*, McGraw-Hill, New York, USA.

Frangopol, D.M., Estes, A.C., Augusti, G. and Ciampoli, M. (1998). *Optimal Bridge Management Based on Lifetime Reliability and Life-Cycle Cost*, Short course on the Safety of Existing Bridges, ICOM&MCS, EPFL, Lausanne, Switzerland, pp 112–120.

Galambos, C.F. and Armstrong, W.L. (1969). Loading history of highway bridges, *National Academy of Sciences – National Research Council – Highway Research Record*, 295, 85–98.

Garus, D., Gogolla, T., Krebber, K. and Schliep, F. (1997). Brillouin optical- fiber frequency-domain analysis for distributed temperature and strain measurements, *Journal of Lightwave Technology*, 15(4), 654–662.

Gastineau, A., Johnson, T. and Schultz A. (2009). *Bridge Health Monitoring and Inspections – A Survey of Methods*, Minnesota Department of Transportation, Report No. MN/RC 2009-29.

Gentile, C. and Bernardini, G. (2010). Radar-based measurement of deflections on bridges and large structures, *European Journal of Environmental and Civil Engineering*, 14, 495–516.

Gerber, M., Weaver, C., Aygun, L.E., Verma, N., Sturm, J.C. and Glisic, B. (2018). Strain transfer for optimal performance of sensing sheet, *Sensors*, 18(6), 1907.

Ghafoori-Shiraz, H. and Okoshi, T. (1986). Fault location in optical fibers using optical frequency domain reflectometry, *Journal of Lightwave Technology*, 4(3), 316–322.

Ghanem, R., Higdon, D. and Owhadi, H. (editors) (2017). *Handbook of Uncertainty Quantification*, Springer International Publishing, Switzerland.

Glisic, B. (2000). *Fibre Optic Sensors and Behaviour in Concrete at Early Age*. Ph.D. Thesis N° 2186, Swiss Federal Institute of Technology Lausanne, EPFL, Switzerland.

Glisic, B. (2009). *Structural Health Monitoring*, Graduate Course CEE539, Princeton University, Princeton, NJ.

Glisic, B. (2011). Influence of gauge length to accuracy of long-gauge sensors employed in monitoring of prismatic beams, *Journal of Measurement Science and Technology*, 22(3), 035206 (13pp).

Glisic, B. and Inaudi, D. (2012). Development of method for in-service crack detection based on distributed fiber optic sensors, *Structural Health Monitoring*, 11(2), 161–171.

Glisic, B. (2019). Comparative study of distributed sensors for strain monitoring of pipelines, *Geotechnical Engineering (Journal of Southeast Asian Geotechnical Society – SEAGS)*, 50(2), 28–35.

Glisic, B. (2022). Concise historic overview of strain sensors used in the monitoring of civil structures: The first one hundred years, *Sensors*, 22, 2397.

Glisic, B. and Inaudi, D. (2003a). Components of structural monitoring process and selection of monitoring system, *6th International Symposium on Field Measurements in GeoMechanics (FMGM 2003)*, Oslo, Norway, pp. 755–761.

Glisic, B. and Inaudi, D. (2003b). Integration of long-gage fiber-optic sensor into a fiber-reinforced composite sensing tape, *Proceedings of the SPIE's*, 5050, 179–186.

Glisic, B. and Inaudi, D. (2007). *Fibre Optic Methods for Structural Health Monitoring*, John Wiley & Sons, Inc., Chichester, UK.

Glisic, B. and Inaudi, D. (2012). Development of method for in-service crack detection based on distributed fiber optic sensors, *Structural Health Monitoring*, 11(2), 161–171.

Glisic, B. and Simon, N. (2000). Monitoring of concrete at very early age using stiff SOFO sensor, *Cement and Concrete Composites*, 22(2), 115–119.

Glisic, B. and Verma, N. (2011). Very dense arrays of sensors for SHM based on large area electronics, *Structural Health Monitoring 2011: Condition-Based Maintenance and Intelligent Structures - Proceedings of the 8th International Workshop on Structural Health Monitoring* 2, pp. 1409–1416.

Glisic, B. and Yao, Y. (2012). Fiber optic method for health assessment of pipelines subjected to earthquake-induced ground movement, *Structural Health Monitoring*, 11(6), 696–711.

Glisic, B., Inaudi, D., Kronenberg, P., Lloret, S. and Vurpillot, S. (1999). Special sensors for deformation measurements of different construction materials and structures, *Proceedings of the SPIE*, 3670, 505–513.

Glisic, B., Inaudi, D. and Vurpillot, S. (2002a). Whole lifespan monitoring of concrete bridges, *IABMAS'02, First International Conference on Bridge Maintenance, Safety and Management,* paper on conference CD, Barcelona, Spain.

Glisic, B., Inaudi, D. and Nan, C. (2002b). Piles monitoring during the axial compression, pullout and flexure test using fiber optic sensors, *Transportation Research Record (TRR)*, 1808, 02-2701, pp. 11–20.

Glisic, B., Posenato, D. and Inaudi, D. (2007). Integrity monitoring of old steel bridge using fiber optic distributed sensors based on Brillouin scattering, *Proceedings of the SPIE - The International Society for Optical Engineering*, 6531, art. no. 65310P.

Glisic, B., Inaudi, D. and Casanova, N. (2010). SHM process as perceived through 350 projects, *Proceedings of the SPIE - The International Society for Optical Engineering*, 7648, art. no. 76480P

Glisic, B., Chen, J. and Hubbell, D. (2011). Streicker Bridge: A comparison between Bragg-gratings long-gauge strain and temperature sensors and Brillouin scattering-based distributed strain and temperature sensors, *Proceedings of the SPIE*, 7981, 79812C-1–79812C-10.

Glisic, B., Inaudi, D., Lau, J.M. and Fong, C.C. (2013). Ten-year monitoring of high-rise building columns using long-gauge fiber optic sensors, *Smart Materials and Structures*, 22(5), art. no. 055030 (15pp).

Glisic, B., Yarnold, M.T., Moon, F.L. and Aktan, A.E. (2014). Advanced visualization and accessibility to heterogeneous monitoring data, *Computer-Aided Civil and Infrastructure Engineering*, 29(5), 382–398.

Glisic, B., Yao, Y., Tung, S.-T., Wagner, S., Sturm, J.C. and Verma, N. (2016). Strain sensing sheets for structural health monitoring based on large-area electronics and integrated circuits, *Proceedings of the IEEE*, 104(8), 1513–1528.

Guan, S., Rice, J., Li, C. and Wang, G. (2014). Bridge deflection monitoring using small, low-cost radar sensors, *Proceedings of the Structures Congress (ASCE)*.

Habel, W.R., Hoepcke, M., Basedau, F. and Polster, H. (1994). Influence of concrete and alkaline solutions on different surfaces of optical fibers for sensors, *Proceedings of the SPIE*, 2361, 168–171.

Hafezolghorani, M., Hejazi, F., Vaghei, R., Jaafar, M.S.B. and Karimzade, K. (2017). Simplified damaged plasticity model for concrete, *Structural Engineering International*, 27(1), 68–77.

Hallaji, M. and Pour-Ghaz, M. (2014). A new sensing skin for qualitative damage detection in concrete elements: Rapid difference imaging with electrical resistance tomography, *NDT and E International*, 68, 13–21.

Hallee, M.J., Napolitano, R.K., Reinhart, W.F. and Glisic, B. (2021). Crack detection in images of masonry using CNNs, *Sensors*, 21, 4929.

Hartog, A.H. (1983). A distributed temperature sensor based on liquid-core optical fibers, *Journal of Lightwave Technology*, 1(3), 498–509.

HBM (2018). *Datasheet of LS31HT Strain Gauge*, HBM, Darmstadt, Germany, https://www.hbm.com/en/3450/ls31ht-weldable-strain-gauge-high-temperatures/ (last accessed on December 25, 2018).

Her, S.-C. and Huang, C.-Y. (2016). The effects of adhesive and bonding length on the strain transfer of optical fiber sensors, *Applied Sciences*, 6(1), 13.

Hibbeler, R.C. (2008). *Structural Analysis* 7th edn, Prentice Hall, Hoboken, NJ.

Hill, K.O. (1974). Aperiodic distributed-parameter waveguides for integrated optics, *Applied Optics*, 13(8), 1853–1856.

Hogg, D., Turner, R.D. and Measures, R.M. (1989). Polarimetric fiber-optic structural strain sensor characterization, *Proceedings of the Fiber Optic Smart Structures and Skins II, SPIE*, 1170, 542–550.

Horiguchi, T., Kurashima, T. and Koyamada, Y. (1993). Measurement of temperature and strain distribution by Brillouin frequency shift in silica optical fibers, *Proceedings of the SPIE - The International Society for Optical Engineering*, 1797, 2–13.

Hou, T.-C. and Lynch, J.P. (2006). Rapid-to-deploy wireless monitoring systems for static and dynamic load testing of bridges: Validation on the grove street bridge, *Proceedings of the SPIE - The International Society for Optical Engineering*, 6178, 61780D.

Hoult, N.A., Bennett, P.J., Middleton, C.R. and Soga, K. (2009). Distributed fibre optic strain measurements for pervasive monitoring of civil infrastructure, *Structural Health Monitoring of Intelligent Infrastructure - Proceedings of the 4th International Conference on Structural Health Monitoring of Intelligent Infrastructure, SHMII 2009*.

Hubbell, D. and Glisic, B. (2013). Detection and characterization of early age thermal cracks in high performance concrete, *ACI Materials Journal*, 110(3), 323–330.

Inaudi, D. (1997). *Fiber Optic Sensor Network for the Monitoring of Civil Engineering Structures*. Ph.D. Thesis N° 1612, Swiss Federal Institute of Technology Lausanne, EPFL, Switzerland.

Inaudi, D. and Glisic, B. (2002). Long-gage sensor topologies for structural monitoring, *The First fib Congress on Concrete Structures in the 21st Century*, Volume 2, 15–16 October, Osaka, Japan.

Inaudi, D. and Glisic, B. (2005). Development of distributed strain and temperature sensing cables, *Proceedings of the SPIE - The International Society for Optical Engineering* 5855 PART I,53, pp. 222–225.

Inaudi, D. and Glisic, B. (2006). Distributed fiber optic strain and temperature sensing for structural health monitoring, *Proceedings of the 3rd International Conference on Bridge Maintenance, Safety and Management - Bridge Maintenance, Safety, Management, Life-Cycle Performance and Cost*, pp. 963–964.

Inaudi, D. and Glisic, B. (2010). Long-range pipeline monitoring by distributed fiber optic sensors, *Journal of Pressure Vessel Technology, Transactions of the ASME*, 132(1), 0117011–0117019.

Inaudi, D., Vurpillot, S., Glisic, B., Kronenberg, P. and LLoret, S. (1999a). Long-term monitoring of a concrete bridge with 100+ fiberoptic long-gage sensors, *Proceedings of the SPIE - The International Society for Optical Engineering*, 3587, 50–59.

Inaudi, D., Glisic, B. and Vurpillot, S. (1999b). Packaging interferometric sensors for civil structural monitoring, *Proceedings of the SPIE - The International Society for Optical Engineering* 3746, p. 37463I.

ISO (1993). *International Vocabulary of Basic and General Terms in Metrology (VIM)*, 2nd edn, Geneva, Switzerland.

ISO/IEC (2007). *International Vocabulary of Metrology — Basic and General Concepts and Associated Terms, VIM*, ISO/IEC Guide 99-12:2007, 3rd edn, International Organization for Standardization, Geneva, Switzerland.

Jacobsen, S.C., Mladejovsky, M.G., Davis, C.C., Wood, J.E. and Wyatt, R.F. (1991). Advanced intelligent mechanical sensors (AIMS), *Transducers, '91*, 969–973.

Jang, S., Jo, H., Cho, S., Mechitov, K., Rice, J.A., Sim, S.-H., Jung, H.-J., Yun, C.-B., Spencer Jr., B.F. and Agha, G. (2010). Structural health monitoring of a cable-stayed bridge using smart sensor technology: Deployment and evaluation, *Smart Structures and Systems*, 6(5–6), 439–459.

Jarosevic, A., Fabo, P., Chandoga, M. and Begg, D.W. (1996). Elastomagnetic method of force measurement in prestressing steel, *Inzinierske stavby (Bratislava, Slovakia)*, 7, 262–267.

Jauregui, D.V., White, K.R., Woodward, C.B. and Leitch, K.R. (2003). Noncontact photogrammetric measurement of vertical bridge deflection, *Journal of Bridge Engineering*, 8, 212–222.

Jin, Y. (2020). *The Influence of Stereotomy on the Overall Behavior of Stone Cantilever Staircases*. Senior Thesis, Department of Civil and Environmental Engineering, Princeton University, Princeton, NJ, USA.

Karbhari, V.M. and Ansari, F. (editors) (2009). *Structural Health Monitoring of Civil Infrastructure Systems*, Woodhead Publishing in Materials.

Keller, T. (2003). *Use of Fibre Reinforced Polymers in Bridge Construction*, IABSE-AIPC-IVBH, ETH, Hönggerberg, Zurich, Switzerland.

Kikuchi, K., Naito, T. and Okoshi, T. (1988). Measurement of Raman scattering in single-mode optical fiber by optical time-domain reflectometry, *IEEE Journal of Quantum Electronics*, 24(10), 1973–1975.

Kingsley, S.A. (1984), Distributed fiber-optic sensors, *Advances in Instrumentation*, 39(1), 315–330.

Kirkup, L. and Frenkel, B. (2006). *Introduction to Uncertainty in Measurement*, Cambridge University Press, Cambridge, UK.

Kliewer, K. and Glisic, B. (2017). Normalized curvature ratio for damage detection in beam-like structures, *Frontiers in Built Environment - Structural Sensing*, 3, art. no. 50 (13pp).

Koo, K.Y., Brownjohn, J.M.W., List, D.I. and Cole, R. (2013). Structural health monitoring of the Tamar suspension bridge, *Structural Control and Health Monitoring*, 20(4), 609–625.

Kurata, N., Spencer Jr., B.F., Ruiz-Sandoval, M., Miyamoto, Y. and Sako, Y. (2003). A study on building risk monitoring using wireless sensor network MICA mote, *Structural Health Monitoring and Intelligent Infrastructure - Proceedings of the 1st International Conference on Structural Health Monitoring and Intelligent Infrastructure* 1, pp. 353–357.

Ladani, R.B., Wu, S., Kinloch, A.J., Ghorbani, K., Mouritz, A.P. and Wang, C.H. (2017). Enhancing fatigue resistance and damage characterisation in adhesively-bonded composite joints by carbon nanofibers, *Composites Science and Technology*, 149, 116–126.

Laflamme, S., Kollosche, M., Connor, J.J. and Kofod, G. (2013). Robust flexible capacitive surface sensor for structural health monitoring applications, *Journal of Engineering Mechanics*, 139(7), 879–885.

Lanticq, V., Gabet, R., Taillade F. and Delepine-Lesoille, S. (2009). Distributed Optical Fibre Sensors for Structural Health Monitoring: Upcoming Challenges. In: Christophe L. (editor), *Optical Fiber New Developments*, InTech, Available from: http://www.intechopen.com/books/optical-fiber-new-developments/distributed-optical-fibre-sensors-for-structural-health-monitoring-upcoming-challenges.

Lanzara, G., Feng, J. and Chang, F.-K. (2008). A large area flexible expandable network for structural health monitoring, *Proceedings of the SPIE - The International Society for Optical Engineering*, 6932, 69321N.

Lee, J. and Fenves, G.L. (1998). Plastic-damage model for cyclic loading of concrete structures, *Journal of Engineering Mechanics*, 124, 892–900.

Lemnitzer, L., Eckfeldt, L., Lindorf, A. and Curbach, M. (2008). Biaxial Tensile Strength of Concrete – Answers from Statistics. In: Walraven J.C. and Stoelhorst D. (editors), *Tailor Made Concrete Structures*, Taylor & Francis Group, London, UK.

Li, S. and Wu, Z. (2005). Structural identification using static macro-strain measurements from long-gage fiber optic sensors, *Journal of Applied Mechanics*, 8, 943–949.

Li, Y., Guo, Z. and Chen, B.-C. (2012). Application of optoelectronic liquid lever sensor in urban bridges deflection monitoring, *Applied Mechanics and Materials*, 198–199, 1184–1189.

Liehr, S., Lenke, P., Krebber, K., Seeger, M., Thiele, E., Metschies, H., Gebreselassie, B., Münich, J.C. and Stempniewski, L. (2008). Distributed strain measurement with polymer optical fibers integrated into multifunctional geotextiles, *Proceedings of the SPIE*, 7003, 700302.

Lim, A.S., Melrose, Z.R., Thostenson, E.T. and Chou, T.W. (2011). Damage sensing of adhesively-bonded hybrid composite/steel joints using carbon nanotubes, *Composites Science and Technology*, 71, 1183–1189.

Lin, Y.B., Chang, K.C., Chern, J.C. and Wang, L.A. (2005). Packaging methods of fiber-Bragg grating sensors in civil structure applications, *IEEE Sensors Journal*, 5(3), 419–424.

Liu, H. and Yang, Z. (1998). Distributed optical fiber sensing of cracks in concrete, *Proceedings of the SPIE - The International Society for Optical Engineering*, 3555, 291–299.

Loh, K.J., Hou, T.C., Lynch, J.P. and Kotov, N.A. (2009). Carbon nanotube sensing skins for spatial strain and impact damage identification, *Journal of Nondestructive Evaluation*, 28, 9–25.

Lubliner, J., Oliver, J., Oller, S. and Onate, E. (1989). A plastic-damage model for concrete, *International Journal of Solids and Structures*, 25, 229–326.

Lynch, J.P., Law, K.H., Kiremidjian, A.S., Kenny, T. and Carryer, E. (2002). A wireless modular monitoring system for civil structures, *Proceedings of the SPIE - The International Society for Optical Engineering*, 4753(I), 1–6.

Lynch, J.P., Sundararajan, A., Law, K.H., Kiremidjian, A.S., Kenny, T. and Carryer, E. (2003). Embedment of structural monitoring algorithms in a wireless sensing unit, *Structural Engineering and Mechanics*, 15(3), 285–297.

Malvern, L.E. (1969), *Introduction to the Mechanics of a Continuous Medium*, Prentice-Hall, Inc., Englewood Cliffs, NJ.

Maraval, D., Gabet, R., Jaouen, Y. and Lamour, V. (2017). Dynamic optical fiber sensing with Brillouin optical time domain reflectometry: Application to pipeline vibration monitoring, *Journal of Lightwave Technology*, 35(16), 3296–3302.

Mason, T.G.B., Valis, T. and Hogg, W.D. (1992). Commercialization of fiber optic strain gauge systems, *Proceedings of the SPIE 1795, Fiber Optic and Laser Sensors X*, pp. 215–222.

McCollum, B. and Peters, O.S. (1924). A new electrical telemeter, *Technologic Paper of the Bureau of Standards*, 17, 737–781, National Institute of Standards and Technology, US Department of Commerce, www.nist.gov.

MCNY (2019). *History of 1220 Fifth Avenue*, Museum of the City of New York, https://www.mcny.org/story/history-1220-fifth-avenue (last accessed on February 27, 2019).

Measures, R.M. (2001). *Structural Monitoring with Fiber Optic Technology*, Academic Press, San Diego, USA.

Meltz, G., Dunphy, J.R., Glenn, W.H., Farina, J.D. and Leonberger, F.J. (1987). Fiber optic temperature and strain sensors, *Proceedings of the SPIE 0798, Fiber Optic Sensors II*, pp. 104–114.

MnDOT (2009). *I-35W St. Anthony Falls Bridge*, Minnesota Department of Transportation – MnDOT, FAQ.

Modjeski, R., Masters, F.M. and Chase, C.E. (1931). *Tacony-Palmyra Bridge over the Delaware River*. Final Report to the Tacony-Palmyra Bridge Company, Modjeski, Masters & Chase Engineers.

Mohamad, H., Bennett, P.J., Soga, K., Mair, R.J., Lim, C.-S., Knight-Hassell, C.K. and Ow, C.N. (2007). Monitoring tunnel deformation induced by close-proximity bored tunneling using distributed optical fiber strain measurements, *Geotechnical Special Publication*, 175, 84.

Moragaspitiya, H.N.P. (2011). *Interactive Axial Shortening of Columns and Walls in High Rise Buildings.* Ph.D. Thesis, Faculty of Built Environment and Engineering, Queensland University of Technology, Australia, p157.

Moreu, F., Bleck, B., Vemuganti, S., Rogers, D. and Mascarenas, D. (2017). Augmented reality tools for enhanced structural inspection, *Structural Health Monitoring 2017: Real-Time Material State Awareness and Data-Driven Safety Assurance - Proceedings of the 11th International Workshop on Structural Health Monitoring, IWSHM 2017*, 2, pp. 2563–2568.

Morlier, P. (1994). *Creep in Timber Structures*, Report of RILEM Technical Committee 112-TSC. E & F N Spon, London, 149 p.

Mufti, A.A. (2001). *Guidelines for Structural Health Monitoring*, Design Manual No. 2, ISIS, Canada.

Mufti, A.A., Oshima, T., Bakht, B. and Mohamedien, M.A. (2003). *SHM Glossary of Terms*, ISHMII – International Society for Structural Health Monitoring of Intelligent Infrastructure.

Muravljov, M. (1989). *Gradjevinski Materijali (Constructuion Materials)*, Naučna Knjiga and Faculty of Civil Engineering, Belgrade, Serbia.

Murnen, G.J. and Laubenthal, P.F. (1985). Instrumentation of The East Huntington Bridge, *Proceedings of the 2nd Annual International Bridge Conference, Pittsburgh, PA, USA*, pp. 135–141.

Myrvoll, F., Bergstrand, B., Glisic, B. and Enckell, M. (2009). Extended operational time for an old bridge in Sweden using instrumented integrity monitoring, *Proceedings of the 5th Symposium of Strait Crossings*, Trondheim Norway.

Napolitano, R., Blyth, A. and Glisic, B. (2017). Virtual environments for structural health monitoring, *Structural Health Monitoring 2017: Real-Time Material State Awareness and Data-Driven Safety Assurance - Proceedings of the 11th International Workshop on Structural Health Monitoring, IWSHM 2017* 1, pp. 1549–1555.

Napolitano, R., Blyth, A. and Glisic, B. (2018). Virtual environments for visualizing structural health monitoring sensor networks, data, and metadata, *Sensors*, 18(1), art. no. 243 (14pp).

NASA (2020). *The Earth*, https://imagine.gsfc.nasa.gov/features/cosmic/earth_info.html (last accessed on December 20, 2020).

Neville, A.M. (1975). *Properties of Concrete*, Pitman International.

Nikles, M., Thevenaz, L. and Robert, Ph. (1994). Simple distributed fiber sensor based on Brillouin gain spectrum analysis, *Optics Letters*, 21(10), 758–760.

Nikles, M., Thevenaz, L. and Robert, Ph. (1997). Brillouin gain spectrum characterization in single-mode optical fibers, *Journal of Lightwave Technology*, 15(10), 1842–1851.

Nilson, A.H. (1987). *Design of Prestressed Concrete*, Wiley, Singapore.

NJDOT (2007). *Standard specification for road and bridge construction, division 900 Materials*, New Jersey Department of Transportation, http://www.state.nj.us/transportation/eng/specs/2007/spec900.shtm#s903 (last accessed on May 14, 2011).

NOAA (2013). *Hourly observations at Trenton Mercer Airport (14792), Quality Controlled Local Climatological Data (QCLCD)*, National Oceanic and Atmospheric Administration, http://ncdc.noaa.gov/land-based-station-data/qualitycontrolled-local-climatological-data-qclcd.

Nonis, C., Niezrecki, C., Yu, T.-Y., Ahmed, S., Su, C.-F. and Schmidt, T. (2013). Structural health monitoring of bridges using digital image correlation, *Proceedings of the SPIE 8695, Health Monitoring of Structural and Biological Systems* 8695, p. 869507.

Oh, B.-K., Glisic, B., Kim, Y. and Park, H.-S. (2020). Convolutional neural network-based data recovery method for structural health monitoring, *Structural Health Monitoring*, 19(6), 1821–1838.

Oh, B.K., Park, H.-S. and Glisic, B. (2021). Prediction of long-term strain in concrete structure using convolutional neural networks, air temperature and time stamp of measurements, *Automation in Construction*, 126, 103665.

Oluokun, F. (1991). Elastic modulus, Poisson's ratio, and compressive strength relationships at early ages, *ACI Materials Journal*, 88, 3–10.

Omega (1998). *Transactions in Measurement and Control*, Vol. 1–3, Putman Publishing Company and OMEGA Press LLC.

Omenzetter, P. and Brownjohn, J.M.W. (2006). Application of time series analysis for bridge monitoring, *Smart Materials and Structures*, 15(1), 129–138.

PAC (2011), *TrendSafe-SCADA*®, User's Manual, Process Automation Corporation, Belle Mead, NJ.

Pereira, M. and Glisic, B. (2022). A hybrid approach for prediction of long-term behavior of concrete structures, *Journal of Civil Structural Health Monitoring*, 12, 891–911.

Perry, C.C. and Lissner, H.R. (1955). *The Strain Gage Primer*, McGraw-Hill book Company, Inc., New York, NY.

Piyasena, R. (2002). *Crack Spacing, Crack width and Tension Stiffening Effect in Reinforced Concrete Beams and One-Way Slabs*. Ph.D. Thesis, Griffith University, Southport, Queensland, Australia pp. 6-1–6-30.

Posenato, D., Kripakaran, P., Inaudi, D. and Smith, I.F.C. (2010). Methodologies for model-free data interpretation of civil engineering structures, *Computers and Structures*, 88(7–8), 467–482

Posey, R. Jr., Johnson, G.A. and Vohra, S.T. (2000). Strain sensing based on coherent Rayleigh scattering in an optical fibre, *Electronics Letters*, 36(20), 1688–1689.

Potocki, F.P. (1958). Vibrating-wire strain gauge for long-term internal measurements in concrete, *Thermal Engineering*, 206(5369), 964–967.

Pozzi, M., Zonta, D., Trapani, D., Amditis, A.J., Bimpas, M., Stratakos, Y.E. and Ulieru, D. (2011). MEMS-based sensors for post-earthquake damage assessment, *Journal of Physics: Conference Series*, 305, 012100.

Psimoulis, P.A. and Stiros, S.C. (2007). Measurement of deflections and of oscillation frequencies of engineering structures using robotic theodolites (RTS), *Engineering Structures*, 29, 3312–3324.

Pyo, S., Loh, K.J., Hou, T.C., Jarva, E. and Lynch, J.P. (2011). A wireless impedance analyzer for automated tomographic mapping of a nanoengineered sensing skin, *Smart Structures and Systems*, 8, 139–155.

Rabinovich, S.G. (2005). *Measurement Errors and Uncertainties, Theory and Practice*, 3rd edn, AIP Press / Springer Science and Media, Inc., New York, NY.

Radojicic, A., Bailey, S. and Brühwiler, E. (1999). Consideration of the serviceability limit state in a time dependant probabilistic cost model, *Application of Statistics and Probability*, Balkema, Rotterdam, Netherlands, 2, 605–612.

Ran, Z., Yan, Y., Li, J., Qi, Z. and Yang, L. (2014). Determination of thermal expansion coefficients for unidirectional fiber-reinforced composites, *Chinese Journal of Aeronautics*, 27(5), 1180–1187.

Ravet, F., Briffod, F., Glisic, B., Nikles, M. and Inaudi, D. (2009). Submillimeter crack detection with Brillouin-based fiber-optic sensors, *IEEE Sensors Journal*, 9(11), 5257462, 1391–1396.

Reilly, J. (2019). *Temperature Driven Structural Health Monitoring*. Ph.D. thesis, Princeton University, Princeton, NJ, USA.

Reilly, J. and Glisic, B. (2018a). Tracking changes in coefficient of thermal expansion through the lifespan of a concrete bridge, *Engineering Mechanics Institute Conference*, Boston, MA, USA.

Reilly, J. and Glisic, B. (2018b). Identifying time periods of minimal thermal gradient for temperature driven structural health monitoring, *Sensors*, 18(3), 734.

Rinaudo, P., Paya-Zaforteza, I., Calderón, P. and Sales, S. (2016). Experimental and analytical evaluation of the response time of high temperature fiber optic sensors, *Sensors and Actuators, A: Physical*, 243, 167–174.

Roberts, G.W., Meng, X. and Dodson, A.H. (2004). Integrating a global positioning system and accelerometers to monitor the deflection of bridges, *Journal of Surveying Engineering*, 130, 65–72.

Rogers, J.D. (2010). Hoover Dam: Evolution of the Dam's Design, *Proceedings of the Hoover Dam 75th Anniversary History Symposium*, Las Vegas, NV, USA, 21–22 October 2010; pp. 1–39.

Rogers, A.J. and Handerek, V.A. (1992). Frequency-derived distributed optical-fiber sensing: Rayleigh backscatter analysis, *Applied Optics*, 31(21), 4091–4095.

Rosin-Corre, N., Noret, C. and Bordes, J-L. (2011). L'auscultation par capteurs à corde vibrante, 80 ans de retour d'expérience (Vibrating wire sensors monitoring, 80 years of feedback), *Proceedings of the Colloque du Comité Français des Barrages et Réservoirs (CFBR)*.

RST Instruments (2022). Carlson Strain Meter, *Datasheet CAB0003H*, www.rstinstruments.com (last accessed on February 28, 2022).

Rytter, A. (1993). *Vibration Based Inspection of Civil Engineering Structures*. Ph.D. Dissertation, University of Aalborg, Denmark.

Saboonchi, H. and Ozevin, D. (2015). MetalMUMPs-based piezoresistive strain sensors for integrated on-chip sensor fusion, *IEEE Sensors Journal*, 15(1), 6879247, 568–578.

Salowitz, N.P., Guo, Z., Kim, S.-J., Li, Y.-H., Lanzara, G. and Chang, F.-K. (2014). Microfabricated expandable sensor networks for intelligent sensing materials, *IEEE Sensors Journal*, 14(7), 6701343, 2138–2144.

Schaefer, O. (1919). Die Schwingende Saite als Dehnungsmesser, *Zeilschrift des Vereines Deutscher Ingenieure*, 63(41), 1008.

Schumacher, T. and Thostenson, E.T. (2014). Development of structural carbon nanotube-based sensing composites for concrete structures, *Journal of Intelligent Material Systems and Structures*, 25, 1331–1339.

Sigurdardottir, D.H. and Glisic, B. (2013). Neutral axis as damage sensitive feature, *Smart Materials and Structures*, 22(7), art. no. 075030 (18pp).

Sigurdardottir, D.H. and Glisic, B. (2015). On-site validation of fiber-optic methods for structural health monitoring: Streicker Bridge, *Journal of Civil Structural Health Monitoring*, 5(4), 529–549.

Sigurdardottir, D., Stearns, J. and Glisic, B. (2017). Error in the determination of the deformed shape of prismatic beams using the double integration of curvature, *Smart Materials and Structures*, 26, art. no. 075002 (12pp).

Sikorsky, C. (1999). Development of a health monitoring system for civil structures using a level IV non-destructive damage evaluation method. F.K. Chang (Ed.), *Structural Health Monitoring*, 2000, 68–81.

Simpson, W. and TenWolde, A. (1999). Physical Properties and Moisture Relations of Wood. In: *Wood Handbook: Wood as an Engineering Material*, USDA Forest Service, Forest Products Laboratory, Madison, WI. General technical report FPL; GTR-113, pp. 3.1–3.24.

Sohn, H., Farrar, C.R., Hemez, F.M., Shunk, D.D., Stinemates, D.W., Nadler, B.R. and Czarnecki, J.J. (2004). *A Review of Structural Health Monitoring Literature: 1996–2001*, Los Alamos National Laboratory, Report No. LA-13976-MS.

Song, H.W., Song, Y.C., Byun, K.J. and Kim, S.H. (2001). Modification of creep-prediction equation of concrete utilizing short-term creep tests, *Transactions SMiRT*, 16, Paper 1151 (Washington, DC).

Straser, E.G., Kiremidjian, A.S., Meng, T.H. and Redlefsen, L. (1998). Modular, wireless network platform for monitoring structures, *Proceedings of the International Modal Analysis Conference – IMAC*, 1, 450–456.

Thevenaz, L., Facchini, M., Fellay, A., Robert, Ph., Inaudi, D. and Dardel, B. (1999). Monitoring of large structure using distributed Brillouin fiber sensing, *Proceedings of the SPIE*, 3746, 345–348.

Tian, L., Pan, B., Cai, Y., Lian, H. and Zhao, Y. (2013). Application of digital image correlation for long-distance bridge deflection measurement, *Proceedings of the SPIE*, 8769, 87692V.

Timoshenko, S.P. and Goodier, J.N. (1970). *Theory of Elasticity*, McGraw-Hill, New York, NY.

Timoshenko, S. and Young, D.H. (1945). *Theory of Structures*, McGraw-Hill Book Company, Inc., New York, NY.

TML (2017). *TML Strain Gauge Catalogue 2017*, Tokyo Measuring Instrument Lab, Tokyo Sokki Kenkyujo Co., Ltd., Tokyo, Japan.

Torfs, T., Sterken, T., Brebels, S., Santana, J., van den Hoven, R., Spiering, V., Bertsch, N., Trapani, D. and Zonta, D. (2013). Low power wireless sensor network for building monitoring, *IEEE Sensors Journal*, 13(3), 909–915.

Tung, S.-T., Yao, Y. and Glisic, B. (2014). Sensing sheet: The sensitivity of thin-film full-bridge strain sensors for crack detection and characterization, *Measurement Science and Technology*, 25(7), 075602 (14pp).

Udd, E. and Kunzler, M. (2003). *Development and Evaluation of Fiber Optic Sensors*, Project 312, Final Report for Oregon Department of Transportation and Federal Highway Administration, Blue Road Research, Gresham, Oregon, USA.

Udd, E., Clark, T.E., Joseph, A.A., Levy, R.L., Schwab, S.D., Smith, H.G., Balestra, C.L., Todd, J.R. and Marcin, J. (1991). Fiber optics development at McDonnell Douglas, *Proceedings of the SPIE 1418, Laser Diode Technology and Applications III*, pp. 134–152.

Varadan, V.V., Varadan, V.K., Bao, X., Ramanathan, S. and Piscotty, D. (1997). Wireless passive IDT strain microsensor, *Smart Materials and Structures*, 6(6), 745–751.

Venuti, F., Domaneschi, M., Lizana, M. and Glisic, B. (2021). Finite element models of a benchmark footbridge, *Applied Sciences*, 11(19), 9024.

Vurpillot, S., Krueger, G., Benouaich, D., Clement, D. and Inaudi, D. (1998). Vertical deflection of a pre-stressed concrete bridge obtained using deformation sensors and inclinometer measurements, *ACI Structural Journal*, 95, 518–526.

Wait, P.C. and Hartog, A.H. (2001). Spontaneous Brillouin-based distributed temperature sensor utilizing a fiber Bragg grating notch filter for the separation of the Brillouin signal, *IEEE Photonics Technology Letters*, 13(5), 508–510.

Wait, P.C. and Newson, T.P. (1996). Reduction of coherent noise in the Landau Placzek ratio method for distributed fibre optic temperature sensing, *Optics Communications*, 131(4–6) 285–289.

Wang, M.L., Wang, G. and Zhao, Y. (2005). Application of EM Stress Sensors in Large Steel Cables. In: Ansari F. (editor), *Sensing Issues in Civil Structural Health Monitoring*, Springer, Dordrecht.

Wang, P., Wittmann, F.H., Lu, W. and Zhao, T. (2016). Influence of sustained load on durability and service life of reinforced concrete structures, *Key Engineering Materials*, 711, 638–644.

Wenzel, H. (2009), *Health Monitoring of Bridges*, John Wiley & Sons, Inc.

Wenzel, H. and Pichler, D. (2005). *Ambient Vibration Monitoring*, John Wiley & Sons, Inc.

Wenzel, H. and Tanaka, H. (2006). *SAMCO monitoring glossary – structural dynamics for VBHM of bridges*, F02 Monitoring Glossary, SAMCO Final Report, www.samco.org.

Widow, A.L. (1992). *Strain Gauge Technology*, 2nd edn, Elsevier Science Publishers, Ltd., UK.

Withey, P.A., Vemuru, V.S.M., Bachilo, S.M., Nagarajaiah, S. and Weisman, R.B. (2012). Strain paint: Noncontact strain measurement using single-walled carbon nanotube composite coatings, *Nano Letters*, 12, 3497–3500.

Xu, B., Wu, Z. and Yokoyama, K. (2003). Parametric identification with dynamic strain measurement from long-gauge FBG sensors and neural networks, *Structural Health Monitoring 2003: From Diagnostics and Prognostics to Structural Health Management - Proceedings of the 4th International Workshop on Structural Health Monitoring, IWSHM 2003*, pp. 1522–1529.

Xu, J., Dong, Y. and Li, H. (2014). Research on fatigue damage detection for wind turbine blade based on high-spatial-resolution DPP-BOTDA, *Proceedings of the SPIE - The International Society for Optical Engineering* 9061, 906130.

Yang, Y., Dorn, C., Mancini, T., Talken, Z., Theiler, J., Kenyon, G., Farrar, C. and Mascarenas, D. (2018). Reference-free detection of minute, non-visible, damage using full-field, high-resolution mode shapes output-only identified from digital videos of structures, *Structural Health Monitoring*, 17(3), 514–531.

Yao, Y. and Glisic, B. (2012). Reliable Damage Detection and Localization Using Direct Strain Sensing. In: *Proceedings of the 6th International IAMBAS Conference Bridge Maintenance Safety Management*, Stresa, Italy, 8–12 July 2012; CRC Press: London, UK, pp. 714–721.

Yao, Y. and Glisic, B. (2015a). Detection of steel fatigue cracks with strain sensing sheets based on large area electronics, *Sensors*, 15, 8088–8108.

Yao, Y. and Glisic, B. (2015b). Sensing sheets: Optimal arrangement of dense array of sensors for an improved probability of damage detection, *Structural Health Monitoring*, 4(5), 513–531.

Yarnold, M.T. (2013). *Temperature-Based Structural Identification and Health Monitoring for Long-Span Bridges*. Ph.D. Thesis, Drexel University, Philadelphia, PA.

Yarnold, M.T. and Weidner, J.S. (2016). Monitoring of a bascule bridge during rehabilitation, *Bridge Structures Journal*, 12(1–2), 33–40.

Yarnold, M.T., Moon, F.L., Aktan, A.E. and Glisic, B. (2012). Structural Monitoring of the Tacony-Palmyra Bridge using Video and Sensor Integration for Enhanced Data Interpretation, *Bridge Maintenance, Safety, Management, Resilience and Sustainability - Proceedings of the 6th International Conference on Bridge Maintenance, Safety and Management*, pp. 2168–2172.

Yarnold, M.T., Moon, F.L. and Aktan, A.E. (2015). Temperature-based structural identification of long-span bridges, *Journal of Structural Engineering*, 141(11), doi: 10.1061/(ASCE)ST.1943-541X. 000127.

Zhou, D., Dong, Y, Wang, B., Pang, C., Ba, D., Zhang, H., Lu, Z., Li, H. and Bao, X. (2018). Single-shot BOTDA based on an optical chirp chain probe wave for distributed ultrafast measurement, *Light: Science and Applications*, 7(1), 32.

Zonta, D., Chiappini, A., Chiasera, A., Ferrari, M., Pozzi, M., Battisti, L. and Benedetti, M. (2009). Photonic crystals for monitoring fatigue phenomena in steel structures, *Proceedings of the SPIE - The International Society for Optical Engineering*, 7292(1), 729215.

Zur, L., Tran, L.T.N., Meneghetti, M., Tran, V.T.T., Lukowiak, A., Chiasera, A., Zonta, D., Ferrari, M. and Righini, G.C. (2017). Tin-dioxide nanocrystals as Er^{3+} luminescence sensitizers: Formation of glass-ceramic thin films and their characterization, *Optical Materials*, 63(1), 95–100.

Appendix A

Structural Health Monitoring Glossary

A.1 Statement

There are several different terminologies used to describe structural health monitoring (SHM) as a consequence of its multi-disciplinary nature: SHM was born at the intersection of civil engineering, metrology, informatics, statistics, etc. The aim of this document is not to define or impose one terminology, but rather to provide a practical glossary of the most frequently used (and sometimes confused) terms in SHM as a support for researchers, students, and practitioners from all the involved branches of engineering.

A.2 Glossary

Absolute error (of a measurement). Difference between measured value and true value of measurand.

Accelerometer. Dynamic sensor that measures acceleration at points in the structure. Depending on the principle of functioning, the most common types are (i) piezo-resistive accelerometers, (ii) capacitive accelerometers, (iii) force balance accelerometers, and (iv) fiber-optic accelerometers. The type of accelerometer to use in a specific application depends on the vibration characteristics of the structure. By numerical integration of time history acceleration, it is possible to determine the time history of displacement and/or velocity. However, important errors can be propagated during the numerical integration, and intensive processing is needed as the accelerometers generate a large amount of data.

Accessories. Part of the DAQ system is used to connect sensors to DAQ hardware and protect all hardware components from exposure to the elements. The synonym is "peripherals." The accessories allow for wired or wireless communication (transfer of signals) between sensors and the reading unit or channel switch, and between the reading unit and data management hardware. They also provide for physical protection of all the components of the monitoring system from environmental or man-made actions (temperature, humidity, vandalism, etc.). Typical components of accessories are cables, junction boxes, enclosures, SIM cards, etc.

Accidental error (of a measurement). See "random error."

Accuracy (of a monitoring system). *Qualitative term*. Closeness of the agreement between the result of a measurement and the actual value of the measurand. From a metrological point of view, "accuracy" is considered only as a qualitative term and quantitative value should not be attributed to this term (e.g., one can say that a strain monitoring system is more accurate than another, but one should not say that the accuracy of the system is, for example, $2\,\mu\varepsilon$). However, in civil engineering practice, a qualitative value is frequently attributed to accuracy by suppliers of monitoring systems. In these cases, the user of the system should ask for more information from the supplier in order to understand how this value is determined and whether it is relevant for his application. The term "accuracy" is frequently confused in practice with the terms "resolution," "error limits," and "precision."

Important: the system that is accurate is not necessarily precise. For example, let us assume that the actual strain at a point in the structure is $100 \mu\varepsilon$, and let us assume that five repeated strain measurements at that point have values of $95 \mu\varepsilon$, $100 \mu\varepsilon$, $105 \mu\varepsilon$, $102 \mu\varepsilon$, and $97 \mu\varepsilon$. The system is accurate because the measurements are, by value, very close to the actual strain (minimal difference is $-5 \mu\varepsilon$ and maximal difference is $5 \mu\varepsilon$), but at the same time the system is not very precise because the measurements are not close to each other (maximal difference is $10 \mu\varepsilon$). Statistically, a precise system has a small difference between the mean value of repeated measurements and the actual value of measurand, assuming the latter experiences no change during the measurement.

Acoustic emission (AE) sensor. Qualitative sensor that detects the acoustic waves (energy) released by a damage (e.g., crack). The acoustic waves (energy) generated (released) by damage (crack initiation, crack propagation, corrosion, etc.), which travel away from the point of interest in a radial fashion, are detected by the piezoelectric sensing element. Typical frequencies detected by the sensors range between 1 kHz and 2 MHz. A dense network of sensors is needed to determine the location of the damage by triangulation. The advantages of the AE sensors are that they detect damage in real time and that they can detect damage that is far from sensors. The disadvantage is that AE due to damage is very weak, and damage detection and location can be obstructed by noise from many other sources and reflections within the structure. Sophisticated real-time data analysis involving triangulation and/or neural networks is needed to detect and locate the occurrence of cracks.

Ambient vibration. Vibration is caused by ambient excitations that are common for structures in specific geographic areas. The most common causes of vibration are wind, traffic, and other human activity, and in some geographic areas, tremors and low-magnitude earthquakes. In a more general sense, vibration caused by natural and man-made hazards can also be considered ambient. Since the hazards are rare and create vibrations with large amplitudes, these vibrations are frequently considered exceptional events. It is important to highlight that the causes of ambient vibrations cannot be controlled, and excitation parameters (e.g., excitation force, frequency, mass, and displacement) are generally considered unknown.

Ambient vibration test. Determination of the dynamic response of the structure and modal parameters (natural frequencies, mode shapes, and damping factors) using ambient vibrations. The most frequent aims of static tests are system identification and detection of unusual behaviors. The most frequently measured quantities during the test are accelerations and strain. The main assumption is that the input is a broadband stochastic process that can be modeled by white noise and thus the system properties can be represented by the power spectral density function of the dynamic response. Measurement sessions are repeated within the timeframe and with the frequency that are required to reliably and accurately capture the dynamic response and determine the modal parameters of the structure. An ambient vibration test is less reliable than a forced vibration test, but it is significantly less expensive to perform because no exiting devices are needed and the bridge is open for traffic during the test.

Bias. See "systematic error."

Computer-aided design (CAD). The use of computer technology for the process of design, creation of geometric models, and establishment of design documentation. Combined with non-contact scanning technologies (e.g., laser scanning or laser total station) that offer the ability to quickly and accurately measure the geometrical properties of a structure, it helps establish drawings and 3D model of structures without documentation, verification of design/as-is, and the construction of a representative 3D model that can be used as an interface to a comprehensive monitoring and heuristic database, aid in visualization, and serve as the foundation of a finite element model.

Channel commutator. See "channel switch."

Channel switch. Part of the DAQ system, which provides the reading unit with the capability of reading multiple sensors. The synonym is "channel commutator." A channel switch can be integrated into the reading unit or a separate piece of equipment. In general, it consists of interfaces to reading units and sensors, switching devices, processors, and firmware, but other configurations exist on the market, depending on technology. Also see "reading of multiple sensors."

Damage. Unfavorable change in the condition of a structure that can adversely affect structural performance. The unfavorable change may refer to the mechanical properties of construction materials and/or the geometrical properties of a structural system (including changes to the structural members, member connections, and supports).

Damage characterization. Process of ascertaining the time of occurrence, physical location, and size of the damage.

Damage detection. Process of ascertaining whether the damage to structure exists or not. Three main approaches to damage detection are visual inspection, nondestructive testing, and SHM. The reliability of visual inspection is challenged by its inherently subjective nature. In addition, it is limited to damage visible on the accessible surface of the structure, while hidden damage remains imperceptible. Reliability of nondestructive testing and SHM is challenging for damage that occurs in locations far from the sensors, since it depends on sophisticated algorithms whose performance is often decreased due to various influences that may "mask" the damage, such as high temperature variations, load changes, outliers, or missing data in monitoring results. In general, innovative approaches that are based on direct damage detection (integrity monitoring, the use of distributed sensors, or sensing cables) are more reliable than classical approaches based on indirect damage detection (the use of discrete sensors combined with sophisticated algorithms).

Damage identification. In addition to damage detection and characterization, damage identification also includes ascertaining the cause of the damage and its consequences.

Damage sensitive feature. A quantifiable property or pattern is sensitive to damage. It can be either directly monitored (e.g., strain) or extracted from monitoring data (e.g., modal characteristics from accelerometer measurements).

DAQ. See "data acquisition."

Data acquisition (DAQ). Sampling and digitizing the signals from sensors. See also "reading of sensors."

Data acquisition hardware. Device or set of devices that consists of a number of electronic components that convert analog signals (encoding parameters) from sensors into digital signals and transmit sensor signals to data management hardware (e.g., computer). There are various configurations of DAQ hardware components found on the market; in general, three components can be distinguished: the reading unit (interrogator or readout unit), accessories (peripherals including cabling, connection boxes, protections, etc.), and the channel switch (channel commutator).

Data acquisition software. Package of programs that control the DAQ system and allow the user to interact with the DAQ system, i.e., to operate and configure it. There are various configurations of DAQ software package, but in general, there are two main components: internal software that operates hardware components (so-called firmware) and user interface software that allows for communication with, control and configuration of the DAQ system. DAQ software supports self-monitoring capability (subsystem for diagnostic and reporting errors in the functioning of the DAQ hardware).

Data acquisition system. Totality of means used for DAQ. It consists of hardware and software.

Data analysis. Making conclusions about structural health condition or performance based on results of monitoring, i.e., transforming the data (monitoring results) into actionable information. Data analysis is crucial to achieving the aims of monitoring. Since different structures are in general exposed to different conditions, there is no universal methodology for data analysis, but rather a broad range of analytical, numerical, statistical, and heuristic approaches are used for this purpose, and they have been continuously researched and developed. The automation of data analysis is an important goal, and significant progress is made to achieve it, in many cases the final conclusions are, however, carried out by engineers.

Data cleansing. The process of identifying and correcting corrupt or erroneous measurements from a data set. Typical examples refer to identifying incomplete records (missing data), incorrect values (outliers), or inaccurate values due to the temporary malfunctioning of the monitoring system or its components (sensors, communication lines, etc.). The corrupted data is mostly removed and sometimes modified or replaced using some pre-defined algorithms.

Data filtering. Process of enhancement of acquired data (measurements) that allows the extraction of specific components of interest by removing noise.

Data interpretation. Association of the measurement results of each sensor to actions and processes that were exerted on the structure during the observed measurement period. For example, a sudden change in strain time history followed by strain oscillations can be associated with a heavy truck crossing the bridge. Data interpretation can be performed at both qualitative (example above) and/or quantitative level (e.g., if a numerical model of the bridge is established, then the truck weight can be determined from strain measurements).

Data management hardware. Device or set of devices used for data management. It usually consists of a computer or server that hosts the data management software or its parts, the data archival mediums (e.g., another computer or server allocated for this purpose, a backup server, and etc.,), and the communication lines between various data management hardware components and DAQ hardware (wired or wireless).

Data management software. Package of computer programs that manage the data. The minimum requirement for the software package is to provide for automatic: data cleansing, filtering, synchronization, quality assurance, visualization, and storage (archival) in a local archive. The advanced software package provides for data export (e.g., allowing export to some other software for further data processing), data interpretation (determining of meaning of individual results from each sensor), data analysis (making conclusions about structural health condition or performance), storage in a searchable database, and communication of actionable information to the user regarding the structure (e.g., sending messages with data analysis conclusions or with the status of the monitoring system).

Data mining. Automatic or semi-automatic analysis of large data sets in order to identify previously unknown patterns in structural behavior. In particular, data mining can help with the identification of clusters of data (measurements related to parts of structures with similar structural behavior that were unknown or not expected), the detection of unusual behaviors (damage), and finding relationships between various variables (e.g., environmental influences and structural behavior).

Deterioration. Processes that adversely affect structural performance, including reliability over time (Wenzel 2009).

Digital image correlation (DIC). A static non-contact, 2D distributed strain measurement system based on the comparison of digital images of a structure. A digital camera with a specific resolution is used to photograph the structure. A baseline image is taken, and a specific distance between the points (gauge length) in a structure is correlated with a certain number of pixels, which determines the accuracy of measurement. Based on this correlation, the deformation of the structure can be determined from images taken at a later date. The measurement is most effective if a pattern is painted on the surface of the structure or some other set of distinguishing marks is present. The advantages of DIC lie in its non-contact and 2D distributed sensing nature. The disadvantages are that the camera must stay immobilized at one specific place to make images from different times comparable, and that multiple cameras are needed to cover complete structure. Thus, DIC may be best suited for measurements over smaller areas of interest, such as specific structural members or connections. An unobstructed view between the digital camera and the area of interest is necessary for proper measurement.

Discrete sensor. Sensor that measures parameters related to a single point in the structure. The synonym for "discrete sensor" is "point sensor." For example, a strain-gauge is a discrete sensor whose reading provides a strain value at the location of the sensor (i.e., point in the structure). A tri-axial accelerometer is a discrete sensor whose reading provides three values of acceleration (acceleration in each of three directions) at the location of the sensor.

Distributed sensor. Sensor or dense arrays of sensors relative to one-dimensional (1D), two-dimensional (2D), or three-dimensional (3D) areas of structure and whose readings provide a large number of measurand magnitudes relative to that area. Examples of one-dimensional distributed sensors are fiber-optic sensors (see "sensing cable") for strain and/or temperature monitoring. Examples of two-dimensional distributed sensors are laser scanning, digital image correlation, etc. Examples of three-dimensional distributed sensors are ground penetration radar (GPR) and ultrasonic scanning.

Drift. Short-term or long-term losing of stability of a monitoring system, i.e., increase of the measurement error in time.

Dynamic monitoring. Monitoring of quantities such as acceleration, strain, displacement, and forces on a structure subjected to dynamic loads. Dynamic loads do change magnitude and/or position in time, and these changes are "fast enough," so the accelerations at points of monitored structure cannot be neglected. In true dynamic monitoring, it is assumed that the dynamic load state of the structure does not change over the period of time needed to complete a measurement session (one measurement over all sensors). This is, however, true only if all the measurements within one session are taken simultaneously. Many monitoring systems feature sequential reading of sensors (sequential DAQ), and in these cases, the speed of reading should be "fast enough" so the differences in dynamic load state over the period of time needed to complete the measurement session can be neglected. In this case, the monitoring is quasi-dynamic. Due to the large amounts of data that may be generated, dynamic monitoring is usually performed in the form of measurement sessions triggered by exceptional events (e.g., heavy truck, earthquake, and strong wind) with the aim of registering the structure's response to these events. The measurement rate sessions depend on the natural frequencies contained in the event (see "Nyquist rate"). Dynamic monitoring is frequently combined with vibration monitoring, as the same monitoring system can be used for dual purposes. Dynamic monitoring combined with vibration monitoring can also be used to assess fatigue.

Dynamic range (of a monitoring system). *Quantitative term*. Maximal interval of variation of the measurand that can be registered by the system. For example, strain-gauge can typically measure the strain from -3000 to $+3000\,\mu\varepsilon$, and this is its dynamic range, while the dynamic range of interferometric fiber optic sensors (SOFO) is -5000 to $+10{,}000\,\mu\varepsilon$. A large dynamic range allows for monitoring of extreme values of measurand, and consequently, a large dynamic range is a desirable specification.

Electrical resistance strain gauge. Static or dynamic discrete short-gauge strain sensor based on the change in electrical resistance in the electrical resistor due to strain. The most frequently used resistors are metallic (Constantan – Cu—Ni alloy, Nichrome V – Ni—Cr alloy, Platinum alloys – usually tungsten, Isoelastic – Ni—Fe alloy, and Karma-type alloys – Ni—Cr alloy), but also foil and semiconductor resistors exist on the market. The typical resolution is $1\,\mu\varepsilon$, the rate of measurement can be higher than $100\,\text{Hz}$ (depending on the reading unit), and the gauge length ranges between 0.4 and $100\,\text{mm}$. Electrical resistance strain gauges are usually bonded to the structure using an appropriate adhesive. The advantages are simple use and a very low cost. Disadvantages are degradation in performance, especially stability and accuracy in the long-term due to exposure to elements, and electro-magnetic interference, which makes the electrical resistance strain gauges suitable for rather short-term measurements.

Encoding parameter. Physical quantity to which the measurand is converted by the sensor. Also see "sensor."

Error limits (of a monitoring system). *Quantitative term*. Minimal and maximal difference between the measured and true value of a measurand. The error limits can be expressed in two manners: (i) as a constant value, independent of the value of the measurement, and (ii) as a percentage of the value of the measurement.

False negative (damage detection). Error in damage detection. The monitoring system does not report damage to the structure, while the damage is actually present. False negatives may be very dangerous because the manager of the structure has a false impression that there is no damage to the structure, and consequently does not perform actions to mitigate the adverse consequences (risk) of the damage in time.

False positive (damage detection). Error in damage detection. The monitoring system reports damage to the structure, but the damage is actually not present. False positives are undesirable since the manager of the structure may undertake some unnecessary and costly actions to mitigate non-existing risks, because she/he has the impression that the structure is damaged.

Fiber optic sensors (FOS). A broad class of sensors that have an optical fiber as a sensing element. Standard telecommunication silica optical fiber is commonly (but not only) used for sensing purposes, and it can be physically modified or not, depending on principle of functioning. FOS covers a large spectrum of parameters that can be monitored (e.g., strain, inclination, temperature, humidity); thus, multiple parameters can be combined on the same network. Compared with conventional electrical sensors, FOS offers two new and unique sensing tools: long-gauge strain sensors and truly distributed strain and/or temperature sensors. The former can be combined in topologies that allow for global structural monitoring, while the latter allows for one-dimensional strain field and integrity monitoring. Commercially available discrete FOS are based on the following functioning principles: (i) extrinsic Fabry–Perot interferometry (EFPI), (ii) intrinsic Michelson and Mach–Zehnder interferometry (SOFO), (iii) Fiber Bragg Grating (FBG), and (iv) intensity based (micro-bending). Commercially available distributed FOS are based on the following functioning principles: (i) spontaneous Brillouin scattering, (ii) stimulated Brillouin scattering, (iii) Rayleigh scattering, and (iv) Raman scattering. FOS feature high precision and long-term stability, and are particularly suitable for long-term monitoring (proven over more than 15 years). The optical fiber can be used for both sensing and signal transmission purposes. It is resistant to most chemicals in a wide range of temperatures and intrinsically immune to any electromagnetic interference, which makes it suitable for long-term applications in harsh environments. Various packaging, especially designed for field applications, made FOS robust and safe to use even in very demanding environments. The ability to measure over distances of several tens of kilometers without the need for any electrically active component is an important feature when monitoring large and remote structures, such as landmark bridges, dams, tunnels, and pipelines. The disadvantage of FOS is their cost, which is still higher compared to conventional sensors, but is still affordable and justified by their superior long-term performance.

Firmware. Internal software that operates hardware components.

Forced vibration. Vibration caused by controllable devices or actions (e.g., shaker, mass release, etc.) with known excitation parameters (e.g., excitation force, frequency, mass, and displacement).

Forced vibration test. Determination of the dynamic response of the structure (acceleration, strain, tilt, displacements, curvature, stiffness, dynamic factor, etc.) and modal parameters (natural frequencies, mode shapes, and damping factors) using forced vibrations with known position, frequency, and magnitude. The most frequent aims of static tests are system identification and detection of unusual behaviors. The most frequently measured quantities during the test are accelerations and strain. A forced vibration test on a bridge is usually performed using some exciting device (e.g., shaker, impulse, mass release) at pre-defined locations. Measurement sessions are repeated within the timeframe and with frequency that are required to reliably and accurately capture the dynamic response and determine the modal parameters of the structure. The bridge should be closed for traffic during the test.

Gauge length. The physical length over which the strain sensor interacts with the structure, i.e., measures the strain. As a consequence the measurement represents an average strain over the gauge-length. In general, the sensors are considered either as a short-gauge or as a long-gauge, depending on the length of the gauge and properties of the monitored material. Traditional sensors, such as strain gauges and vibrating wires, belong to the group of short-gauge sensors. Depending on their type and packaging, optical-fiber sensors can function as short-gauge or long-gauge sensors. General principles for the selection of an appropriate sensors' gauge length, depending on application type and construction material, are presented in Section 4.5 (Glisic 2011). See also "short-gauge sensors" and "long-gauge sensors."

Global Positioning System (GPS). World-wide radio navigation system made available by US government consists of three components: satellites orbiting the Earth, control stations, and GPS receivers. The GPS receiver can be used as a discrete sensor for position and/or displacement monitoring. GPS receivers continuously communicate with satellites that are used as references to calculate their own global position using triangulation. The accuracy of the system depends on the constellation of satellites "visible" to the GPS sensor at the moment of measurement and, to a certain extent, on the weather conditions. Sub-centimeter reproducibility can be achieved, depending on the system and the strategy applied. Due to its limited precision, GPS is mostly used to monitor the global displacement of the structure, e.g., due to the settlement of foundations or tilt. The advantages of GPS are that it has practically unlimited dynamic range, the sensors need no mutual connection or visibility, and it features excellent long-term stability, so it can be used for long-term monitoring, assuming the US government will keep the system available for civil uses.

Ground Penetrating Radar (GPR). Qualitative 3D distributed static sensing system for the detection and location of discontinuities in mechanical and geometrical properties in a structure. Typical applications are related to the assessment of bridge concrete deck condition, i.e., detection of cracks or zones of delamination from rebars. The principle of functioning of GPR is based on analysis of a radio signal sent from the source into the structure and received after reflection from different layers of the structure's material. By measuring the time and intensity of the reflected radio waves and comparing these with historical records, it is possible to assess the condition of the structure's material. For example, the radio waves reflect in a similar fashion off the rebars, unless the concrete is non-uniform, implying deterioration. By analyzing the data, the user is able to determine the possible locations of cracks, voids, delamination, and corrosion in the concrete. The advantages of GPR are the ability to detect defects deep under the surface of the structure (i.e., invisible from the surface) and the relatively high speed of measurements. The disadvantages are that traffic lanes must be closed during the measurements, and the data analysis can be subjective, depending on the operator who analyzes the data. Computer programs that aid in the analysis of the GPR data make the data analysis less subjective. GPR can be either operated manually or located on a vehicle to increase the speed of measurements.

Hazard. A situation that creates a threat (e.g., to lives, goods, society, etc.). There are two broad types of hazards in bridge management:

Heuristics. Any technique or approach for problem solving, discovery, and education based on experience. Typical examples are an engineering judgment, an intuitive judgment, a proven repeated solution, "common sense".

Inspection. On-site non-destructive examination to establish the present condition of the structure (Wenzel 2009).

Integrity. Quality of being whole and complete, or the state of being unimpaired.

Integrity monitoring. SHM uses the system with capability of direct detection, location, and quantification (or rating) of local strain changes generated by damage or deterioration. Direct detection of local strain changes implies the use of distributed sensors that provide for coverage of all sections of the monitored structure or monitored member (see "distributed sensor" and "sensing cable"). Integrity monitoring features very high reliability in damage detection and location since the sensor presence is practically ensured in the sections of structure subjected to damage or deterioration.

Interrogating (of a sensor). See "reading (of a sensor)."

Interrogator. See "reading unit."

Laser scanning. Non-contact quasi-static geometry measurement system, which consists of one or more laser scanners that are installed on tripods. The laser scanner emits light toward the structure and receives back the light reflected from the structure. Depending on the type, (i) the time of flight of the reflected light or (ii) the phase difference of the peak amplitude in the reflected light are used to measure the distance to the points in a structure. The laser can rotate to accommodate large structures. However, if multiple profiles of a bridge need to be monitored, more scanners would be necessary. The error limits of laser scanners typically range between 1 and 10 mm. The main advantages of the system are that an entire structure can be surveyed with no physical access to it and that entire views of a structure may be obtained. The disadvantages are that differences in surface types on the same structure can have an impact on accuracy and unfavorable weather conditions (e.g., rain or fog) as well. An unobstructed view between the laser scanner and the structure is necessary for proper measurement.

Laser total station. Static discrete displacement measurement system. A stationary laser sends the light, which reflects back from prism targets manually installed at points of interest in a structure. The total station is able to automatically measure the distance and relative angles at all points. This data can be converted into tri-dimensional displacements using appropriate software. An unobstructed view between the total station and all the measurement points is necessary for proper measurement.

Life-cycle monitoring. See "lifetime monitoring."

Lifespan monitoring. See "lifetime monitoring."

Lifetime monitoring. Long-term monitoring that extends over the entire lifespan of the structure, from the construction until the decommission or dismantling. Synonyms for lifetime monitoring are "lifespan monitoring" and "life-cycle monitoring."

Long-gauge sensors. If monitored material is homogeneous (e.g., steel): strain sensors whose gauge length is longer than 100 mm (4 in.); if monitored material is inhomogeneous (e.g., concrete): strain sensors with a gauge length several times longer than the maximal distance between two discontinuities in material mechanical properties (e.g., distance between two cracks or maximal size of aggregate). For example, in the case of cracked reinforced concrete, the gauge length of a long-gauge sensor is several times longer than both the maximum distance between cracks and the diameter of the aggregate. The main advantage of long-gauge sensor measurement is in its nature: since it is obtained by averaging the strain over a long measurement basis, it is not influenced by local material discontinuities and inclusions. Thus, the measurement contains information related to global structural behavior rather than local material behavior. Fiber optic sensors based on low coherence interferometry (SOFO) and specially packaged FBG belong to the group of long-gauge sensors.

Long-term monitoring. Monitoring performed over an extended period of time (typically several months to several years) with the aim of obtaining information related to structural behavior and performance that cannot be achieved by short-term monitoring. Typical examples include (but are not limited to) tracking the structural behavior and detection of unusual behaviors and degradation in performance due to temperature variations, rare events such as strong wind or an overloaded truck, slow deterioration processes (e.g., corrosion, scour, and fatigue), rheologic effects (e.g., creep, shrinkage, and differential settlements).

Man-made hazard. Any hazard situation that is caused by humans, such as improper design or construction, hazards present at a construction site, and terrorist act.

Measurand. Physical quantity being measured (e.g., strain, temperature, acceleration, etc.). In a broader sense, a property that may have only qualitative attributes can also be included under the definition of a measurand. For example one can "measure" whether the crack is active by gluing the glass patches over the cracks; if after a certain time the patch is broken, then the crack is active; this "measurement" is not performed in the traditional sense where the magnitude of the crack is quantified (measured), but rather in a broader sense where qualitative information relative to measurand is acquired ("the crack is active").

Measurement session. Set of measurements of different sensors associated to one moment in time, i.e., to one event. It contains one measurement per sensor. If the measurements are taken simultaneously, then the time of measurement corresponds to the time associated with the session. If the measurements are taken sequentially, then each sensor is measured at (slightly) different times, and the time associated with each session is defined as either the time of the measurement of the first sensor, the time of measurement of the last sensor, or the mean value between all the measurements.

Monitoring system. The set of all the means destined to carry out measurements or observations and to register them is called a monitoring system. Monitoring system has two subsystems: measurement and observation subsystem and data management subsystem. The measurement and observation subsystem consists of sensors, observation tools, and the DAQ system – DAQ hardware (including reading units, accessories, and channel switch) and DAQ software. The data management subsystem consists of data management hardware and data management software. Nowadays, there are a large number of monitoring systems based on different functioning principles, and which system will be used in a specific application depends on specifications of the monitoring system, such as resolution, measurement error, maximal measurement rate, and data management software.

Natural hazard. Any hazard situation that is caused by a natural process, such as earthquake, hurricane, tornado, volcano, landslide, corrosion, and fatigue.

Nyquist frequency. Half the measurement frequency. Nyquist frequency is a property of the monitoring system, and it is the highest frequency beyond which the time history of measurand (e.g., strain, acceleration, displacement, etc.) cannot be properly represented by measurements due to aliasing.

Nyquist rate. Two times the highest frequency in the structure's dynamic response. Nyquist rate is property of measurand (i.e., the structure's dynamic response), and it is a lower bound for the measurement rate that eliminates aliasing. For example, if the structure's dynamic response has frequencies of 1 Hz, 3 Hz, and 7 Hz then, to avoid aliasing, Nyquist rate should be higher than $2 \times 7 = 14$ Hz and the rate of measurement rate should not be lower than this value. However, due to imperfections in data pre-processing and post-processing, it is recommended that the measurement rate be several times (10 times if possible) higher than the Nyquist rate of the structure's response.

Peripherals. See "accessories."

Piezoelectric material. Sensing element are most frequently made of ceramic (e.g., Barium titanate – the first discovered, PZT – Lead zirconate titanate, the most used); single crystals (e.g., Gallium phosphate, quartz, and tourmaline) or piezo-polymers (PVDF – Polyvinylidene Fluoride), which convert dynamic change in their shape (i.e., strain) into an electrical field or electric potential, and vice versa. It is used as a sensing element in many sensors for dynamic measurements, such as accelerometers, strain sensors, and AE sensor.

Point sensor. See "discrete sensor."

Precision (of a monitoring system). *Qualitative term*. The degree of mutual closeness of repeated measurements with no change in measurand. The definition of precession is similar to definitions of repeatability and reproducibility, and that is why it is often confused in practice with these two specifications. It is important to notice that "precision" does not specify conditions of measurement (see "repeatability" and "reproducibility"). From a metrological point of view, "precision" is considered only as a qualitative term, and quantitative value should not be attributed to this term (e.g., one can say that a strain monitoring system is more precise than another, but one should not say that the precision of the system is, for example, $2\,\mu\varepsilon$). However, in civil engineering practice, a qualitative value is frequently attributed to precision by suppliers of monitoring systems. In these cases, the user of the system should ask for more information from the supplier in order to understand how this value is determined and whether it is relevant for his application. The term "precision" is frequently confused in practice with the terms "resolution," "error limits," and "accuracy."

Important: the system that is precise is not necessarily accurate. For example, let us assume that the actual strain at a point in the structure is $100\,\mu\varepsilon$, and let us assume that five repeated strain measurements at that point have values of $119\,\mu\varepsilon$, $120\,\mu\varepsilon$, $118\,\mu\varepsilon$, $121\,\mu\varepsilon$, and $117\,\mu\varepsilon$. The system is very precise because the measurements are, by value, very close to each other (maximal difference is $4\,\mu\varepsilon$), but the system is at the same time inaccurate because the measurements are not close to the actual strain (minimal difference is $17\,\mu\varepsilon$ and maximal difference is $21\,\mu\varepsilon$). Statistically, a precise system has a small dispersion of repeated measurements, assuming no change in measurand during the measurement.

Prognosis. Coupling of historic and current information from SHM, nondestructive evaluation, environmental and operational conditions, heuristics, and numerical modeling to estimate the remaining serviceability of the structure or reliability of the structure.

Pseudo-dynamic testing. A hybrid experimental method based on the simultaneous combination of numerical simulation and active control (e.g., by means of displacement sensors and actuators) for evaluation of dynamic properties (typically non-linear stiffness) of structures subjected to dynamic actions. In general, the mass is lumped and the structure is discretized; thus, displacements and actuation forces are imposed and monitored at a finite number of nodes defined by the degrees of freedom created by lumped mass discretization; the inertia and damping properties are simulated; and restoring forces are acquired from the structure under the test. The stiffness properties are determined by the restoring forces using numerical integration procedures. Pseudo-dynamic testing is mostly performed in laboratories on large-scale physical models of structures (Aktan 1986). With some restrictions, it can also be applied on-site (e.g., the critical part of the structure is tested while a reminder of the structure is numerically modeled).

Qualitative sensor. See description under "sensor."

Quasi-static monitoring. See "static monitoring."

Quasi-dynamic monitoring. See "dynamic monitoring."

Random error (of a measurement). Error of measurement caused by uncontrollable and small changes in environment that affect the measurement system, or measurand, or both. Repeated measurements of the same quantity under unchanging conditions will give different values due to random errors. The random error of a specific measurement can be defined as the difference between this measurement and the mean of repeated measurements. Thus, random errors can be reduced by repeating measurements. Synonym for "random error" is accidental error.

Reading (of a sensor). Acquiring the magnitude of measurand or qualitative information relative to measurand from a sensor at a given moment in time is part of DAQ. It usually consists of the following phases: initiation of measurement, sending encoding signals to the sensors (i.e., sensing elements), receiving the signals from the sensors that are decoded by measurement, decoding the signal, and retrieving the magnitude of or qualitative information relative to the measurand. The reading of sensors is performed by the DAQ system, most commonly by a reading unit. See more under "reading unit" and "reading of multiple sensors."

Reading of multiple sensors. Reading of more than one sensor. It is performed using a reading unit combined with a channel switch. There are three possible ways of reading multiple sensors: (i) simultaneous (all sensors are read at the same time), (ii) sequential (sensors are read one after another), and (iii) combined (sensors are first divided into several groups; the sensors belonging to the same group are read simultaneously, but the groups are read sequentially).

Reading unit. Part of DAQ system that initiates and performs the reading of sensors. Synonyms frequently used in practice are "interrogator" and "readout unit." In general, the reading unit is a single piece of equipment that contain the source of encoding signals, the set-up for generating the encoding signals, the receiver for decoded signals from sensors, a set-up that retrieves from decoded signals the magnitude of or qualitative information relative to measurand, processor, and firmware. Nevertheless, since the components of the reading unit depend on the functioning principle of the sensors, various other configurations can be found on the market. The reading unit has the capability of reading a limited number of sensors, and usually, the capability of reading multiple sensors is augmented by the channel switch (commutator). The communication between the reading unit and sensors and between the reading unit and data management hardware is enabled by accessories.

Readout (of a sensor). See "reading (of a sensor)."

Readout unit. See "reading unit."

Regular inspection. Inspection performed as part of a regularly scheduled program (e.g., once every year or two).

Relative error (of measurement). Ratio between absolute error and true value of measurand.

Reliability (of a monitoring system). Ability of the system to perform its required functions under stated conditions for a specified period of time. This implies that system specifications must not change during the specified period of time. A quantitative conventional measure of reliability is mean time to failure (MTTF), which estimates how long the system is expected to perform its required functions under stated conditions. High reliability is crucial for long-term monitoring.

Reliability (of a structure). Probability that a structure will perform its intended functions during a specified period of time under stated conditions (Wenzel 2009).

Repeatability (of a monitoring system). *Quantitative term.* Closeness of the agreement between the results of repeated measurements of an unchanged measurand (i.e., value of measurand is constant while measuring) carried out under the unvarying conditions of measurement. For example, strain measurement is repeated on a deformed beam, whose strain does not change, in a room with constant temperature and humidity; all the measurements are carried out by one operator using the same reading unit and the same sensor. In summary, the measurements were repeated under the most favorable conditions. The repeatability is determined by applying statistical analysis to a large number of measurements repeated in the laboratory, and the most frequently calculated is as a dispersion of measurements or as a multiple of dispersion (e.g., two times dispersion or three times dispersion). In practice, the term "repeatability" is often confused like terms with "resolution," "accuracy," and "precision."

Reproducibility (of a monitoring system). *Quantitative term.* Closeness of the agreement between the results of measurements of an unchanged measurand (i.e., value of measurand is constant while measuring) carried out under varied conditions of measurement. For example, strain measurements are carried out on a deformed beam, but conditions (e.g., temperature and humidity) surrounding the beam and the reading unit are varied during the measurements; several operators were taking the measurements one after another, and they also exchanged the reading unit during the measurement. In summary, the measurements were performed under the most unfavorable conditions. The reproducibility is determined by applying statistical analysis to a large number of measurements performed under varying conditions, and the most frequently calculated is as a dispersion of measurements or as a multiple of dispersion (e.g., two times dispersion or three times dispersion). In practice, the term "reproducibility" is sometimes confused with terms like "resolution," "accuracy," and "precision."

Resolution (of a monitoring system). *Quantitative term.* Minimum detectable change Δx of the measurand x that can be perceived by the system. If the value of resolution is small, then the system is said to have high resolution, and it is able to monitor the measurand with higher accuracy and precision. Thus, a small value of resolution is a desirable specification. In practice, term "resolution" is often confused with the terms "repeatability," "accuracy," and "precision."

Risk. *Qualitative:* A situation that exposes (e.g., lives, goods, and society) to threat. *Quantitative:* Expected value (e.g., number of injuries/lives lost, costs, and economic losses for society) of an undesirable outcome (e.g., failure of structure or structural element) caused by external or internal vulnerabilities combined with natural or man-made hazards. The risk $R(H)$ due to a hazard H is calculated as the product of the probability of the hazard $p(H)$, vulnerability $p(f|H)$ to the hazard H (probability of failure given the hazard), and the quantified consequences $v(f)$ (number of injuries/lives lost, costs, and economic losses for society) of outcome f (e.g., failure). This is expressed by the following equation:

$$R(H) = p(H) \times p(f|H) \times v(f) \qquad (A.1)$$

The risk is calculated as the sum of risks due to all the hazard scenarios:

$$R = \sum_H R(H) \tag{A.2}$$

Risk may or may not be mitigated through preventive or corrective actions.

Risk assessment: Process of analysis or quantification of risk for different hazard scenarios.

Sensitivity (of a monitoring system). *Quantitative term.* Ratio dy/dx between variations of the encoding parameter y and the measurand x. If small change in the measurand x generates a big change in the encoding parameter y, then the monitoring system is able to monitor the measurand with higher accuracy and precision. Thus, high sensitivity is a desirable specification.

Sensing cable. A distributed fiber-optic sensor is represented by a single cable that is sensitive to measurand (strain and/or temperature) at any point along its length. Hence, one distributed sensor can replace a large number of discrete sensors. There are four main physical principles used in distributed sensing: Rayleigh scattering in optical fibers (allows strain and temperature measurements), spontaneous Brillouin scattering (allows strain and temperature measurements), stimulated Brillouin scattering (allows strain and temperature measurements), and Raman scattering (allows temperature measurements). Since the cable is continuous, it provides for monitoring of a one-dimensional (1D) strain and/or temperature field, i.e., it provides a distribution of the measured strains and/or temperature along the sensor. Sensing cables can be installed along the whole length of the structure, and in this manner, each cross-section of the structure is practically instrumented. The sensing cable is sensitive at each point of its length, and it provides for direct damage detection, avoiding the use of sophisticated algorithms. In this manner, integrity monitoring of structures can be reliably performed. Also see "damage detection" and "integrity monitoring."

Sensing element. Part of a sensor that converts measurand into encoding parameter (e.g., electrical resistor, fiber-optic Bragg grating, and piezo-electric material).

Sensor. Sensor is a device that interacts with some physical quantity called the measurand and converts it into another physical quantity called the encoding parameter, which can be read by an observer or by a reading unit (and DAQ system). In the traditional sense, the sensor "measures" the magnitude of measurand (e.g., strain-gauge "measures" the magnitude of the strain). In a broader sense, the sensor can also provide qualitative information relative to measurand; see the example presented under "measurand." This type of sensor is called a qualitative sensor. Depending on the physical principle behind conversion of the measurand into the encoding parameter (principle of functioning), the sensors can be in general considered electric, fiber-optic, acoustic, GPS-based, laser scanning, etc., and each of these groups can be subdivided based on specific physical principle, e.g., electric sensors can be resistive, capacitive, inductive, piezo-electric, etc. The main components of a sensor are (i) sensing element, i.e., device that converts measurand into encoding parameter (e.g., electrical resistor, fiber-optic Bragg grating, piezo-electric material, etc.), (ii) packaging that protects the sensor during handling, installation, and in long-term (e.g., from environmental influences), and (iii) interface to connect with reading unit (and DAQ system). Depending on principle of functioning, some other components can be present, such as equipment for conditioning and processing of electrical signals in the case of electric sensors, coupling and multiplexing in the case of optical sensors, etc. The sensors that should be in direct contact with the structure also have anchoring pieces, i.e., the parts whose purpose is to anchor the sensor to the structure.

Short-gauge sensors. If monitored material is homogeneous (e.g., steel): strain sensors with a gauge length shorter than 100 mm (4 in.); if monitored material is inhomogeneous (e.g., concrete): strain sensors with a gauge length comparable to or shorter than the distance between two discontinuities in material mechanical properties (e.g., distance between two cracks or the maximal size of aggregate). While short-gauge sensors perform well in homogeneous materials, the measurement performed with short-gauge sensors in inhomogeneous materials is strongly influenced by discontinuities in material mechanical properties. Thus, in inhomogeneous materials, short-gauge sensor measurement provides information related to local material properties and is not suitable for global structural monitoring. Traditional sensors, such as strain gauges and vibrating wires, belong to the group of short-gauge sensors. Depending on their type and packaging, fiber-optic sensors can function as short-gauge sensors as well (e.g., Fabry–Perot interferometers and FBG).

Short-term monitoring. Monitoring performed over a limited period of time with the aim of obtaining specific information related to structural behavior and performance. The period of time usually ranges from a few minutes to a few days, but, depending on the specific objectives of monitoring, it can be extended to longer periods (e.g., from few weeks to few months). Typical examples include (but are not limited to) monitoring during static, ambient vibration, or forced vibration tests, tracking short-term fatigue growth, or monitoring the response of a bridge for a permit vehicle.

Specifications (of a monitoring system). Qualitative and quantitative properties of the measurement subsystem, namely: sensitivity, resolution, dynamic range, accuracy, stability, precision, repeatability, reproducibility, and error limits.

Stability (of a monitoring system). *Qualitative term*. Property of having no change in encoding parameter, i.e., no change in measurement, if there is no change in actual value of measurand over a certain period. Stability is crucial for long-term monitoring. The main sources of instability in the monitoring system are aging and wear of components, and environmental influences such as temperature, humidity, corrosion, and electro-magnetic interferences due to proximity of power lines and lightning. From a metrological point of view, "stability" is considered only as a qualitative term, and quantitative value should not be attributed to this term (e.g., one can say that a strain monitoring system is more stable than another, but one should not say that the stability of the system is, for example, 2% per year). However, in civil engineering practice, a qualitative value is frequently attributed to accuracy by suppliers of monitoring systems. In these cases, the user of the system should ask for more information from the supplier in order to understand how this value is determined and whether it is relevant for his application.

Static monitoring. Monitoring of quantities such as strain, displacement, and forces on a structure subjected to static loads. Static loads do not change magnitude or position in time, or these changes are "slow enough," so the accelerations at points of monitored structure can be neglected. In true static monitoring, it is assumed that the static load state of the structure does not change over the period of time needed to complete a measurement session (one measurement over all sensors). However, the bridges in service are almost never subjected to static loads only (unless the traffic is closed or the bridge is empty during the measurements), and for practical reasons, measurement sessions are performed under the traffic load. Moreover, many monitoring systems feature sequential reading of sensors (sequential DAQ). In these cases, the monitoring is quasi-static. Static monitoring (true or quasi) is usually performed in the mid-term or long-term, in order to observe slow changes in structural behavior. That is why only monitoring systems that feature long-term stability should be used for this type of monitoring. The measurement pace ranges between a few minutes and a few days.

Static test. Determination of the static response of the structure (strain, tilt, displacements, curvature, influence lines, and stiffness) caused by a static load with known position and magnitude. The most frequent aims of static test are load rating, system identification, and detection of unusual behaviors. A static test on a bridge is usually performed using several trucks with known loads that are placed at pre-defined locations. Measurement sessions are registered for each load case once or repeated several times within pre-defined time interval (e.g., to assess rheologic effects). The bridge should be closed for traffic during the test.

Strain gauge. Synonym for "strain sensor"; in jargon, the synonym for "electrical resistance strain gauge."

Strain sensor. Sensor designed to measure the strain. There is a large variety available on the market: static or dynamic, discrete or distributed, short-gauge or long-gauge, embeddable or surface mountable, etc. There are various sensing elements used for strain gauges: electrical resistive material, vibrating wire, piezoelectric material, and optical fibers. The summaries of best performances are for different technologies given in Sections 3.5 and 3.6. See also "strain-gauge," "vibrating wire sensor," and "fiber optic sensor."

Structural health monitoring (SHM). The process of implementing an identification strategy for unusual structural behaviors (e.g., damage) is referred to as SHM (modified Farrar and Worden 2007). It is a process aimed at providing accurate and real-time information concerning structural health conditions and performance (Glisic and Inaudi 2003a). The information obtained from monitoring is used to increase safety, plan and design maintenance activities, optimize rescue actions, verify hypotheses, reduce uncertainty, and widen the knowledge concerning the structure being monitored. It is characterized by four "dimensions": monitored parameters (and damage-sensitive feature), timeframe, level of sophistication, and scale.

Structural identification (St-Id). See "system identification."

System identification. Describes the building of mathematical representations of a structure using either mechanistic (physics-based) or statistical methods based on the results of measurements. More commonly, an initial approximate mathematical model of a structure is created independent of measurements (e.g., based on structural analysis), and measurements are used to determine critical parameters of the model; if the initial model cannot be parameterized with satisfactory accuracy using the measurements, the model is abandoned and a new model that better describes the structure is built. Less commonly, an initial mathematical model does not exist, and mathematical model is built purely using statistical methods. There are numerous approaches to how system identification can be performed, and system identification is typically performed either in the time or frequency domains.

Systematic error (of a measurement). Error that remains constant when measurement is repeated under the same conditions. It is caused by an offset in the measurement system, the operator (person) who performs the measurement, etc. Systematic error can be additive (adds a constant to absolute error) or multiplicative (adds a constant to relative error). It cannot be identified or reduced by statistical methods, but other methods are to be used, e.g., by tracing calibration to the primary standard (etalon) or by changing the experimental set-up (e.g., exchanging the reading unit, having a different operator of measurement, etc.). The synonym for "systematic error" is "bias."

Unusual structural behavior. Structural behavior that is not in accordance with design of the structure, structural analysis, construction method, previously recorded behavior, or engineering experience. Unusual behavior points to some alteration of structural health condition and/or performance due to (i) damage or deterioration induced by environmental degradation, wear, and episodic events like earthquake or impact, (ii) changes in operational and environmental conditions, or (iii) unintentional design, construction, and maintenance imperfections or errors.

Vibrating wire (VW). Sensing element that converts the natural frequency of a tensioned wire into an electrical signal by means of an exacting and pick-up coil. As the natural frequency of tensioned wire is related to strain in the wire, which depends on the relative position of the extremity of the wire, the vibrating wire is used as a sensing element in strain sensors, but also in other types of sensors for which a change in length can be used as a functioning principle, such as pressure sensors, and load cells.

Vibrating wire strain sensor. Static and/or dynamic discrete short-gauge or long-gauge strain sensors based on vibrating wire functioning principle. Typical resolution is ranged between 0.35 and 1 µε, and the gauge length is ranged between 50 and 150 mm. Vibrating wires can be either surface mounted on an existing structure (welded, bolted, or bonded) or embedded in fresh concrete during construction. They feature an excellent stability in long-term, and are proven in on-site conditions. As they are sensitive to temperature, usually a temperature sensor for compensation purposes is placed in the packaging of the sensor. Their disadvantage is their sensitivity to electromagnetic interferences, especially lightening, which can damage the sensors.

Vibration. Repetitive variation in time of some mechanical quantity (strain, stress, force, and displacement) about a point of equilibrium. The vibrations can be periodic (e.g., motion of a pendulum) or random (e.g., motion of points on a bridge under the traffic load).

Vibration monitoring. Monitoring of mechanical quantities related to vibration, such as accelerations, strains, displacements, and forces. Vibration monitoring is performed using dynamic monitoring systems with accelerometers, strain sensors, and displacement sensors. Due to the large amounts of data that may be generated, vibration monitoring is usually performed in the form of scheduled periodic short-term measurements sessions (e.g., for monitoring ambient vibrations), or tests (e.g., for monitoring forced vibrations). The rate of measurement sessions depends on natural frequencies of vibration (see "Nyquist rate"). Vibration monitoring (both ambient and forced) is used to determine modal parameters of the structure: natural frequencies, corresponding mode shapes, and damping factors, and to observe changes in these parameters in the long-term. Vibration monitoring is frequently combined with dynamic monitoring as the same monitoring system can be used for dual purposes. Vibration monitoring, combined with dynamic monitoring can also be used to assess fatigue.

Vulnerability. *Qualitative*: The susceptibility of a structure or its members to experience an undesirable outcome (e.g., failure). *Quantitative*: Probability $p(f)$ of occurrence of an undesirable outcome f (e.g., failure). If several hazards H can cause the undesirable outcome f, then the vulnerability $p(f)$ is calculated as the sum of the products of the probabilities $p(H)$ of the hazard H and the vulnerability to this hazard $p(f|H)$:

$$p(f) = \sum_{H} p(H) \times p(f|H) \tag{A.3}$$

Index

Note: *Italic* page numbers refer to *figure* and **Bold** page numbers reference to **tables**.

a

absolute elastic strain   182, **197–198**, 212
absolute error   80
absolute relative error   80
accuracy
  of compensated strain measurement   90
  of integration methods   266
  of measurement   22–23, 79, 80, 88
  of strain measurement   109
ACI 318-02 (ACI 2002)   285
acoustic systems   21
aging, of infrastructure   1
air temperature measurements   125, 126, *126*
American Society of Civil Engineers (ASCE)   1
ASCE (American Society of Civil Engineers)   1
automatic data analysis   12
average strain   93

b

bare optical fiber   290
bascule span   34
BCBC (Burlington County Bridge Commission)
  34
beam
  linear cross-sectional thermal gradients on
    161, *161*
  stress and strain in   152–156, *153*
bilinear distribution of curvature   266
BOFDR (Brillouin Optical Frequency Domain
  Reflectometry)   57
BOTDA (Brillouin Optical Time Domain
  Analysis)   57, 69, 70, *108*

BOTDR (Brillouin Optical Time Domain
  Reflectometry)   57, 69
Brillouin effect   45
Brillouin Optical Frequency Domain
  Reflectometry (BOFDR)   57
Brillouin Optical Time Domain Analysis
  (BOTDA)   57, 69, 70, *108*
Brillouin Optical Time Domain Reflectometry
  (BOTDR)   57, 69
Brillouin scattering   57
  distributed fiber optic strain sensors   99
  strain sensors on   66–70
buckling stress   186
Building 166A   207–212
Burlington County Bridge Commission (BCBC)
  34

c

calibration constants   91
Carlson strain meters   49, *50*
Cauchy–Schwarz inequality for covariance   205
CEB-FIP Model Code 1990 (MC90)   169, 176,
  285
centroid of stiffness   *243*, 243–244
clamping error   107
cleansing of data   120–121
CNNs (Convolutional Neural Networks)   75
coating function   57
commercial strain sensors   42
compensated strain sensor   90
compensation function   90
compressive strength   249, **250**

Introduction to Strain-Based Structural Health Monitoring of Civil Structures, First Edition. Branko Glišić.
© 2024 John Wiley & Sons Ltd. Published 2024 by John Wiley & Sons Ltd.

compressive stress   186
connecting cables   46
Constant Current Circuit   54
constitutive law   179
construction materials   138
   strain constituents in   **139**
   thermal expansion coefficients of   **158**
Convolutional Neural Networks (CNNs)   75
coordinates mapping, distributed sensors
      292–294, *294*
crack detection principle
   algorithm   295
   crack openings   283, 284, **284**, 287, *287*
   direct and indirect detection   279
   initiation times   **284**
   long-gauge strain sensors   279–282
   short-gauge strain sensors   279–282
   soft-packaged short-gauge sensors   274–279
crack identification   294–296
crack quantification
   long-gauge strain sensors   282–284
   short-gauge strain sensors   282–284
creep   169–174, 204
   elastic strain and   173
   estimation   **170**
   evolution   *170*
   of reinforced concrete   172
   and stress relationship   173
crossed topology   251–252, 255–260
cross-section   *188*
   bending moment in   220
   geometrical properties of   154, 189, **189**,
      239
   mechanical properties of   **239**
   normal force in   219–220
   Southeast Leg of Streicker Bridge   238–239,
      239
   thermal gradients   158–166, *159*

d

damage-sensitive features   4, 10, 33, 39
data analysis algorithms   33, 241
data analysis and management   11, 21, 29–31
   basic and advanced   *29*
   components of   31, *31*, 32
   integration and visualization   30
   subsystem   *31*

data-driven approach/methods   123, 126, 173
data exploitation   12
data interpretation   30
Data Management Component (DMC)   35
data visualization   16, *16*, 30
deformed shape and deflection   *263*, 263–272,
      271, **271**, *271*
deterministic approach   82
dimensionless partial shear coefficient   254
direct deflection measurement   40
direct detection approach   274
direct determination of error   100
direct measuring system   *22*
direct monitoring   39
direct sensing approach   71–73, *72*
discrete fiber optic sensors (FOS)   45
discrete sensors   46, *113*
   and distributed sensors   92
   gauge length of   *103*
   inherent source of error   92–94, *93*
   mechanical interlocking   113
   strain measurements with   108
displacement sensors   217
distributed fiber optic sensors (FOS)   45, *68*
distributed fiber-optic strain sensors (FOSS)   69,
      112
distributed sensors   66–70, *70*, 107, 109, 177, *177*,
      178
   coordinate mapping   292–294, *294*
   crack detection and localization using. *see*
      crack detection principle
   crack identification   294–296
   inherent source of error   104–109
   for integrity monitoring   289–297
   mechanical interlocking   113
   one-dimensional (1D)   289
   packaging of   289
   principle of   *105*
   "Profile" strain sensor   *113*
   sampling interval of   105, 292–294
   sensor packaging, installation, survivability
      289–292
   spatial resolution for   105, 294–296
distributed "Tape" sensor   *114*
DMC (Data Management Component)   35
drifting sensor   *127*
drift measurement   24, *24*, 125

dynamic monitoring   7
dynamic strain sensors   34

e

early age thermal gradient   247, **247**
EC26 (Punggol East Contract 26) project   203,
    207–212
EFPI (Extrinsic Fabry–Perot Interferometry)   57
elastic curvature   219, 236, 238, 240, 241, 243,
    247, 247–248, 285
    due to thermal gradient   247
elastic strain   195
    absolute elastic strain   **197–198**
    model parameters and   **193**
    relative uncertainty of   204
    and stress   152
elastomagnetic sensor   39
electrical power   33
electrical sensors   42, 49
electrical systems   21
electrical telemeters   43, *43*, 44
electromagnetic interference (EMI)   49
embedded sensors   119
EMI (electromagnetic interference)   49
endpoint error   107
environmental noise   33
epistemic error   179
epistemic model error   268
equivalent shear modulus   254
erroneous measurement values   121
error and uncertainty
    analysis   85–86
    propagation formula   **83**, 83–85, **84**
    rounding and significant figures   86–88
    to SHM system   80–83
    thermal compensation of sensor   88–92
error of model. *see* model error
error propagation formula   83, **83**, 134, 136, 152
expandable sensor networks   45
Extrinsic Fabry–Perot Interferometry (EFPI)   57

f

Fabry–Perot interferometry   279
false positive detections   33
FBG. *see* Fiber Bragg-grating
Federal Highway Administration (FHWA)   188
FHWA (Federal Highway Administration)   188

Fiber Bragg-grating (FBG)   279
    strain sensors   58, *63*, 63–66, *64*, 188
FiberMetrics Corporation   45
fiber-optic (FO) displacement sensors   264
fiber optic sensors (FOS)   42, 44–45, 111, 137
fiber-optic strain sensors (FOSS)   56, *58*
fiber-optic technologies   44
firmware   21
first-generation strain sensors   42
    advantages and challenges   48–49
    best performance of   **50**, 56
    resistive strain gauge   53–56
    vibrating wire strain sensor   49–53
FOS (fiber optic sensors)   42, 44–45, 111, 137
FOSS (fiber-optic strain sensors)   56, *58*
Four-Wire Ohm Circuit   54
frictional stress distribution   261–262
full-bridge strain gauges   *112*
fundamental frequency   51

g

gauge factor (GF)   53
gauge length
    of discrete sensors   92–94
    discontinuities in material   98–99
    inclusions in material   96–97
    homogeneous part of material   94–96
    for monitoring of beam-like structures   104
    scale of monitoring   102–103, *103*
    strain measurement   99–102
    and theoretical strain distributions   *101*
Gaussian distribution   82, 83
geometrical properties, of beam elevation   154
GF (gauge factor)   53
global-scale SHM approach
    average curvature   231–232
    average normal force   225
    axial and bending stiffness   235–238
    axial stiffness of cell   223–224
    bending moment   238
    change in length of cell   222–223
    change in relative rotation   231–232
    crossed topology   251–252, 255–260
    deformed shape and deflection   *263*, 263–272
    examples of   260–272
    implementation guidelines   213–215
    neutral axis   233–235

global-scale SHM approach (*contd.*)
  overview   213
  parallel topology   230–231, *231*
  physics-based interpretation   218–219
  pile subjected to axial loading   *215*, 215–216
  shear force and shear stiffness   253–255
  shear strain and shear stress   252–253
  simple topology   221–222
    measurements   225–230
  strain and normal force   255
  strain at centroid   235
  of Streicker Bridge   216–218, *217*
  topologies used in   213–214, *214*
  topology of sensors   219–221
Götaälvbron. *see* Gota Bridge project
Gota Bridge (Götaälvbron) project   *13*, 13–18
  crack detection   17, *17*
  damaged flange of   *14*
  sensors   *17*
  SHM system   **15**

h

hardening time of concrete   241
HDB (Housing and Development Board)   203
heterogeneous SHM system   19, 34–38
homogeneous materials   94–96, 104, 106, 131,
    177
homogeneous system   135
horizontal thermal gradients   285
Housing and Development Board (HDB)   203
hyper-static (indeterminate) structures   172

i

indirect approach   71
indirect measuring system   *22*
indirect sensing approach   71–73
inherent error, in gauge length   96–99
inhomogeneous materials   104, 106, 157
instrumental measurement uncertainty   25
integration methods   266, **267**
integrity monitoring   8
  on direct and indirect detection   279
  distributed sensing for   289–297
  implementation of   296–297
integrity scale SHM
  crack identification   274–297
  direct detection approach   274

overview   273–274
International Bridge Study (IBS)   188
interpretation
  of data   30
  of single-sensor measurement   213
  of standard error   82
  of strain measurements. *see* strain
      measurements
  of strain sensor measurements   131

l

L-brackets installation   114–115, *115*
level 1 SHM   7–8
level 1.5 SHM   7–8
level 2 SHM   7–8
level 3 SHM   7–8
level 4 SHM   7–8
light propagation   57
limited maintenance costs   3
limits of error   25, 80, 82–83
  of elastic strain   193
  evaluation of   185–186
  propagation formula   **83**, 136
  reference time and comparisons with   *199*
  for thermal strain   192
linear beam theory   179, 193, 203
linear cross-sectional thermal gradients   *161*, 162
  short-term   162–166, *163*
linear distribution of curvature   266
linear theory of beams   255, 268, 269
linear thermal compensation function   91
linear variable differential transformers (LVDTs)
    40
Live Portal Software   36, *37*
local-scale SHM   8
  mechanical strain and stress   140–156
  overview of   131
  physics-based interpretation   176–212
  principal strain components   134–136
  reference (zero) measurement   137–138
  reference time   137–138
  rheologic strain   169–176
  strain constituents   138–140
  strain tensor transformations   132–134
  thermal strain and thermal gradients   156–169
  total strain   138–140

long-gauge strain sensors   56, 92, *93*, 103, 116, 177, *177*, 213
  crack detection and localization   279–282
  crack diagnosis and prognosis using   284–289
  crack quantification   282–284
  crossed topology of   220
  discontinuities on   *98*
  FBG sensors   64–65, *65*, 108
  SOFO strain sensors   *114*
longitudinal thermal gradients   159, 166–169
Long-term Bridge Performance (LTBP) Program   188
long-term reliability
  of sensors   48
LTBP (Long-term Bridge Performance) Program   188
LVDTs (linear variable differential transformers)   40

m
machine learning (ML) methods   74, 75
Mach–Zehnder Interferometry   57, 59–63
Main System Software   36, *37*
malfunction of sensors   128
maximum permissible measurement error   25
maximum reading frequency   26
MCNY (Museum of the City of New York)   255, *256*
measurement error   80, 82–83
  of monitoring system   102
measurement fiber   61
measures of accuracy   22–23, 79, 80, 88, 90
measuring interval   25
measuring subsystem   22–25
  accuracy   22–23, *24*
  components   *35*
  drift   24, *24*
  precision   23, *24*
  repeatability conditions   23
  reproducibility conditions   23
  resolution   25
  sensitivity   25
mechanical interlocking   113
mechanical strain   40
  absolute value   181
  and stress   140–156, 172
  thermal strain and   160

MEMS (Micro-Electro-Mechanical Systems)   46
metadata   4
Michelson interferometry   59–63
Micro-Electro-Mechanical Systems (MEMS)   46
minimal tensional elastic strain   246
missing data handling   123–124, *124*
ML (machine learning) methods   74, 75
model error   179
modern sensors   32
monitoring authority   12
monitoring parameters, SHM process   6, **6**
monitoring strategy   9, 12, 214
monitoring system   86
  components of   11
  measurement error of   102
  measure of accuracy of   91
  uncertainty of   239
Museum of the City of New York (MCNY)   255, *256*

n
Newton–Cotes formulae   264
Nikon D90 digital camera   75
non-drifting sensor   *127*
non-linear temperature distribution   155, 160
non-linear thermal gradients   129, 158, *160*
normal strain components   41, 42, 132
numerical modeling   285

o
one-dimensional (1D) distributed sensors   289
optical fibers   56, 111
optical systems   21
ordinary optical fiber   56, *56*
outliers   121–123, *122*

p
pace   6–7
packaging (or housing)   46, 48, 51
  of distributed sensors   66, 289
  sensor   109–120, *112*
  of strain gauge   *54*
  of VW strain sensor   52
parabolic distribution of curvature   266
parallel topology   232–250
peak-tracking software   121
periodic monitoring   11

physics-based approach   123
physics-based interpretation   176–212
    corresponding model values   179–184
    global-scale SHM approach   218–219
    of strain measurements   176–212
    stress, evaluation of   184–187
    of total strain and stress   176–179
piezo-electric sensing elements   49
pile subjected to axial loading   *215*, 215–216
planar beams   152, *155*
Playback Software   36, *38*
point sensor. *see* discrete sensors
Poisson's coefficient   255
Poisson's ratios   152, 171
polyimide-coated fibers   57
polymer-based FOSS   57
post-tensioning force   100, 101, 102
precision measurement   23, *24*
prestressing force   139, 232, 237, 240–241, 244,
        245, **245**, *245*, **247**, 291
    maximal and minimal stresses   **249**
principal axes   134, 136
principal strain components. *see* principal strains
principal strains   41, 134–136
probabilistic approach   82
"Profile" sensor   292
    distributed strain sensor   *113*
    packaging   119–120
prognosis   5
propagation formula
    elementary   85
    error   83, **83**, 134, 136
    uncertainty   84, **84**, 84, 207
Punggol East Contract 26 (EC26) project   203,
        207–212

q
quantification of error   111
quasi-continuous monitoring   11
quasi-distributed sensors   45

r
Raman scattering   57
random (accidental) error   80–81, **81**
raw (unprocessed) data   30
Rayleigh scattering   45, 57, 106
    strain sensors   66–70

recording monitoring parameters   7
rectangle rule   264, *265*, 266, **267**
reference fiber   61
reference (zero) measurement   137–138
reference quantity value   25
reference reading   137
reference time   137–138, 239
refractive index   63
reinforced concrete beam   98
relative error   80
    in strain measurement   *97*, **207**
relative humidity (RH)   204
relative systematic error   203
reliability
    of measurement values   124–129
    of sensor measurements   201
    of SHM system   33
    of strain measurement   120–129
resistive strain gauge   53–56, *54*, *55*
resistive strain sensors   42, 43, 73
resistor   53
rheological strains   138, 169–176
    concrete slabs   194
    creep   169–174, *170*, **170**
    model for   183
    shrinkage   174–176, *175*
rosette   54, *54*
    principal strains using   136
    sensors measurements   133–134
rounding   86–88

s
Saint-Venant's principle   100, 101, 104
sampling interval   67
    of distributed sensors   105, 292–294
SCADA (Supervisory Control and Data
        Acquisition)   36
SCC (System Control Component)   35
second-generation strain sensors
    advantages and challenges   56–59
    best performances of   **59**, **60**, 70–71
    distributed sensors   66–70
    Fiber Bragg-grating (FBG)   *63*, 63–66
    SOFO sensors   59–63, *60*
self-weight load   281, *288*
sensing element   46
Sensing Sheet   73–74, *74*, 119

sensitivity of measuring subsystem   25

sensor-based SHM   4

sensor measurements   179, 203

  reliability   201

sensor measuring (reading) time   25, *26*

sensor networks   15, 33

sensor packaging   109–120, *112*

  geometrical and mechanical properties of
        115–117

  Profile   119–120

  strain in   115

sensor reading   *66*, 137, **138**

sensors   19

  and reading unit   *66*

  and sensor packaging   109–120

  survivability of   117–120

  types of   56

sensor topologies   213–214, *214*

  crossed topology   251–252, 255–260

  parallel topology   232–250

  simple topology   221–230

  and unusual structural behaviors
        218–219

session measuring (reading) time   26

  for combined reading of sensors   *28*

  for sequential reading of sensors   *28*

  for simultaneous reading of sensors   *27*

shear force

  distribution   155

  and shear stiffness   253–255

shear stiffness   253–255

shear strain   132, 258

  and change in average shear stress
        252–253

  component   41, 42, *153*

  conjugate property of   133

SHM. *see* structural health monitoring

SHM data   29

SHM metadata   29

SHM system

  components   19, *20*, 32

  data analysis and management subsystem   21,
     29, 29–31, *31*

  deployment   34

  error and uncertainty to   80–83

  heterogeneous   19, 34–38

  maintenance   29, 32–33

operational conditions   32–33

principle of functioning of   21–22

properties of   22–34

  measuring subsystem   19, 22–25

  reading type and frequency   25–28

reliability   33

structure and environment   32–33

subsystems of   19, *20*

short-gauge strain sensors   92, 116, *177*,
    178

  crack detection and localization   279–282

  crack diagnosis and prognosis using
    284–289

  crack quantification   282–284

  FBG   64–65, *65, 66*

short-term linear cross-sectional thermal
    gradients   162–166, *163*

shrinkage   174–176, *175*, 204

significant figures   86–88

simple topology   221–230

single-sensor measurement   213

SMARTEC   15, 16, *16*

SOFO sensors   59–63, *60, 62, 114*, 203

soft-packaged short-gauge sensors   274–279

soft polyethylene   119

software packages   31, 35, 36, *36*

sources of error   79, 179

  analytical model   179

  discrete sensors   92–94

  gauge length of. *see* gauge length

  thermal compensation   88–92

southeast leg of Streicker Bridge   217, 238, **239**,
    286

spatial resolution   67, 105

  for distributed sensors   105, 294–296

"spider configuration"   *50*

spontaneous Brillouin scattering   57

standard error of measurements   82–83

standard uncertainty   82–83, 82–86, 92, *210*

static monitoring   7

static reading unit   62

static strain sensors   35

steel–concrete beam

  elastic strain constituents in   184

  expressions for   178

  unusual structural behaviors in   176–212

  Young's modulus of   188–189

Stevenson Creek Experimental Dam   43, *44*
stimulated Brillouin scattering   57
strain
   in beam-like structures   263
   and change in temperature   **190–191**
   in concrete structure   203–212
   definition of   41–42
   measurements   100
   sensing technologies   42
   in steel structure   187–203
   and temperature coefficients   64
strain-based SHM   39–41, 299. *see also* structural
     health monitoring (SHM)
   of civil structures and infrastructure   48
   technologies   300
strain components   *41*, 41–42
strain constituents   138–140, 138–140, **139**, *160*,
     205
   estimation of   **206**
strain distribution   94, 105
strain gauges   42–43, 53
   full-bridge   *112*
   for long-term static monitoring   55
   reading units for   *55*
   resistive   53–56, *54*, *55*
   rosette of   133
   sensing element and packaging of   *54*
strain measurements   105
   with distributed BOTDA   *297*
   errors/uncertainties in   79, 109–120
   physics-based interpretation of   176–212
   relative error in   **207**
   reliability of   120–129
      "cleansing of data"   120–121
      handling missing data   123–124
      "outliers"   121–123, *122*
   and temperature   *240*
   uncertainty of   239
strain-measuring FBG   128
strain meter   44
strain monitoring   39–41, 43, 45, *206*
strain sensors   15, 32, 41, *76*
   Brillouin and Rayleigh Scattering
     66–70
   compensated   90
   damage detection capabilities   *76*
   desirable properties of   46–48

discrete. *see* discrete sensors
distributed. *see* distributed sensors
first-generation. *see* first-generation strain
     sensors
installation of   111
mapping functions   294
and normal force   255
resistive   42
within rosette measures   133–134
second-generation. *see* second-generation
     strain sensors
shrinkage   174
and strain-based sensing techniques   46, *46*
structural health monitoring   39–41
technology   42, 48, 300
and temperature measurements   294
third-generation. *see* third-generation strain
     sensor
types   177
vibrating wire (VW)   43
strain state/simply strain   41, 132
   and strain components   41, *41*
strain tensor transformations   132–134
strain transfer   109–111, *110*
   quality of   117–120
Streicker Bridge project   107, *108*, 125,
     279
   curvature constituents   *241*, *242*
   curvature distribution in   **271**
   deformed shape of   271
   global-scale SHM of   216–218, *217*
   hardening time   *245*
   southeast leg of   217, 238, **239**, 286
   strain and temperature taken at   *240*
   tensional and compressive strength of   **250**
   vertical thermal gradient   *286*
stress, evaluation of   184–187, **199**, **200**
stress measurement   184
stress–strain relationship   40
   in beams   152–156, *153*
   instantaneous multidirectional   145–152
   instantaneous uniaxial   140–145
   least-square linear interpolation of   216
stress tensor   171
structural deficiency   7
structural engineering   152

structural health monitoring (SHM). *see also*
    SHM system
  concept   2
  core activities   *9*, 9–12, *13*
  critical challenges   4–5
  definition   5–6
  dimensions   6–9, **8**
  Gota Bridge (Götaälvbron) project   *13*, 13–18,
    14, **15**
  level of sophistication   7–8, 10
  monitoring parameters   6, **6**
  as nervous system   2–3, *3*
  overview   1–5
  parties involved in   12–13, *13*
  potential benefits   3
  process   5–13
  scale of application   8–9
  strain-based. *see* strain-based SHM
  timeframe   6–7, 15
Supervisory Control and Data Acquisition
    (SCADA)   36
surface mountable sensors   52, *52*, 54
systematic error/bias   80–81
System Control Component (SCC)   35

t

Tacony–Palmyra Bridge   *34*, 34–38
teletensometer   43
temperature compensation   49
temperature measurements   64, *64*
  thermal expansion coefficient and   90
temperature-sensing optical fiber   292
temperature sensors   35, 188, 294
tensional strength   249, **250**
tensional stress   160, 186, 249, 285
TensorFlow   75
theory of elasticity   100
thermal compensation   203, 289–292
thermal expansion coefficient   182,
    212
thermal gradients   156–169
  and boundary conditions   201
  cross-section   158–166, *159*
  early-age   **247**
  elastic curvature due to   247
  longitudinal   159, 166–169
thermal strain   156–169, 212

determination of   157
  at point   156–158
  thermal curvature and   162
thin-walled cross-sections   255
third-generation strain sensor   45, 76
  advantages and challenges   71–73
  contact-based   73–74
  non-contact based   74–75
threshold value   40
timeframe, of SHM   6–7, 15
time-stamp indicators   241
total strain   138–140, *160*
  measured and estimated   *192*
  measurement   184
  physics-based interpretation   176–179
Trafikkontoret   14, 16, 18
trapezoid rules   264, *265*, 266, **267**
Trenton Airport measurements   125–126,
    126
truly distributed sensors   56
2D sensing techniques   45
2D strain sensing techniques   73

u

ultimate stress   186
Ultrasonic Pulse Velocity test   129
uncertainty
  error and. *see* error and uncertainty
  of measurement   82–83
  of model. *see* limits of error
  model and measurements   **208**
  of monitoring system   239
  propagation formula   84, **84**, 84, 207
  of strain measurement   239
uncorrected error   268
uniaxial compression   116
uninterruptible power supply (UPS) devices
    33
unit sensors   74
unshielded electrical sensors   32
unusual structural behaviors   3, 4, 74
  characterization   39
  direct and indirect detection of   71–73,
    246
  identification   184, 185
  proposed model in   *202*
  quantification of   184

unusual structural behaviors (*contd.*)
  sensor topologies and   218–219
  in steel and concrete   176–212
  in Wayne Bridge   200–203
UPS (uninterruptible power supply) devices
      33

v

validation, of measurements   125
vertical thermal gradients   285
vibrating wire (VW) displacement sensors   264
vibrating wire (VW) strain sensor   43, 49–53,
      52
  frequency   51
  functioning principle   49
  packaging   52
  reading units   *52*
vibration frequency   121
virtual sensors   101, 102
visual inspection   250

visualization of data   30
VW Spectrum Analyzer   35

w

Wayne Bridge   187, *187*, *188*, 189–203
Whitestone Bridge, uniaxial strain measurement
      54
whole lifespan monitoring   7
wired SHM systems   *20*, 21
wireless sensing network   28
wireless SHM systems   *20*, 21
wireless strain sensing   46
working interval   25

y

yield strain values   183
yield stress   186, 200
Young's modulus   40
  of sensor   117
  for steel and concrete   157, 188–189